AF598679

Three

Large-scale pile tests in clay

Proceedings of the conference
Recent large-scale fully instrumented pile tests in clay
held at the Institution of Civil Engineers,
London, on 23–24 June 1992

Thomas Telford, London

Published by Thomas Telford Services Ltd, Thomas Telford House,
1 Heron Quay, London E14 4JD.

The conference was organized by Thomas Telford Services Ltd and co-sponsored by BP Engineering and the Health & Safety Executive.

Organizing Committee: J. Clarke, BP Engineering, BP International Ltd, UK (Chairman); W. Cox, Fugro-McClelland Marine Geosciences Inc., USA; T. Inge Tjelta, Statoil, Norway; J. Little, Ground Board, Institution of Civil Engineers, UK; M. Thompson, Health and Safety Executive, UK.

First published 1993

Distributors for Thomas Telford books are
USA: American Society of Civil Engineers, Publications Sales Department, 345 East 47th Street, New York, NY 10017-2398
Japan: Maruzen Co. Ltd, Book Department, 3–10 Nihonbashi 2-Chome, Chuo-ku, Tokyo 103
Australia: DA Books and Journals, 648 Whitehorse Road, Mitcham 3132, Victoria

Classification
Availability: Unrestricted
Content: Original research
Status: Refereed papers
Users: Geotechnical engineers and researchers, offshore engineers

A CIP catalogue record for this book is available from the British Library.

ISBN 0-7277-1918 1

Papers or other contributions and the statements made or the opinions expressed therein are published on the understanding that the author of the contribution is solely responsible for the opinions expressed in it and that its publication does not necessarily imply that such statements and/or opinions are or reflect the views or opinions of the organizers or publishers.

Printed and bound in Great Britain by Redwood Books, Wilts.

Acknowledgements

The catalyst for the two major research projects on which the conference 'Recent large scale fully instrumented pile tests in clay' was based was John Rigden, Manager of Exploration Facilities Engineering, BP Exploration. His contribution is greatly acknowledged. The project reports for the Magnus foundation monitoring, Large diameter pile test and the Tilbrook Grange tension and lateral test projects are available from HMSO report OTH 92 392. The authors of all papers related to the various projects wish to thank the joint venture participants for permission to publish their papers. The participants in the various projects are listed below:

Magnus foundation monitoring

- British Petroleum Development Limited
- Britoil Plc
- Brown & Root Inc
- Commission of the European Communities
- Den Norske Stats Oljeselskap AS (Statoil)
- Engineers India Limited
- Exxon Production Research Company
- John Brown Offshore Structures Limited
- Marathon Oil Company
- Total CPF
- UK Department of Energy
- Large Diameter Pile Test Project
- Amoco (UK) Exploration Company
- British Petroleum Development Limited
- Britoil Plc
- Commission of the European Communities
- Esso Exploration and Production UK limited
- Lloyd's Register of Shipping
- Marathon Oil (UK) Limited
- McClelland Limited
- Shell UK
- Statoil Netherlands BV
- UK Department of Energy

Tilbrook Grange tension and lateral test

- British Petroleum Development Limited
- Britoil Plc
- Esso Exploration and Production UK Limited
- Fugro-McClelland Limited
- Statoil Netherlands BV
- UK Department of Energy

NGI pile test at Pentre and Tilbrook (paper 17)

(part of 'Design of offshore piles in clay: field test and computational modelling')

- Agip SpA
- A/S Norske Shell
- British Petroleum Development (Norway) Limited
- Conoco Norway Inc
- Elf Aquitaine Norge A/S
- Esso Norge AS
- Marathon Oil Company
- Mobil Research and Development Corporation
- Norsk Hydro AS
- Saga Petroleum AS
- Statoil AS
- UK Department of Energy

Pile load tests in stiff clay (paper 22)

- Amoco (UK) Exploration Company
- British Petroleum Development (Norway) Limited
- UK Department of Energy

Degradation of skin friction for driven piles in clay (paper 23)

- Amoco (UK) Exploration Company
- British Petroleum Exploration Inc
- Exxon Production Research Company
- Shell UK Exploration and Production Limited
- UK Health & Safety Executive

Contents

Magnus foundation monitoring project

1. Magnus foundation monitoring — an overview. D. E. SHARP 1

2. Magnus foundation monitoring project instrumentation data processing and measured results. R. M. KENLEY 28

3. Design implications arising from the Magnus foundation monitoring project. C. BOWLEY 52

4. Magnus foundations: soil properties and predictions of field behaviour R. J. JARDINE and D. M. POTTS 69

5. Mudmat interaction and foundation analysis. M. R. HORSNELL, V. A. NORRIS and B. IMS 84

6. Magnus FMP — dynamic response of foundations. T. I. TJELTA, E. SKOMEDAL and O. SKJASTAD 98

Discussion 116

Large diameter pile test project

Investigation and instrumentation

7. Large diameter pile test project — overview. C. R. MULLIS 128

8. Investigation and interpretation of Pentre and Tilbrook Grange soil conditions. M. D. LAMBSON, D. G. CLARE and R. M. SEMPLE 134

9. The development of instrumentation for drivability testing and load testing of large diameter piles. I. J. SOLOMON, W. R. COX, J. CLARKE and T. J. POSKITT 197

10. Instrumentation and calibration of two 762 mm diameter pipe piles for axial load tests in clays. W. R. COX, I. J. SOLOMON, and K. CAMERON 217

Installation and testing

11. Large diameter pile tests data acquisition systems. C. CURRIE and I. SOLOMON 237

12. Measurements of axial strain and pile wall displacement, and the determination of skin friction during the driving of instrumented piles. T. POSKITT, K. YIP-WONG and W. COX 250

13. Static and cyclic axial load tests on two 762 mm diameter pipe piles in clays. W. R. COX, K. CAMERON and J. CLARKE 268

14. Reduction of field data and interpretation of results for axial load tests of two 762 mm diameter pipe piles in clays. C. E. GIBBS, J. MCAULEY, U. A. MIRZA and W. R. COX 285

Discussion 346

Additional large-scale tests

15. The axial tension test of an instrumented pile in overconsolidated clay at Tilbrook Grange. J. CLARKE, M. M. LONG and J. HAMILTON 362

16. Cyclic lateral loading of an instrumented pile in overconsolidated clay at Tilbrook Grange. J. HAMILTON, M. M. LONG, J. CLARKE and M. D. LAMBSON 381

17. NGI's pile tests at Tilbrook and Pentre — review of testing procedures and results. K. KARLSRUD, S. B. HANSEN, R. DYVIK and B. KALSNES 405

Discussion 430

Additional analysis of test data

18. Analysis of stress-wave data from pile tests at Pentre and Tilbrook. M. F. RANDOLPH 436

19. Analysis of behaviour of the Tilbrook Grange lateral test pile. J. M. HAMILTON and T. W. DUNNAVANT 448

20. LDPT as calibration for new pile models. G. SVANO, T. I. TJELTA and A. EIDE 463

Discussion 485

Design implications

21. The impact of axial pile load tests at Pentre and Tilbrook on the design and certification of offshore piles in clay. R.HOBBS 491

22. Comparison of recent tests on OC clay and implications for design. F. NOWACKI, K. KARLSRUD and P. SPARREVIK 511

23. Degradation of skin friction for driven piles in clay.
T. R. ALDRIDGE, and F. SCHNAID 538
Discussion 554

Panel discussion

Panel discussion 563

Closing address

Professor J. B. Burland 590

1. Magnus foundation monitoring — an overview

D. E. SHARP, BP Engineering

1. Introduction

BP's Magnus oil-field was discovered in 1974 in Block 211/12 of the UK sector of the North Sea as shown on Fig. 1. The field was developed at a total cost of £1,300 million by a single production platform installed in 186m of water in 1982. The Magnus structure is shown on Figs 2 and 3. When the structure was being designed, in 1979, it was to be located in deeper water and further north than other structures in the North Sea, resulting in particularly severe environmental design criteria. As a result of very high deck loads to be supported, the tower was to be the heaviest steel structure installed off-shore. At float out it weighed 34,000 tonnes.

At the time, design methods used for off-shore structural foundations were extrapolated conservatively from onshore experience although pile sizes employed off-shore were very much larger than land piles. There was a need to confirm pile group capacity, the actual loads imposed on off-shore pile groups, the distribution of those loads within pile groups and the conservatism inherent in design. As a result, in addition to monitoring environmental conditions and structural response, BP proposed to monitor the performance of the foundation system of one leg of the platform, leg A4, in order to obtain some basic information to verify current design methods. This paper describes that project, believed to be the most comprehensive pile monitoring project attempted off-shore, the Magnus Foundation Monitoring Project (FMP). At approximately the same time, the structure and a pile of one of the relatively small platforms in the Ekofisk field, off-shore Norway were being instrumented as reported by Spidsoe et al (1980). The foundations of many of the large gravity structures deployed in the North Sea have been instrumented.

2. Objectives

Much of the required capacity of off-shore piles is attributable to wave loading but there were few measurements to confirm how much of this loading is actually transmitted to the piles. Furthermore, little work had been done onshore and none off-shore on the distribution of loads within a pile group. The overall objective of FMP was to compare actual foundation behaviour with design. Specific objectives were to measure

- the variation of static load distribution with time;
- the environmental load acting on a pile group during storms;
- the proportion of this load taken on the mudmat and on the pile group;
- the distribution of this load within the pile group and along the length of one pile;
- the effective stiffness of a pile group during storms;
- the variation of pore and total pressure along a pile with time;
- the short and long term absolute vertical settlement/movement of a leg.

To achieve these objectives, the lower section of one of the four legs of the platform and the piles supporting that leg were instrumented to determine the actual loads imposed on the piles and seabed by the structural and environmental forces. However, pressure sensors on the piles were lost in an incident during upending and not replaced and settlement gauges did not perform well. As a result the last two objectives were not achieved.

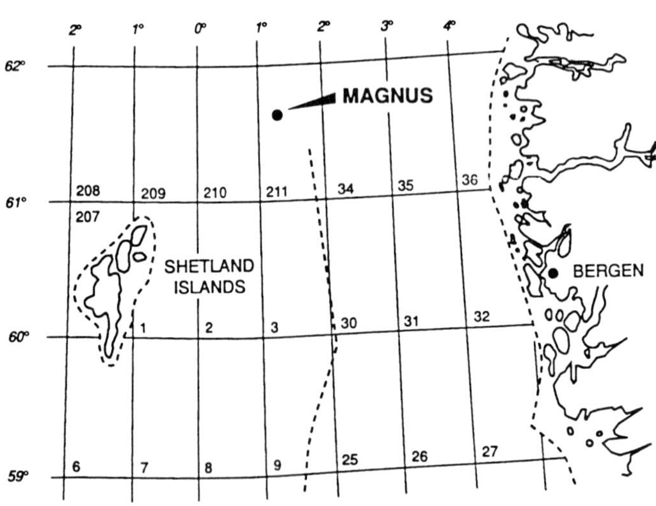

Fig. 1. Vicinity map

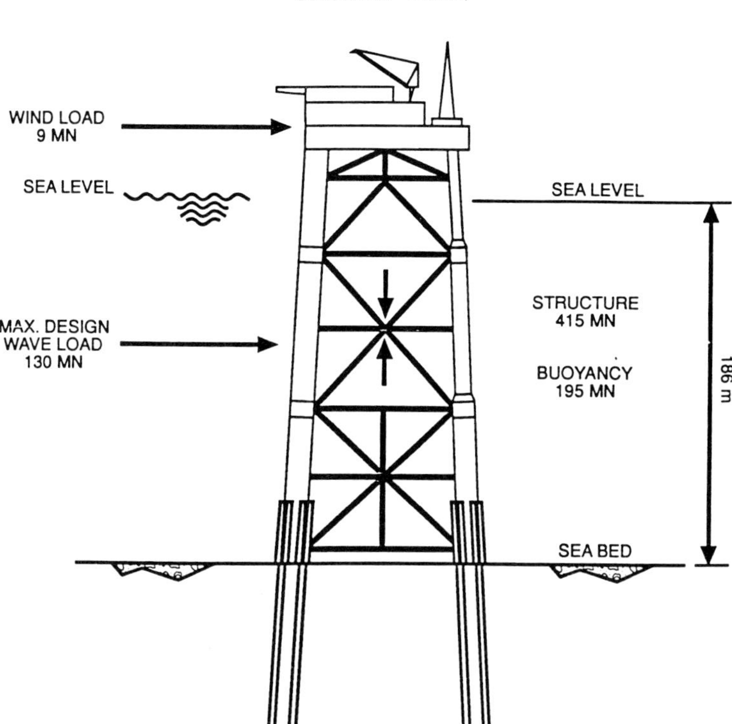

Fig. 2. Magnus' structure

3. Structure and foundations

Constructed at the Nigg Bay yard in Scotland, the Magnus platform is a four-legged tower type self-floating structure. The tower was floated outfrom its graving dock at the end of March 1982, towed to the Magnus site and upended. During upending, ten of fourteen piles which had been pre-installed on the tower before float-out were released prematurely. This incident was a major set-back to the FMP; as a result, approximately half of the original sensors were lost. It was evident that the three instrumented piles which survived would not provide adequate data to meet project objectives. Retrieval of those piles which had released prematurely was concluded to be impractical and replacement piles were fabricated. The platform was successfully placed on the seabed in early April 1982.

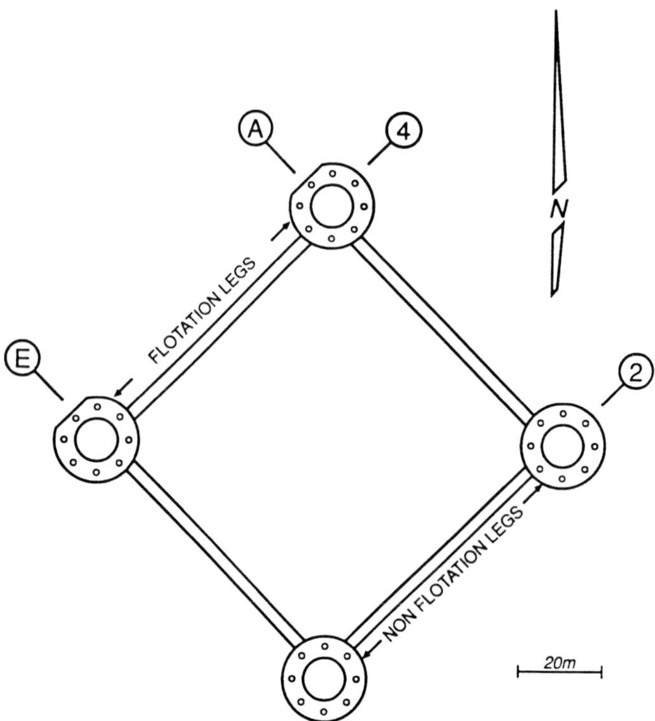

Fig. 3. Base of structure

The structure is supported by four groups of nine 2.134m outside diameter open-ended pipe piles. The piles were provided with an externally flush 1400mm long driving shoe having a wall thickness of 80mm. Apart from this shoe the piles have a uniform wall thickness of 63.5mm. The Pitch Circle Diameter of each group is 13.8m. Details of the design and installation of the Magnus foundations are given by Rigden and Semple (1983). Immediately after upending, the structure was supported on four mudmats, each approximately 17m in diameter. After a storm, but before piling commenced, the mudmat on leg A4 was observed to have penetrated about 1m. The 36 piles were driven to penetrations of 80 to 87m using Menck 1700 underwater hammers and a spread of Menck MRBS 8000 and 12500 steam hammers with a conventional follower arrangement. On leg A4, two piles, piles 7 and 9, were driven during April and grouted in mid May. Remaining piles were driven in two groups a few days apart in early June and pile sleeve grouting completed later that month. Soil levels measured in seven piles after installation were within 1m of the seabed indicating that the piles did not plug during driving.

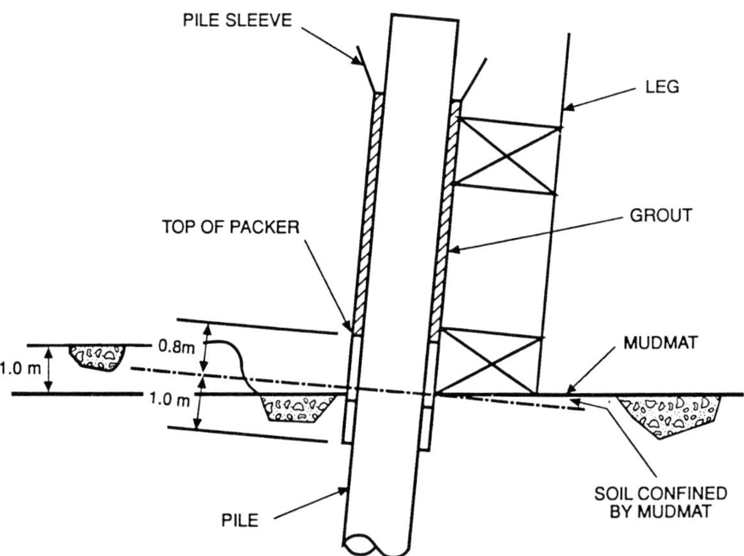

Fig. 4. Detail of pile/sleeve connection

The design detail of the pile/sleeve connection is shown in Fig. 4. However, installation reports show that on only one of the nine pile sleeves was the packer inflated. On the remainder, grout plugs were cast. In July 1982 platform topsides, weighing approximately 25,000 tonnes, were lifted onto the tower. Hook-up and commissioning of the platform continued for the next two years. The platform continues as an operational installation.

4. Soil conditions

Details of the site investigations performed at Magnus and the soil conditions disclosed by those investigations are described by Semple and Rigden (1983). A boring log and design shear strength profile is presented as Fig. 5. Except for a thin surface layer which is not always present, the predominantly cohesive soils are relatively strong and incompressible, particularly the upper 20m. In situ cone resistances in a 2.5m thick sand layer which occurs about 15m below the seabed indicate a dense to very dense condition.

Review of the installation records suggests that soils in the upper 20m, an area of particular interest for lateral behaviour, may be somewhat stronger than assumed during design. Penetrations of piles under combinations of pile/follower/hammer were less than predicted whilst pile driving between 7 and 20m was somewhat harder than predicted.

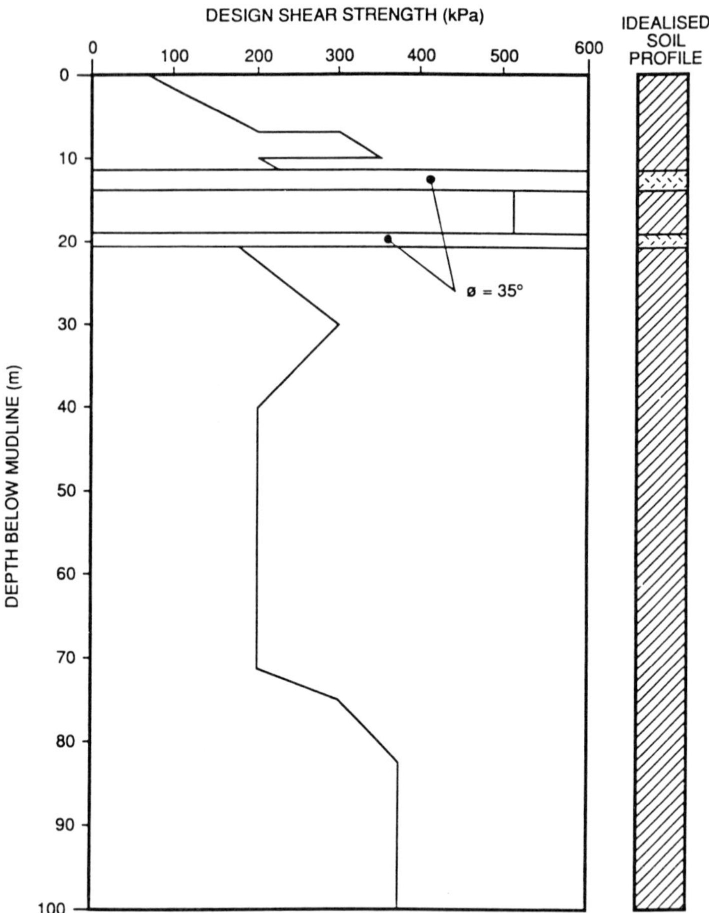

Fig. 5. Design shear strength

5. Project organisation

Total cost of the programme was £2.8 million. Whilst underwritten by BP, FMP was partly funded as a joint industry project. Participants included other oil companies, engineering contractors and the UK Department of Energy. Assistance was also provided by the European Community. The commercial negotiations which led to the involvement of the many participants were both complex and of a lengthy duration (close to 18 months) and as such required considerable senior management time.

FMP was managed as a stand alone programme by BP Engineering. A summary of the project programme is presented in Fig. 6. The project's successful execution required a major collaborative effort between academia and many parts of industry. Future practitioners must ensure appropriate support for such a lengthy research programme exists throughout their and other participating organisations to ensure the inevitable differences and interfaces between so many diverse parties, can be managed smoothly. Continuity of some personnel was key to success.

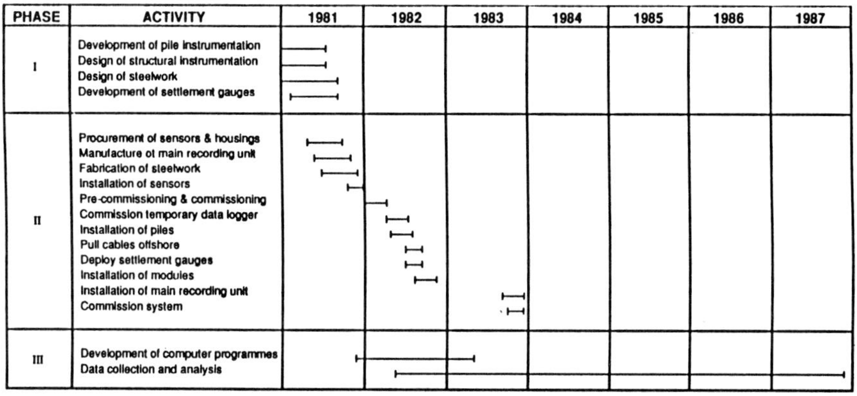

PHASE	ACTIVITY	1981	1982	1983	1984	1985	1986	1987
I	Development of pile instrumentation Design of structural instrumentation Design of steelwork Development of settlement gauges							
II	Procurement of sensors & housings Manufacture of main recording unit Fabrication of steelwork Installation of sensors Pre-commissioning & commissioning Commission temporary data logger Installation of piles Pull cables offshore Deploy settlement gauges Installation of modules Installation of main recording unit Commission system							
III	Development of computer programmes Data collection and analysis							

Fig. 6. FMP programme

6. Instrumentation system and performance

The foundation monitoring system is one of three sub-systems which together comprise the complete platform monitoring system. Environmental and structural systems have been described by Webb and Corr (1989). Of the environmental system, signals from infra-red type wave height sensors installed on two diagonal corners of the platform have been utilised to define wave height conditions incident on the platform. Originally, structural and foundation monitoring instrumentation was to consist of strain gauges, accelerometers, pressure gauges and settlement gauges fitted to the tower structure and strain gauges and pressure gauges fitted to the piles which would support leg A4. However, pressure gauges were only installed on one pile which was one of those lost. No pressure gauges were installed on replacement piles.

The principal instrumentation finally installed on the platform and piles is shown on Fig. 7 and comprised:

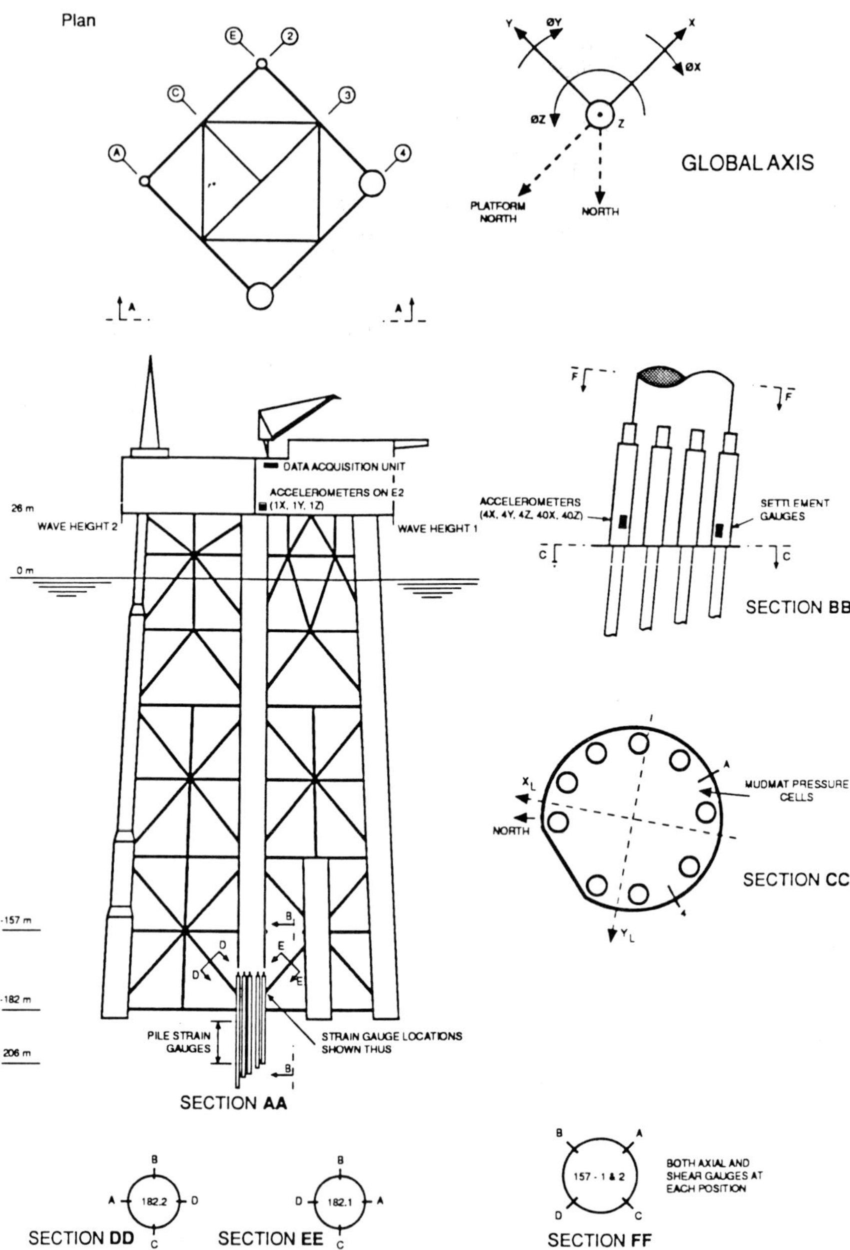

Fig. 7. Location of sensors

6.1 Accelerometers

A triaxial set of linear and a biaxial set of angular accelerometers were installed at the –182 m level to determine the dynamic displacements and rotations of the bottom of the leg adjacent to the top of the piles. Linear and angular accelerometers were Sundstrand type QA-1300-KA00-12 and Systron Donner type P/N 4590 AF-1-AG respectively.

The accelerometers performed well except that the gauge electronics of an angular accelerometer failed 30 months after installation. By this time, excellent correlations had been established between the accelerometers installed at the base of leg A4 and similar instrumentation installed at deck level.

6.2 Structural strain gauges

Ailtech weldable strain gauges were installed on the platform leg at -157m, just above the top of the pile sleeves, and at -181m on each of two diagonal braces which meet the leg near its base. At each location, four longitudinal gauges were installed at 90 degree intervals around the outside circumference of the member so that axial loads and bending moments could be derived. On the main leg additional gauges were installed at the same positions as the longitudinal gauges to enable shear forces and a torsional moment to be derived.

These gauges, which provide dynamic information only, had been used extensively on other off-shore platforms and again proved reliable.

6.3 Mudmat pressure cells

Nine total pressure cells, designed and supplied by the Norwegian Geotechnical Institute (NGI), were installed on the underside of the A4 mudmat to measure pressures from which the static and dynamic loads transmitted through the mudmat were derived. The pressure cells incorporate two modified NGI/Geonor type P-100 vibrating wire sensors, a primary and a back-up.

These instruments performed extremely well. Strong correlations were achieved between each pair of sensors and between readings recorded by both recording systems and manufacturer-supplied test equipment.

6.4 Settlement gauges

Two early versions of settlement gauges described by Jardine, Hight and McIntosh (1988) were developed and manufactured by Imperial College, London but data obtained is not judged to be reliable. Similar gauges were successfully deployed later on the Hutton TLP foundations.

6.5 Pile strain gauges

Following a long term development programme at Queen Mary College, London, pile strain gauges and the methods for their installation were selected on the basis of the long term drift and stability characteristics of the gauges and the ability of the system to survive the high accelerations and strains imposed on piles during driving. Pile gauge systems were tested on an impact rig, described by Solomon *et al* (1992), capable of producing suitably high accelerations and strains.

Originally, sets of strain gauges were installed at four levels on the inside of one of the piles to be installed on leg A4, Pile 3, and a single level approximately 75m from the pile toe on each of the remaining eight piles. At each level the sets comprised four pairs of HBM model LG6/350 (modified) foil gauges spaced at 90 degree intervals to enable pile forces and bending moments to be derived. The two gauges of each pair were installed as the opposite active arms of a full bridge and completed at the gauge location with a CIL SGA 220 strain gauge amplifier. The dummy gauges were mounted on a strain free plate attached to the amplifier board. Additionally, at each location a SINCO 52621 model vibrating wire strain gauge was installed to provide redundancy. Gauges and associated downhole electronics were installed within special cast steel housings protected from mechanical damage by steel ploughs.

Following a review of the design, fabrication and installation of the original system it was concluded that it would be impractical to duplicate it for replacement piles. No delay in platform installation schedule due to instrumentation of replacement piles was acceptable. An alternative system was designed based upon the use of SINCO vibrating wire gauges only which was compatible with hardware installed on the tower and capable of being installed either in a fabrication yard or off-shore. Four replacement piles were instrumented at one level and a further pile, again Pile 3, was instrumented at three levels. Instrumentation levels were revised on the basis of actual pile driving experience and a sensor level 82m from the toe selected for piles instrumented at a single level. These five piles were installed and together with the three remaining original piles, Piles 2, 6 and 9, and one uninstrumented pile, Pile 7, formed the pile group monitored as part of FMP.

Whilst 10 of the 12 foil gauges commissioned showed an excellent response to dynamic conditions, their long term stability has not been as good as expected. Circumstantial evidence suggests the cause to be microscopic scale damage to gauge housings. However, this has not reduced the overall effectiveness of the system which was designed to be tolerant of just such a generic fault. Monitoring of the long term behaviour of the foundations has relied on data from vibrating wire sensors.

Occasionally, the output from some vibrating wire gauges was erratic and it appeared that these instruments were susceptible to noise, a phenomenon which had not been seen during the detailed selection trials. Despite this approximately 75% of the 40 vibrating wire gauges installed on the piles provided useful data and, as the loss of a single gauge from a set of four was not significant, both axial forces and bending moments were computed from up to 9 of the 10 instrumentation levels installed. The only instrumentation level from which no data was recorded was the lowest level of sensors on the most instrumented pile.

6.6 Cabling, cable terminations and off-shore hook-up

Wherever possible, all cabling was enclosed within steel conduit for mechanical protection. Considerable attention was given to the design of both subsea cabling, in particular its long term water resistance, and underwater connections. Traditional underwater connectors were considered insufficiently reliable for long term applications and as a result high quality hot injection moulded connections were used. Structural strain gauges, accelerometers and mudmat pressure gauges were hard wired back to junction boxes inside the top of the leg in the fabrication yard. Cabling from pile sensors was coiled inside the top of each pile in the fabrication yard before float out. After piling, the cables were retrieved by divers and pulled through pre-installed conduits on the jacket and connected to junction boxes inside the top of the leg.

6.7 Data recording

Mudmat pressure gauges were first recorded approximately 24 hours after touchdown of the tower. A month later, BP commissioned a temporary data logger (TDL) which automatically sampled data from mudmat pressure gauges and, once they were hooked-up, pile strain gauges at a frequency of 2 Hz for five minutes every six hours. Summarised results were stored on cartridge tapes. Additionally, manual recordings were made to supplement the data base: whilst topsides modules were being installed; during periods when the TDL was unserviceable; and at other times to provide an independent check on data recorded automatically. Towards the end of 1983, the TDL was replaced by a sophisticated data acquisition unit (DAU) which was connected to all sensors from both the structural and foundation monitoring systems as well as selected environmental sensors. Routinely, the DAU powered up and recorded data for 40 minutes every six hours whilst during stormy weather data could be recorded continuously. Performance of this unit has been excellent.

7. Analytical techniques and results

Time series data, examples of which are shown in Fig. 8, were inspected visually and analysed statistically. The dynamic behaviour of the structure and foundations is dominated by a quasi-static response to wave action. Time series signals are strongly correlated with virtually identical signals from many sensors. The effects of a single wave are seen in forces in the structure, mudmat and piles. Results from over 500 events were similar, statistically well behaved and none showed unusual behaviour. Consequently, a reasonably representative understanding has been obtained by examining in detail single waves from selected recordings. To demonstrate that the results obtained from this analysis were generally applicable further analyses were performed.

Firstly, long term relationships between structure and foundation forces and displacements were examined by comparing statistical data from each days recordings. Secondly, every wave in the selected recordings was analysed to obtain relationships between structure and foundation forces and displacements. Furthermore, a frequency analysis was undertaken on selected recordings in order to investigate structural response as a function of frequency and to obtain relationships between displacements and structural and foundation forces. Data analysis procedures are described by Kenley and Sharp (1992).

Overall data quality has been exceptionally high and consistent results have been derived from the various analytical methods.

8. Design predictions

In order to compare measurements with design predictions the original design contractor undertook a structural and foundation analysis of the Magnus tower under installation and environmental conditions close to those actually experienced off-shore (Bowley, 1992). Analyses were performed in accordance with the original design techniques. Account was taken of:

- the actual soil conditions disclosed by a boring drilled close to leg A4;
- the as-built details of the structure;
- the marine growth on structural members appropriate for the monitoring period;
- the actual topside loads;
- the probable pile head conditions;
- the actual number of well conductors and other appurtenances installed.

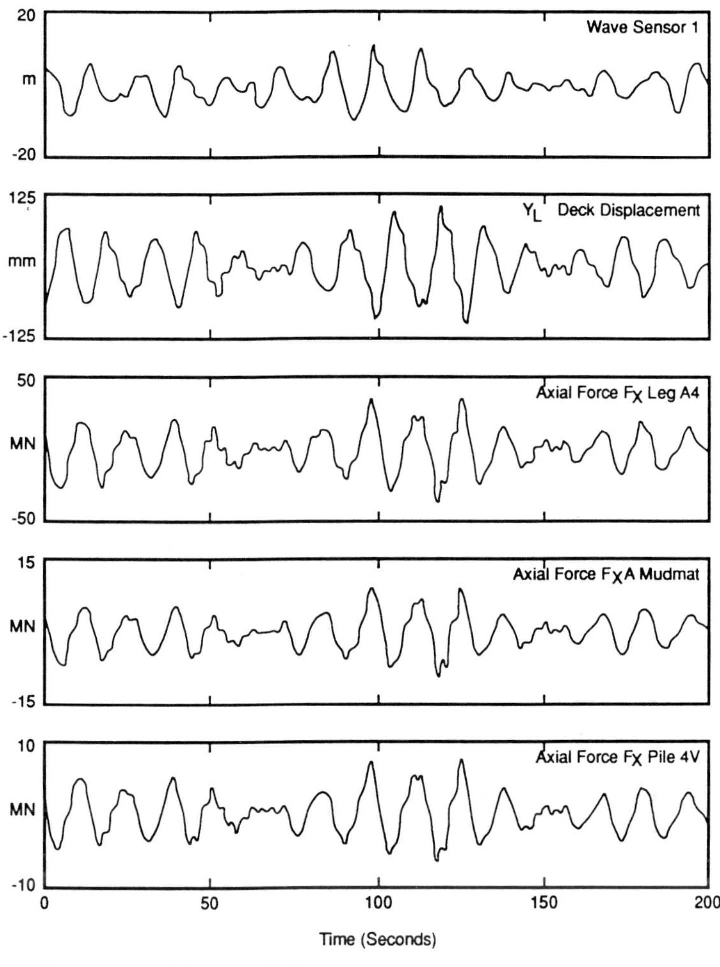

Fig. 8. Examples of time history signals

Estimates of the global dynamic amplification factors (DAFs) were made on both calculated and measured fundamental periods. However, in deriving design predictions for comparison with actual measurements, DAFs based on the calculated period were used.

9. Review of project objectives

In general, the objectives of FMP have been achieved. The TDL recorded data from soon after platform installation until mid 1983. The DAU was commissioned towards the end of 1983 and operated until April 1986. The system was then shut-down until December 1987 when the DAU was replaced by a simpler system which analyses data only from the accelerometers and structural strain gauges. Specific comment on each of the project objectives is given below.

9.1 Variation of static load distribution with time

Kenley and Sharp (1992) present details of the long term foundation behaviour, a summary of which follows. Before piling all load was carried by the mudmat. After driving of just two piles on leg A4, but before grouting, there is evidence of substantial load transfer from the mudmat to the piles, maybe through friction within the beaded pile sleeves, and later back to the mudmat, possibly as a result of slippage due to vibration during piling on other legs. The phenomenon was repeated when the remaining seven piles on leg A4 were piled. Additionally, pile sleeve extensions and a 'cookie cutter' (which connects these extensions close to the perimeter of the mudmat), both of which extend 1m below the mudmat, may also carry some load. Penetration resistance of these elements was estimated during installation studies to be between 7 and 14MN. At least initially, the distribution of static load within the pile group appears to have been influenced strongly by a complex and disrupted installation sequence. It is beyond the scope of this overview to attempt to rationalise all aspects of the observed behaviour during this period.

During module installation, a period of eight days, pile and mudmat loads increased by between 5 and 7MN per pile and 4 to 5MN, respectively, from which a total measured load change of around 59MN is derived. This compares with the predicted increase in axial load on leg A4 during module installation of 64MN. Throughout this period measurements show load transferring from the mudmat.

During the next 20 months, load on the mudmat slowly decayed from about 19MN to a steady state of about 9MN. Meanwhile pile loads increased by an average of about 3MN per pile, ranging from 1.5 to 4.5MN per pile, reflecting both load transfer from the mudmat to the piles and a small increase in topsides loads. By early 1985, the average still water static loads measured in the Magnus piling in service were approximately 15MN per pile. The total static load acting measured on the piles and mudmat was computed to be 140MN which again compares well with design which predicted the total static axial load acting on leg A4 in early 1985 to be around 130MN.

Surprisingly, the largest increase in load was recorded on the lower level of gauges on Pile 3 and overall there was little evidence of a reduction in static axial load over the instrumented depth, that is the top 25m. While there is no reason to suspect instrument error, possible explanations for this do include:

- the development of negative skin friction on the upper part of a pile due to consolidation of near surface soils following pile driving and the transfer of loading from the mudmat to the piles;
- that the interaction between piles is not adequately accounted for in the load shedding analysis down the pile;
- that the cyclic nature of the loads allows negative skin friction to become established in the upper section of the piles;
- the possible effect of residual stresses from pile driving or earlier loading.

Over the long term, changes in pile bending moments were expected to be, and have been measured to be, small; less than 1MNm.

9.2 The environmental load acting on the pile group

A wealth of data have been recorded during storms. The most significant event recorded by the DAU was in January 1986 when the significant wave height reached 12.5m with a maximum individual wave of about 20.5m, equivalent to a storm with a return period in excess of a year. The original 1 in a 100 year design wave height was 30.5m although this was later reduced to 27.5m. The maximum environmental axial and shear loads acting on the leg A4 during this storm were deduced to be 89MN and 17MN peak to peak, respectively. The bending moment applied to the foundation of leg A4 was deduced to be 200MNm peak to peak.

During two further storms, in December 1988 and December 1990, maximum individual wave heights of 23.5 and 26.0m respectively were recorded. Significant wave heights were 14 and 12.8m respectively. The direction of the December 1990 storm was similar to one in January 1986 referred to above. In each of the three large storms wave periods were about 11.5 seconds. During the most recent event, the most severe recorded, whilst the maximum peak to peak axial load measured in leg A4 above the pile sleeves was approximately 30% greater than that measured in January 1986, the maximum peak to peak axial force in the two diagonal brace members were similar. Furthermore, standard deviations of all measured data in the later event were lower. Thus it is judged unlikely that the overall forces applied to foundation of leg A4 during the December 1990 event were more than 20 to 30% greater than those measured during the January 1986 storm. More probably they were of a similar magnitude.

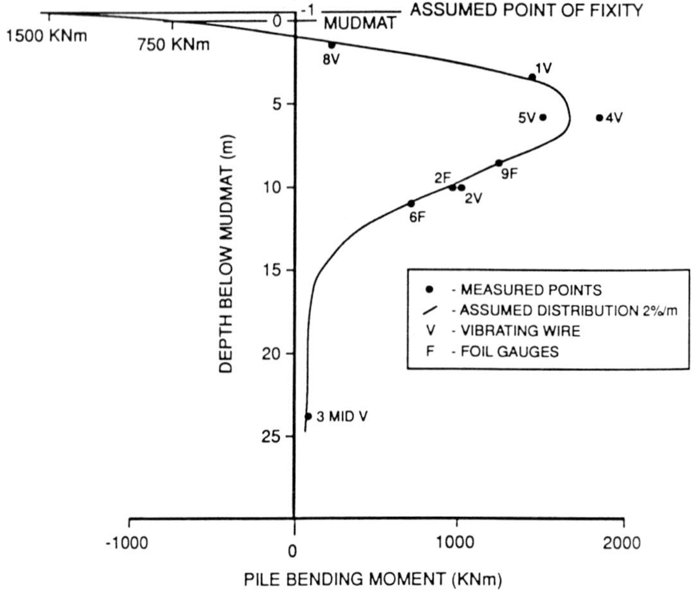

Fig. 9. Derived bending moment distribution down a pile

The predicted applied loads were between 40 and 90% higher than measured for equivalent wave conditions. This has little to do with foundation behaviour although the high soil stiffness measured, see below, will contribute to higher platform natural frequencies and reduced dynamic amplification factors. The fundamental natural frequencies of the platform were measured to be some 23% higher than those predicted during design. The effect of the higher natural frequency of the platform is to lower DAFs by about 5-10% for wave periods in the range of interest. As a further consequence, pile fatigue lives estimated from measured data are many times greater than predicted.

9.3 The proportion of the environmental load taken by the mudmat and on the pile group

During storm loading the mudmat carries between 15 and 20% of the environmental axial load acting on the foundation and about 10% of the applied bending moment. No attempt was made to measure shear load carried by the mudmat but Rigden and Semple (1983) showed that for a mudmat penetration of 1m, the shear capacity of the mudmat was about 20MN. Axial capacity of the mudmats was predicted during design to be in excess of 100MN. However, because it was considered prudent to allow for possible scour the mudmat was assumed not to contribute to either the load carrying capacity or the stiffness of the foundation. In practice, in the absence of the scourable sands, it made a substantial contribution dynamically which was consistent with theoretical expectations.

The maximum environmental load measured in an individual pile was less than 15MN peak to peak, resulting in a maximum compressive load of less than 25MN in the most heavily loaded pile. Originally, it was hoped that FMP might provide some indications of the ultimate capacity of off-shore piles. However, in the event environmental loading has been much lower than expected and as a result individual pile loads have been relatively modest in comparison with an original ultimate design capacity of more than 60MN. The specific requirement for a programme of large diameter pile tests loaded to failure was recognised; the results of which are presented in the second part of this conference.

9.4 The distribution of environmental load within the pile group and along the length of one pile

9.4.1 Behaviour under lateral loading. Since design calculations predict similar lateral loading on all piles, to within 5%, an estimate of the bending moment distribution down a pile was obtained by combining results from all piles and assuming piles would behave in a similar manner. This distribution, shown in Fig. 9, was used to derive shear forces, rotations and displacements, results of which were consistent with values measured on the platform. The bending moment distribution reached a maximum about 6m below the mudmat and was approximately zero about 1m below the mudmat. The maximum pile bending moment measured was about 2.5MNm peak to peak. Below about 20m pile bending moments are very small. As shown in Fig. 10, Kenley and Sharp (1992) observed a slight difference in the shape of the bending moment distribution in high energy environments, that is during inclement weather.

As shown in Fig. 11, predicted bending moments are significantly different from those measured with the predicted maximum negative bending moment occurring at a much greater depth. This result is consistent with the findings from Ekofisk. There is also a particularly wide discrepancy between the predicted and measured pile head bending moment. As discussed later, laterally, the pile group is approximately 2 to 3 times as stiff as assumed during design.

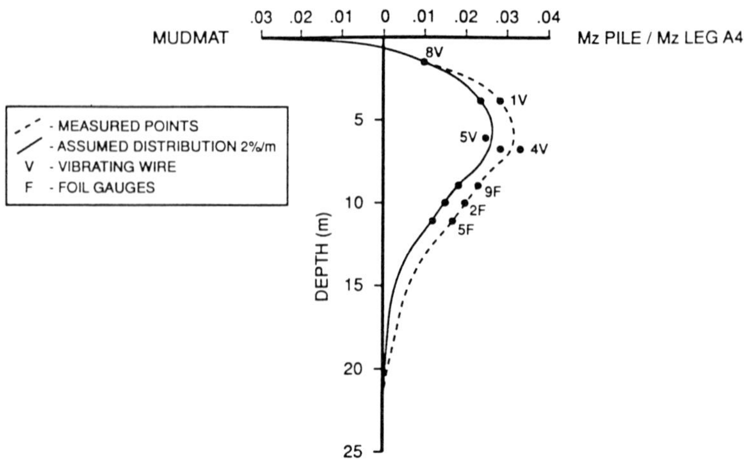

VALUES AT MUDMAT	LOW	HIGH
SHEAR FORCE (N/Nm) BENDING MOMENT (Nm/Nm) ROTATION (μ rad/MNm) DISPLACEMENT (mm/MNm)	0.028 0.031 4.3 0.036	0.028 0.031 5.5 0.048
VALUES AT -182m		
ROTATION (μ rad/MNm) DISPLACEMENT (mm/MNm)	3.5 0.047	4.7 0.062

Fig. 10. Bending moment distributions in different wave conditions from long term plots

In part, the values measured may reflect different constraint conditions at the pile head in practice from those assumed in design. Analysis carried out during design assumed 1m of scour, i.e. no contact between soil and mudmat. Furthermore, installation records show only one grout packer operated successfully on leg A4 and as a result grout plugs were cast on the remaining eight piles. Horsnell, Norris and Ims (1992) have shown that a more realistic analysis of lateral behaviour using p-y data derived from Reese, Cox and Koop (1975) and which takes due regard of pile head fixity and includes the mudmat in the analysis, yields both a reasonable bending moment profile and a lateral stiffness that is comparable with that measured. Kenley and Sharp (1992) suggest that measurements indicate that the effective soil modulus assumed in design was too low, although they are unable to isolate the contribution from the actual soil modulus from interaction effects.

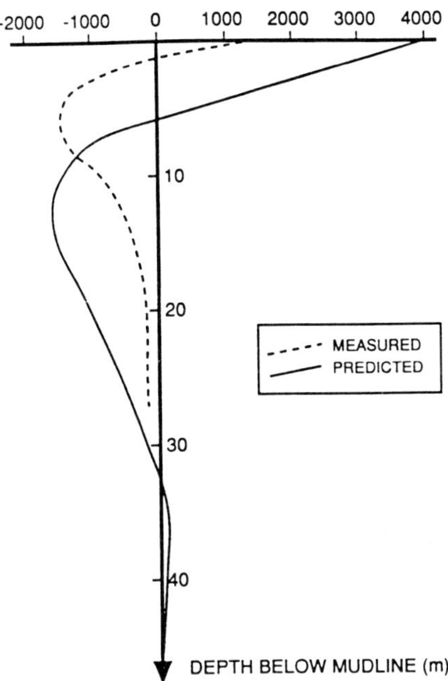

Fig. 11. Comparison of measured and predicted bending moment distribution

9.4.2 Behaviour under axial loading. Due to the failure of the lowest level of gauges on the most instrumented pile only limited data has been collected on the load distribution down the length of one pile. Since the axial force in an individual pile depends both upon the axial displacement and the rotation of the base of the leg it is not possible to estimate the axial load distribution in a similar manner to that adopted for the bending moment distribution.

An approximate indication of the axial load distribution down a pile was obtained initially by assuming it decreased linearly with depth. It was then possible to estimate the axial loads in each pile at mudmat level from the measured results. These results were plotted against pile position from the axis of rotation assuming this axis was perpendicular to the direction of wave approach. A best fit straight line was then drawn through the data points and a revised estimate obtained from this of the axial load at the top of each pile. This was used in turn to obtain an improved estimate for the load distribution down a pile. The process, shown in Fig. 12, was then repeated until no improvement could be obtained in the estimated values. Whilst an initial linear decrease of about 2%/m, as shown in Fig. 13, appears to be the most appropriate there is sufficient scatter in the data to enable other interpretations, particularly in the upper few metres, to be assumed. This is a more rapid decrease than seen during static loading. As shown on Fig. 14, during environmental loading the measured and predicted distribution of axial load down a pile compare well.

The maximum environmental axial load and bending moment acting on the pile group have been deduced to be 82MN and 197MNm peak to peak, respectively. The distribution of pile and mudmat loads is consistent with the application of an overall load and bending moment to a rigid pile cap and may be described by a linear variation along the predominant storm direction.

No evidence of non-linear behaviour has been observed, although overall loading has been much less than assumed in design.

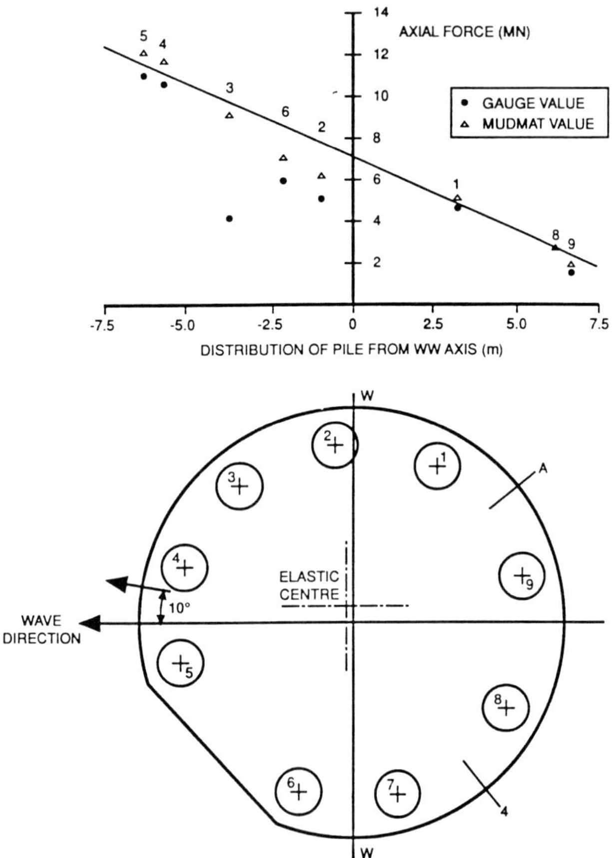

Fig. 12. Derivation of pile axial load distribution

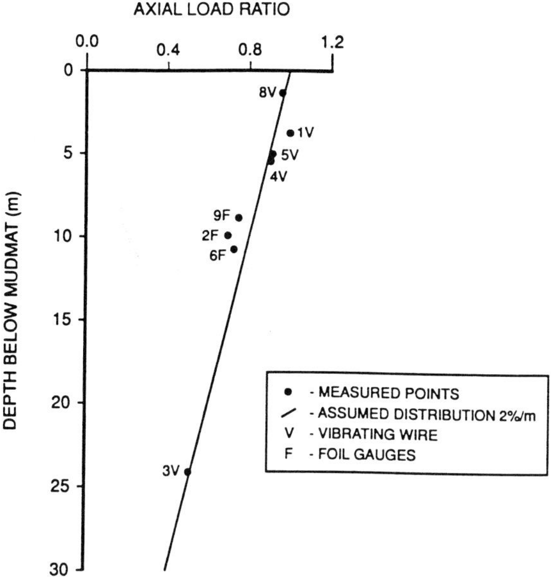

Ratio is the ratio of axial load at the gauge position compared with the estimated axial load at mudmat level.

Fig. 13. Derived axial load distribution down a pile

9.5 The effective stiffness of a pile group.

Estimates of the stiffness of the pile group have been deduced from displacements, derived both from measured force data and from accelerometer data.

9.5.1 Axial stiffness. In the case of axial loading, stiffness estimates are derived from the cyclic component of axial load only, not the total axial load, and thus are equivalent to a tangent rather than a secant modulus. The pile group has proved to be considerably stiffer than predicted, more axially than rotationally. Estimates of the measured dynamic axial stiffness range from 20 to 27 MN/mm, derived from a spectral analysis and reported by Skjastad and Skomedal (1992) up to 32MN/mm, derived from analysis of individual waves and reported by Kenley and Sharp (1992). There was no evidence of non-linear axial behaviour.

Using today's analytical tools, allowing for the contribution from the mudmat and increasing the Young's modulus of the soil, to the upper end of the range considered appropriate for Magnus, Horsnell, Norris and Ims (1992) predict an axial stiffness of less than 10MN/mm which is far lower than that measured. They suggest that conventional analyses may treat interaction effects incorrectly at load levels experienced by the Magnus piling. Using a non-linear finite element analysis and soil parameters derived from a series of laboratory tests performed at Imperial College, Jardine and Potts (1992) predict a dynamic axial stiffness for the Magnus foundations of up to 32MN/mm which compares well with measurements. Kenley and Sharp (1992) suggest that differences between measured and predicted axial stiffness are mainly associated with interaction effects. Further detailed work is warranted to understand more fully pile interaction mechanisms and to devise improved modelling techniques.

The load deflection behaviour of the mudmat was linear over the range of dynamic loads measured and the vertical stiffness was about 2700MN/m. This compares well with an assessment made during design of about 2400MN/m.

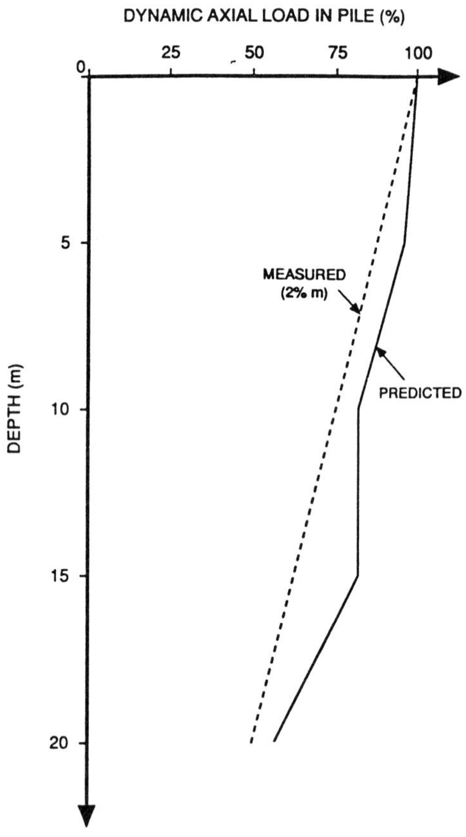

Fig. 14. Comparison of predicted and measured dynamic load distributions

9.5.2 Lateral stiffness. Estimates of the lateral stiffness of the pile group range from 2.4MN/mm up to about 3.5MN/mm which is 2 to 3 times that predicted at the design stage. The foundation behaved in a linear manner throughout any particular storm but there appears to have been a reduction in overall lateral stiffness of about 20% in stormy conditions compared with calm conditions. The lateral stiffness thus appeared to be dependent more upon the overall energy level of a storm and the average period of the waves during the storm than the severity of an individual wave. Evidence from the recent, more severe, storms confirms this tendency which suggests either that there is some softening due to pore pressure accumulation during a storm or that the foundation response is influenced by the rate of loading.

Increasing foundation stiffness by a factor of three is computed to increase the fundamental natural frequencies by about 8% and thus is clearly not the only factor contributing to the difference between the measured and predicted natural frequencies of the platform. This aspect is discussed in detail by Webb and Corr (1984).

9.6 Comparison of actual foundation behaviour with design predictions
In general form, the measured and predicted behaviour is similar which gives reasonable confidence in the current design methods for off-shore structures. There were however significant differences in the numerical values of certain measured and predicted parameters as described above. In particular, the results have demonstrated that it is important to distinguish between the behaviour of the foundation in the long term and during wave loading. This is perhaps most obvious with the loading carried by the mudmat. Initially, on placement, it had to carry all of the applied the load. Most of this transferred to the piles over a period of about a year such that in the long term the mudmat carried only a small amount of static load. However, during wave loading, the mudmat again carried a significant percentage of the transient loading.

10. Design implications

10.1 Overall
The design and analysis procedures used to predict the environmental loads applied to the foundation were conservative. Structural designers may wish to examine the assumptions of the structure's natural period, the inertia coefficient (Cm), wave spreading and the design idealisation of a train of identical, as opposed to random, waves. Generally, techniques used to design the foundations were also slightly conservative.

10.2 Axial capacity and load distribution

On Magnus, as with most structures of its era, the number of piles per leg was determined by the maximum expected axial load on the foundations. The contribution of the mudmat, particularly during storm loading, was significant. If operators are prepared to ensure that mudmats make good contact with the seabed, design can derive benefit from the mudmat contribution, particularly on strong cohesive soils.

As shown on Fig. 12, which plots peak to peak load variation against pile position, during storms, piles on the leading edge of the pile group tended to see the largest fluctuation in load whilst those on the trailing edge experienced the smallest fluctuations. The prediction of the highest individual pile load is used to design the pile/structure connection and is a factor in determining pile wall thickness. Design methods appeared to predict both the distribution of axial load between piles and the maximum individual pile load reasonably well.

At the load levels experienced, the reduction in axial load down a pile has few design consequences for a structure like Magnus. However it is reassuring that during storm loading the load variation measured down the piles was broadly as predicted.

10.3 Behaviour under lateral load

Whilst significant differences were found between measured and predicted pile head bending moments this was of little consequence to Magnus because pile bending moments did not affect design of Magnus's battered piling. Pile wall thicknesses were determined by drivability considerations. However, on platforms with high ratios of lateral load to axial load, for example when vertical piles are employed, pile head bending moments are very important in determining a pile's section properties and thus the differences, which are likely due to inaccurate modelling of pile head fixity and lateral foundation stiffness, are also important.

10.4 Foundation stiffness

The foundation was measured to be at least twice as stiff in any mode as that predicted. This will impact on the predicted natural frequency of the platform, but will only account for about one-third of the difference between the natural frequency measured and that predicted. Even if it was to be assumed that the foundations were rigid, analysis has shown that it would still not be possible to account for all of that difference. Other factors which will affect the natural frequency of the platform include:

- jacket member stiffness properties;
- jacket and topside loads;

- marine growth;
- added hydrodynamic mass.

Of these factors all except the added hydrodynamic mass are known to a high degree of confidence. The added hydrodynamic mass depends on as-built member properties and an inertia (added mass) coefficient (Cm). Adoption of the measured natural frequency in design leads directly to lower dynamic amplification factors.

Increasing the foundation stiffness would result in significantly lower pile head bending moments. The overall axial loading on pile groups on a structure like Magnus is relatively independent of foundation stiffness but on less structurally determinate structures, such as those with more than four pile groups or with piles arranged around the base of the structure, the foundation stiffness will influence structural load paths strongly and will have particular implications for the design of compliant towers proposed for development in certain deep water situations.

11. Conclusions

11.1. An overall objective of FMP was to reduce the uncertainty associated with the design of off-shore piling. This it has certainly achieved whilst unexpected benefits have accrued to improving structural and environmental design. FMP provides valuable support during regular platform re-certification.

11.2. The Magnus instrumentation system can serve as a model for future instrumentation projects of piled off-shore structures. It is expected that developments in instrumentation and in the integrated circuit technology would lead to changes in downhole amplifier and signal conditioning systems.

11.3. The project benefited from the involvement of a few key staff in all stages from system development through to data analysis. Attention to detail during design and installation was vital.

11.4. In general, the long term loads and load changes in the foundation loads are consistent with expectations and compare well with the applied loads. However, the implications of the observation that there appears to be little or no take out in static axial load over the top 25m of the piling should be explored in the context of the similar observations from pile tests at Tilbrook, described by Gibbs *et al* (1992) and Clarke *et al* (1992) later in this conference.

11.5. Dynamically, the structure and foundation behave like a single integrated system at all frequencies. Response of the system is quasi-static and the stiffness mechanism of the soil-structure interaction is predominantly linear.

12. Acknowledgements

The Author wishes to thank The British Petroleum Company plc for permission to publish this paper.

FMP was a joint industry funded project and the contribution, both financially and technically, of the following organisations is gratefully acknowledged:

- BP Petroleum Development Limited
- Britoil plc
  (Both of the above are now part of BP Exploration)
- Brown and Root Inc.
- Engineers India Limited
- Exxon Production Research Co.
- Total CFP
- Marathon Oil Co.
- Den Norske Stats Oljeselskap AS (Statoil)
- John Brown Offshore Structures Ltd
- UK Department of Energy
- Commission of the European Communities.

I would like to acknowledge the contribution of the myriad of consultants and contractors, and specific individuals, whose talents and commitment were so crucial in ensuring firstly that FMP was initiated and secondly that it was implemented so successfully. In particular, the dedication of staff from Queen Mary College, London, the consistent involvement of key staff from Structural Monitoring from conceptual planning through to data analysis and the contribution of many colleagues within BP was invaluable.

References

BOWLEY, C. (1992) Design implications arising from the Magnus foundation monitoring project. This volume.

CLARKE, J., LONG, M. M. and HAMILTON, J. (1992) Tilbrook Grange tension tests. This volume.

GIBBS, C. E., MCAULEY, J., COX, W. R. and MIRZA, U. (1992) Reduction of field data and interpretation of results. This volume.

HORSNELL, M. R., NORRIS, V. A. and IMS, B. (1992) Mudmat interaction and foundation analysis. This volume.

JARDINE, R. J., HIGHT, D. W. and MCINTOSH, W. (1988) Hutton tension leg platform foundations: measurement of pile group load–displacement relations. *Géotechnique* **38,** No. 2, 219–230.

JARDINE, R. J. and POTTS, D. M. (1992) Magnus foundations: soil properties and predictions of field behaviour. This volume.

KENLEY, M. and SHARP, D. E. (1992) Magnus FMP — Instrumentation, data processing and measured results. This volume.

REESE, L. C., COX, W. R. and KOOP, F. D. (1975) Field testing and analysis of laterally loaded piles in stiff clay. Offshore Technology Conference, Houston

RIGDEN, W. J. and SEMPLE, R. (1983) Design and installation of the Magnus foundations: prediction of pile behaviour. *Design in off-shore structures*, London: Thomas Telford.

RIGDEN, W. J. and SEMPLE, R. (1983) Design and installation of the Magnus foundations: installation studies and platform installation. *Design in off-shore structures*, London: Thomas Telford.

SEMPLE, R. and RIGDEN, W. J. (1983) Site Investigation for Magnus. Offshore Technology Conference, Houston.

SKJASTAD, O. and SKOMEDAL, E. (1992) Magnus FMP - Dynamic response of foundation. This volume.

SOLOMON, I. J., COX, W. J., CLARKE, J. and POSKITT, T. J. (1992) The development of instrumentation for drivability testing and load testing of large diameter piles. This volume.

SPIDSOE, N., BERG, S., HOEN, C. and BECK, G. (1980) Measured behaviour of platforms on the Norwegian continental shelf. European Offshore Petroleum Conference and Exhibition. London.

WEBB, R. and CORR, R. B. (1989) Full scale measurements at Magnus. The oil industry international exploration and production forum workshop: Waves and current kinematics and loading, IFP.

2. Magnus foundation monitoring project instrumentation data processing and measured results

R. M. KENLEY, Structural Monitoring Division, Fugro McClelland Limited, and D. E. SHARP, BP Engineering

Introduction

The Magnus Foundation Monitoring Project was a major instrumentation project involving the measurement of the actual performance of the foundations of the Magnus platform under service conditions. The overall objective of the project was to obtain basic information to verify foundation design methods.

The intention of the instrumentation system was to obtain as much information as possible about the behaviour of the foundation on one of the legs, leg A4. This included the total load in the leg, the mudmat load, the loads in each pile, motion of the pile head, long term settlement, and pore and total pressures along one pile.

Strain gauges, accelerometers and settlement gauges were installed at the base of leg A4, total pressure cells were installed in the mudmat, strain gauges were installed at one level on each pile, and one pile was fitted with additional strain gauges and with pore and total pressure cells. In addition, wave sensors were installed to give wave height information and accelerometers were installed at deck and base level to give platform motion.

The sensors were connected to recording units situated at deck level. Initially, a Temporary Data Logger (TDL) was installed inside the top of the leg. This recorded static data only from the strain gauges on the piles and the mudmat pressure cells. The object was to provide information on the variation in static load distribution between the mudmat and piles immediately after platform installation.

The TDL was later replaced by a permanent Data Acquisition Unit (DAU) which collected time series data from all the sensors on a regular basis. The main objective was to provide information on dynamic performance.

The sensors were initially installed on the platform and piles during fabrication in 1981 and 1982. The platform was floated out at the end of March 1982. During upending, six of the seven pre-installed piles on leg A4 were released prematurely and lost. Replacement piles were fabricated but within the timescale it was not possible to re-install all of the instrumentation. In particular, the pore and total pressure cells could not be installed and only one type of strain gauge was fitted instead of the two types on the original piles.

The platform was successfully placed on the seabed on 4 April 1982. The piles were driven between 6 April and 17 June. Grouting of pile sleeves was completed on 4 July. The instrumentation from five of the piles was connected to the TDL before module loading which began on 9 July and finished on 17 July. The remaining pile instrumentation was connected up on 6 August. The TDL was operated until July 1983. The DAU was commissioned in December 1983 and operated continuously up to the end of April 1985.

Data from both the TDL and DAU were analysed in some detail onshore. Information on the static performance of the foundations was mainly based on the TDL data. Information on the dynamic performance was obtained from the data collected by the DAU between December 1983 and April 1985. Data was also collected over the 1985/86 winter but only one recording, during a severe storm in January 1986, was subsequently analysed.

Information obtained from the data analysis was compared with theoretical predictions made previously during design. These comparisons were not exhaustive and were intended primarily to identify areas of interest for possible further investigation. Details of the design of the Magnus foundations were given by Rigden and Semple (refs 1 and 2)

The DAU was replaced by a smaller computer based data acquisition and analysis system in 1987 which was connected to the structural strain gauges, accelerometers and wave sensors only. No data has been obtained from the pile and mudmat sensors since 1986 although they are still present.

Instrumentation systems

Figure 1 shows the principal instrumentation which was finally installed on the platform and piles.

Wave sensors

A laser-type wave sensor, manufactured by EMI, was installed on both the platform North-East (leg A2) and South West (Leg E4) corners.

Accelerometers

A triaxial set of linear accelerometers (x,y,z) were installed at deck level (+26m) and at the base (-182m) of leg A4, along with a bi-axial set of angular accelerometers about the horizontal axes at the base of leg A4. These were used to measure platform displacements.

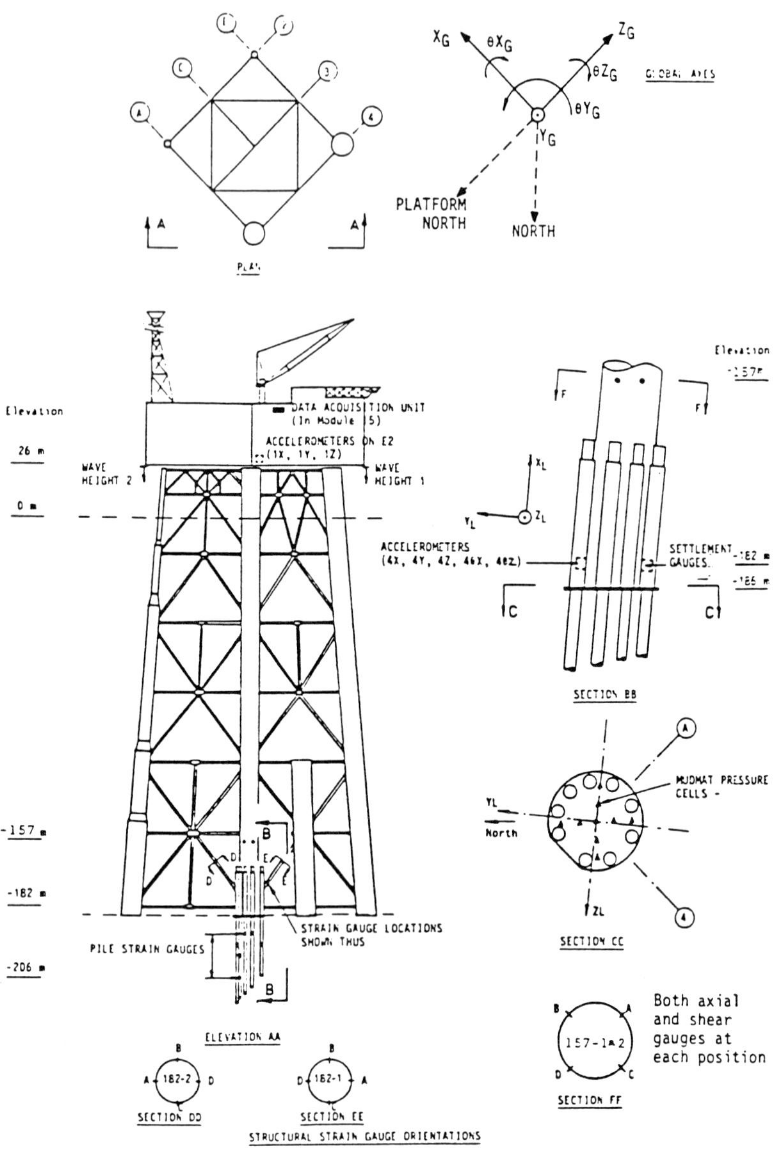

Fig. 1. Location of sensors

Settlement gauges

Two settlement gauges were installed at the bottom of leg A4. They were of special design and measured pressure difference between a point on the platform and a reference point on the seabed nearby.

Structural strain gauges
Ailtech weldable strain gauges were installed on leg A4 at -157m, which is just above the top of the pile sleeves, and at -182m on the two diagonal braces which meet the base of leg A4. At each location there were four longitudinal gauges at 90 degree intervals around the member circumference to give axial loads and bending moments. In addition, on the main leg only, there were shear gauges to give shear loads.

Mudmat pressure cells
Nine total pressure cells were installed underneath the mudmat of leg A4 in order to determine the static and dynamic loads transmitted through the mudmat. These use vibrating wire type sensing elements.

Pile strain gauges
In the final as-built configuration, strain gauges were installed on all the piles of leg A4 except for pile 7. There were gauges at 3 levels on pile 3 and at one level on the other piles. On piles 2, 6 and 9, two types of strain gauges were installed at each location-a Sinco vibrating wire type and a HBM foil resistance type. On the remaining piles, there were vibrating wire type only. Initially it was intended to install both types of gauges at each location on each pile but it was only possible to replace the vibrating wire gauges following the loss of 6 of the instrumented piles during upending.

Two different types of gauges were used for redundancy and to guard against any possible unforeseen generic fault. It was anticipated that the vibrating wire gauges would provide better long-term stability and the foil gauges would be better dynamically.

Gauge protection and cable connections
All the sensors installed below the waterline and their cabling were well protected against mechanical damage and water ingress. Cable connections under water were minimised and, where essential, high quality moulded connections were used. The structural strain gauges, base accelerometers and mudmat pressure cells were hard wired back to junction boxes inside the top of leg A4 in the fabrication yard.

The pile strain gauges were hard wired to a length of cable which was coiled inside the top of each pile. This was subsequently pulled through conduits pre-installed on the jacket and connected to junction boxes inside the top of leg A4.

Temporary data logger (TDL)
The TDL collected and processed data from all the mudmat pressure cells, all foil strain gauges, 24 vibrating wire gauges and all settlement gauges. It sampled data for 5 minutes every 6 hours from all connected sensors, com-

puted mean, maximum, minimum and standard deviation values and stored results on a cartridge tape. The TDL was intended primarily to give static information obtained from the mean values.

Data acquisition unit (DAU)

The DAU was commissioned in December 1983, approximately 5 months after the TDL was removed. There was little change in the static sensor values over this time so the gap in the data was not significant. The DAU consisted of signal conditioning for all the sensors, a microprocessor to sample data from all sensors and a tape recorder to store time series data.

In normal operation, data was recorded for 40 minutes every 6 hours but, during stormy weather, recording was continuous. The DAU operated entirely automatically apart from tape changes which were made manually when the tapes were full. The tapes were then sent ashore for analysis.

System performance

The instrumentation system performed well throughout the project with high quality information being obtained from the majority of the sensors. The accelerometers, structural strain gauges and mudmat pressure cells gave the best performance; high quality data was obtained and only one sensor out of 32 failed. A number of pile strain gauges either failed during installation or failed before the end of the main data analysis in April 1985. However, 34 of the original 52 gauges were working then and, because of redundancy, little information was lost.

Data quality was good overall except that, after the first few months, the foil gauges drifted and could not be used for long term information; the vibrating wire gauges had high long term stability. One of the settlement gauges failed during pile driving and the other never gave reliable data. Unfortunately, therefore, no information on long term settlement was obtained.

Table 1 summarises the status of the sensors throughout the project.

Table 1. Status of sensors

Sensor	No. orig. intended	No. finally installed	No. working after floatout	No working June 1984	No. working April 1985
Structural strain	16	16	16	16	16
Accels	8	8	8	8	7
Mudmat pressure	9	9	9	9	9
Settlement	2	2	1	0	0
Pile Pressure	10	0	0	0	0
Pile VW strain	48	40	33	30	28
Pile foil strain	48	12	10	10	6

Data analysis procedures

The TDL obtained mean, maxima and standard deviation values for each signal and stored them on tape. The tapes were replayed onshore and the statistical values stored in a data base. The data was used to determine long term behaviour mainly by plotting values against time.

The DAU recorded only time series signals offshore. Data was analysed by replaying data into a computer onshore and initially a routine analysis was undertaken. This analysis obtained mean, maxima, standard deviation and period values for each signal. In addition, the acceleration signals were integrated to give displacements and wave direction was estimated from the direction of platform motion. The strain signals were combined to give axial and bending loads in the structure and piles. The mudmat gauges were analysed to give axial forces and bending moments taken by the mudmat.

Normally a routine analysis was undertaken on one recording each day, although additional recordings were analysed during stormy conditions. The results were stored in a database and were examined primarily by plotting one parameter against another or against time. They were used to determine general behaviour of the platform and to identify interesting periods for more detailed analysis.

In particular, 9 recordings were selected and subvjected to a detailed analysis. This included 5 storms and 4 recordings, when the significant wave height was about 4m for 4 different wave directions. Table 2 summarises the characteristics of the recordings selected for detailed analysis.

Table 2. Recordings selected for detailed analysis

Date	Sign wave Ht (m)	Period (secs)	Direction (degs)	Standard deviation Disps (mm)	Standard deviation A4 leg force (MN)
17/01/84	4.2	7.6	178	10.6	3.0
22/01/84	10.7	8.0	168	27.5	8.2
05/02/84	9.0	N/A	347	21.7	3.6
21/02/84	4.8	N/A	151	11.6	3.3
22/11/84	7.1	8.8	157	19.2	5.4
27/11/84	9.2	9.6	198	21.9	5.7
22/01/85	4.8	7.1	51	9.3	2.4
01/02/85	5.4	8.3	90	13.5	2.3
10/01/86	12.6	11.4	160	35.2	10.1

The long term behaviour of the foundations was examined by plotting the mean values of sensor signals from both the TDL and DAU data against time. Results are given in the section headed 'Long-term foundation behaviour'.

The dynamic response of the structure and foundation was found by undertaking a number of separate but related analysis methods. Initially the analysis concentrated on a detailed study of single selected waves. This is the simplest approach and can be directly related to a specific applied static load. It yields a considerable understanding of the behaviour of the foundation.

The general applicability of the dynamic results was then demonstrated by a statistical analysis of selected recordings and by examining the largest waves in all recordings. Examples of results are presented in the section headed 'Dynamic behaviour'.

Considerable care was exercised to ensure that maximum possible accuracy was obtained from the measurement and analysis system. Where possible all instrumentation was calibrated before installation. Checks on accuracy were carried out in service by comparing similar results from different gauges. Axial loads, for example, could be computed in two separate ways from opposite pairs of gauges; the foil and vibrating wire gauges on the piles were at the same location and could be directly compared.

It was considered that the static mudmat loads were accurate to about 2.5 MN, and static pile loads to about 1MN. Dynamically the accuracy was considered to be higher, typically better than about 15 per cent.

Long-term foundation behaviour

Information from the sensors was combined to obtain the total load carried by the mudmat and the axial loads in the piles. The mean values were plotted against time to determine the long term variations in load.

The left-hand plot in Figure 2 shows the variation in total mudmat load from touchdown in April 1982 up to April 1985. Data up to April 1983 was obtained from the TDL and after December 1983 from the DAU. No data were obtained until early May 1982, except for a manual reading just after touchdown. On touchdown the measured load was about 20MN (i.e. on top of the load due to hydrostatic pressure of 350MN). This corresponds closely to the load due to self weight of the platform.

Soon after touchdown, the legs were flooded andit is expected that the load on leg A4 increased to 50MN. The next measurement was made in early May, after two piles had been driven but not grouted, and only showed a load of 25MN.This suggests about 25MN of load was transferred to the piles through friction between piles and guides.

During the first half of May the measured mudmat load increased to 35MN during pile driving operations on the other legs. This suggests slip between the platform and the ungrouted piles on leg A4, resulting in load transfer back into the mudmat. In the second half of May, after the 2 piles had been grouted, the load remained constant.

During early June the mudmat load decreased to about 20MN during driving of the remaining piles on leg A4. Load may have been transferred either to the two grouted piles or to the new ones in friction. The remaining piles were grouted between 9 June and 4 July.

During module loading, the mudmat loads increased by between 4 and 5MN compared with an applied load increase of about 65MN. The total load on the mudmat above hydrostatic at this time was about 25MN. The net final load on the mudmat when a stable position was reached in mid 1983 was about 9MN.

The right-hand plot of Fig. 2 shows the variation in axial load in one pile. Similar variations in form occurred in all piles, although there were differences in the magnitudes of loads from pile to pile.

In this Figure, the load is plotted to a common zero reference of 6 August; this was when all piles were connected to the TDL. Estimates of absolute loads were available based on readings taken in the fabrication yard when the piles were horizontal. Changes in load of between 5 and 7MN occurred in piles for which measurements were available during module loading, which is consistent with an anticipated increase of about 60MN on the pile group.

There was a steady increase in pile load from after module loading until about mid 1983. The overall load on the pile group at this time was estimated to be about 130 MN. The increase after module loading was partly due to load transfer from the mudmat and partly due to increases in deck weight as the platform was being commissioned.

Table 3 summarises the estimates made of measured and applied loads at three stages of the project. The total measured load (mudmat and pile group) agreed reasonably well with the applied load giving confidence in the accuracy of the measurements. It should be noted that, by April 1985, the mudmat load is small.

Table 3. Comparison of measured and applied loads

	Jacket only 7/82	Modules and jacket 6/8/82	Final 22/4/85
Pile group	52	107	132
Mudmat	15	19	9
Total measured	67	126	141
Applied load	9	114	130

All values are leg A4 loads in MN.

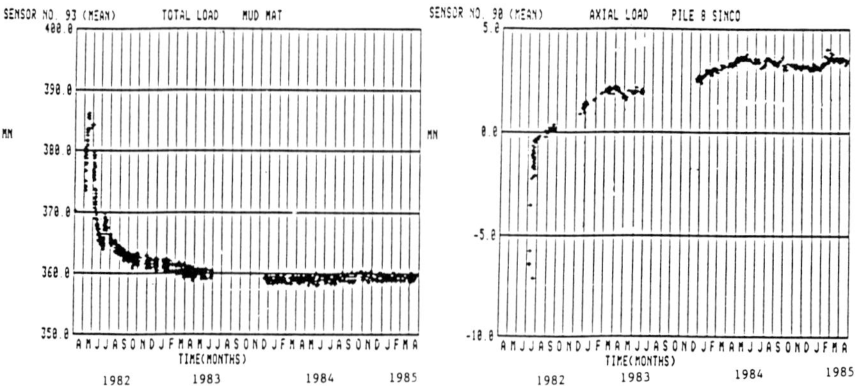

Fig. 2. Long-term mudmat and pile variations

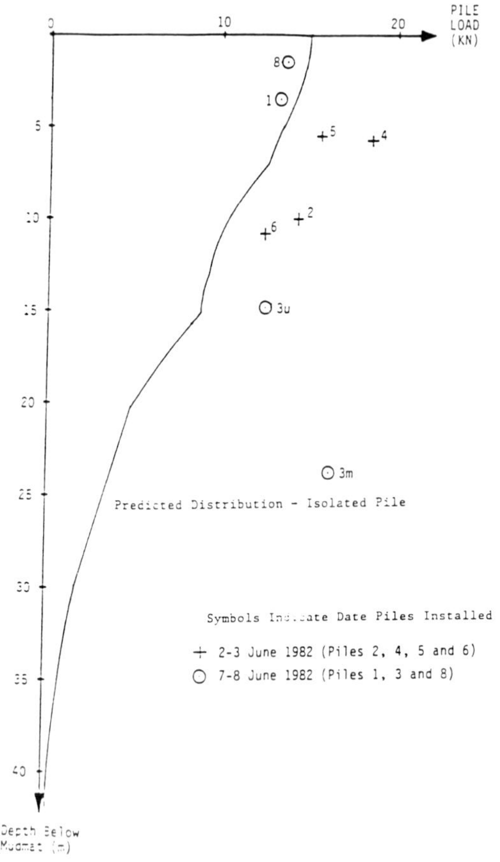

Fig. 3 Static pile axial load distribution

Figure 3 plots the absolute axial loads at each gauge location against the depth of the gauges below mudmat level for the final measurements in April 1985. If it is assumed that all piles are equally loaded and behave in a similar manner, then the plot will approximate the static axial load distribution down a pile. There is no obvious decrease in axial load with depth in the measured data. The solid line of Figure 3 is the theoretical distribution of axial load for an isolated pile at Magnus carrying a 15MN load (cf the design pile capacity of 60 MN). This was calculated using the Coyle and Reese method (ref. 3). The theoretical distribution does not allow for interaction effects between the piles of the group which may account for the poor agreement with the measured data. Also, as the piles were installed at different times and may be unequally loaded, it is strictly not valid to directly compare the measured and theoretical distributions.

Because of the failure of the settlement gauges, no information was obtained on the long term settlement of the platform.

Dynamic behaviour

Time history signals and general response characteristics

Figure 4 shows examples of time series signals. They were from the largest storm recorded by the system on 10 January 1986, when the significant wave height reached 12.6m with an individual wave of 21m. The plots show a simultaneous 200-second portion of signal which includes the largest wave. Only the dynamic portion of the signals is shown; the static portion was removed prior to plotting.

The upper plot is the wave signal; it is random in form with a zero-crossing period of about 12 seconds. The next plot is deck displacement of the platform; this is similar in form to the wave signal, indicating that most of the motion of the platform is in direct response to wave action. The peak-to-peak displacement was about 230mm.

The third plot is the axial load in leg A4 at minus 157m. This is virtually identical to the displacement signal and has a maximum peak-to-peak value of about 70MN. The fourth plot, mudmat load, is also similar with a maximum peak to peak value of about 17MN. The final plot is axial force in one pile, which has a maximum peak-to-peak value of about 13MN. The main feature of the plots is the high degree of correlation between the various signals which suggests that the platform and its foundations are responding in a quasistatic manner to the wave loading.

Short-term distributions of signal amplitude were calculated for most sensor signals. The upper plot in Fig. 5 shows a typical example. The histogram bars are measured data and the solid line is a Gaussian distribution of the same standard deviation. The measured data is approximately Gaussian in form. This was found to be the case for all signals examined.

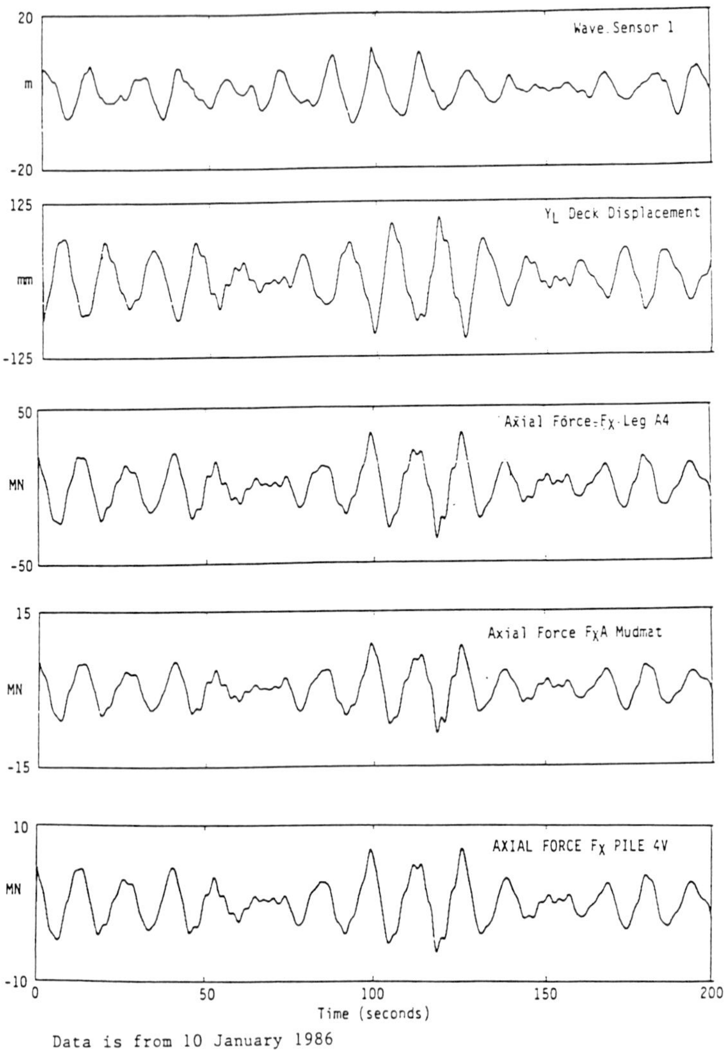

Fig. 4. Examples of time history signals

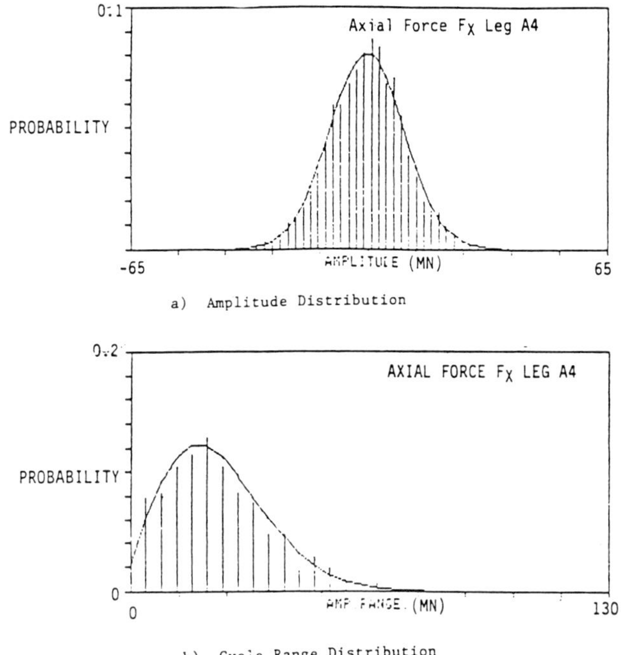

a) Amplitude Distribution

b) Cycle Range Distribution

Fig. 5 Examples of short term distributions

Short-term distributions of cycle ranges were also calculated. The lower plot of Figure 5 shows a typical example. Again, the histogram bars are measured data; in this case, the solid line is a Rayleigh distribution of the same standard deviation. It was found that all the measured cycle count distributions were approximately Rayleigh.

The distributions obtained from the detailed analysis showed that the signals were well behaved and could be simply described in statistical terms. The form of the distributions corresponds to those typically assumed in design analysis.

Selected individual waves

The largest wave in each of the recordings analysed in detail was selected along with the corresponding cycles of displacement, leg A4 load, mudmat load and axial forces and bending moments in the piles. The cycle ranges between upcrossings were obtained as illustrated in Fig. 6a and behaviour was studied by comparing the equivalent individual cycles for the various signals. The wave direction for a cycle was estimated by plotting the time histories of the horizontal deck displacements against each other; Fig. 6b shows an example for the largest recorded cycle. The load and bending moment cycles were then resolved in this direction.

Table 4 shows the maximum individual cycles for various signals from the two largest storms on 22/1/84 and 10/1/86. This data is used as examples of the results in the following sections.

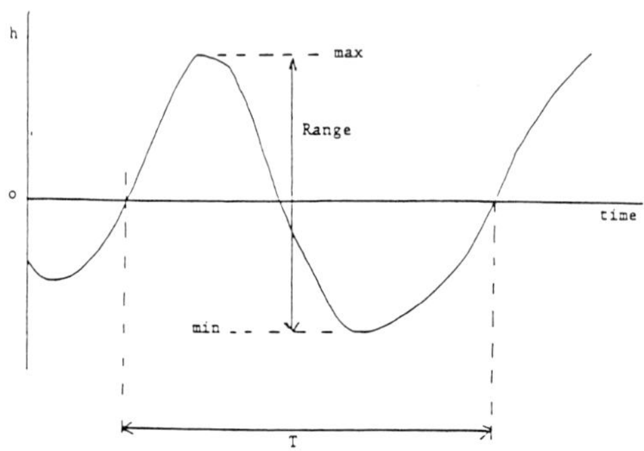

Fig. 6a Selection of a cycle

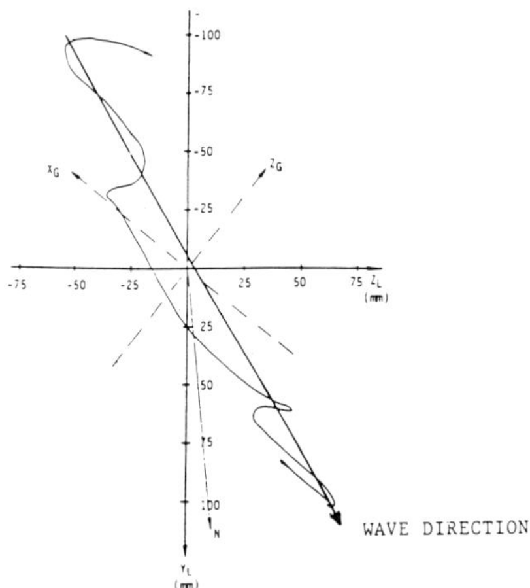

Fig. 6b Direction of largest cycle – 10 Janaury 1986

Table 4. Selected cycle values

	22.1.84	10.1.86
Wave (m)	17	21
Direction (degrees)	175	155
Deck disp– lateral (mm)	168	230
Base disp– lateral (mm)	5.7	8.4
Base rotation (mrads)	0.25	0.42
A4 axial force (MN)	54.9	68.3
Diag. 1 axial (MN)	10.4	10.8
Diag. 2 axial (MN)	9.6	18.4
Mudmat force (MN)	12,8	17.2
Mudmat moment (MNm)	16.4	28.3
Pile axial forces (MN)		
1	4.84	4.87
2	5.20	5.34
3	4.96	5.92
4	10.2	13.4
5	10.7	—
6	6.1	9.7
8	—	3.4
9	1.7	—

It should be noted that the maximum amplitude (max relative to zero) of the measured force in leg A4 was about 35MN compared to a 100 year design value of total A4 load of about 130MN. Consequently, the loads experienced by the foundations were well short of their ultimate capacity. The 21m maximum wave height compares to a design value of 30m.

Data has been obtained from the wave and leg A4 structural sensors since 1987. The highest waves experienced to date were in a storm on 22 December 1990 when the individual wave height exceeded 25m and the maximum amplitude force in leg A4 was over 45MN.

Lateral loading of piles

The behaviour of the piles under lateral loading was examined from individual cycle data. Ideally this would be done by obtaining a bending moment distribution along the pile. Such a distribution cannot be measured directly in this case as there are, at most, only two instrumented levels on each pile. If it can be assumed, however, that all piles behave in a similar manner under lateral loading results from the instrumented locations on each pile could be combined and a distribution inferred. It should be noted that design calculations predict similar lateral loadings on all piles to within 5 per cent.

Figure 7a shows an example of results obtained from the largest cycle for the 10 January 1986 storm. The pile bending moments are plotted against the depth of the strain gauges from mudmat level. The curve is the best fit to the data points, drawn by eye. It is postulated that this curve represents the bending moment distribution down a pile.

The pile bending moments have very low values below 20m depth. They increase in magnitude above 20m to reach a maximum of about 2500kNm at about 6m and then decrease rapidly to zero just below mudmat level.

Figure 7b compares measured and predicted bending moment distributions from 22 January 1984. The moments were predicted using the method of Matlock and Reese (ref. 4) with non-linear p-y curves appropriate to repeated loading (ref. 5). The interaction effects between piles were estimated using the Poulos and Davies method (ref. 6). The predicted maximum value of bending moment occurs at a much lower depth than that measured and the predicted bending moment at mudline is much greater than measured.

It is possible to use the measured bending moment distribution to estimate the pile displacement and rotations which are shown in Figures 7c and 7d for the 22 January 1984 data. Table 5 compares measured and predicted values of mudline displacements, rotations and bending moments for similar applied load levels. Displacements and rotations were obtained from the accelerometers at the base of leg A4 and these values were consistent with estimates obtained from pile displacement and rotation distributions. The predicted lateral displacement is about twice that measured.

The differences in the measured and predicted bending moment and displacement distributions are entirely consistent with differences between

Table 5. Comparison of pile bending moments and displacements

	Measured	Predicted
Lateral displacement (mm)	2.85	6.0
Rotation (m rads)	0.125	0.18
Pile bending moment at mudline (KNm)	750	4000
Max sub-soil bending moment (KNm)	900	1390
Depth of max below mudline (m)	7	14

Predictions based on the same force in leg A4 as measured. Values are maxima relative to zero.

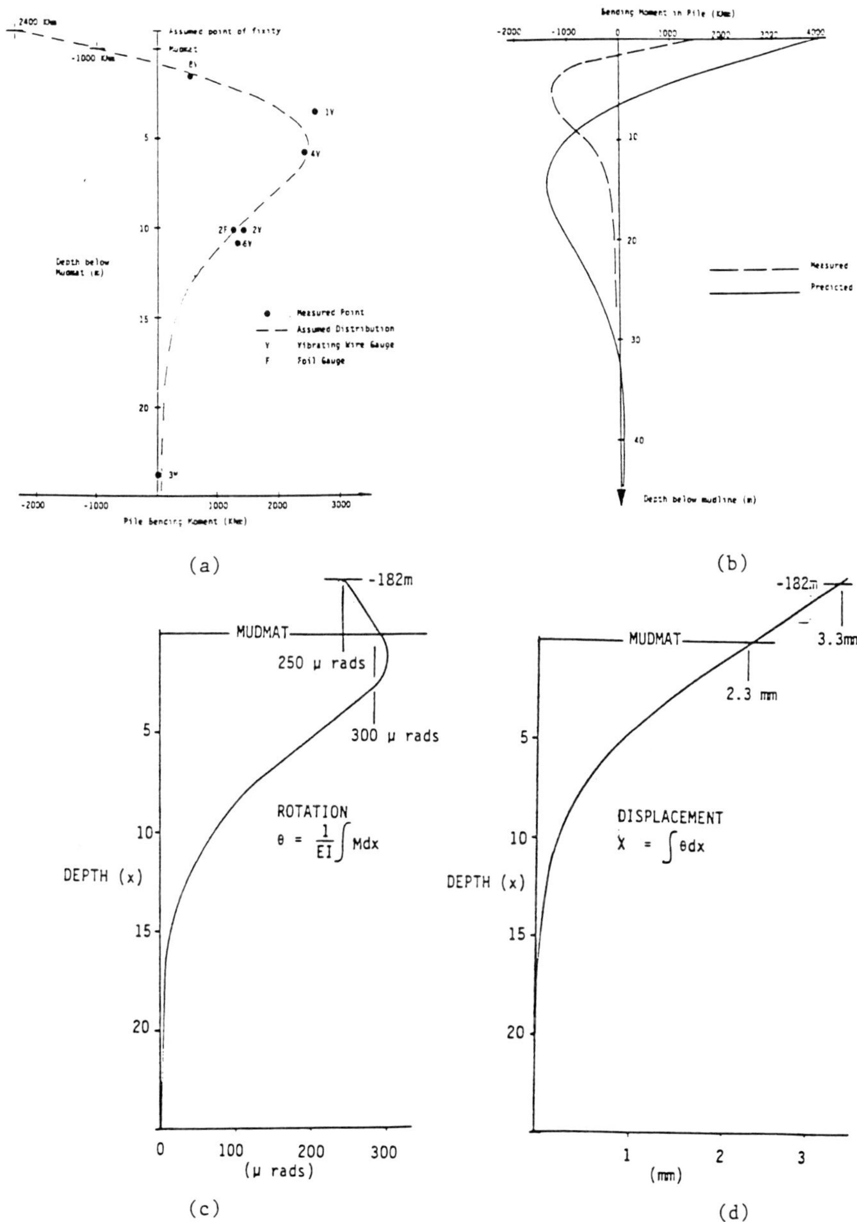

Fig. 7. Bending moment, rotation and lateral displacement distributions

measured and predicted soil stiffness. An increase in soil stiffness will raise the position of the maximum bending moment, reduces the mudline bending moment and reduces deflection. The differences between the measurements and predictions can be explained by an increase in effective soil modulus by a factor of about 4.

It is not possible from the measurements to isolate whether the apparent difference in effective soil modulus is due to the actual modulus of the soil, interaction effects or other factors. Interaction, however, was estimated to reduce lateral pile group stiffness by a factor of 2.8 and the differences could be explained if interaction was much less than expected.

Axial loading of piles

The behaviour of the piles under dynamic axial loading was also examined from the individual cycle data. Under dynamic loading there is bending of the leg, which means that each pile in the group will see a different applied axial load. The method used to estimate the bending moment down a pile cannot therefore be used to estimate an axial load distribution down a pile.

An approximate indication of load distribution was however obtained by initially assuming that the load would decrease linearly with depth below the mudmat. An initial estimate of the rate of decrease was made and used to estimate the axial loads in each pile at mudmat level. From these the axial load and bending moment in the pile group was obtained.

Then, assuming that each pile is rigidly attached to the leg, a revised estimate was obtained of axial load at the top of each pile and of the load distribution down a pile. This was repeated until no further improvement could be obtained in the estimated values.

Figure 8 shows an example of the results from the largest cycle from 10 January 1986. The upper plot shows measured forces at each gauge location, estimated forces at mudmat level, and a best fit line to the mudmat forces. The data is plotted against pile position relative to an axis perpendicular to the wave direction. The lower plot shows the ratio of axial force at the gauge position compared with the estimated force at mudmat level, against gauge depth. The solid line is the estimated distribution of force with depth, a decrease of 2%/m. An overall axial load and bending moment taken by the pile group can be obtained from the estimated pile forces at mudline.

It should be noted that the measured axial load distribution under dynamic loading is different to the measured distribution under static loading (Figure 3).

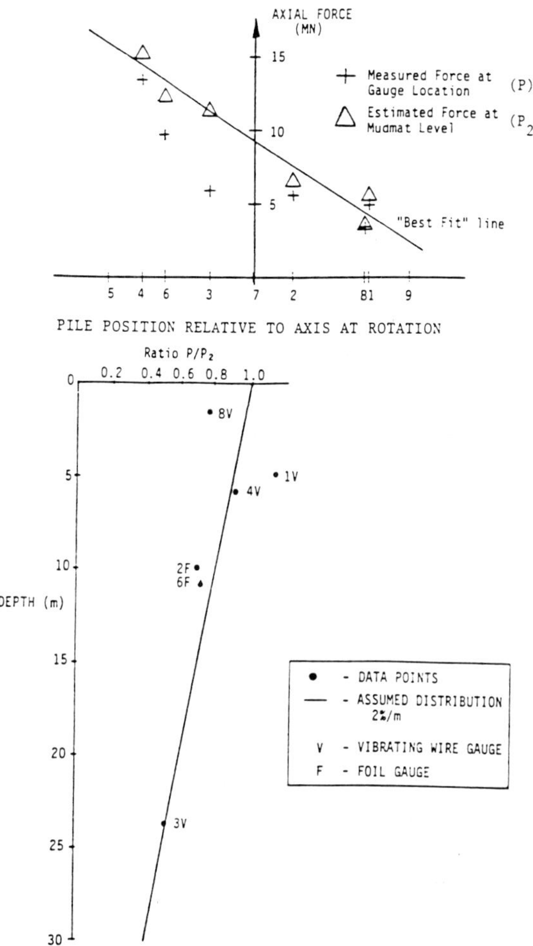

Fig. 8. Axial force distribution– dynamic loading

Table 6. Comparison of measured and predicted axial pile group behaviour

	Measured	Predicted
Pile group axial force (MN)	32	32
Pile Group moment (MNm)	82.5	92
Vertical displacement (MM)	1.0	4.0
Rotation (m rads)	0.125	0.16
Axial stiffness (MN/m)	32 000	8000
Rotational stiffness (MNm)	660 000	575 000
Rotational: axial ratio (sqm)	21	72

Predictions are for the same axial force as that measured. Values are amplitudes (maxima above zero)

Table 6 compares measured and predicted pile group forces, moments and displacements for one cycle. The measured displacements and rotations were obtained from the accelerometers at the base of the leg. The predicted displacements are approximately four times those measured.

Table 6 also compares measured and predicted axial and rotational stiffness and the rotational to axial stiffness for the pile group. This ratio should be independent of the actual soil stiffness, and so the difference cannot be explained by differences in measured and predicted soil stiffness.

The predicted pile group stiffness in Table 6 included the effects of interaction between piles and were used in the design process. Interaction tends to significantly reduce the predicted axial stiffness of a pile group but makes very little difference to the rotational stiffness. The ratio of rotational stiffness to axial stiffness can be used as an indicator of the magnitude of the interaction. It was estimated, if interaction was excluded from the design predictions, that the axial stiffness would rise to about 20,000 MN/m whilst the rotational stiffness would decrease slightly to about 450,000 MN/m. This gives a ratio of rotational to axial stiffness of about 22 which is similar to that measured. This suggests that dynamically there is little interaction between the piles of the group. If the effect of interaction is removed both the predicted axial and rotational stiffnesses are about 50 per cent greater than measured. This may be because the soil itself was stiffer than predicted.

Overall force balance

The instrumentation was designed to measure the total applied loads on leg A4, the mudmat load and the pile group load. Table 7 summarises all the measured loads in the structure and foundations for the selected cycle on 10 January 1986.

Table 7. Force balance on Leg A4

	Structure		Foundation		% diff.
Axial force					
	Main leg	68.3	Mudmat	15.6	10
	Diag.	7.6	Pile group	82.5	
	Diag. 2	13.0			
	Total	88.9	Total	98.1	
Lateral force (MN)					
	Main leg	2.2	Pile group	12.6	25
	Diag. 1	5.4			
	Diag. 2	9.2			
	Total	16.8	Total	12.6	
Bending moment					
	Leg moment	74.4	Mudmat	24.7	2
	Leg force	63.8	Pile group	197.0	
	Diag. 1	21.6	Local pile moments	-21.6	
	Diag. 2	36.8			
	Total	196.6	Total	200.1	

The axial forces in the structure were obtained from the axial forces in the leg and the two diagonal bracing members at the bottom of the leg. The axial forces in the foundations were obtained from the pile group axial force and the mudmat axial loads. The foundation and structural forces agree to within 10 per cent which is considered good in view of the approximate nature of the estimate of pile group axial loads. The axial force taken by the mudmat was approximately 15% of the total load, the remainder being taken by the piles. Dynamically, the mudmat has a significant load-carrying capacity.

The lateral forces in the structure and foundation agree to within about 25 per cent. The assessment of lateral load in the foundation excludes any lateral load taken by the mudmat. This was not measured but is likely to be significant as the mudmat has a 1m skirt. The bending moments in the structure and foundations are very similar.

The results in Table 7 indicate that there is a good force balance between the applied load and that transmitted to the foundations. This increases confidence in the accuracy of the measured results.

General applicability of results

The examples presented above were based on only one wave cycle, the maximum one measured. The general applicability of the results was demonstrated by repeating the analysis for one wave in each of the nine selected recordings and also by statistical analysis of all waves within one recording. Inevitably there was some variation from cycle to cycle because of the random nature of the sea. The underlying feature, however, was that all results were similar and entirely consistent with each other.

Finally the dynamic behaviour was examined by plotting all results from the routine data analysis which covered a wide range of wave heights, periods and directions. Figure 9 shows some examples of results. Figure 9a plots standard deviation of deck displacement measured in the same direction as the wave direction against significant wave height. Each + symbol represents data from one 40 minute recording. The points virtually all lie on a straight line. This suggests that, over the range of the measurements, the applied loading on the structure is linear; there is a very slight indication of non-linearity for the largest storm.

Figure 9b plots maximum axial force cycles against deck displacement cycles in the direction of the A4/E2 diagonal. Each point is obtained from the maximum measured cycle in each recording. There is a strong linear relationship between leg force and displacement.

Figure 9c shows mudmat load cycles against axial force cycles. Figure 9d shows axial force in one pile against axial leg force. Once more, all points lie close to a straight line. All results suggest, that under axial loading, the foundations behave in linear manner for all conditions measured.

Figures 9e and 9f plot the maximum bending moments in piles 4 and 6

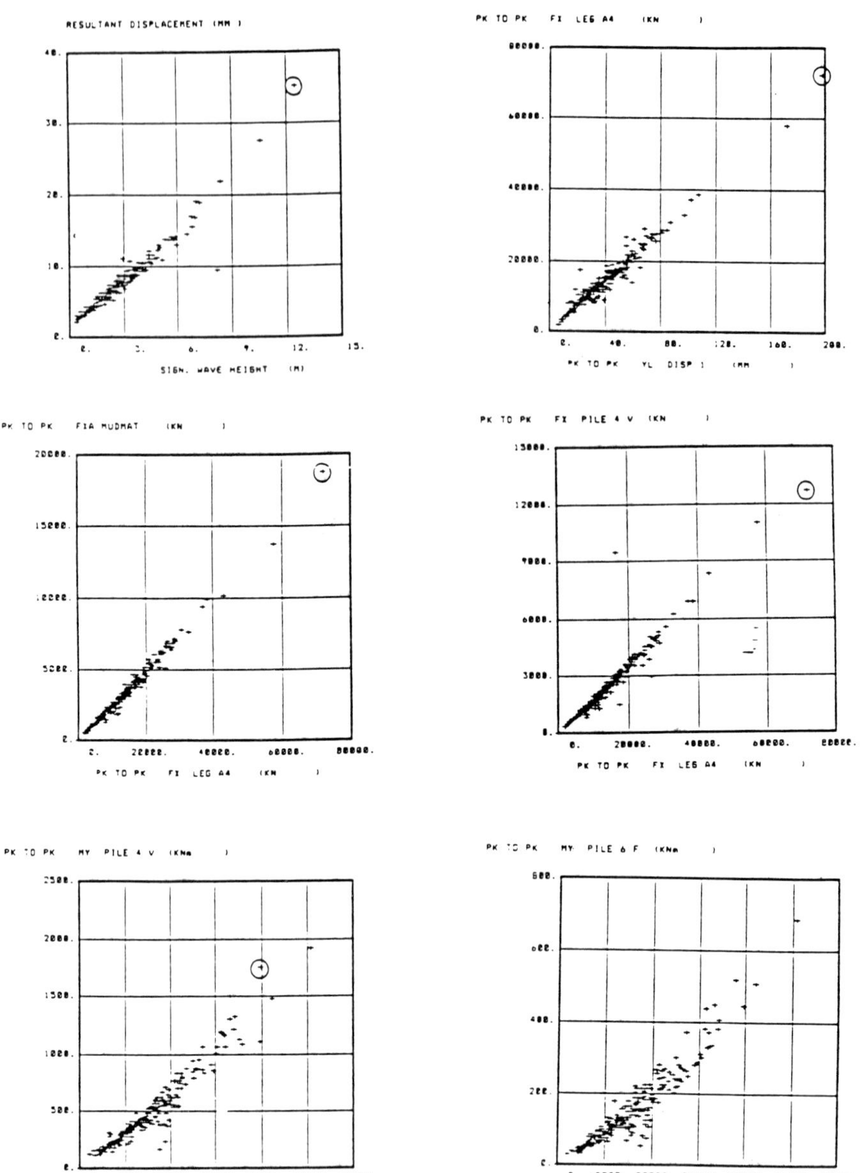

Fig. 9. Dynamic relationships from many recordings

against maximum bending moment in leg A4. All points lie close to a line, although in this case, the line is slightly curved. This suggests slight non-linear behaviour of the piles under lateral loading.

Values of bending moments, normalised relative to the leg A4 bending moments, were obtained for each pile from the long term plots for low and high wave conditions: They were used to estimate bending moment distributions down a pile in a similar manner to the selected cycle values. Figure 10 summarises the results for the low and high wave conditions in each direction. It also tabulates estimated shear force, bending moment, rotation and displacement at the top of a pile.

The bending moment distributions are similar in shape although higher bending moments (per unit applied force) occur in the 2 to 15 m depth range for the higher wave condition. The rotations and lateral displacements are

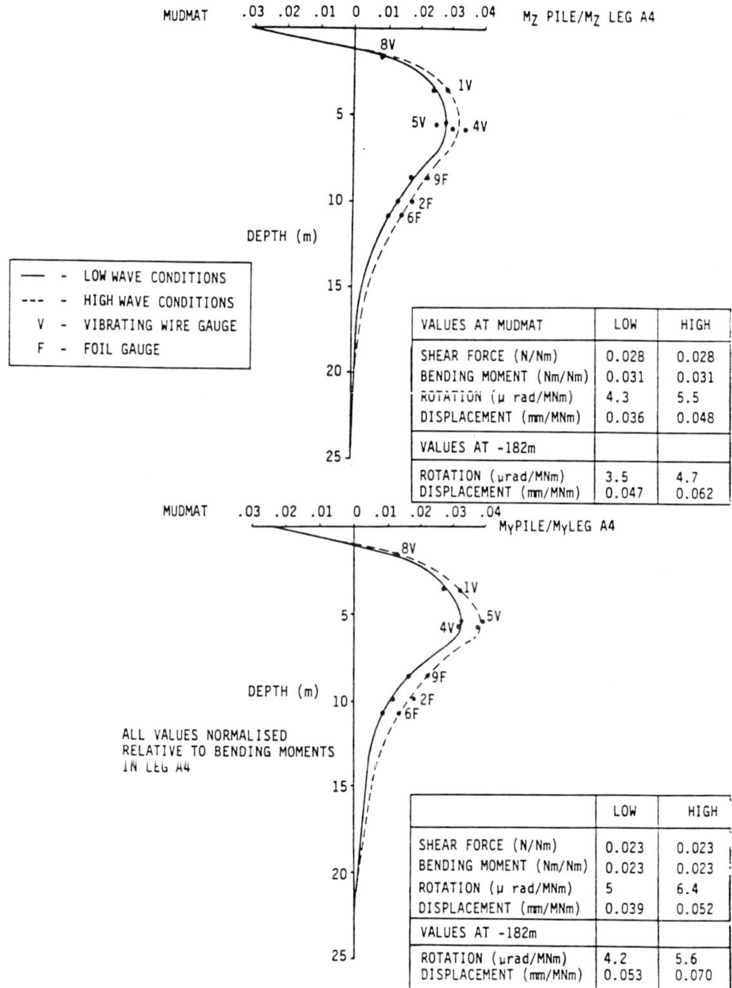

VALUES AT MUDMAT	LOW	HIGH
SHEAR FORCE (N/Nm)	0.028	0.028
BENDING MOMENT (Nm/Nm)	0.031	0.031
ROTATION (μ rad/MNm)	4.3	5.5
DISPLACEMENT (mm/MNm)	0.036	0.048
VALUES AT -182m		
ROTATION (μrad/MNm)	3.5	4.7
DISPLACEMENT (mm/MNm)	0.047	0.062

	LOW	HIGH
SHEAR FORCE (N/Nm)	0.023	0.023
BENDING MOMENT (Nm/Nm)	0.023	0.023
ROTATION (μ rad/MNm)	5	6.4
DISPLACEMENT (mm/MNm)	0.039	0.052
VALUES AT -182m		
ROTATION (μrad/MNm)	4.2	5.6
DISPLACEMENT (mm/MNm)	0.053	0.070

Fig. 10. Bending moment distributions form long term data

about 30 per cent greater in the higher wave conditions. It should also be noted that there are slight differences between the two directions.

Conclusions

The Magnus Foundation Monitoring Project was successful and met its main objectives.

The instrumentation performed well, even though some of it was novel and had not previously been used under such difficult conditions. Very high quality data was obtained from virtually all sensors.

Information was obtained from the measurements and directly compared with design predictions. In general form the measured and predicted behaviours were similar. The predictions were conservative which gives reasonable confidence in design methods.

Information was obtained on the long term transfer of loads from the mudmat to the piles and on the static load distributions along a pile. Virtually all load was transferred to the piles within one year.

An estimate of the static axial load distribution down a pile was obtained. This appeared to be significantly different from a theoretical distribution for an isolated pile. This may be because of interaction effects or because the piles were unequally loaded.

Information was obtained on the dyriamic loads due to wave action taken by the mudmat and piles, on the axial and lateral distributions along a pile and on the dynamic stiffness of a pile group. It was found that about 15 per cent of the dynamic load was carried by the mudmat.

The measured overall dynamic axial stiffness was about 4 times that predicted although the measured and predicted rotational stiffnesses were reasonably similar. This is consistent with interaction effects between the piles of the group being much less than expected.

There was a significant difference in the dynamic predicted and measured bending moment distributions down a pile and the measured pile group lateral stiffness was about twice that predicted. The differences are consistent with the soil modulus being greater than predicted which again may be because interaction effects were less than expected.

There was no evidence of any non-linear behaviour under axial loading of the foundations but evidence of slight non-linear behaviour under lateral loading.

The dynamic loads actually measured were less than 30% of the ultimate design loads; consequently no information was obtained on the ultimate load capacity of the foundations.

References

1. RIGDEN W. J., and SEMPLE R. M. Design and installation of the Magnus foundations: prediction of pile behaviour. Design in offshore structures, Thomas Telford, London, 1983.
2. RIGDEN W. J., and SEMPLE R. M. Design and installation of the Magnus foundations: installation studies and platform installation. *Design in offshore structures*. Thomas Telford, London, 1983.
3. COYLE H. M. and REESE L. C. Load transfer for axially loaded piles in clay. *Journal, Soil Mechanics and Foundation Division, American Society of Civil Engineers*, 1966.
4. MATLOCK L. H. and REESE L. C. Generalised solutions for laterally loaded piles. *Journal, Soil Mechanics and Foundation Division, American Society of Civil Engineers*, 1960.
5. REESE L. C., COX W. R. and KOOP F. D. Field testing and analysis of laterally loaded piles in stiff clay. *Proceedings, Seventh Offshore Technology Conference*, Houston 1975.
6. POULOS H. G. and DAVIES E. H. *Elastic solutions for soil and rock mechanics*. John Wiley & Sons, New York, 1974.

3. Design implications arising from the Magnus Foundation Monitoring Project

C. J. BOWLEY, John Brown Engineers & Contractors Ltd

Introduction

The BP Magnus structure is the most northerly platform in the North Sea and stands in 186 m of water. In 1979 during the design of the structure the proposed piles were larger, deeper and were required to carry heavier loads than any previous development. It was proposed that the structure, including the mudmat and piles at one leg, should be instrumented to provide valuable information on the actual loading regime and foundation behaviour.

The aim of the Magnus Foundation Monitoring Project was to verify current offshore piled foundation design methods. The data collected has also allowed wave parameters to be compared with foundation forces giving valuable data on wave loading.

The maximum operating design topside weight is 35 000 tonnes and the structure and piles weigh a further 45 000 tonnes. Thirty-six piles of 84% (2.134 m) diameter support the structure and are arranged in groups of nine around the four legs. The average penetrations of the piles is in the order of 85 m. A general arrangement of the structure is shown in Fig. 1.

The foundation monitoring project was conducted in three stages:

(*a*) The collection and analysis of raw data from the instrumentation fitted to the jacket and piles.
(*b*) The prediction of the structure and foundation response to topside and environmental loads in the range of the measured values following the methods used during the detailed design.
(*c*) A comparison of the predicted and measured values.

This paper presents a summary of the work carried out under stages (2) and (3) and highlights a number of implications for the design of foundations for offshore structures.

Summary of foundation instrumentation

A detailed account of the instrumentation and the measured results is reported by Kenley and Sharp (1993). Here follows, for completeness, a brief

Large scale pile tests in clay. Thomas Telford, London, 1993

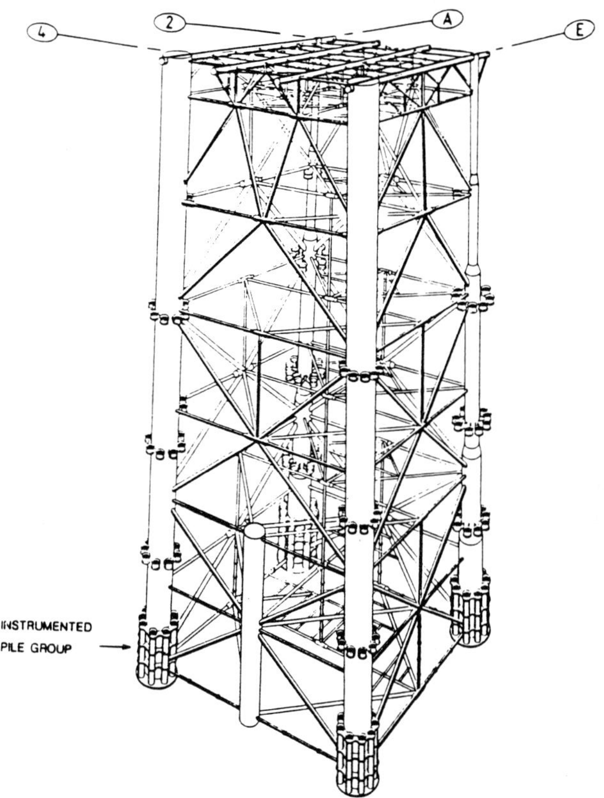

Fig. 1. BP Magnus

summary of the foundation instrumentation which was fitted to Leg A4:

- A settlement gauge was fitted at mudmat level with a reference post some metres away.
- Accelerometers were attached at the bottom of the leg at the -182 m level. Double integration allowed displacement in five degrees of freedom (all except torsion) to be calculated.
- Strain gauges were installed on the leg at the -157 m level, which is just above the pile sleeve cluster, and at -182 m on the two diagonal braces which frame into the leg. At each location the gauges were arranged to give the axial load and bending moment in the member.
- Additional gauges were arranged on the leg to measure the shear load and torsional moment.

Nine pressure cells were fitted to the underside of the mudmat in order to determine the loads carried by the mudmat.

Eight of the nine piles were fitted with strain gauges at a point between 2 m and 15 m below the mudmat. In addition, Pile No. 3 had gauges at two stations lower down the pile.

A summary of pile penetrations and pile instrumentation is given in Table 1.

Table 1. Pile penetrations and instrumentation on Leg A4

Pile No.	Pile Length (m)	Final Penetration Below Mudmat (m)	Dist. of Inst. from Tip (m)	Instrument below mudmat (approx) (m)	Instrument * below mudline (approx) (m)
1	106.5	84.2	81	3	4
2	110.1	85.2	75	10	11
3	106.5	84.6	40, 60, 70	45,25,15	46, 25, 15
4	110.1	87.0	81	6	7
5	110.1	86.9	81	6	7
6	110.1	86.0	75	11	12
7	104.1	82.9	-	-	-
8	104.1	82.6	81	2	3
9	110.1	82.8	75	8	9

Note* Mudmat A4 settled approximately 1m below the sea bed.

Design predictions

In order that a comparison could be made with the measured forces it was first necessary to develop a number of design predictions for the structure and foundations to match the selected prevailing static and dynamic load conditions. A number of analyses were performed, as far as possible in accordance with the original design techniques. These analyses used the as-installed computer model of the jacket structure with appropriate foundation stiffness models based on the soil conditions recorded for the borehole nearest leg A4; borehole B11. The design conditions considered are briefly described below. A summary of the design parameters is given in Table 2.

Table 2. Design parameters

PARAMETER	PREVIOUS ANALYSES		FOUNDATION MONITORING PROJECT			
	TOWER2 (1)	TOWER6 (2)	MODULE PLACEMENT	STATIC ANALYSIS	SPECTRAL ANALYSIS	FUNDAMENTAL PERIOD
ENVIRONMENTAL:						
1) Water depth	186m	185m	185m	185m	187.2m	185m
2) Settlement	none	0.7m av.	1m @ A4	1m @ A4	none	0.7m av.
3) Scour	2m	2m	2m	2m	2m	2m
4) Max wave height	30.3m	30.3m	n/a	16.0m (4)	17.5m (10)	n/a
5) Max current speed						
Surface	1.52m/s	1.52m/s	n/a	0.67m/s	n/a	n/a
Seabed	0.62m/s	0.62m/s	n/a	0.31m/s	n/a	n/a
6) Marine growth thk.						
Maximum	75mm	75mm	n/a	10mm (6)	75mm	75mm
General	25mm	25mm	n/a	zero	25mm	25mm
7) Max wind speed	44m/s	44m/s	zero	35.5m/s (7)	n/a	n/a
APPURTENANCES:						
1) No. of conductors	20	20	n/a	7 (8)	20	7 (8)
2) No. of exp/inp risers	4	4	n/a	2 (4)	4	4
3) No. of flowlines	23	23	n/a	23	23	23
4) No. of caissons	9	9	n/a	9	9	9
DECK LOADS:						
1) Calm sea conditions	33,068t(11)	35,009t	various	n/a	30,000t	32,872t
2) Extreme storm	31,616t	33,545t	various	n/a	n/a	n/a
GENERAL:						
1) Drag coeff. (Cd)						
Maximum	0.85	0.85	n/a	0.85	0.85	n/a
General	0.65	0.65	n/a	0.65	0.65	n/a
2) Inertia coeff. (Cm)	2.00	2.00	n/a	2.00	2.00	2.00
3) Wave theory (9)	Stokes V	Stokes/Airy	n/a	Airy	Airy	n/a
4) Fundamental period	3.86sec	n/a	n/a	3.86sec	3.86sec	various
5) Module installation	Skidding	Placement	Placement	n/a	Placement	n/a
6) Stiffness model	n/a	n/a	"TOWER6"	"TOWER6"	"TOWER2"	"TOWER6"
SOILS DATA:						
1) Borehole	B/H 4	B/H 4	B/H 11	B/H 11	B/H 11	various

Notes:

(1) - Detailed design second inplace analysis and spectral fatigue (1980)
(2) - As installed inplace analysis and design check (1983)
(3) -
(4) - A range of wave heights considered.
(5) -
(6) - Based on actual marine growth survey.
(7) - One year return period value
(8) - Appropriate for monitored period
(9) - Differences between Stokes V and Airy insignificant
(10) - Measured wave spectra used
(11) 30,000t used for the modal and spectral fatigue analyses

Static loads under calm sea conditions

The increase in the load at leg A4 and the distribution of the load into the piles was predicted for two stages of module placement and three topside weights under calm sea conditions. The three topside weights covered the period from end of module installation in July 1982 (24979T), July 1983 (26352T), and February 1985 (31571T).

Static wave loading

The forces in leg A4 due to wave and wind forces on the structure and the distribution of the pile group load to individual piles were calculated for 3 wave heights, 3 wave periods and 5 wave directions.

The wave loading analysis used the geometry and member properties of the as-built jacket and the hydrodynamic properties from the detailed design. The wave direction, heights and periods which were considered are given below:

- directions (relative to true north): NE, SE, SW, S, SSE
- wave heights (H_{max}): 4, 10 and 16 m
- wave periods for each wave height: 6, 9 and 13 seconds

These wave parameters were selected because they covered the full range of results reported by the environmental monitoring project and would allow reasonable interpolation/ extrapolation for particular waves to be made. As the instrumentation on the jacket measures the variation between the maximum and minimum stresses rather than absolute values the waves were stepped through the jacket over a full wave length to predict the corresponding variation in load.

Spectral analysis of storms

The forces in leg A4 were calculated for six measured wave spectra for which the corresponding, but at that time undisclosed, pile forces were available.

Natural frequency study

The fundamental natural period of the structure was calculated for three different foundation stiffnesses:

- the original foundation stiffness used for detailed design
- a stiffness 3 times that used for the as-installed analyses
- rigid foundations

Figure 2 shows the sign convention adopted for the analyses and to which all predicted and measured forces were referenced.

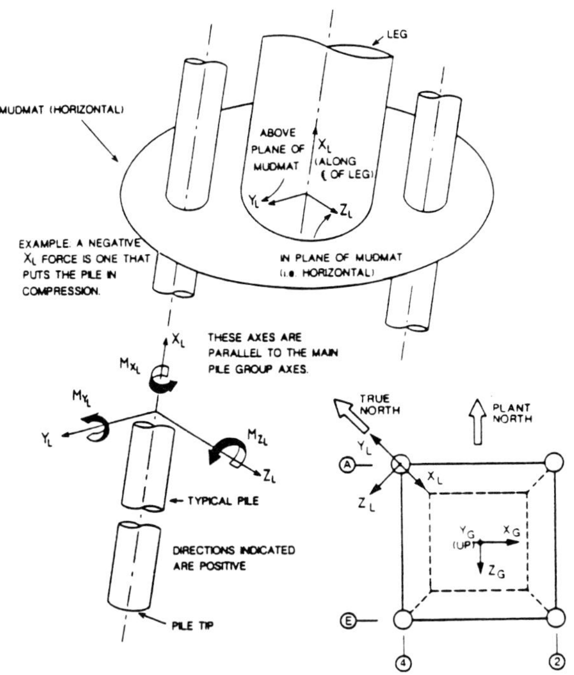

Fig. 2. Sign convention for forces and deflections

Foundation modelling

For the structural analyses the restraint provided by the foundations was modelled by a 6 × 6 stiffness matrix at the base of each leg, developed from consideration of the various modes of load-deflection behaviour as set out below.

Axial load deflection behaviour of an isolated pile was predicted using the *t-z* curve approach (Coyle and Reese, 1966). *T-Z* and *q-z* curves were generated following the recommendations of Vijayvergiya (1977)

The lateral load deflection behaviour of an isolated pile was investigated using *p-y* curves calculated from the method of Reese, Cox and Koop (1975). The behaviour of a pile in a group was predicted by softening the *p-y* curves with a *y*-modifier so that the lateral deflection of the individual pile matched that of the group.

The axial and lateral behaviour of the pile group was predicted using the methods proposed by Poulos and Davies (1974) modified to allow for non-linear behaviour after Focht and Koch (1973).

Comparison of predicted and observed forces on Leg A4

Loads from individual wave cycle

This section presents the comparison of the predicted and observed forces on Leg A4 for an individual wave cycle. The selected wave cycle was from the data recording for a storm on 22 January 1984. This storm was chosen as it was one of the more severe storms experienced during the monitoring period. A section of the wave height record is shown in Fig. 3. The selected wave cycle was interpreted to have a wave height of 12.5 m (crest to trough), it was chosen as it was that which best fits a sinusoidal profile and was one of the larger peak to peak responses during the recording. The wave train is fairly typical of the waves measured during the monitoring period and comes close to the design idealisation of a train of regular waves. As the wave is asymmetrical periods of 8 s and 13 s were chosen for the comparison.

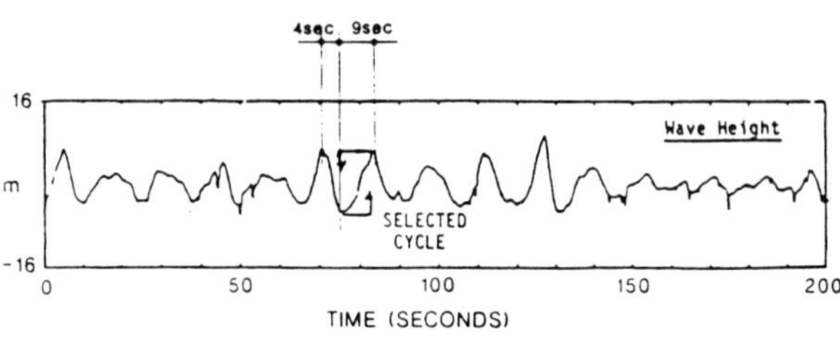

Fig. 3. Example of individual cycle selection (after W. A. Fairhurst)

For this exercise the effects of a 12.5 m high wave were linearly interpolated from the results for the 10 and 16 m waves with 9 s and 13 s periods described earlier. Table 3 compares the forces on the structure for the selected wave. The measured loads presented represent the most positive value minus the most negative value in the selected cycle. The predicted loads include dynamic amplification factors (DAFs).

The measured wave is believed to approach the platform at an angle of approximately 175 degrees to true north. This was compared to a calculated wave approaching from the south.

This difference means that the predicted axial leg loads can be considered to be about 5 per cent high.

The measured forces and moments at the bottom of the leg were developed from the measured forces in the diagonal braces and in the leg at elevation -157 m. Detailed examination of the force balance on the leg bottle in the analytical model shows that the shear at the bottom of the leg is 5 per cent greater than that which would be calculated from the shear at elevation

Table 3. Comparison of measured and predicted force for 22/01/84

QUANTITY		MEASURED	PREDICTED (Tz=9s)	PREDICTED / MEASURED	PREDICTED (Tz=13s)	PREDICTED / MEASURED
Wave Ht (m)		12.5	12.5	N/A	12.5	N/A
Period (s)		8-13	9	N/A	13	N/A
Direction (degs)		175	180	N/A	180	N/A
Leg Loads	FX (MN)	-54.9	-59.7	1.09	-67.9	1.24
at EL	FY (MN)	1.72	0.45	0.26	1.0	0.58
-157M	MZ (MN-M)	57.4	30.2	0.53	39.7	0.69
Brace Axial	(1720)	-10.4	-5.6	0.54	-10.2	0.98
Loads (MN)	(1150)	-9.6	-12.9	1.34	-16.2	1.69
Leg Loads	FX (MN)	-68.4	-66.1	0.97	-86.9	1.26
at EL	FY (MN)	12.6	9.6	0.76	15.4	1.22
-186	MZ (MN-M)	147.9	111.4	0.75	169.0	1.14
Pile Group Moment	MZ (MN-M)	129.2	93.4	0.72	145.5	1.13
Base	X (MM)	-4.42	-9.7	2.19	-11.6	2.62
Displacements	Y (MM)	-5.67	-10.0	1.76	-14.8	2.61
	Z (MRAD)	0.25	0.29	1.16	0.46	1.84

-157 m and the axial load in the diagonal braces. The measured shear at elevation -186 m was, therefore, increased by this amount. The measured moment at top of piles, and the theoretical moment in the leg at -186 m, were estimated assuming a leg-to-pile group eccentricity equal to that used in the computer model.

The over-prediction of wave loads can be attributed to a number of causes as follows:

(*a*) It is difficult to accurately measure the wave height, period and direction, and to compare this with the idealised wave considered in design. It is possible that the structure disrupts the water surface near the measurement location.

(*b*) The DAFs used in design were calculated using the design value of the structure natural period; as will be discussed later, the measured value is somewhat smaller, and consequently the values used are too high. The overestimate is 10 per cent for a 9 s wave period, and 5 per cent for a 13 s wave period.

(*c*) No account is taken of shielding, and the coefficient of inertia (C_M) may be too high.

Spectral wave loads

Table 4 compares the standard deviation axial load at the bottom of leg A4 for the various storm spectra analysed. The predicted forces are higher than the measured by a substantial margin. It is noticeable that the over-prediction is greatest at the smaller wave heights.

Table 4. Comparison of predicted and measured spectral axial forces

DATE	Hs (m)	DIRECTION (Deg)	STD DEVIATION FORCE (KN) (EL -186m)		PREDICTED / MEASURED
			PREDICTED	MEASURED	
21/02/84	5.2	151	8850	4160	2.1
22/01/84	10.7	168	17850	10240	1.7
27/11/84	9.2	218	11450	7180	1.6
22/11/84	7.1	151	11400	6730	1.7
17/01/84	4.2	178	7500	3710	2.0
22/01/85	4.8	47	5450	3040	1.8

Notes: 1 Wave direction is relative to true North.
2 Measured values are for El -157m adjusted to give values at el -186m.

The effect of natural period was found to be small when investigated using the following method. The spectrum for 22/01/84 was factored by the DAF at each frequency for both the measured fundamental period and for the original calculated period of 3.86 s. The degree of over-prediction was found to be 2 per cent if the wave loads were inertia dominated, and 5 per cent if they were drag dominated.

Wave spreading also causes an over-prediction and this was not considered in the theoretical work. Assuming a '$\cos^2 \theta$' spreading function it can be shown that over-prediction will be 15 per cent if the forces are drag dominated and 33 per cent if they are inertia dominated. Interestingly, the

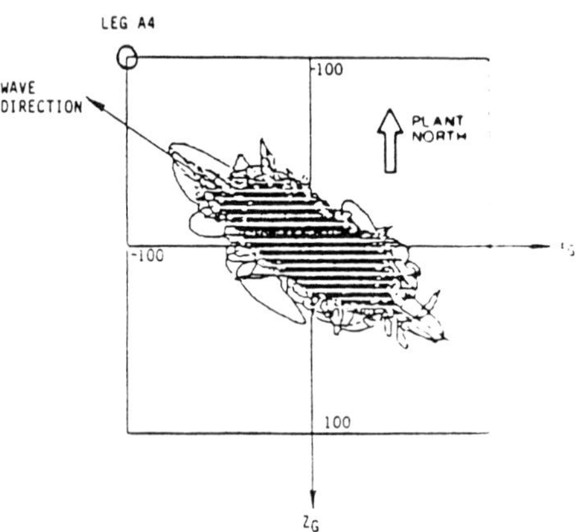

Fig. 4. Trace of platform motion (after W. A. Fairhurst)

plan platform trace illustrated in Fig. 4 can be partly explained by wave spreading. If the forces were drag dominated the length of the trace would be about 4 times the width, and if the forces were inertia dominated the length would be about 1.7 times the width (Sarpkaya and Isaakson, section 7.6.4). The trace shown has an aspect ratio of about 2.

Dynamic response: fundamental period

Table 5 compares measured and predicted fundamental natural period of the structure in the plan north-south direction.

Table 5. Platform fundamental natural period – plant N–S direction

REF	PERIOD (S)	TOPSIDE MASS (TONNES)	COMMENT
1	2.82	25000	JULY 82
2	2.86	25500	JANUARY 83
3	2.94	APPROX 30000	JANUARY 84
4	2.97	31300	MARCH 84
5	2.98	31600	APRIL 85
6	3.49	32872	ORIGINAL STORM FOUNDATION SPRINGS
7	3.20	32872	3 X ORIGINAL STORM FOUNDATION SPRINGS
8	2.95	32872	FIXED BASE
9	3.11	25175	3 X ORIGINAL STORM FOUNDATION SPRINGS
10	2.96	25175	3 X ORIGINAL STORM FOUNDATION SPRINGS REDUCED CM VALUE
11	3.86	30000	PREDICTED: ORIGINAL SPECTRAL ANALYSIS

Notes Ref 1 to 5: Measured
Ref 6 to 11: Predicted

This table illustrates the influence of a variety of factors:

(*a*) Differences in topside mass. Measurements 1 to 5 indicate the variation in period with topside mass. Similarly, prediction 9 is different from prediction 7 due to its topside mass.
(*b*) Differences in foundation stiffness. Predictions 6 to 10 present the variations investigated. Prediction 7, 9 and 10 are based on foundation stiffnesses close to those measured.
(*c*) Differences in structure added mass. In prediction 10 the C_M value for the legs of the tower was reduced from 2.0 to 1.6.

Comparison of predicted and measured foundation behaviour

Static pile loads

The predictions for static loads described earlier were made ignoring the presence of the mudmat. Measurements showed that the load at the end of module placement was 3 MN higher than at the start and that this load decayed by 8.6 MN from end of module placement to February 1985. Furthermore, Kenley and Sharp (1993) report that at that time, the mudmat was still carrying a residual load of about 9MN.

The gauges on pile 8 are only 2.6 m below mudline and so it is not unreasonable to ignore any depth effects. The predicted increase in mudline load on this pile during module placement was 7.91 MN and the measured increase was 5.32 MN (Table 6). This difference can only be partly explained by considering that part of the load is being carried by the mudmat as the increase in measured load on the mudmat for the same period is 3 MN or 0.33 MN per pile. It is possible that further load is being carried by the 1 m deep skirts beneath the mudmat.

Table 6. Static pile loads in calm sea conditions

Stage	July 82		July 83		Feb 85	
Load (MN) Pile	Predicted Mudline	Measured Gauge	Predicted Mudline	Measured Gauge	Predicted Mudline	Measured Gauge
1	6.73		7.26		8.17	
2	6.22	4.60	6.70	5.6	7.55	6.4
3	6.19		6.67		7.52	
4	6.49		6.99		7.90	
5	7.37	6.25	7.94	8.3	8.98	9.9
6	8.14	4.98	8.77	7.5	9.92	8.7
7	8.02		8.65		9.76	
8	7.91	5.32	8.53	8.5	9.61	10.0
9	7.50	4.65	8.09		9.10	

Table 6 also presents the behaviour of instrumented piles up to February 1985. For pile 8, remarkably good agreement is seen at July 1983 and February 1985, suggesting that the additional load carried by the mudmat has transferred to the pile in the long term.

Similar comparisons are difficult to make for other piles as the gauges are deeper below mudline. However, piles 2 and 5 both have gauges at a similar depth (11 m and 12 m respectively) so a comparison of their behaviour is of interest.

Two things are apparent from the results. Firstly, the measured load at instrument level, as a percentage of the predicted mudline load, is almost identical on both piles in the long term. This suggests that the piles are behaving in a reasonably consistent manner. Secondly, the piles gain load in the long term in a similar way to pile 8.

Storm pile loads – axial forces

The approach taken here was to determine the measured load on the leg A4 foundations for a particular storm and then to extract a suitable load case from the predictions that had similar total foundation forces. The measured pile forces were then compared with the predicted distribution of forces within the group. The storm considered was that of 22 January 1984 (see Table 1). The values in Table 1 are peak to peak so mean to peak values will be approximately half.

Table 7 presents the comparison of predicted and measured forces in each pile. The general trend of load distribution between piles is correct, i.e. maximum load on pile 5, followed closely by pile 4, with pile 9 the most lightly loaded. However, in general, the predicted loads at instrument level are rather higher than those measured. This must be due in part to the fact that the mudmat (which was ignored when making the predictions) is taking a significant proportion of the foundation load.

The measured peak to peak axial load on the mudmat was 11.7 MN, adjusting the predicted values by half this value gives much closer agreement (Table 7).

Storm pile loads – bending moments

The predictions for bending moment in the piles due to storm loads showed that there was little variation in the lateral loading between piles. This behaviour meant that the measured bending moments for each pile could be combined to give a single bending moment diagram for the top 25 m

Table 7. Axial load on piles during 22/01/84 storm

		No Load on Mudmat			11.7MN Load Taken by Mudmat	
Pile	Gauge Depth (m)	Predicted Mudline (MN)	Predicted Gauge (MN)	Measured (MN)	Predicted Mudline (MN)	Predicted Gauge (MN)
1	4	2.44	2.39	2.42	1.95	1.91
2	11	3.70	3.17	2.52	2.96	2.54
3m	26	5.01	1.81	2.48	4.01	1.44
4	7	6.04	5.53	5.10	4.84	4.42
5	7	6.79	6.22	5.35	5.43	4.98
6	12	5.54	4.51	2.92	4.43	3.61
7	-	3.44	-	-	2.75	-
8	3	1.79	1.72		1.43	
9	9	1.44	1.26	0.86	1.15	1.01

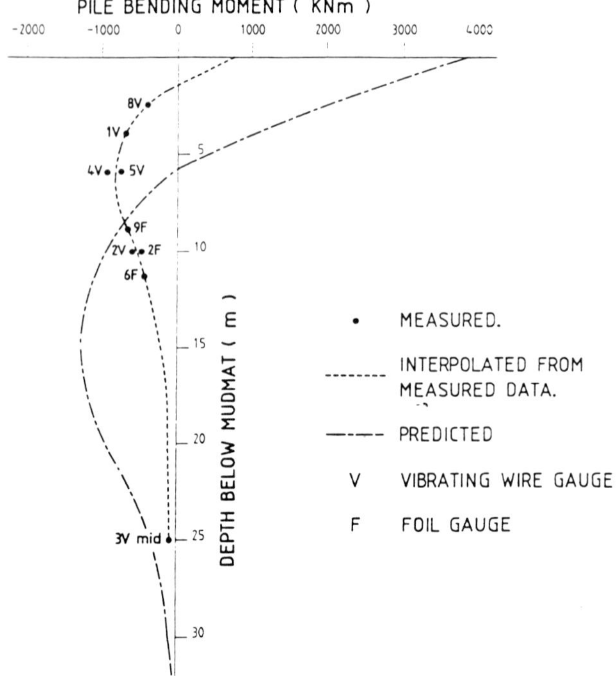

Fig. 5. Bending moment diagram for single pile – 22/01/84 storm

of the piles. The measured and predicted curves for the storm of 22/01/84 are compared in Fig. 5. There is a significant difference in the values at the point of fixity, the shape of the bending moment curve is also quite different with the predicted maximum reverse moment occurring much deeper. It is also interesting to note that the largest measured moment rather than being at the point of fixity occurs lower down the pile. This suggests less fixity at the pile head than assumed in design.

Foundation stiffness

From the measured forces and deflections, foundation stiffness values were deduced:

$$\begin{bmatrix} Fx \\ Fy \\ M_z \end{bmatrix} = \begin{bmatrix} 14600 & 0 & 0 \\ 0 & 2880 & -25200 \\ 0 & -25200 & 1100000 \end{bmatrix} \cdot \begin{bmatrix} X \\ Y \\ \theta z \end{bmatrix}$$

kN, M, radians

When compared with the predicted foundation stiffness, below, it can be seen that the measured behaviour is generally twice as stiff.

$$\begin{bmatrix} F_x \\ F_y \\ M_z \end{bmatrix} = \begin{bmatrix} 7729 & 0 & 0 \\ 0 & 1337 & -8519 \\ 0 & 8519 & 670300 \end{bmatrix} \cdot \begin{bmatrix} X \\ Y \\ \theta z \end{bmatrix}$$

kN, m, radians

Design implications

Design axial load on the foundation

On Magnus, and for most other jackets also, it is the maximum predicted axial load on the foundation that determines the required number of piles per leg. This load is a combination of gravity loads, both dead weight and live operating loads, with environmental loads due to wave, wind and current.

In a quasi-static analysis, such as is usually used in design, the environmental load will be factored by a DAF to account for the forces induced due to the dynamic response of the structure, the value of which depends on the period of the dynamic loading in relation to the natural period of the structure.

The environmental component of load on leg A4 was examined in detail for a selected cycle in one storm and statistically for six storms. The selected cycle was compared with the static predictions of leg load and, depending on the period that the actual wave was deemed to have (which is difficult to determine), the design prediction ranged from 97% to 126% of the actual. Making different assumptions about the natural period, the inertia coefficient (C_M), and wave directions would reduce the design prediction by up to 10%, 25% and 5% respectively.

Turning to the assessment of the 6 storms, the spectral predictions of leg load ranges from 160% to 210% of the measured. Again, different assumptions about natural period, C_M, and wave spreading would reduce the design predictions by up to 5%, 25% and 33% respectively.

These factors could well account for the larger part of the discrepancy and indicates the sensitivity of the design prediction to choice of parameter and suggests that changes to the method of design should only be made after careful consideration.

Mudmat contribution to axial capacity

It has been shown that in 1985, 3 years after installation, the mudmat carries about 9 MN of the measured calm sea load (total 139 MN). During short period, cyclic loading, the mudmat appears to carry 11.7 MN, about 17% of the fluctuation in total leg load (69 MN). This is for a wave of approximately 12.5 m trough to crest. Assuming that the mudmat's behaviour remains constant up to the 30 m design wave (estimated fluctuation in leg load of 390 MN) this is obviously a very valuable contribution to the foundation's axial capacity. However, even allowing for over-estimation of the wave loading during design the 30 year wave will cause a leg load 4–5 times that investigated here and extrapolation of the mudmat's behaviour should be carried out with care.

The disadvantages of including the mudmat's capacity in design are:

(*a*) it must be ensured that all mudmats make good contact with the sea floor and that the structure does not stand on 3 legs;
(*b*) the operator must include scour measurements in the routine inspection surveys and be prepared to take remedial measures should it become apparent that the mudmat is being undermined;
(*c*) it must be checked that the pile group efficiency calculations are compatible with a load carrying mudmat.

Distribution of axial load between piles

In calm seas, the variation in predicted load between piles in a group is small. Because the instrument depth varied between piles, it is difficult to identify whether this variation in predicted load exists. Its effect on design will anyway be small.

There was an obvious trend in the storm case considered for the piles on the 'leading edge' of the pile group to see the largest fluctuation in load and for the piles on the trailing edge (the crotch piles in this instance) to experience the smallest fluctuation in load.

After adjustment for the load carried by the mudmat, it was found that the total axial load predicted at instrument level agreed with the total measured load within 8%. The predicted load on the most heavily loaded pile was within 7%. In design, the prediction of the highest individual axial load is used to design the grouted connection and to determine the required wall thickness of the pile (when combined with the predicted bending stress). It is therefore an important quantity and one which design techniques seem to predict within acceptable limits.

Distribution of axial load along a pile

The reduction in axial load down a pile is of little importance in design. However, because the instruments on the piles were a varying distance below the mudmat, it was necessary to estimate load decay for prediction purposes.

The measurements show two things. Firstly, for the permanent gravity loads, there appears to be little or no decay over the instrumented depth (top 25 m). Secondly, during storm loading the load variation in the top 25 m of pile is broadly as predicted using design techniques.

Pile bending moment

Significant differences were found to exist between measured and predicted pile bending moments, with the maximum measured bending moment in a pile due to storm loading being only 24% of the maximum predicted. This discrepancy is most probably due to inaccurate modelling of lateral foundation stiffness. Although bending moments did not affect the design of the Magnus piles, for platforms with higher ratios of lateral load/vertical load on foundations and especially those with vertical piles the bending moments become very important in determining the wall thickness of the piles.

Foundation stiffness

It was found that the measured foundation stiffness is generally twice that predicted, whether axially, laterally or rotationally. Calculations show that this will have only small impact on the predicted natural frequency of the structure (less than 8%) but will result in substantially lower pile bending moments.

Because Magnus is supported on only four discrete pile groups, the axial load applied to each pile group is virtually independent of soil stiffness. However, on a jacket that has either a greater number of pile groups, or skirt piles distributed around the jacket perimeter, the foundation stiffness plays a larger part in determining the distribution of load. Further analysis of the axial behaviour revealed by the Magnus Foundation Monitoring Project should therefore benefit this type of jacket.

Conclusions

The Magnus Foundation Monitoring Project has shown that, in general, the techniques used to design the foundations are valid, if slightly conservative.

The measured dynamic loading on the structure has highlighted a number of possible conservatisms in the methods used to calculate environmental loads on the structure.

Both axial and lateral foundation stiffnesses are significantly higher than anticipated during design. The latter is particularly significant as it results directly in reduced bending moments in the piles as shown by the measured data.

The measurements show that the mudmats can play a significant role in the capacity and behaviour of the foundations.

References

COYLE, H. M., and REESE L. C. (1966). Load transfer for axially loaded piles in clay. *Journal of the Soil Mechanics and Foundation Division, Am. Soc. Civ. Engrs*, March.

FOCHT, J. A., and KOCH, K. J. (1973). Rational analysis of the lateral performance of offshore pile groups. *Proc. Offshore Technology Conference*, Houston. Paper 1896.

KENLEY, M. and SHARP, D. E. (1993). Magnus FMP – instrumentation, data processing and measured results. This volume.

POULOS, H. G., and DAVIES, E. H. (1974). *Elastic solutions for soil and rock mechanics*. John Wiley & Sons, Chichester.

REESE, L. C., COX, W. R., and KOOP, F. D. (1975). Field testing and analysis of laterally loaded piles in stiff clay. *Proc. Offshore Technology Conference*, Houston. Paper 2312.

SARPKAYA and ISAAKSON. *Mechanics of wave forces on offshore structures*.

VIJAYVERGIYA, V. N. (1977). Load-movement characteristics of piles. *Proc. Ports '77 Conf.*, California.

4. Magnus foundations: soil properties and predictions of field behaviour

R. J. JARDINE and D. M. POTTS, Imperial College, London

The Imperial College soil mechanics group made three contributions to the Magnus Foundation Monitoring Project: the development and manufacture of sea-bed settlement gauges, a laboratory study of the soils and a numerical analysis of the platform's piled foundations. The settlement gauge deployment was not entirely successful; finer resolution and greater overload protection were required than had been anticipated. Gauges incorporating these revised features were deployed later on the Hutton TLP foundations with much greater success, Jardine, Hight and McIntosh (1988). This paper is concerned with the more positive results obtained from the Magnus laboratory and theoretical studies. Part I describes the characterization of the Magnus ground profile and soil properties. This involved synthesizing the conventional site investigations with results from specialist tests performed at Imperial College (IC). Part II reports how the foundations were modelled using non-linear (elasto-plastic) finite-element analyses which captured the key features seen in the laboratory studies. Predictions are presented from the analyses of the axial load-displacement curves, skin friction distributions and pile-group interactions expected for the Magnus piles undergoing monotonic axial loading.

Part 1 – Ground profile and soil properties

Ground profile

Rigden and Semple (1983) summarize the geotechnical surveys performed at Magnus, which identified a thick and relatively consistent sequence of Quaternary soils. The Imperial College research concentrated on the zone around the platform's most heavily instrumented pile group (Leg A4) and involved experiments on samples from Borehole 11 – which were taken using Fugro's Seacalf unit in April 1981. Following the contractors' reports and the fabric studies described by Jardine (1985), the profile was divided into the five main geotechnical units identified in Fig. 1. Stratum I may have been deposited as a lodgement till, but Strata II to V were considered to be of glaciomarine origin; the relatively thin interglacial sand layers encountered at ≈ 14 and 19 m depth were not thought to have a significant bearing on foundation behaviour. A summary of relevant index parameters is given in Table 1, whilst Fig. 2 presents the trend line with depth of undrained shear strength, using data obtained by Fugro in standard UU triaxial tests on 38mm diameter specimens. Results from Push-in-pressuremeter (PIP) in-situ tests are also shown; the Imperial College data are discussed later.

Overall, Stratum 1 is slightly leaner and coarser grained than the deeper layers. It also has a lower intrinsic compressibility and exists in a denser, and apparently more overconsolidated, state. This is thought to be due to wave compaction, rather than mechanical overconsolidation (Hight 1983, Jardine 1985). Yield Stress Ratios (YSR - equivalent to apparent OCR) fall typically between 15 and 25 within the top stratum. The more plastic Strata (III to V) appear to be only lightly overconsolidated, with a minimum YSR of ≈ 1.0 being found at 70m depth. Stratum 2 appears to be transitional, with intermediate properties; the YSRs also fall from ≈ 6 at 20m to ≈ 2 at 40m depth.

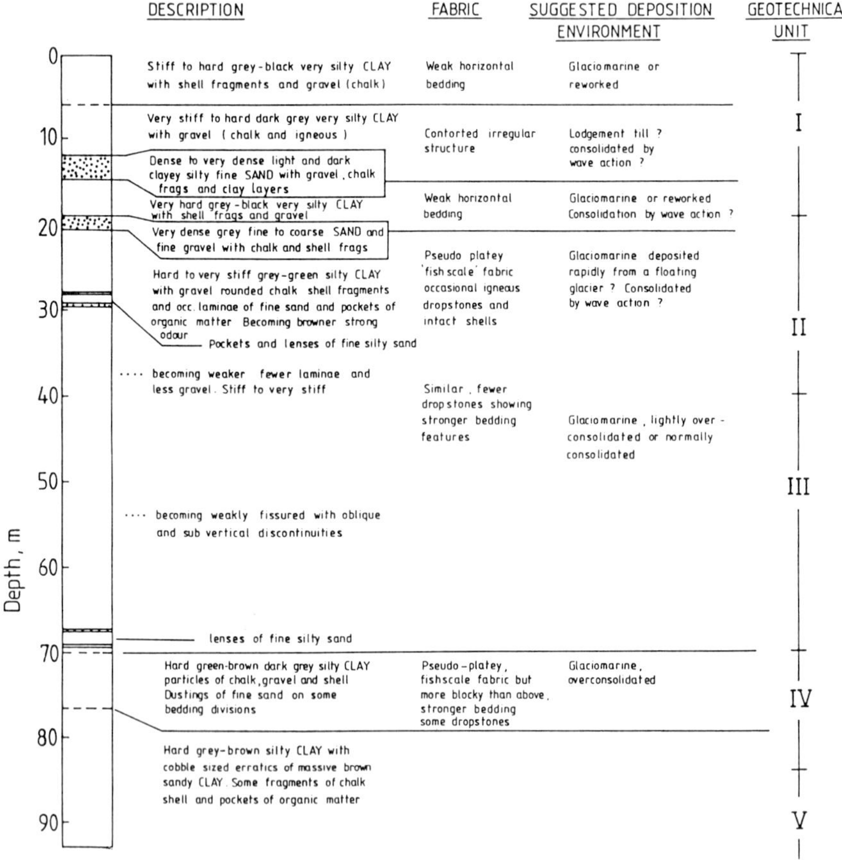

Fig. 1. Magnus soils profile, after Jardine (1985)

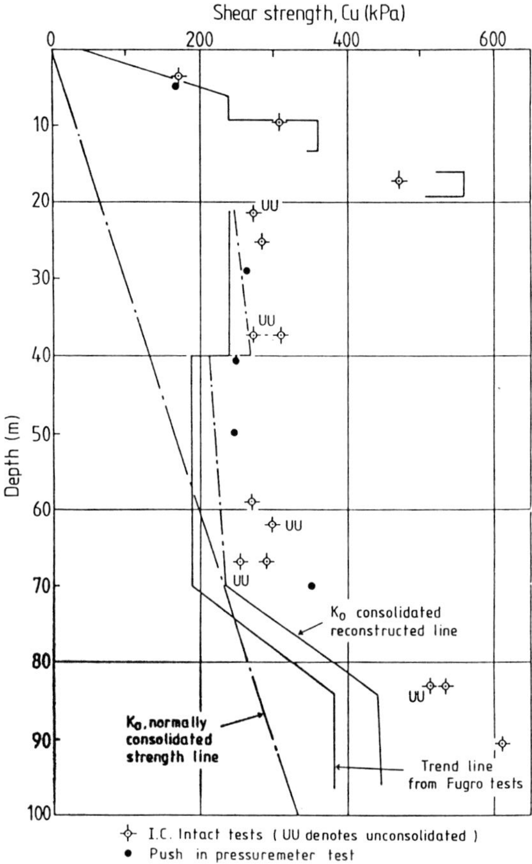

Fig. 2. Undrained shear strength profile from Imperial College and Fugro tests

Table 1. Mean index properties for main strata

Unit	Water content: range %	γ: kN/m^3	I_p: %	I_L	Oedometer parameters e_1	C_c	% clay
I	14-20	21.8	17	-0.07	0,75	0.12	23
II	16-21	21.1	19	0.09	0.95	0.15	27
III	21-24	20.5	22	0.16	1.25	0.23	34
IV	18-23	20.9	25	0.10	1.10	0.20	34
V	18-21	20.9	24	0.03	1.00	0.16	32
Reconstituted soil			18	N/A	1.25	0.21	36

Note: C_c and e_1 apply to the range 2MPa $<\sigma'_v<$3MPa, e_1 is void ratio projected for σ'_v= 1kPa.

Stress path tests

The objective of the stress-patch testing programme was to obtain detailed information on the soil's stress–strain behaviour. Particular emphasis was placed on measuring the response at small strains, using the local strain instrumentation described by Jardine *et al.* (1984)

Reconstituted test series

Parallel test programmes were performed on intact and reconsituted samples. The reconstituted work allowed assessments of the effects of sampling; variations in properties with OCR; differences in compression and extension and the effects of different consolidation procedures. Dried samples taken from between 30 and 70m depth, were crushed and mixed to a slurry (at I_L - 1.3) and then compressed to σ'_v= 200 kPa in a large oedometer. Identical specimens were cut from the resulting clay cakes which could then be subjected to stress path testing. Fig. 3 illustrates the series of effective stress paths obtained in a suite of undreained compression tests on samples which were K_0 consolidated to σ'_v= 400kPa and then swelled back to a range of OCRs. Contours are shown on Fig. 3 of the axial strains developed during undrained shearing: the initial response is very stiff—with only small strains being developed before large scale volumetric yielding occurs. A companion test serries showed that the clay was both weaker and softer when extended to failure.

Test interpretation is helped by defining a reference small-strain secant modulus at 0.01 per cent axial strain ($Eu_{0.01}$). Fig. 4 shows how $Eu_{0.01}/p'$ (p' is the pre shear mean effective stress) changes by less than ± 15 per cent in compression tests on reconstituted clay at OCRs between 1.2 and 20. Stress-strain non-linearity is quantified by the rate of change with strain of the

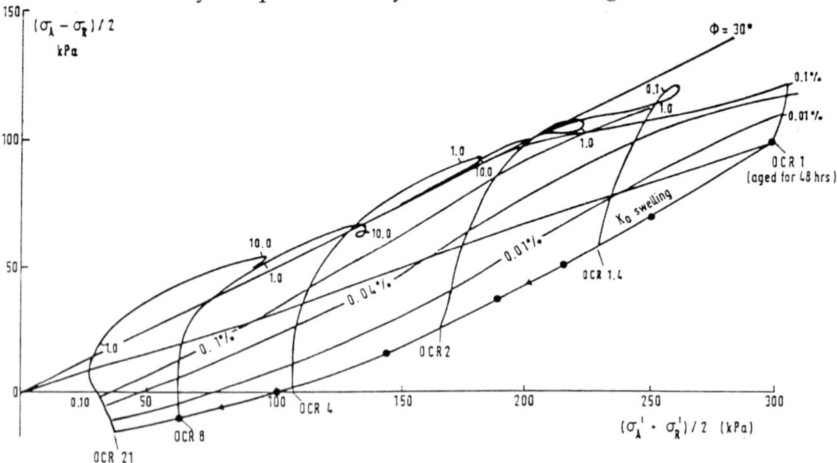

Fig. 3. Undrained effective stress paths and axial strain countours: compression test series on reconstituted clay

normalised secant stiffness $L(\varepsilon)$; $L(\varepsilon)$ is defined as $Eu_{(\varepsilon)}/Eu_{0.01}$. As shown in Fig. 5, $L(\varepsilon)$ reduces most rapidly in tests on normally consolidated clay; the OCR=2 curve appears to represent the median trend.

A parallel study of the effects of perfect and real sampling allowed the undisturbed in-situ triaxial compression c_u profile to be reconstructed, as shown in Fig. 2. The largest corrections apply in the low OCR layers as these tills are expected to suffer the greatest strength losses as a result of tube sampling, Hight *et al.* (1985).

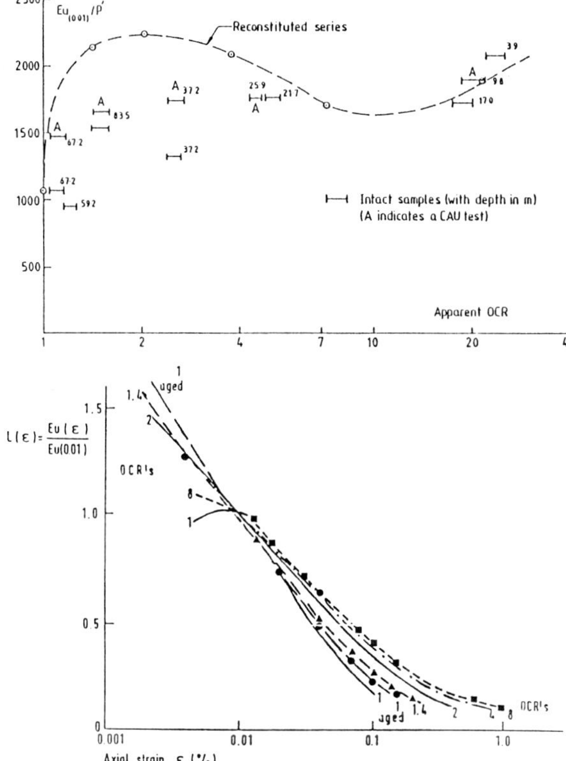

Fig. 4. Undrained compression stiffness index and OCR

Fig. 5. Undrained compression normalised stiffness strain curves: reconstituted series

Intact test series

The reconstituted test programme was complimented by a shorter series of undrained compression tests on intact samples. Fig. 2 shows how the undrained shear strengths measured in CAU and UU tests fall closest to the pressuremeter (PIP) c_u data points. Reasonable agreement is also found with the re-constructed in situ profile, except that the intact samples tested at IC tended to be stronger than expected at low YSR's. This might have been due to in situ cementation or ageing. The effective stress paths followed by typical intact samples from either end of the YSR spectrum are shown in Fig. 6.

The stiffness data from the intact UU and CAU tests are summarised in Figs 4 and 7 using the parameters $Eu_{(0.01)}/p'$ and $L(\varepsilon)$. Overall there is remarkable agreement between the intact and reconstituted sereis, although the $Eu_{(0.01)}/p'$ data diverge from the reconstituted trend line by 25 per cent at YSR = 2.

The intact $Eu_{(0.01)}$ data are plotted against depth in Fig. 8, along with profiles projected from: the reconstituted data shown on Fig. 4 and results from the companion extension test series.

Pressuremeter data are also shown from the five Mangus PIP tests. When interpreted conventionallly, the PIP and laboratory data seem incompatible. But a non-linear interpretation of the field tests reveals far closer agreement, Jardine 1992.

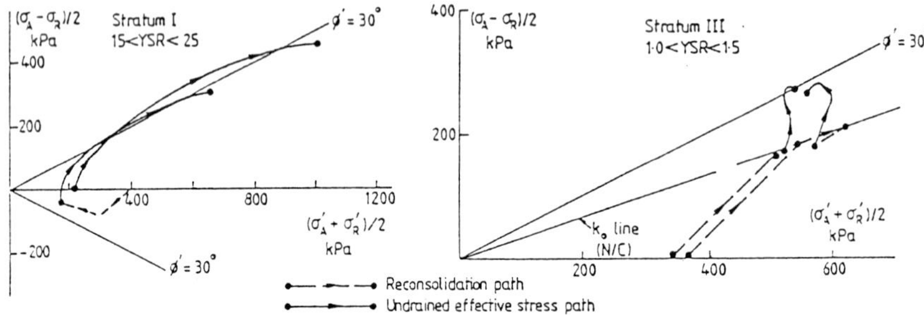

Fig. 6. Undrained effective stress paths, intact samples

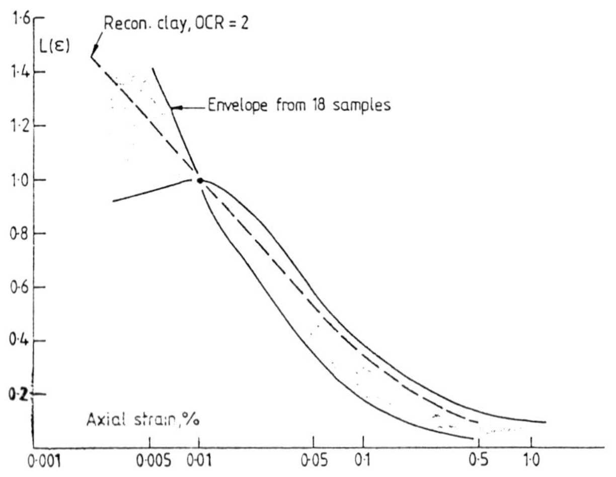

Fig. 7. Undrained compression normalised stiffness–strain curves, intact series

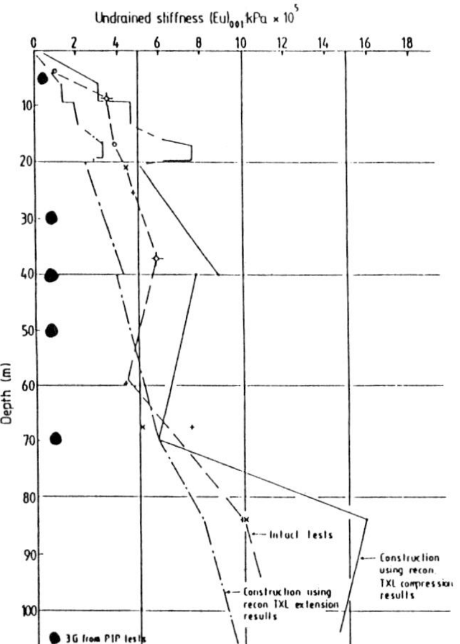

Fig. 8. Profiles of $Eu_{0.01}$ from Magnus tests, Jardine (1985)

Ring shear experiments

Field research with instrumented piles by Bond and Jardine (1991) and others has proven that interface residual fabric can be crucially important with displacement piles in clay. Lemos (1986) performed ring-shear tests to determine the residual properties of reconstituted Magnus clay. He found that the soil-soil $\phi'_{residual}$ was only marginally less than the critical state value ϕ'_{cs}, with angles of 30° and 27° applying respectively. However, markedly lower angles were obtained when the clay was sheared against a sand-blasted steel interface. Fig. 9 shows data from simulated pile tests where the soil was sheared for several metres at a rate of 133mm/minute (to match pile installation), then reconsolidated and sheared again slowly — giving the indicated results. Also shown are the parameters adopted for the numerical analyses; following other research by Lemos, higher δ′ values were assumed in the leaner Strata (I and II).

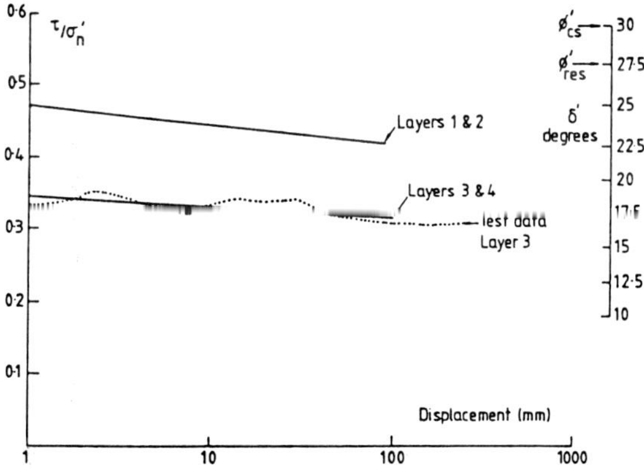

Fig. 9. Ring shear interface test data

Part II – Numerical analyses

General approach

The approach adopted was similar to that used in the later Hutton TLP studies reported by Jardine and Potts (1988). The loading analyses simulated undrained static pile tests and were made with the non-linear finite element code ICFEP (Potts 1985). The piles were taken as being solid elastic bodies , 82m long and 2.13m in diameter, with equivalent $E=24 \times 10^3$ MPa and $\nu = 0.3$. The axisymmetric FE mesh extended 150 to 170m depth, and a 155 m radius and included more than 150 eight noded isoparametric elements, thin elements were used close to the shaft to ensure solution accuracy. Overall group behaviour was assessed through an approximate (non-linear) superposition method that utilized the displacement fields obtained from the finite element calculations.

As tabulated below, three case were analysed. Run A was not considered realistic, but was included so that the potential effects of reduced, residual, interface frictional strength could be assessed directly.

Consolidation studies indicated that the ground stresses might take more than 10 years to stabilize after pile driving, and Runs B and C represented best estimates of typical medium and long term characteristics respectively. In both analyses limiting ratios were specified between the shear stresses (τ_{rz}) and radial effective stresses (σ'_r) that could be developed close to the pile shaft. This allowed interface slippage, and mild displacement softening, without failure necessarily developing in the soil continuum.

Run	Time elapsed after driving	Frictional strength close to pile shaft
A	6 months	No reduction
B	6 months	δ' as in Fig. 9
C	10 years	δ' as in Fig. 9

Soil models

A variant of the modified cam clay model was adopted for Strata I to V in which elastic shear modulus G varies with the level of invariant shear strain.

$$G_{sec}/p' = A + B \cos(\alpha \log)$$

$$G_{tan}/p' = d(\varepsilon_s\, G_{sec}/p')/d\varepsilon_s$$

where $\varepsilon_s = \sqrt{}\,(2/9((\varepsilon_1 - \varepsilon_3)^2 + (\varepsilon_2 - \varepsilon_3)^2 + (\varepsilon_1 - \varepsilon_2)^2))$

The elastic volumetric behaviour was simulated using a non-linear bulk modulus K' which varied with volume strain ε_{vol}. The $K'/p' = f(\varepsilon_{vol})$ function was identical to that relating G/p' to ε_s. The equations apply between limits $\varepsilon_{smax} > \varepsilon_s > \varepsilon_{smin}$ and evolmax> $\varepsilon_{vol} > \varepsilon_{volmin}$, outside which fixed tangents apply that are taken from the functions at their limits.

The basic material parameters (λ, v_1, ϕ'_{cs}, etc) were selected for the five clay layers on the basis of Table 1 and the data shown in Figs 3 and 6. Only one set of non-linear shear stiffness parameters was specified, the normalised shear stiffness characteristic was based on the G/p' curve from the undrained (reconstituted) compression tests at OCR/2 whilst the K'/p' expression was derived from K_0 swelling tests.

The effective stress paths and axial strain contours predicted by the composite model are illustrated in Fig. 10, which should be compared with Fig. 3. The model does not reproduce the brittleness seen at low OCRs nor does it mimic differences between compression and extension. However, these deficiencies are unimportant when failure is controlled by interface residual strength. The model is relatively simple and captures the way in which the soil dilates or contracts, depending on YSR. The small-strain equations simulate monotonic soil behaviour very well and have been used widely and successfully practice, Jardine *et al* (1991)

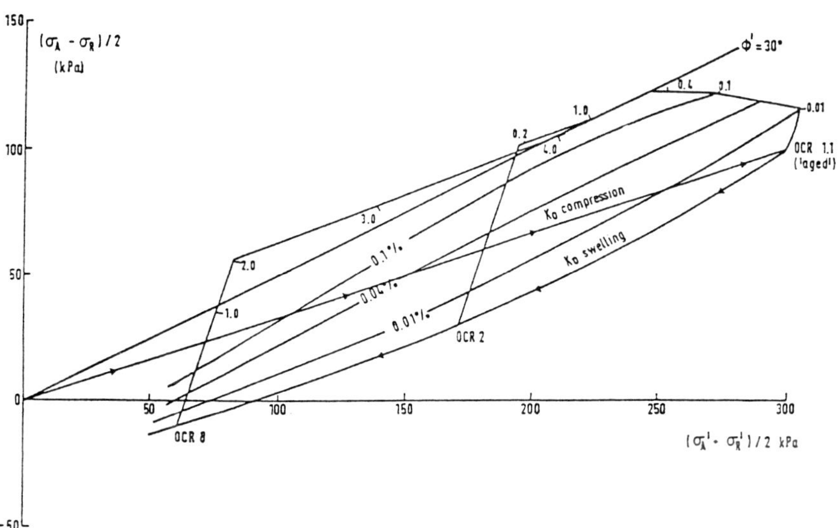

Fig. 10. Predicted undrained effective stress paths and strain contours (compare with test data on Fig. 3)

The stiffness coefficients specified for Magnus clays were

Case	A	B	C	α	γ	ε_{min}	ε_{max}
G/p'	623	643	0.0007%	1.349	0.6385	0.003%	0.25%
K/p'	687	412	0.0007%	1.400	0.797	0.003%	0.25%

Initial stress conditions

Jardine and Potts (1988) describe how the effective stress conditions around the driven Magnus piles were assessed by combining critical reviews of published theories with analyses of the then sparse data base of instrumented piles. The stress ratios specified in the anlyses are summarised in Table 2, a semi-logarithmic radial variation in initial stresses was assumed between the pile face and undisturbed conditions 20 radii from the pile. Overconsolidation ratios were taken as shown in Fig. 11, no account was taken of residual driving shear stresses and soil beneath the pile tip was also treated as undisturbed. The finite element program made small iterative changes before any load increment was applied to ensure overall stress equilibrium.

Much more information is now available on the stresses around displacement piles, following instrumented field tests such as those described at this Conference and by, for example, Bond and Jardine (1991).

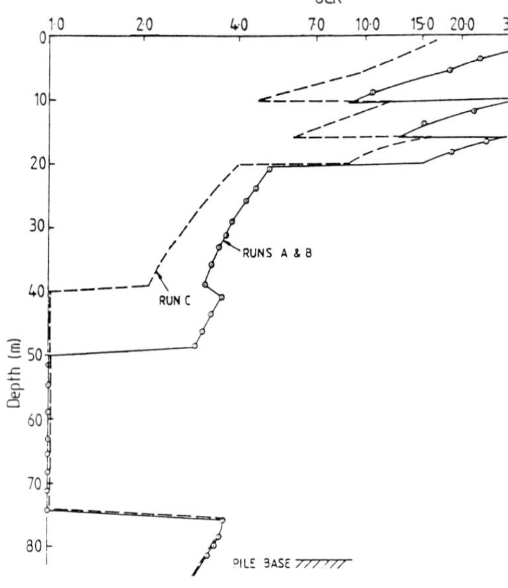

Fig. 11. OCR depth profiles specified in finite element analyses

Table 2. Effective stresses assessed around Magnus piles

Stratum	σ'_v/σ'_{vo} runs A & B	C	σ'_r/σ'_{vo} runs A & B	C	$\sigma'_\theta/\sigma'_{vo}$ runs A & B	C	K_0 Free field
I 0-10 m	1.65	1.80	1.40	2.5	1.2	1.60	1.65
I 10-20 m	1.15	1.80	1.1	2.0	1.0	1.40	0.95
II	1.30	1.70	0.72	1.08	0.6	0.90	0.72
III	1.0	1.4	0.5	0.7	0.45	0.46	0.55
IV	1.0	1.4	0.5	0.7	0.45	0.46	0.55
V	1.0	1.0	0.55	0.55	0.45	0.55	0.55

Results of single pile analyses

The computed single-pile load displacement curves are presented in Fig. 12. Comparing Run A with the other two analyses shows immediately how important the residual interface strength can be. Local slippage starts at a much earlier stage in Runs B and C, with the strength cutoff leading to smaller shaft resistance through two mechanisms. Firstly there is the direct effect associated with the lower δ′ values. Secondly the cut-off effectively truncates dilation. The effective stress paths reported by Jardine (1985) show that, in Run A, large (undrained) post yeild increases in σ'_r were noted at most depth. In contrast, Runs B and C showed only small changes in σ'_r before and after local slippage had started.

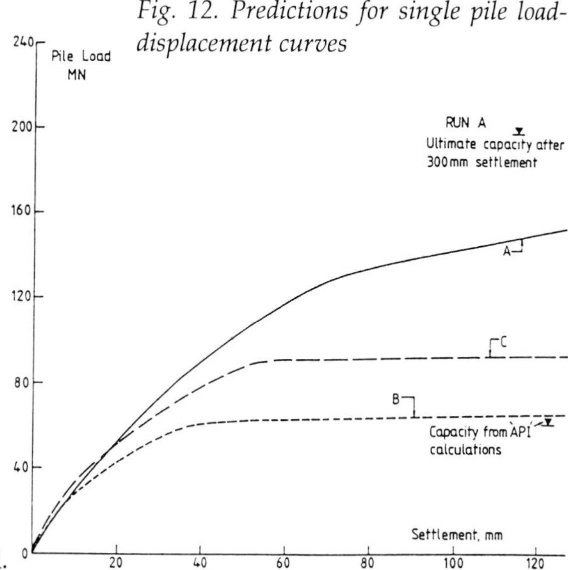

Fig. 12. Predictions for single pile load-displacement curves

Runs B and C predict far more plausible capacities and displacements to failure than A. It is interesting that the medium term anlysis (B) gave a similar capacity to the API calculation reported by Rigden and Semple (1983). Run C indicated that consolidation over the ten years following driving might lead to significant gains in capacity.

The shaft shear stress predicted in Runs B and C are summarised on Figs 13 and 14. These diagrams show the piles failling progressively and give results that may be compared with the load distributions measured on site.

The vertical displacement fields developed in the soil surrounding the piles are illustrated in Fig. 15, where data from two stages of Run B are shown as profiles of normalised surface settlement plotted against logarithm of radium. Linear elastic analyses of long rigid piles predit that settlement should vary almost linearly with log (r/r_0). In this case movements are concentrated close to the shaft by three factors, the compressibility of the pile, the non linearity of the soils continuum behaviour, and local slippage at the pile shaft. The two latter factors become progressively more important as the load factor increases. One important result is that the settlement interaction between piles is far less than would be expected from conventional pile group analyses.

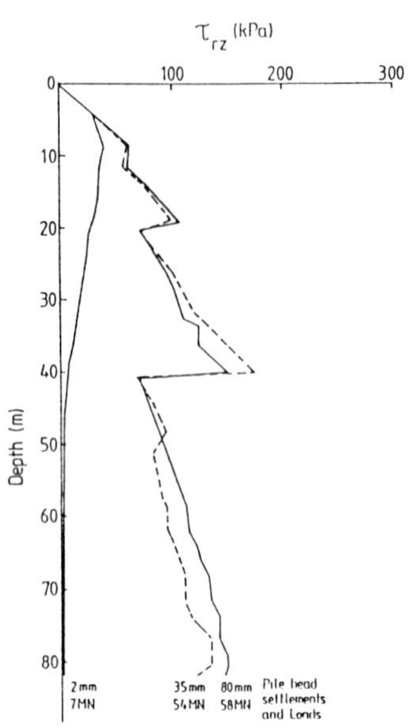

Fig. 13. *Prediction for shaft shear stress distributions*

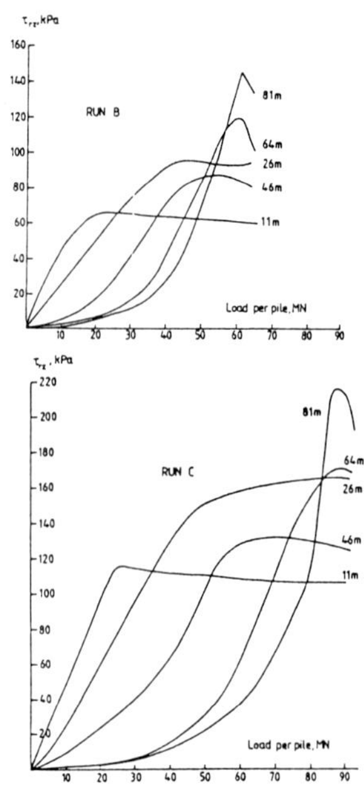

Fig. 14. *Predictions for variations of shaft shear stresses with pile head loads*

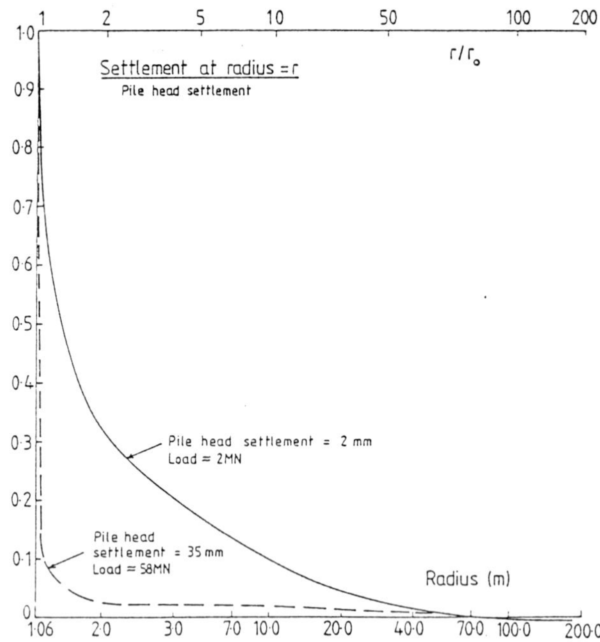

Fig. 15. Radial distributions of settlement, single pile analysis B

Group load-displacement response

Approximate, non-linear, superposition calculations were performed to predict the axial load displacement response of an evenly loaded group of nine Magnus piles. The surface displacement profiles computed at 20 or more stages in Runs A, B and C were used to develop the curves shown in Fig. 16, which also reproduces the prediction form the more conventional analysis reported by Rigden and Semple (1983).

It is interesting that, despite the highly non linear soil properties assumed for analyses A, B and C the initial parts of the curves are almost linear, giving global stiffnesses between of 30 (±2) MN group load per mm settlement. Curves A and B coincide initially, but differences in the local shaft resistances causes divergence when loads exceed ≈ 30MN per pile; curve C has a steeper initial slope - but this prediciton only applies to the artificial condition of first time loading 10 years after platform installation. The non-linear finite element analyses predict that the initial behaviour will be three times stiffer than is expected from the conventional calculations.

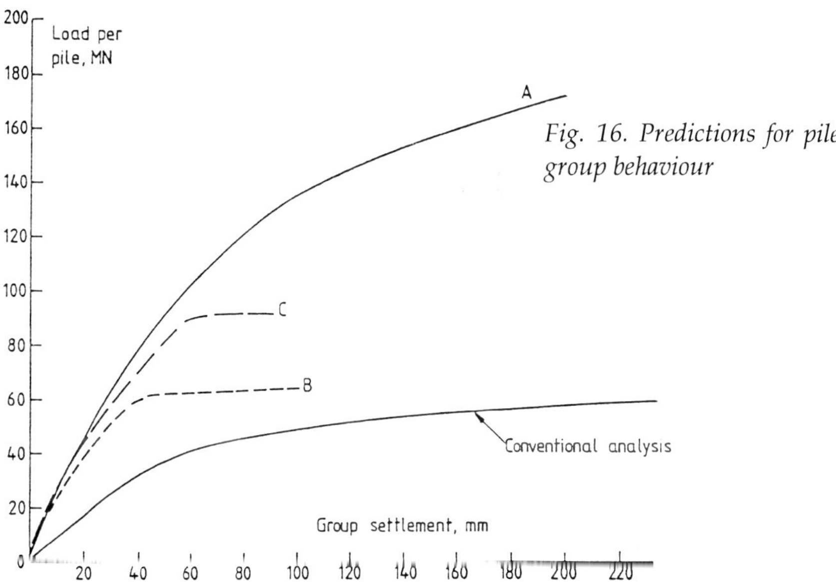

Fig. 16. Predictions for pile group behaviour

In each case the foundation stiffness reduces as loads increase. The static group stiffness predicted (in the medium term analysis, B) during topside placement is 23MN per mm (compared to an initial gradient of 27MN/mm). Storms which apply loads fluctuating between 15 and 25MN per pile group give medium and long term group stiffnesses of 17 and 20 MN/mm respectively. The effects of load level and time since installation are counteracting, so it is possible that a large storm developed some years after installation may give a comparable stiffness to a lesser event occurring at an earlier date. Other ageing processes and kinematic effects on stiffness exist which might cause the foundation response to relatively light dynamic loading to be stiffer than is implied by the appropriate portions of curves B or C. Even so, the group dynamic stiffness is unlikely to exceed the initial slope of curve C (i.e. 32MN/mm).

Field measurements exist of the platforms dynamic foundation behaviour. As reported by Sharp (1992) the Magnus pile groups have turned out to be far stiffer than was expected from the conventional calculations. But the available filed data are in good agreement with the range of predictions from the non linear analyses summarised above. Overall, the comparison indicates similar trends to those reported by Jardine and Potts (1988) from parallel studies of the Hutton Tension Leg Platform foundations.

Summary and conclusions

The paper summarises the laboratory and numerical research performed at Imperial college as part of the Magnus FMP. Stress-path tests, involving local strain measurement allowed the soils stiffness characteristics to be evaluated over a wide range of strain, with programmes on reconstituted and intact samples. Ring-shear interface tests provided data on the residual frictional resistances at the pile shaft.

The observed soil behaviour could be modelled relatively simply using a versatile finite element code. Critical reviews were undertaken to estimate how pile installation might affect the stress conditions around the piles and the various strands of research were drawn together in numerical analyses of the Magnus foundations. Four main conclusions may be drawn from the results

- Realistic predictions can be made without resorting to empirical design rules.
- It is extremely important to allow for clay residual fabric close to the pile shaft.
- Soil stiffness non-linearity plays a key role in controlling load displacement behaviour.
- Conventional analyses underestimate pile group axial stiffness considerably. They may also be conservative with regard to long term axial capacity.

The potential benefits of a soundly based effective stress design method are evident from the Magnus studies. Field research is elucidating the stress conditions associated with pile installation and loading to failure, whilst the above numerical approach is being extended to consider pile groups experiencing horizontal and turning, as well as axial loads.

Acknowledgements

The authors thank the Science and Engineering Research Council (SERC) for financial support, BP for permission to publish the results, and Professor J. B. Burland, Dr D. Hight, Dr L. Lemos and Dr P. Smith for their guidance and assistance.

References

A. J. BOND and R. J. JARDINE (1991) The effects of installing displacement piles in a high OCR clay. *Géotechnique* **41,** No. 3, pp. 341-363.

D. W. HIGHT (1983) *Laboratory investigations of sea bed clays.* PhD Thesis, University of London, Imperial College.

D. W. HIGHT, A. GENS and R. J. JARDINE (1985) Evaluation of geotechnical parameters from triaxial tests on offshore clay. *Proc. Soc. Underwater Technology, conf. on Offshore Site Investigations,* Graham & Trotner, London, pp. 253-268

R. J. JARDINE, M. J. SUMES and J. B. BURLAND (1984) The measurement of soil stiffness in the triaxial apparatus. *Géotechnique* **34,** No. 3, pp. 323-340.

R. J. JARDINE (1985) *Investigations of pile-soil behaviour, with special reference to the foundations of offshore structures.* Phd Thesis, University of London, Imperial College.

R. J. JARDINE, D. W. HIGHT and W. MCINTOSH (1988) Hutton Tension Leg Platform Foundations, measurement of pile group load-displacement relations. *Géotechnique* **38,** No 2, pp. 219-230.

R. J. JARDINE and D. M. POTTS (1988) Hutton tension leg platform foundations, an approach to the prediction of pile behaviour. *Géotechnique* **38,** No. 2, pp. 231-252.

R. J. JARDINE, H. ST. JOHN, D. HIGHT and D. POTTS (1991) Some practical applications of a non-linear ground model. *Proc. Xth ECSMFE, Florence,* Vol. 1, pp. 223-228.

R. J. JARDINE (1992) Non-linear stiffness parameters from undrained pressuremeter tests. *Canadian Geotechnical Testing Journal,* June 1992.

L. J. LEMOS (1985) *The effects of rate on the residual strength of soil.* PhD Thesis, University of London, Imperial College.

D. M. POTTS (1985) *A user's manual for ICFEP.* Internal report, Imperial College.

W. J. RIGDEN and R. M. SEMPLE (1983) Design and installation of the Magnus foundations; prediction of pile behaviour. *Design and construction of offshore structures,* ICE, London, pp. 29-52.

D. E. SHARP (1992) Magnus foundation monitoring – an overview. This volume.

5. Mudmat interaction and foundation analysis

M. R. HORSNELL and V. A. NORRIS, Fugro-McClelland Ltd, and B. IMS, Fugro Douglas Pty Ltd

Fugro-McClelland Limited have carried out analyses of the foundation stiffness at the BP Magnus platform and compared results with those derived from site measurements. Data from the pile installation and the platform monitoring programme have been used in conjunction with the site investigation data to derive as constructed soil and foundation conditions for input to the analyses. A special feature of the analyses is the inclusion of the mudmats (see Fig.1). A conventional pile group analysis, ignoring the influence of the mudmat, had resulted in stiffness values which were of the order of 50% lower than the values derived from jacket instrumentation measurements. This paper describes the ways in which the effect of the mudmat has been included in the analyses as well as other measures taken to obtain better agreement with site measurements.

Foundation response of the Magnus platform foundation

Instrumentation at the Magnus Platform, concentrated on leg A4 (as shown in Fig. 1), has provided the following information for calculating the foundation stiffness:

- leg loads from strain gauges on the main leg and diagonal braces
- pile loads from strain gauges at varying levels on eight of the nine piles
- mudmat loads from total pressure cells on the underside of the mudmat.

Recordings made for an individual storm wave on 22 January, 1984 have been used for the analysis discussed in this paper. For a peak to trough wave height of 17 metres the loads at mudline derived by W. A. Fairhurst and Partners (Ref. 1) from the recorded data are given in Table 1. The pile group moment terms are the components due to (1) axial pile loads and (2) pile bending moments.

The values derived by Fairhurst result from the assumption that axial pile load decreases linearly with depth at a rate of 2% per metre. The actual

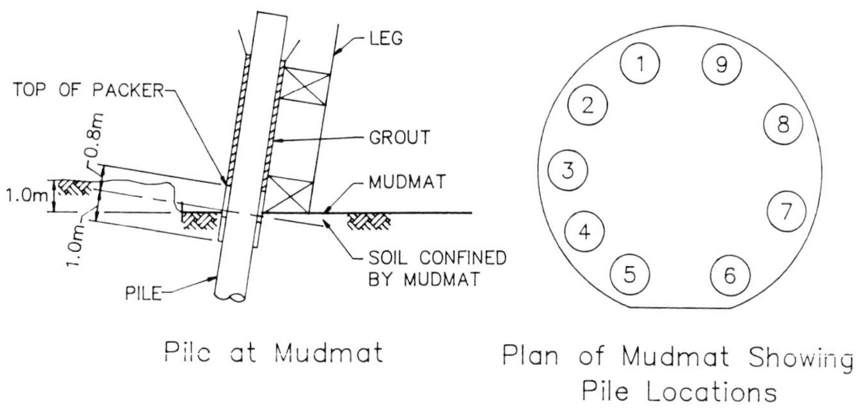

Fig. 1. Leg A4 pile group and mudmat

Table 1. Foundation loading for maximum storm wave, 22/1/84

FORCE	LEG(1)	PILE GROUP(2)	MUDMAT(3)	(2)+(3)-(1)
AXIAL	69.1	64.3	11.7	6.9 (10%)
LATERAL	11.7	6.8	-	-
MOMENT*	146.7	165-13.5	17.0	21.8(15%)

*Moments are given relative to pile group geometric centre
Units: MN,m

Table 2. Foundation displacement for maximum storm wave 22/1/84

AXIAL DISPLACEMENT	LATERAL DISPLACEMENT	ROTATION
2.0mm	5.7mm	0.25mRads

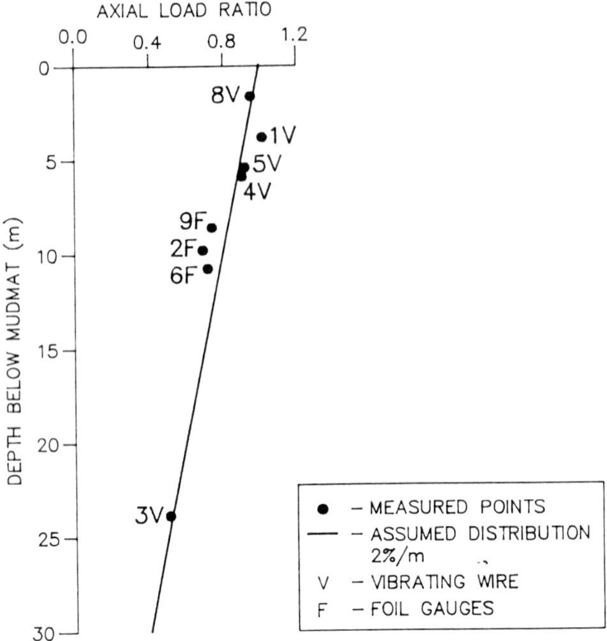

Fig. 2. Derived axial load distribution down a pile

measurements show, however, that there may be an increase with depth at the top of the piles (see Fig. 2). Numerical analyses, described in this paper, show that such an increase could be attributed to interaction between the mudmat and foundation piles. The apparent imbalance of forces presented in the right hand column of Table 1 may therefore be indicative of the level of interaction between the mudmat and the foundation piles. The values used for the moment M and lateral load F of a single pile at mudline have been derived using the pile bending moment diagram shown in Fig. 3 and assuming that $F = dM/dz$ at mudline.

The corresponding dynamic displacements, obtained from accelerometers at the base of the leg, are given in Table 2. It should be noted that these displacements have been measured on the leg at -182 m. They do not necessarily apply directly to the piles at -185 m (mudline) since the leg–pile connection provided by the grout in the pile sleeve is not necessarily rigid to this depth. Allowance is later made for relative movement between pile and leg when examining the foundation behaviour analytically.

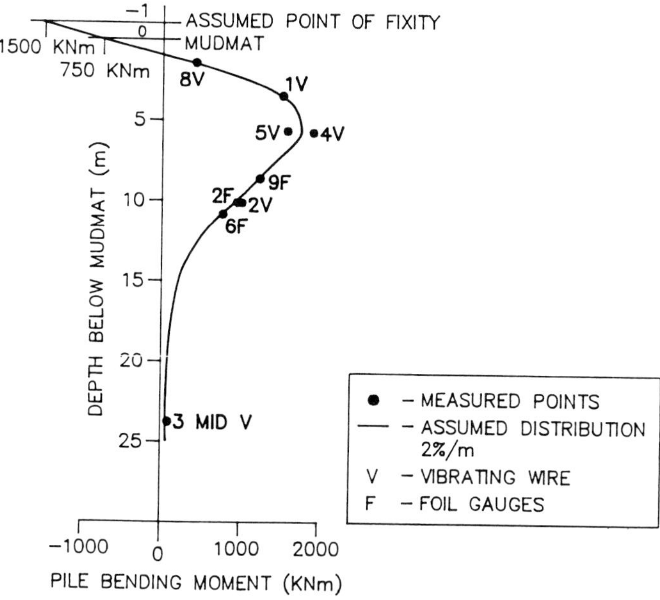

The distribution has been derived by combining the bending moment values from different piles.

Fig. 3. Derived bending moment distribution down a pile

Foundation stiffness matrix

The measured load and displacement data can be used to derive a stiffness matrix for the leg foundation at mudline. The equations used to define the pile group behaviour in the plane of loading take the form:

$$\begin{bmatrix} F_z \\ F_y \\ M \end{bmatrix} = \begin{bmatrix} A_{11} & 0 & 0 \\ 0 & A_{22} & A_{23} \\ 0 & A_{32} & A_{33} \end{bmatrix} \begin{bmatrix} Z \\ Y \\ \Theta \end{bmatrix} \qquad (1)$$

where F_z = axial load, F_y = lateral load, M = moment, Z = axial displacement, Y = lateral displacement, and Θ = rotation

These equations assume no interaction between axial load and moment, which is not strictly true for non-linear behaviour, but is reasonable for the relatively low loading herewith being considered. The axial stiffness A_{11} can thus be calculated simply by dividing the axial force by the axial displacement. Similarly the rotational stiffness coefficient A_{33} can be obtained by dividing the associated group moment by the rotation. However, the moments due to pile bending interact with the lateral displacements and the unknown stiffness terms are too many to be derived from the site measurements. An approximate method has been used by Fairhurst (ref. 1, Appendix 5) using Roark's method for displacements and rotations at the top of a semi-infinite beam. Using this method for a single pile gives the relationship:

$$\begin{bmatrix} F_{y,p} \\ M_p \end{bmatrix} = \begin{bmatrix} 266 & -2,330 \\ -2330 & 40,900 \end{bmatrix} \begin{bmatrix} Y \\ \theta \end{bmatrix} \qquad (2)$$

Units : MN, m, Rads

Multiplying the single pile stiffness matrix by the number of piles, 9, and combining with the axial stiffness (64.3 MN/2.0 mm), and the rotational stiffness due to axial loading, (165 MN m/0.25 mRad), results in the overall stiffness equation:

$$\begin{bmatrix} F_z \\ Fy \\ M \end{bmatrix} = \begin{bmatrix} 32,000 & 0 & 0 \\ 0 & 2,400 & -21,000 \\ 0 & -21,000 & 1,028,000 \end{bmatrix} \begin{bmatrix} Z \\ Y \\ \theta \end{bmatrix} \qquad (3)$$

Units: MN, m, Rad

Design stiffness matrix

The pile group stiffness matrix used for platform design is shown below. As may be seen, the stiffness values

Table 3. Foundation loading computed using modified FUGROUP

CASE NO.	LATERAL LOAD (MN)					
	LEG LOAD		PILE GROUP		MUDMAT	
	COMP.	MEAS.	COMP.	MEAS.	COMP.	MEAS.
1	8.3	11.7	2.6	6.8	5.8	4.9*
2	10.4	11.7	4.6	6.8	5.3	4.9*
3	12.7	11.7	6.9	6.8	5.8	4.9*
4	11.9	11.7	6.6	6.8	5.3	4.9*

*Implied load from measured leg load and corresponding pile group load

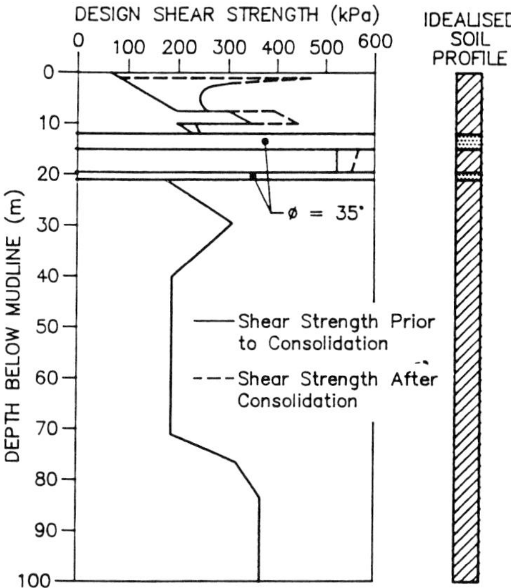

Fig. 4. Design shear strength

$$\begin{bmatrix} F_z \\ F_y \\ M \end{bmatrix} = \begin{bmatrix} 7{,}000 & 0 & 0 \\ 0 & 1{,}270 & -9{,}000 \\ 0 & -9{,}000 & 600{,}000 \end{bmatrix} \begin{bmatrix} Z \\ Y \\ \Theta \end{bmatrix} \qquad (4)$$

Units: MN, m, Rad

obtained from site measurements are approximately twice the design values. The remainder of this paper describes analyses which have been performed to assess the influence of the following factors on pile group stiffness, i.e.

- interaction between mudmat and foundation piles, and
- basic formulations used in 'conventional' pile group analysis.

Pile group analysis with mudmat included

The effect of the mudmat has been included in the lateral loading analysis by incorporating an additional group member with appropriate Winkler springs. Interaction between mudmat and piles was modelled by adopting a '*y*-shift' approach, similar to that commonly adopted for pile–pile interaction. Pile group analyses for lateral loading were performed using the computer

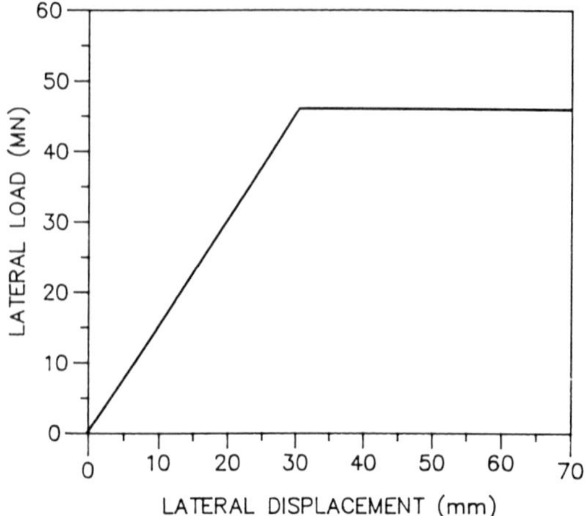

Fig. 5. Mudmat lateral load versus displacement

program FUGROUP (ref. 2), with the coding modified to incorporate mudmat–pile interaction using the following techniques:

- mudmat–mudmat interaction with piles — based upon elastic finite element analysis of mudmat, and
- pile interaction with mudmat — based upon averaging pile–pile interaction at mudmat level.

For axial loading the mudmat has not been included with the pile group as separate finite element analyses had shown that interaction under axial loading was less significant than for the lateral case.

Prior to carrying out the analyses the soil strength profile had been modified to take account of consolidation by the mudmat. Applying an ultimate effective pressure of 50 kN/m^2 due to a flexible mudmat and assuming that the increase in soil strength is proportional to the increase in vertical stress had resulted in the modification shown in Fig.4.

The effect of the mudmat on lateral loading response

Mudmat behaviour under lateral load was modelled using a bilinear load-displacement relationship consisting of an elastic portion derived using the Bycroft formula (ref. 3) for a homogeneous soil and a failure load calculated assuming full soil strength at the soil-mudmat interface (see Fig. 5). In order to model the mudmat-pile interaction a Finite Element elastic analysis was performed to determine the soil displacements due to lateral displacement of the mudmat. The lateral displacements of the soil have been found to be similar at all pile locations and can be represented by the displacement-

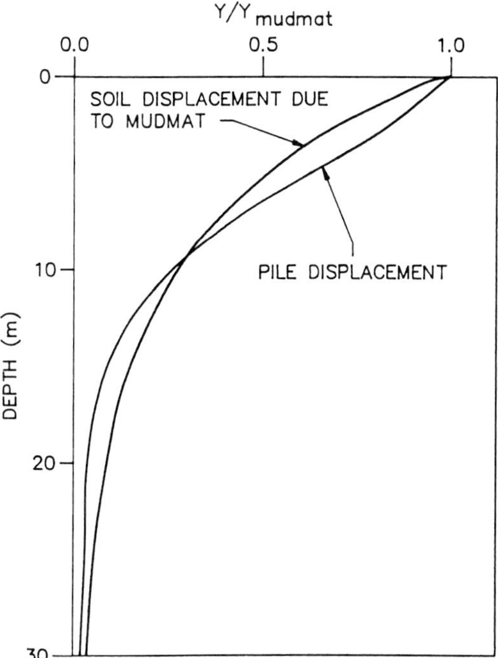

Fig. 6. Displacement of soil due to lateral displacement of mudmat compared with pile displacement

depth profile shown in Fig. 6. Also included is the displacement of a single pile. Because the lateral displacements due to the mudmat are similar to those of the pile one would expect them to have a significant effect on the pile behaviour.

A series of pile group analyses were performed using FUGROUP, modified to incorporate mudmat/pile interaction, to assess the influence of the model used to represent single pile behaviour in stiff clays, i.e.

- Case 1 — Single pile behaviour based upon modified Matlock *p-y* data (ref. 4) with $\varepsilon_{50} = 1\%$
- Case 2 — As for case 1 with $\varepsilon_{50} = 0.4\%$
- Case 3 — Single pile behaviour based upon Reese, Cox and Koop (ref. 5) with $\varepsilon_{50} = 0.5\%$.

Analyses were performed corresponding to a leg rotation (8) of 0.25 mRad and lateral displacement of 5.7 mm at an elevation of minus 182 m. Assuming complete leg pile fixity, these conditions correspond to a rotation of 0.25 m Rad and lateral displacement of 5 mm at mudline elevation of -185 m. Results from the analysis are presented in Table 3 and corresponding bending moment profiles are presented in Figs 7 and 8.

As can be seen from Table 3, application of Reese, Cox and Koop *p-y* data results in the best comparison between computed and measured lateral loads but, as shown in Fig. 8, the comparison of bending moment distribution is not convincing. A further series of analyses was performed to investigate the influence of leg fixity. These analyses identified that best overall comparison between measured and predicted forces and bending moments was obtained for a rotation at mudline of 0.34 m Rad and mudline deflection of 4.1 mm (shown as Case 4 in Table 3). This indicates that in reality leg–pile fixity reduces gradually with depth and is a function of the position of grout packers and group strength and compressibility. The equivalent single pile stiffness matrix corresponding to the revised pile–leg fixity now becomes:

$$\begin{bmatrix} F_p \\ M_p \end{bmatrix} = \begin{bmatrix} 277 & -1,194 \\ -1,194 & 10,290 \end{bmatrix} \begin{bmatrix} Y \\ \theta \end{bmatrix} \quad (5)$$

The effect of the mudmat on vertical loading response

In the case of vertical loading it has not been possible to model the mudmat and pile group together. However, some idea of the significance of their interaction has been obtained by modelling a single pile and the mudmat separately using the Finite Element program harmony. The vertical soil displacement due to the mudmat at the location of the pile is compared with the vertical displacement of the pile in Fig. 9. It may be seen that the soil displacements due to the mudmat are greater than the displacements of the pile between depths of 0 and 8 metres. The axial force in a single pile beneath the mudmat will therefore increase to a depth of 8 metres and decrease from thereon. This finding supports the assumption made previously for pile group loading when deriving axial pile loads from site measurements using Fig. 2.

Because the axial displacements due to the mudmat decrease with depth far more rapidly than those due to the pile one would not expect the interaction effect of the mudmat to be as great as is the case for lateral loading (see Fig. 6), its effect being concentrated on the soil close to the mudrnat. It is therefore considered reasonable to analyse the pile group without the mud-mat and reduce the resultant pile loads by 10%, as was proposed previously to obtain good agreement between leg, pile group and mudmat loads derived from site measurements.

The pile group analyses have been carried out using programme FEMAX to determine the single pile load–displacement relationship and program

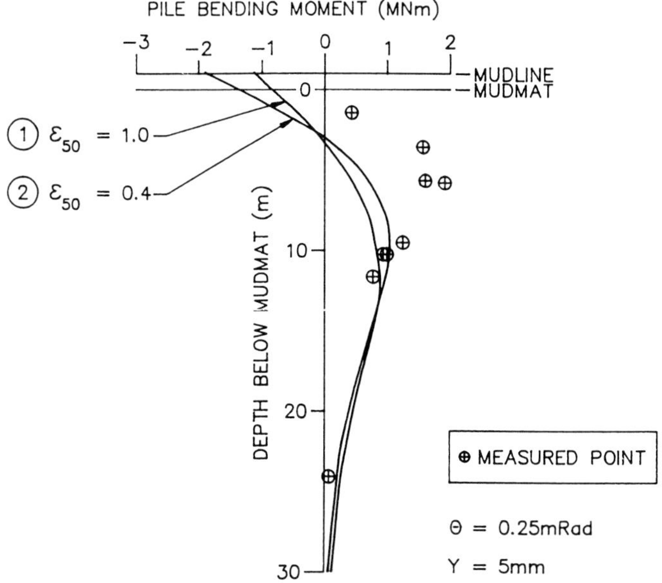

Fig. 7. Single pile bending moment versus depth (Matlock method)

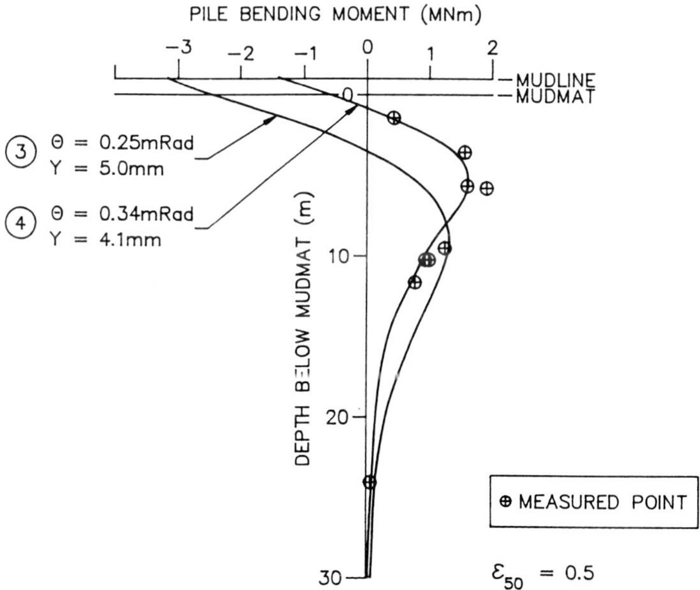

Fig. 8. Single pile bending moment versus depth (Reese, Cox and Koop method

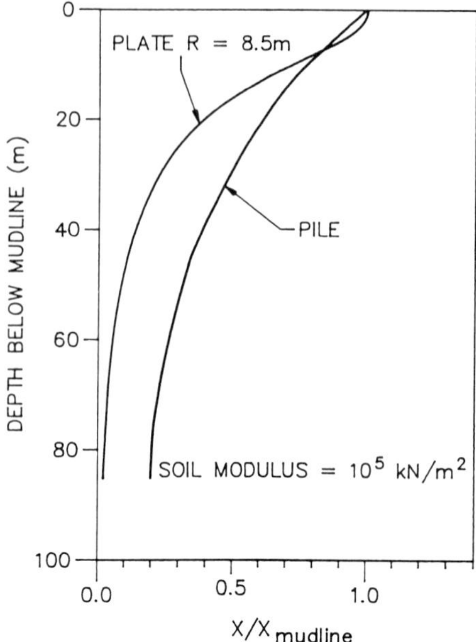

Fig. 9. Vertical displacement of soil at pile location due to pile and plate displacement

ROPRO1 to model pile–pile interaction (ref. 6). The soil strength profile used was that shown in Fig. 5 for soil consolidated by the mudmat. Pile–soil friction has been calculated using API RP2A recommendations (ref. 7).

The load–displacement curve obtained for pile group axial displacement is shown in Fig. 10. For the selected individual wave of 22/1/84 the dynamic axial load amplitude calculated by Fairhurst (ref. 2) is 64.3 MN. The static axial load is 130 MN. For this combination of loading the loaddisplacement curve gives an axial tangential stiffness of 9340 MN/m. An analysis of pile group rotation, starting at the static axial load of 130 MN and applying a rotation of 0.125 m Rad, i.e. half the amplitude of the wave motion, gives a rotational stiffness of 694,000 MN m/Rad. Comparison with the stiffnesses derived from site measurements, i.e. an axial stiffness of 32,000 MN/m and a rotational stiffness of 660,000 MN m/Rad, shows that the analytical axial stiffness is 71% lower whilst the rotational stiffness is 5% higher.

The level of agreement between computed and measured rotational stiffness is encouraging. For a large pile group with near symmetrical configuration, as at Magnus, rotational stiffness at low level will be governed primarily by single pile behaviour with any interaction effects between the piles tending to cancel each other out. The difference between the measured and predicted axial stiffness of 71% indicates the level of interaction which

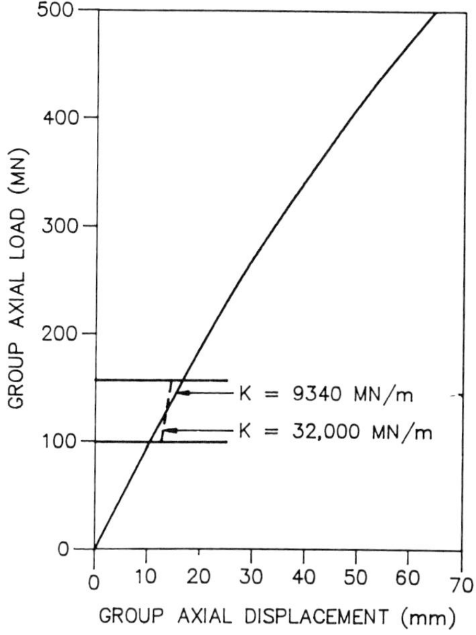

Fig. 10. Pile group axial load versus displacement

results from the assumptions that the soil can be modelled as on equivalent homogeneous, isotropic elastic half-space may be too high at the load level analysed. As discussed previously, and as shown in Fig. 3, axial load distribution for the storm conditions under construction dissipates at a rate of approximately 2% per metre penetration below mudline. This suggests that the bottom half of the pile carries little or no load and, as a consequence, any pile-to-pile interaction is confined within the soil above the mid-point of the piles any may only influence the distribution of shear stresses without inducing additional pile movement. This observation differs significantly from the assumptions upon which conventional Poulos-type interaction factors used in computer analyses, are based. It should be noted, however, that these conclusions are based upon a load levels, which is approximately 18% to 22% of the computed ultimate static capacity of the pile group. At higher load levels, pile to pile interaction may be more significant. In addition, the measured axial stiffness is for dynamic loading over an incremental load range, compared with psuedo static loading used in predictions.

Stiffness of the mudmat

The stiffness of the mudmat with respect to lateral loading may be seen from the analyses involving pile group mudmat interaction. Whilst there are no site measurements for direct comparison with the analytical results it has

been shown that the pile group lateral load agrees well with that obtained from pile measurements whilst the pile group plus mudmat lateral load agrees well with that from leg measurements.

In the case of axial loading the mudmat has not been included in the pile group analyses and no attempt has been made to model the effects of interaction by the piles. However, for comparative purposes it is useful to predict the mudmat behaviour using simple elastic formulae. For a circular footing of radius r on homogeneous soil the axial stiffness is given by:

$$K_z = (2\,Er) \,/\, (1 - \upsilon^2) \tag{8}$$

where υ = Poisson's ratio for the soil.

For $E = 10^5$ kN/m^2, $\upsilon = 0.5$ and $r = 8.5$ m, $K_z = 2{,}270$ MN/m, i.e. about 15% of the measured leg stiffness. This figure compares with a value of 17% given by the measured mudmat load in Table 1. It therefore seems likely that interaction by the piles has no significant effect on the stiffness of the mudmat.

For pure moment loading the mudmat rotational stiffness is given by

$$K_m = (4\,Er^3)/3\,(1-\upsilon^2) \tag{9}$$

For the E, υ and r values used previously $K_m = 131{,}000$ M m/Rad, i.e. about 22% of the measured leg rotational stiffness. This figure compares with a value of 12% given by the measured mudmat rotational stiffness. It may be that a lower value of E is appropriate for rotational stiffness, or that the mudmat pressure cells do not provide an accurate estimate of the moment.

Conclusions

(a) Inclusion of the mudmat in the analysis of lateral loading of a pile group at Magnus gives good agreement with site measurements if leg–pile fixity is assumed to stop at some appropriate height above mudline.

(b) For the small displacements being considered at Magnus it has been found that the Reese, Cox and Koop method for constructing p-y curves for overconsolidated clay is more suitable than the 'modified Matlock' method. This is not necessarily true at larger displacements such as those corresponding to a 100 year storm wave.

(c) In the case of axial pile loading the pile group analysis shows poor agreement with site measurements. The low predicted axial stiffness may be attributed to the method used for modelling pile–pile interaction being unsuitable for the relatively small loads under consideration.

(d) The contribution of the mudmat in the case of a stiff surface clay is very significant in the case of lateral loading, less significant in the case of vertical loading.

References

1. W. A. FAIRHURST AND PARTNERS. Magnus Foundation Monitoring Project, Final Report on Measurements submitted to BP International Ltd, Report No. 450, 1986.
2. HORSNELL, M. R., ALDRIDGE, T. R. and ERBRICH, C. Lateral group behaviour of piles in offshore soil conditions. OTC 6246, 1990.
3. POULOS, H. G. and DAVIS, E. H. *Elastic solutions for soil and rock mechanics*, John Wiley & Sons, 1974.
4. MATLOCK, H. Fugro Ltd internal memorandum, 1979.
5. REESE, L. C., COX, W. R and KOOP F. D. Field testing and analysis of laterally loaded piles in stiff clay, OTC 2312, 1975.
6. TOOLAN, F. E. and HORSNELL, M. R. Analysis of load deflection behaviour of offshore piles and pile groups. *Proc. Int. Conf. on Numerical Methods in Offshore Piling*, ICE, London, May 1980.
7. AMERICAN PETROLEUM INSTITUTE. Recommended Practice 2A, 19th Edition. *Recommended Practice for Planning, Designing and Constructing Fixed Offshore Platforms*, 1991.

6. Magnus FMP — dynamic response of foundations

O. SKJÅSTAD, Sintef, Norway and E. SKOMEDAL, Statoil, Norway

Introduction

The dynamic response of a piled offshore platform is governed by the interaction between the load-carrying structure and the foundation. To compute the dynamic response it is therefore important to model the interaction mechanism as accurately as possible.

The purpose of this study is to improve the understanding of the behaviour of the interaction mechanism between the load-carrying structure and the foundation. The study is based on analysis of full scale measurement data recorded on the Magnus platform, and has been directed against

- dynamic properties of the platform
- stiffness and damping properties of foundation
- load distribution in foundation
- load transfer from structure to foundation

The instrumentation system comprises several types of sensors for measuring environmental conditions and platform and foundation behaviour. The study is based on analyses of the following data types which are either measured directly or derived from measured data

- measured accelerations, strains in structure, strains in piles and mudmat pressures
- leg displacements at base derived from the measured accelerations
- forces in leg and braces derived from measured strains
- pile forces derived from the measured strains
- mudmat forces derived from the measured mudmat pressures

Dynamic properties of the structure and foundation

As a basis for understanding the soil–structure interaction mechanism it is important to investigate the dynamic properties of the platform and foundation. The dynamic properties will be governed by the vibration char-

Table 1. Vibration characteristics from acceleration data.

The direction is given relative to the x-axis.

Recording period: 840122 00:02						
Frequency [Hz]	Damping [%]	Direction deck base [deg]		Phase deck base [deg]	Mode ratio base/deck	Remark
.105		65	57	-7	.078	Quasti-static
.337	1.0	116	108	3	.0268	1.order Y
.339	1.4	26	23	1	.0284	1.order X
.520	1.7	106	132	7	.256	Torsion
.817	.7	-3	7	171	.119	2.order X
.830	.9	85	87	-175	.140	2.order Y
.842	2.9	12	7	-179	.106	2.order X
.973	1.3	-16	-34	-169	.159	
1.108	1.1	132	116	178	.399	2.order torsion
1.426	2.9	9	9	20	.335	3.order X
1.480	4.1	115	112	29	.335	3.order Y
1.848	8.1	69	41	168	.173	
1.992	3.5	11	-8	-170	.387	4.order X

acteristics, i.e. the natural frequencies, mode shapes and modal damping. These are determined by a method which employs the correspondence between a multidegree-of-freedom linear dynamic system and a multichannel Auto-Regressive Moving Average (ARMA) model (Hoen, 1987).

The linear horizontal acceleration data at deck and base of leg A4, shown in Fig. 1 (taken from ref. 1), are applied for estimation of the vibration characteristics of the structure in four recording periods with different sea state severity. Representative results are given in Table 1. The third column of this table gives the direction or vibration plane of the mode at deck and base, while the fourth column gives the phase angle between the modal displacements at deck and base. This information is valuable for interpretation of the mode shape. A phase angle of about 0 deg. indicates a global mode of first, third, fifth etc. order, while a phase angle of about 180 deg. indicates a mode of second, fourth etc. order. The fifth column gives the ratio between the modal displacements of the base and deck, which are used to evaluate the contribution from the foundation in the mode, and also to indicate the order of the mode.

The corresponding mode shapes are shown in Fig. 2. The structure is shown from above and the mode shapes are indicated at leg A4. The elliptic shape, the rotation of the ellipses and the indicated phase angles are calculated from the eigenvector of the modes.

The natural frequencies and modal damping are also estimated from the structural strain data, Table 2. As shown it is possible to clearly identify 5 frequencies from the axial and shear strain data of the leg, which agrees with the frequencies estimated from the acceleration data. The same calculations are performed based on mudmat pressure data, Table 2. Three natural frequencies are identified. The mode shapes, presented in Fig. 3, show that

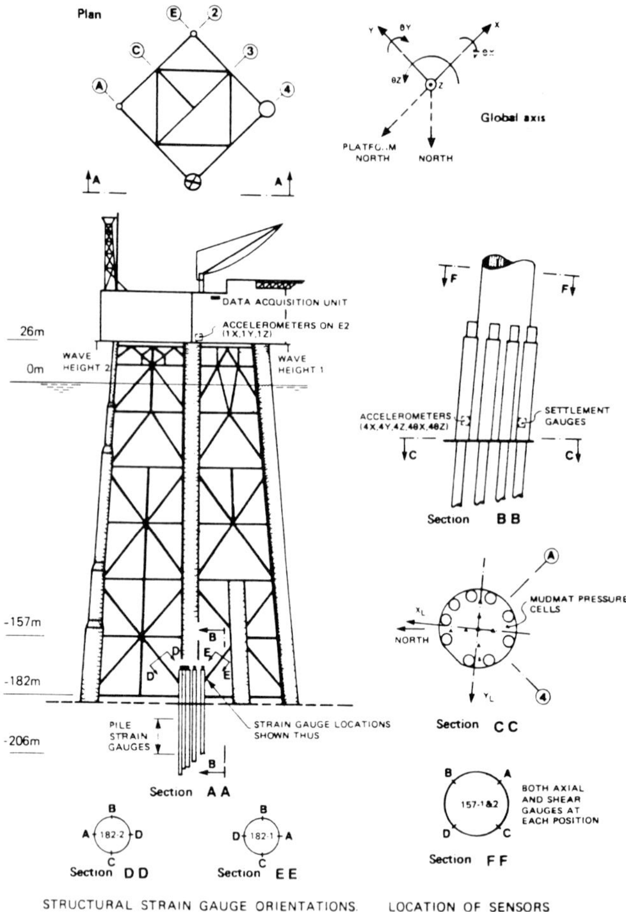

Fig. 1. Sensor locations

the quasistatic mode at 0.076 Hz and the sway mode at 0.340 Hz are quite similar. The magnitude of the sway mode seems almost constant along the line between Cells 1 and 7, while for the perpendicular line, the mode has the greatest magnitude at Cell 3 and decreases towards Cell 9. The torsional mode at 0.51 Hz has its main response along the Cells 1, 2, 5, 6 and 7, where Cells 1 and 2 are about 180 deg. out of phase with Cells 6 and 7, which is indicated by the opposite sign in the figure. The second order bending mode at 0.83 Hz has the largest magnitude at Cells 7 and 9, and Cells 1 and 7 and Cells 3 and 9 are about 180 deg. out of phase.

Five sets of pile strain data have been analysed with respect to natural frequencies and modal damping, see Table 2. The frequencies up to the

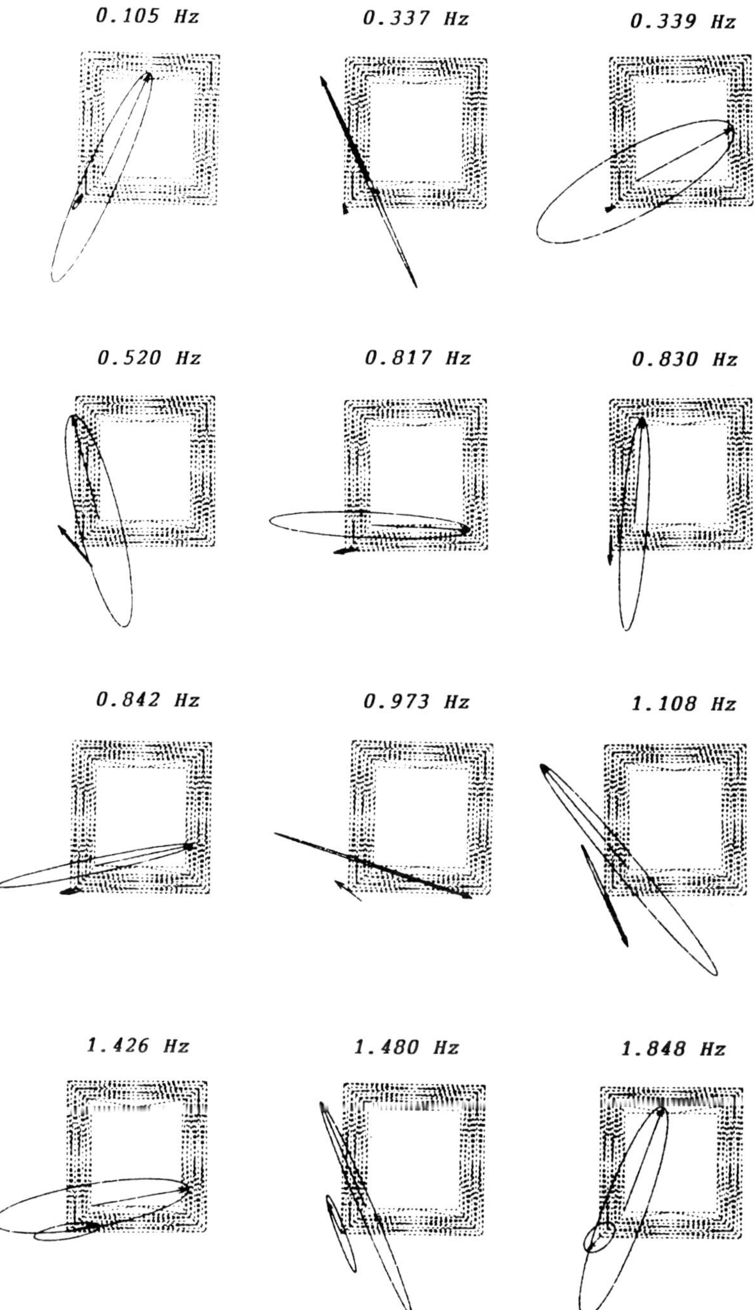

Fig. 2. Estimated mode shapes: recording period 840122 00:02

Table 2. Vibration characteristics from strain and pressure data

	Recording Period: 840122 08:04							
Sensor group	Frequency [Hz] / Damping [%]							
LAS	.075	.337 2.2	.340 1.1	.522 1.3	.829 2.2	- -	(.983) (6.5)	- -
LSS	.077	.339 1.6	.340 1.0	.522 1.2	.818 1.6	.842 2.5	- -	1.131 4.6
B1(x)	.073	- -	.340 1.0	.522 1.4	.823 2.3	- -	- -	- -
B2(y)	.073	.339 1.2	.340 4.9	.523 1.3	- -	.846 3.5	(.99) (6.7)	1.110 2.4
MP	.076	- -	.340 1.5	.506 5.6	- -	.830 6.5	- -	- -
P1V	.065	.335 3.6	.340 1.0	.523 1.3	.818 3.3	- -	- -	1.095 10.0
P2V	- -	- -	.340 1.2	.515 4.3	- -	- -	- -	- -
P2F	.063	.326 6.6	.340 1.1	- -	.819 2.0	- -	- -	- -
P3M	.074	.337 1.1	.342 1.5	- -	- -	- -	- -	- -
P4V	.076	.330 5.2	.340 1.1	.522 1.4	.820 2.3	- -	- -	- -

LAS: axial strain in leg A4
LSS: shear strain in leg A4.
B1 : axial strain in brace 1.
B2 : axial strain in brace 2.
MP : mudmat pressure.
P1V: axial strain pile 1 (-3.50 m v.w).
P2V: axial strain pile 2 (-9.94 m v.w).
P3M: axial strain pile 3 (-23.75 m v.w).
P4V: axial strain pile 4 (-5.72 m v.w).
P2F: axial strain pile 2 (-9.94 m foil).

second order bending mode are present in the pile strain data. One may also note that the strain data sets closest to the base and the structure contain 3-4 natural frequencies. The strain set from Pile 3, which is located at a depth of 23.75 m below mudmat, contains, however, only the quasi-static peak frequency and the sway mode.

Stiffness and damping properties of the foundation

Nature of stiffness and damping mechanisms

The nature of stiffness and damping mechanisms is evaluated on the basis of statistical analysis of the base response; in this context forces and displacements in three global directions (x, y and z). The response consists of a quasistatic part which is controlled by stiffness, and resonant part which is controlled by damping. The two parts are therefore conveniently studied separately. The base response is separated into a quasistatic part (0.05–0.25 Hz), a first resonance part (0.3-0.4 Hz) and a higher order resonance part (0.75–0.9 Hz).

The quasistatic base forces will balance the wave forces and will not be influenced by the soil–structure interaction mechanism, and will thus only include the wave loading effects. The quasistatic base displacements will, however, include both effects from the soil–structure interaction mechanism and the wave loading. A comparison of probability distributions of the quasistatic forces and displacements will thus expose the effects connected to the soil–structure interaction mechanism.

Representative probability distributions of quasistatic base forces and displacements are shown in Fig. 4. The quasistatic forces follow the Gaussian

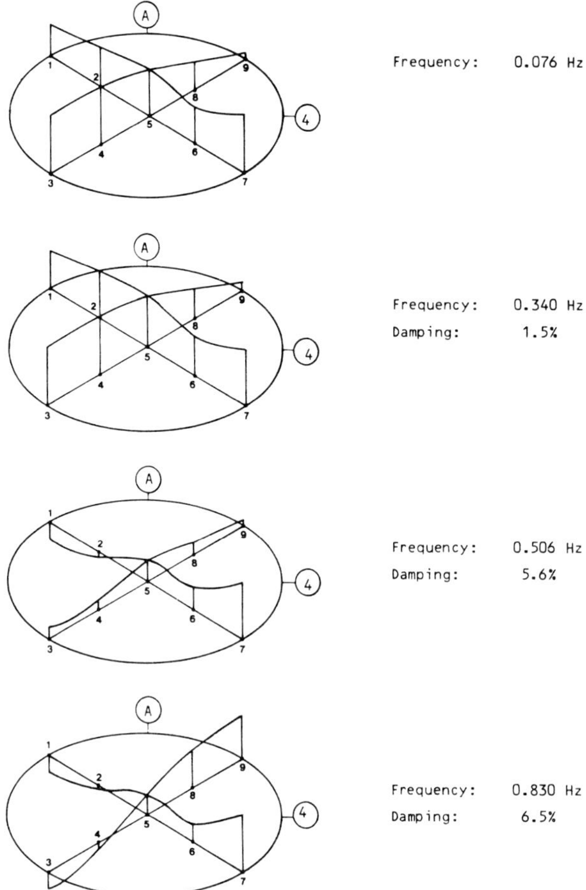

Fig. 3. Mode shapes of mudmat: recording period 840122 08:04

distribution fairly well. No nonlinear hydrodynamic load effects are thus observed. There is no indication of a different nature of the quasistatic base forces as the significant wave height varies nor is the nature different in the global force directions. The probability distributions of quasistatic base displacements have a quite similar Gaussian trend as the forces. The probabilistic nature of the quasistatic base forces and displacements seems thus to be similar and close to Gaussian. This means that there is a linear relation between them. This relation expresses the stiffness mechanism which thus can be regarded as linear.

The nature of the resonant forces and displacements are controlled by the damping of the soil–structure interaction mechanism. This means that nonlinearities in the damping mechanism should be recognized in the resonant

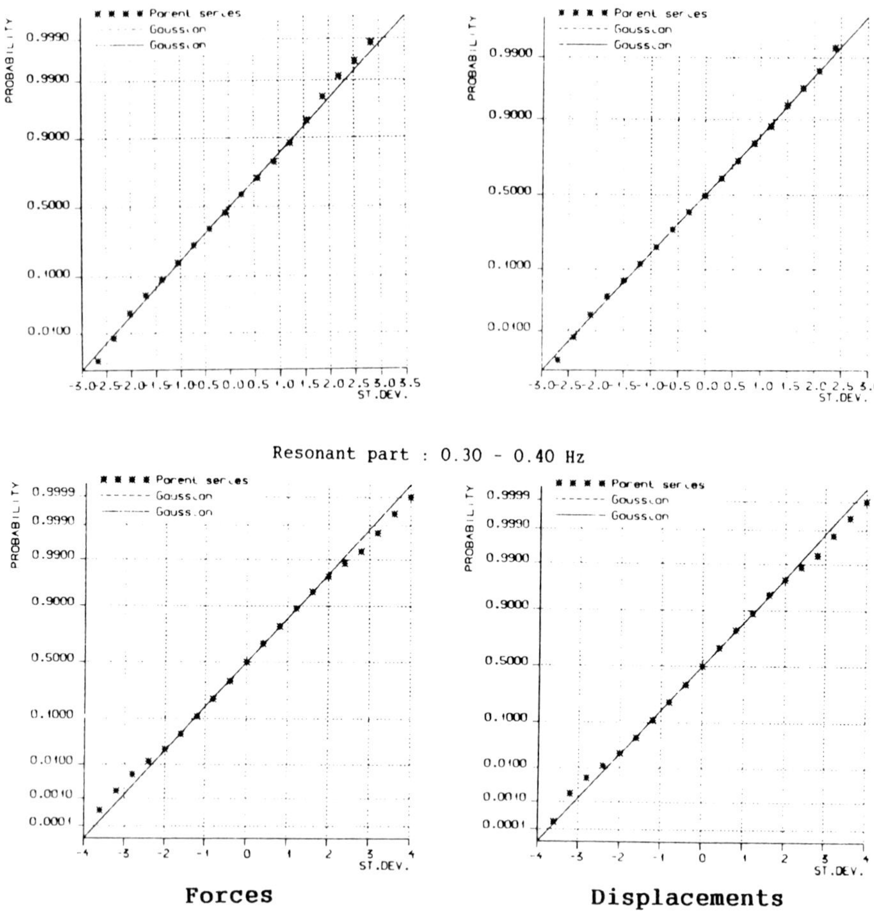

Fig. 4. Probability distributions: recording period 840122 08:04

forces and displacements by a probabilistic nature different from the Gaussian. Two resonant frequency regions have been investigated. The first region (0.3–0.4 Hz) corresponds to the first order sway modes. The second region (0.75–0.9 Hz) corresponds to the second order bending modes.

Representative probability distributions of resonant forces and corresponding displacements are shown in Fig. 4. The results do not indicate any clear probabilistic nature for the resonant response. In some cases the probability distributions deviate from the Gaussian. The reason for this is probably transient peaks in the time series caused by impact loading, and not

Table 3. Stiffness estimates

Recording period	Stiffness (MN/mm)		
	Z-direction	X-direction	Y-direction
840121 00:02	27	3.8	3.0
840122 00:02	27	3.5	2.8
840122 08:04	27	3.5	3.1
860110 11:22	20	2.8	2.8

a nonlinear damping mechanism. The probabilistic nature of resonant forces and displacements are identical, which indicates that there is a linear relation between them. The stiffness mechanism is thus linear at the resonant frequencies as well.

Stiffness properties

The stiffness properties are found by investigating the relation between base forces and displacements. The dynamic stiffness is defined as

$$K(\omega) = | S_{DF}(\omega)/S_D(\omega)|$$

where $S_{DF}(\omega)$ is the cross-spectral density between displacement and forces and $S_D(\omega)$ is the auto-spectral density of displacements. The coherence- and phase-spectra between forces and displacements are also estimated. The dynamic stiffness is only valid when a high coherence is present.

The vertical stiffness, in Fig. 5a, seems as expected to be constant for frequencies where the coherence is high, which are the dominating wave frequencies and natural frequencies.

The relations between forces and displacements *i* x-direction, in Fig. 5b, is characterized by a high coherence in the frequency region up to 0.4 Hz, which includes the quasistatic part and the sway mode frequencies. The stiffness function seems apparently to be frequency dependent in this region, such that the stiffness decreases as the frequencies decreases. The same trend is observed for the y-direction. However, theappliedhorizontal displacements have not been corrected for the influence of the acceleration of gravity. An attempt is made to correct for this, and new dynamic stiffnesses are estimated based on the corrected displacements. Fig. 5c shows the corrected stiffness function *i* x-direction, and indicates that the dynamic stiffnesses in the horizontal directions are constant for frequencies with high coherence, and that the apparent frequency dependent stiffness probably is due to the influence of the acceleration of gravity. Calculations have shown that the quasistatic horizontal base displacements will be reduced by a factor of 2 to 4 when they are corrected for the influence of the gravity. In the resonant region, the

Table 4. Mudmat forces

Recording period	Wave height H_s [m]	Mudmat force [MN]				
		Method	Maximum value	Minimum value	Average	Standard deviation
840121/00:02	3.0	1	1.294	-1.241	0.10	0.334
		2	1.279	-1.227	0.14	0.329
		3	1.267	-1.219	0.16	0.327
840122/00:02	7.6	1	3.979	-4.610	0.17	1.212
		2	3.921	-4.561	0.15	1.197
		3	3.872	-4.524	0.13	1.188
840122/08:04	10.7	1	6.194	-6.600	0.45	1.822
		2	6.079	-6.516	0.42	1.794
		3	6.030	-6.502	0.42	1.782
860110/11:22	12.6	1	9.225	-8.065	-0.03	2.401
		2	9.059	-7.932	-0.07	2.361
		3	8.949	-7.904	-0.04	2.342

displacements and thus the dynamic stiffnesses are only slightly influenced by this effect. Average stiffness values for frequencies with high coherence in the quasistatic and resonant region are listed in Table 3 for the four recording periods. As shown the vertical stiffness level is the same for the recording periods of January 1984, but apparently reduces for the recording period in January 1986. The same trend is observed for the horizontal stiffnesses. There is, however, no tendency of reduced stiffness as the sea state increases as in the January 1984 storm.

There are some matters to point out concerning the stiffness estimates from January 1986. Firstly, the stiffness estimate is somewhat uncertain as it is based on one single recording period. Secondly, the magnitude of the forces is not quite as expected for the high sea state with H_s=12.6 m. Fig. 6 shows the base forces and displacements in terms of standard deviation as a function of the significant wave height. The displacements have a trend as expected. However, for the base forces the trend is not as intuitively expected for highest sea state. Considering the corresponding displacements, one should expect these quanties to increase, which is not the case. This may indicate that the estimated base forces and consequently the estmated stiffnesses are unreliable for the actual recording period of January 1986. Further information or confirmation of findings regarding the stiffness mechanism can be obtained by establishing a time domain relation between base forces and displacements. The time series are again separated in a quasistatic part, a first resonance part and a second resonance part. Representative plots are shown for the horizontal x-direction in Fig. 7. The stiffnesses can be obtained as the main slope of the plotted relation between forces and displacements. The

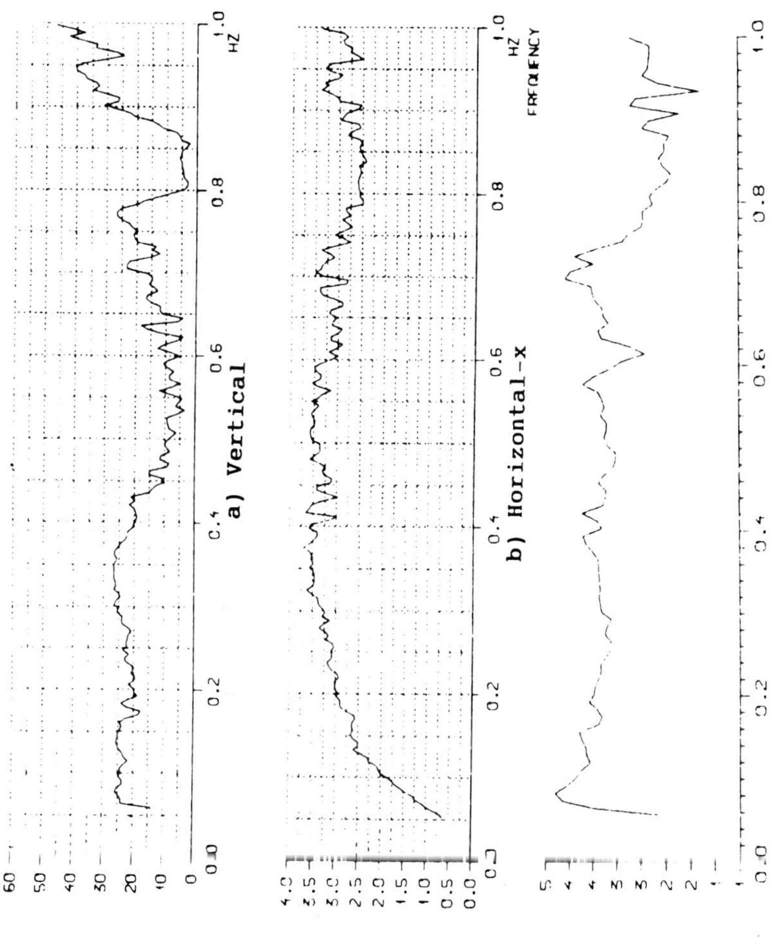

Fig. 5. Estimated stiffnesses: recording period 840122 08:04

stiffnesses estimated this way are in good agreement with the frequency domain estimates.

Damping properties

The statistical analysis discussed in the first section gave no clear indication of the nature of the damping mechanism. The hysteretic curves of the damping controlled resonant response have a typical elliptic shape as shown in Fig. 7. The wider ellipses for the second resonance part than for the first is caused by frequency increases. This is in good agreement with a damping mechanism described by a linear viscous damping model.

Load distribution in the foundation

Analyses have been carried out to study the vertical load distribution in the foundation. The analyses were based on data from pile No. 1, 2, 3, 4 and 6 and mudmat earth pressures recorded by the nine mudmat gauges. A sketch of the mudmat with piles is shown in Fig. 8 (taken from ref. 2).

Table 5. Action and reaction forces

Wave height H_s[m]	Pile force distribution	Pile force [MN]	Mudmat force [MN]	Mudmat + piles [MN]	Leg force [MN]
3.0	1	1.67	0.33	2.00	1.90
	2	1.56		1.89	
7.6	1	5.48	1.20	6.68	6.64
	2	4.91		6.11	
10.7	1	8.87	1.80	10.67	10.61
	2	7.76		9.56	
12.6	1	11.37	2.37	13.74	10.88
	2	7.89		10.26	

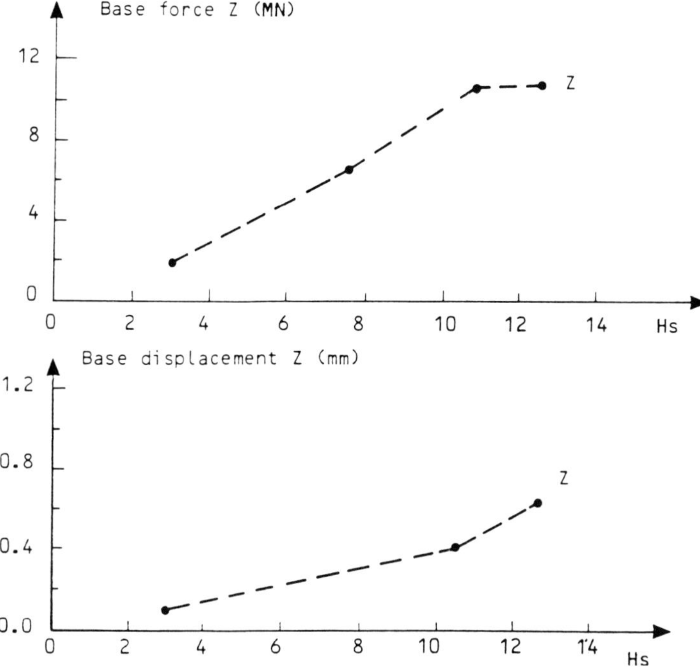

Fig. 6. Base force and diplacement

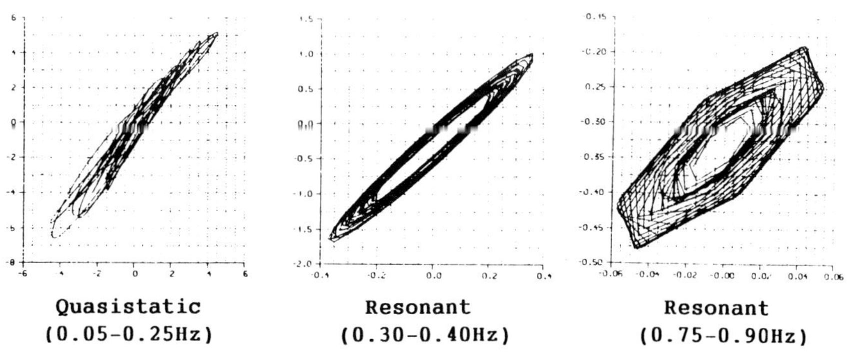

Fig. 7. Horizontal base force versus displacement: recording period 840122 8:04

Mudmat forces

The measured mudmat pressures have been distributed over the mudmat in three ways to check the influence of possible edge concentrations in the various sea states

(a) Average pressure from all cells multiplied by total area.
(b) Average pressure from Cells 1, 3, 7, 9 multiplied by area C, plus average of other cells multiplied with remaining area.
(c) average pressure from Cells 1, 3, 7, 9 multiplied by area C, plus average of Cells 2, 4, 6, 8 multiplied by area B plus Cell 5 multiplied by area A.

The methods for calculation of mudmat forces give very close results in all sea states, Table 4, indicating that any edge concentration of mudmat stresses have apparently no significant influence on total mudmat forces.

Representative mudmat pressure distribution in the main storm directions are shown in Fig. 9. The pressure increases approximately linearly over 60–70 percent of the mudmat and remains constant with maximum pressure over the remaining area. The reason for the constant pressure is probably that piles 4 and 5 (Fig. 8) carry relatively higher vertical loads than the mudmat in the region where pressure cell No. 3 is located.

Pile forces

Five of the piles have all four strain sensors around the 'circumference intact and are therefore considered. The axial pile forces are plotted in the cross section along the storm direction to provide a basis for force extrapolation to piles with no force estimates. Representative force distribution over the pile group is shown in Fig.10.

Two estimates of pile force distribution are presented. Distribution No. 1 is obtained with engineering judgement. Distribution No. 2 is obtained from a linear regression of the estimated pile forces. Pile force distribution No. 1 may cause too large forces in piles 8 and 9, whereas pile force distribution No. 2, may cause too low forces in piles 8 and 9. It is anticipated that the true pile force distribution is somewhere between the two estimated, with somewhat smaller forces in piles 8 and 9 than the average of piles 1, 2, 3 and 6, and with significantly larger forces in piles 4 and 5 than the average.

To have a better indication of the pile force distribution, an attempt is made to estimate the pile forces in piles 5, 8 and 9. Only two or three strain gauges are intact for these piles. These estimates are marked in Fig. 10 and indicate that pile distribution No. 1 slightly overestimates the total pile forces.

The load distribution over the mudmat and pile group have been indicated, and seem to be similar along the main storm directions. This further indicates that the pile group and mudmat behave as an integrated system. Coherence analysis between mudmat pressure and pile forces confirms this by a high coherence at the dominant load frequencies.

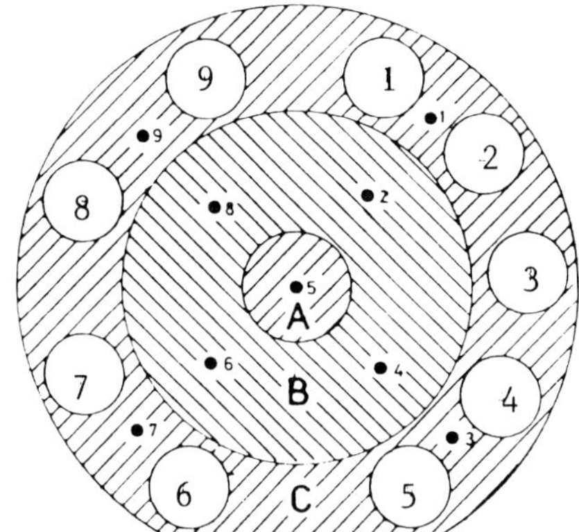

Figure 8 Piles and mudmat sensors.

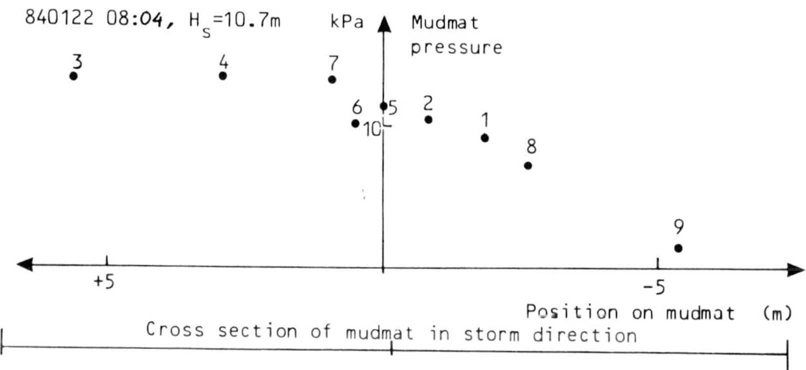

Fig. 9. Estimated mudmat pressure

Load transfer from structure to foundation

The vertical loads from the platform leg to the foundation are carried by the pile group and mudmat, and these forces (action and reaction) should balance each other. A comparison of these forces has been made in terms of standard deviations, and since the response processes are shown to be close

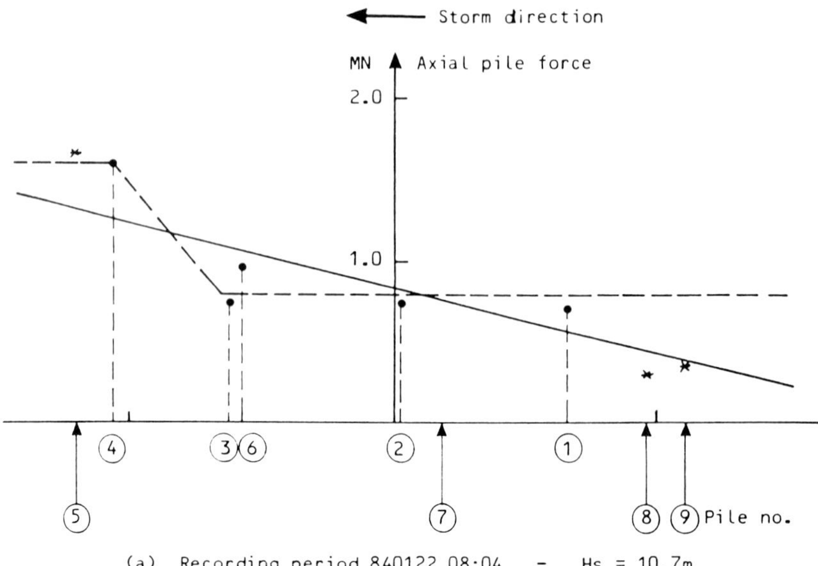

Fig. 10. Pile force distribution

to Gaussian, conclusions drawn on this basis will also be valid for the extreme response.

Load transfer to piles

The vertical forces of the leg and the foundation are shown in Table 5. Pile force distribution No. 1 results in good agreement between the action force (in the leg) and the reaction force (mudmat + piles) for the three recording periods in the January-84 storm. For the recording period from the January-86 storm, the estimated leg force is significantly lower than the estimated total reaction force. The trend of the mudmat force, the pile force as well as the total reaction force seems reasonable with respect to significant wave height. This is not the case for the leg force which seems to be too low for the January-86 storm with H_s=12.6 m, refer to discussion in the section Stiffness Properties. The total reaction forces estimated with pile force distribution No. 2, is lower than the estimated action force for all recording periods.This indicates that linear regression underestimates the forces in piles 8 and 9.

Pile force distribution No.l, which seems to be most accurate, indicates that about 82–84% of the total reaction load is carried by the piles, and 16-18% by the mudmat. The recording periods include different sea state levels, and the results show no tendency of a percentage increase of mudmat forces.

The above discussion of the load transfer from the structure to the foundation has dealt with the magnitude of the vertical forces.More information about the load transfer can be obtained by investigating the coherence between the loads in the platform leg and loads in the foundation, for vertical as well as horizontal forces. Representative coherence between the vertical leg force and the pile forces, are shown in Fig. 11a. High coherence is observed for the quasistatic part and the first order resonant part at 0.34 Hz. This indicates that the platform and piles are acting like an integrated system at these frequencies.

Representative coherence between the horizontal leg forces and the axial pile forces in Fig. 11b shows high coherence both in the quasistatic region, the sway mode region (0.43 Hz) and the torsional mode region (0.52 Hz). This means that some of the horizontal forces of the leg is transferred to axial forces in the piles, and is probably due to inclined piles and bending of the piles.

Load transfer to mudmat

The transfer of vertical leg forces to the mudmat is demonstrated in Fig. 11c which shows the coherence between vertical leg forces and mudmat pressure. The coherence is high in the quasistatic and the first resonant region which confirms that the mudmat carries vertical loads from the structure.

Conclusions

The stiffness mechanism of the soil–structure interaction is linear, both in the horizontal and vertical directions. The damping mechanism of the soil–structure interaction has not been fully understood.The results indicate, however, that a linear viscous damping model can be applied.

The mudmat pressure distribution over the mudmat area is described by a variation along the main storm direction such that the mudmat carries lowest load closest to the centre of the platform. The load distribution over the pile group seems to be similar to that for the mudmat indicating that the pile group and the mudmat behave as an integrated system. Coherence analysis between mudmat pressure and pile forces confirms this.

The pile group carries about 80% and the mudmat about 20% of the vertical loads from the platform. No indication of redistribution is observed as the significant wave height varies.

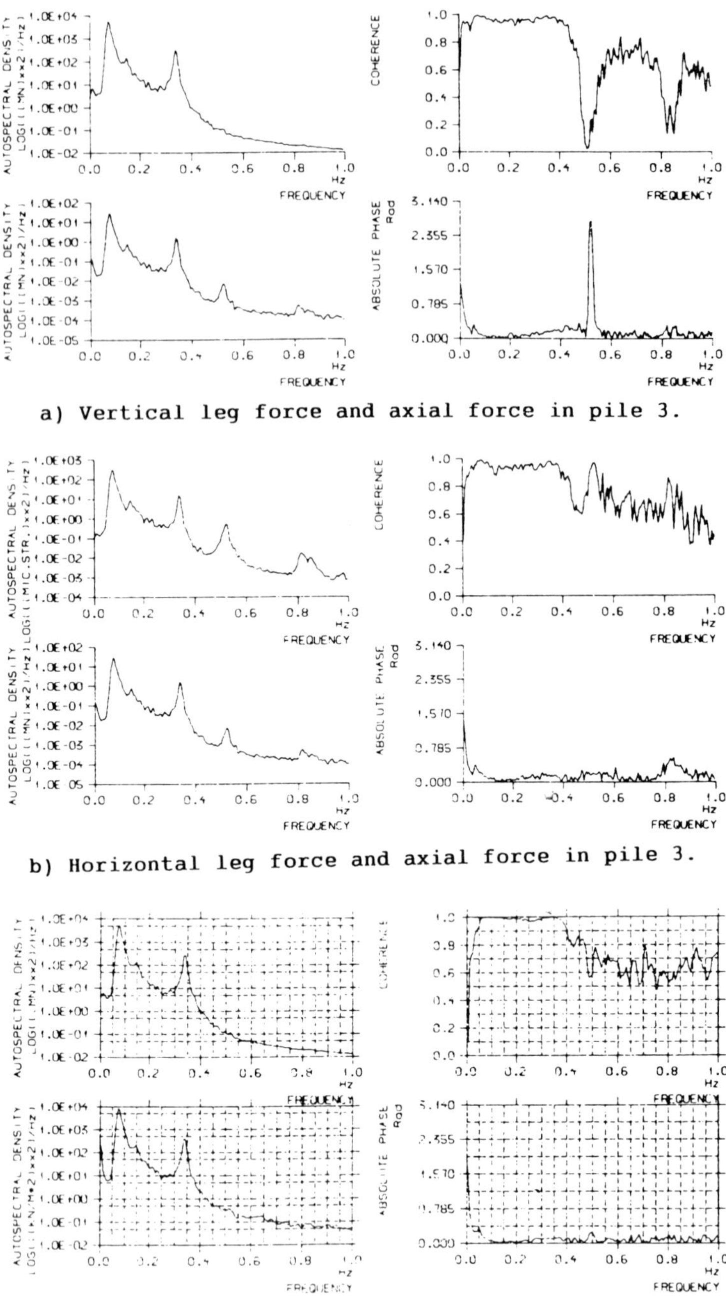

a) Vertical leg force and axial force in pile 3.

b) Horizontal leg force and axial force in pile 3.

c) Vertical leg force and mudmat pressure cell 4.

Fig. 11. Auto- coherence- and phase spectra: recording period: 840122 08:04

Acknowledgements

This paper is based on results from a research project initiated and founded by Statoil, Skjåstad *et al.* (1988). The authors appreciate the permission to publish these results.

References

1. W. A. FAIRHURST & PARTNERS. *Magnus Foundation Monitoring Project Data Aquisition Unit. Analysis Procedure.* Report No. 383, July 1984.
2. STRUCTURAL MONITORING LIMITED. *Magnus Foundation Monitoring, Temporary Data Logger, Analyses procedure.* Report No. 306, Rev. 1, March 1983.
3. HOEN, C. *MARCO – A Program System for Estimation of Structural System Modal Parameters from Vibration Measurements.* Theory Manual, SINTEF Report STF71 F87057, Trondheim 1987.
4. SKJÅSTAD, O., BRENNODDEN, H., LIENG, J. and SPIDSOE, N. *Soil–structure interaction of a piled offshore platform.* SINTEF Report STF71 F88015, Trondheim 1988.

Discussion

Part 1

Dr R. L. WOOTON, Atkins Oil & Gas Engineering

This session constitutes a description of an experiment, a measurement programme that demonstrates meticulous preparation and management and execution over a very long period and I congratulate all those who have been involved in it in their determination to see the work through to fruition and to present the results. Mr Kenley mentioned to me that he was involved in the first meeting to discuss this experiment in 1979, and that is a long time to follow something through, and I congratulate him and others on their determination over that period. We have heard a description of a joint industry-funded project that went on a long time and, clearly, has been very successful in terms of producing results and in verifying the design methodology used in the design of Magnus, which was, I suppose, based on the technology of more than a decade ago. How does BP view this as a project? I would also be interested to hear from other sponsors about how they viewed the experiment as it went along and how they also viewed the results since seeing them when they were originally presented.

D. E. SHARP, BP Exploration

I have not been involved in the geotechnics field for a number of years now, but I understand from my colleagues in BP Engineering that the results from the Magnus programme have been used to support recertification and additional loading on the platform. Additional wells have been drilled when, on the basis of the original design loads, it was not thought possible. The accumulative benefits of this and other actions has been estimated in-house at something like $50 million, so in that sense I guess we are quite pleased with the results. I think it is fair to say that those benefits arose not just from monitoring foundations, but from the environmental, structural and foundation monitoring projects combined as a whole instrumentation system. An understanding of the behaviour of these platforms is clearly a benefit in future recertification and in upgrading the capacity of the platform production capacity in future years.

[In response to a question from the floor.]

The stiffnesses presented in Mr Kenley's paper actually relate to the stiffnesses predicted using the original design methods at the loads that were actually seen on the platform, so they are comparable.

R. M. KENLEY, Structural Monitoring, Fugro McClelland Ltd

I can only present the results and present an opinion and leave it for others to interpret them, but the results were very, very clear. There was a very substantial decrease in the load recorded on the mudmat when two piles were driven. Those were the first two piles that were driven on the platform so there seems to be no explanation other than that, during driving, load had to be transferred from the mudmat into those two piles. Perhaps the pile was partly jammed in the guides.

D. E. SHARP, BP Exploration

The pile sleeves on Magnus are about 18–20 m long, and on the inside of those sleeves there are a number of weld beads. On the top of the piles there are also a number of weld beads to improve the strength of the grouted connection. A hypothesis may be, that as the pile leans against the inside of the sleeve, which in the case of Magnus is battered, there must be substantial transfer of friction through these weld beads. The beads are about 12 mm high, I think, both on the outside surface of the pile and inside surface of the sleeve.

Professor G. T. HOULSBY, Department of Engineering Science, Oxford University

Could I offer a different explanation for the reduction of load on the mudmat? We have evidence from research sponsored by BP at Oxford that when a pile is driven through a mudmat, that triggers downward movement of the mudmat (S. S. Gue, 1984 *Ground Heave Around Driven Piles in Clay*, DPhil thesis, Oxford University). Since the foundations of the Magnus platform have four legs, it is therefore a statically indeterminate structure, and if there was a tendency for one mudmat to move downwards, that would simply transfer the load to the adjacent legs. I think it is a much more likely mechanism, that the load was being transferred to the other legs, not that the load was being transferred to the piles.
[D. E. Sharp accepted that]

Dr J. D. MURFF, Exxon Production Research Co. USA

Please explain the basis for determining the pile stiffness. In our experience, the inference of pile stiffness based on platform dynamic behaviour has not always been reliable. The stiffness of the structure itself is also highly uncertain and simply using the foundation to tune the response may be overly simplistic.

R. M. KENLEY, Structural Monitoring, Fugro McClelland Limited

The whole of the stiffness predictions came from the dynamic behaviour due to wave loading. There was a set of accelerometers at the very bottom of the leg which were taken to be the movements of the pile head. By double integrating those accelerometer signals we could obtain displacements. We had load signals which were obtained by combining the various loads and so we got displacement signals and load signals, as signals which you use and can more or less see how strongly correlated they are, and from these individual cycle ranges, for those signals, we can work out the stiffness by just taking the load over deflection.

A delegate from the floor gave his interpretation of how load is transferred on to the pile. Unlike modern day structures, which all have vertical sleeves and piles, the Magnus structure had raked piles which makes it easier for jamming to occur.

Dr R. J. JARDINE, Imperial College of Science, Technology & Medicine

Dr Murff commented that dynamic foundation stiffness measurements might be misleading. An attempt was made to monitor static displacements (settlement) on the Magnus jacket using Imperial College settlement gauges. This did not work: they were either insufficiently sensitive or were damaged during installation. However, gauges with far finer resolution (0.03 mm) were developed by Imperial College for the Hutton TLP and used with much greater success. The TLP static measurements confirm the trends shown by the dynamic measurements at Magnus (i.e. stiffness under-predicted by 400%). See Jardine & Potts 1988, *Géotechnique* **38**, No. 2, 231-252, "Hutton TLP Foundations: An approach to predicting driven pile behaviour". The reasons for this under-prediction are addressed in paper 4.

Dr R. L. WOOTON, Atkins Oil & Gas Engineering

Regarding the fact that the mudmat is taking a significant part of the load, presumably for very low costs compared with piles: if you are optimising the design of this foundation system based on these results, would you in fact go for a larger mudmat?

D. E. SHARP, BP Exploration

I think the size of the mudmat was partly determined by the depth available in the actual fabrication yard, so I am not sure if you could have significantly increased mudmat size. Clearly, the point is well made in the sense that the mudmat is taking the equivalent loading of one or two piles. That, in eight piles per platform and at 300 tonnes of steel per pile, one is talking about a lot of money.

M. RANDOLPH, University of Western Australia

Would either of the authors like to comment on the deduced relationship between environmental loading ratios and wave heights and whether the magnitude of the environmental loading was in line with predictions?

D. E. SHARP, BP Exploration

I believe one of the main reasons why the loads on the platform were considerably less than that we had been expecting, was because there appears to be substantial conservatism in the wave loading and environmental loading predictions. I think there are several reasons for that, which I have noted in the paper, but predominantly, we are looking at the drag factor, which we believe is significantly less than was used in design. I think they used a C_d of 2 in design and I think the measurements suggest that a factor nearer 1.6 is closer to the mark. Another factor is that even up to wave heights very close to design conditions, we observed virtually no non-linear behaviour, either in the structure or the foundations. Design predictions had been for substantial non-linear behaviour, so there appears substantial conservatism here. I hope that a little later we can summarise the overall conclusions and that is clearly a significant one.

Professor G. T. HOULSBY, Department of Engineering Science, Oxford University

The evidence of linearity which was presented involved the plotting of one load against another, but according to my understanding of linearity it usually involves plotting loads against deflections. Is there any equivalent evidence which the load-deflection behaviour revealed?

R. M. KENLEY, Structural Monitoring, Fugro McClelland Ltd

Yes. All the various plots were done, the examples shown there were just a few of them. There was no evidence of non-linearity at all in axial load behaviour, even plotting load against displacements, and the evidence in terms of the bending moment showed that in the maximum wave conditions, stiffness was perhaps 10% less than it was in the low wave height conditions. So even in bending, where there was some evidence, it was really very small. It certainly was not of the order of perhaps a factor 2, 3 or 4 that you might well expect if you were taking a pile fairly close to its ultimate capacity. So we can conclude that we are a long way from the ultimate capacity of the piles.

D. E. SHARP, BP Exploration

The one point that has not come out is that during one set of analyses, there was indication that in high energy environments the response of the platform laterally was slightly softer than the response of the platform in lesser energy environments.

During an individual event, that is a single storm, the response was always linear no matter whether the particular wave height was 2 m, 4 m or 12 m.

T. ALDRIDGE, Fugro McClelland Limited

On the structural response, you showed a range of wave heights against the load in the structure. I am interested to know whether the frequency of the waves as measured was close to the calculated natural frequency of the structure, and whether there was the expected dynamic amplification in the structural response.

R. M. KENLEY, Structural Monitoring, Fugro McClelland Ltd

The amount of response at the platform natural frequencies was reasonably small. Perhaps it accounted for something like 10% of the total response. It did vary with wave heights: it was a slightly bigger percentage at low wave heights, but at the high wave heights it was perhaps only as much as 10%. So really from a design point of view, one could almost ignore it. Perhaps one cannot ignore it from a fatigue point of view, but from a basic maximum wave design the behaviour at the natural frequencies can be ignored.

A. R. BIDDLE, Independent Consultant

I have a query on why different gauges are used at level 3 (25 m), on figures 3 and 8 (3M and 3V) and why the distribution of actual load with depth differs between static and dynamic cases.

R. M. KENLEY, Structural Monitoring, Fugro McClelland Ltd

The question was relating to data presented for pile 3 where there are two results presented, 3V and 3M. The question was also why were they different and why was there different behaviour dynamically and statically. There are two gauges on pile 3. As mentioned earlier on, there was to have been a fully instrumented pile with gauges at several levels. This was one of the piles that was lost, but on the replacement piles an attempt was made to reproduce this fully instrumented pile by installing gauges at three separate levels. One was fairly close to the top, the other one was 25 m down and the final one was at the pile tip. Unfortunately the one at the pile tip did not survive pile driving, so we were left with two gauges on pile 3. There was different behaviour statically and dynamically. Dynamically it is more or less what you would expect, with a predicted fall off with depth, the lower level one being significantly less than the load level and upper level ones. Statically there was relatively little difference in load at the two levels.

Part 2

M. ENGLAND, Cementation Piling & Foundations

We have been doing considerable research on the behaviour of footings and piles under load and in relation to time, and I would wish to challenge some of the conclusions that have been presented, simply because they have not taken time to account when assessing how the mudmat is actually operating.

If we look at an illustration of the behaviour of a standard 0.75 m diameter pile in time under constant load, in which the pile head settlement is recorded. One may notice the scale of the time effects, and these increase with the size of the foundation.

The mudmats are of the order of 17 m diameter and this compares with the back analyses we have performed on water tanks of 15 m diameter, which have been settling in time on soil materials with a modulus of elasticity of approximately 10 000 kN/m^2. The results indicate that many days are needed in order to get half way to the final displacement behaviour.

In the light of this, it is clear that the inflated capacity of the mudmat in the short term needs to be correctly addressed. Otherwise engineers may make serious misjudgements regarding the final characteristics of the piles and mudmats in computing their designs.

Another interesting thing to note is that the shaft component of pile behaviour will dominate the initial reaction of the foundation system to the applied load. In this case, an example is shown in which the shaft behaviour and the elastic shortening are the first components to be mobilised, as illustrated in Fig. 1. The final base characteristic takes a significant time to develop.

It should be noted that the duration of the force induced by a wave is approximately seven seconds. This is well into the region where very little of the total consolidation/creep effect is taking place, and hence it needs to be looked at very carefully to ensure that the inflated resistance does not lead to incorrect conclusions.

J. PELLETIER, Shell Development, USA

I was interested in seeing the work of Imperial College on the effect of soil non-linearity on decreasing the group interaction effects, and the resultant increases in predicted foundation stiffnesses. We have been interested in this in the past when doing feasibility studies for deepwater compliant towers, particularly the effect of foundation stiffness on natural period. In some of our work in the late 1980s, we came across a reference in the *Proceedings of the 3rd International Conference on Numerical Methods of Offshore Piling* (Chow, 1986). Chow published interaction factors of Poulos type, but took into account soil non-linearity, which showed this same rapid decrease in inter-

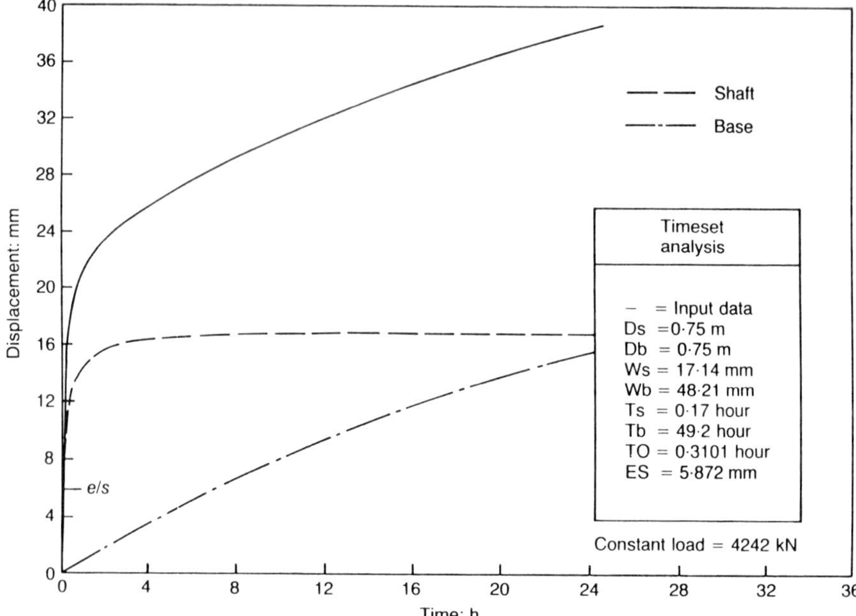

Fig. 1.

action effects due to soil non-linearities. My question to BP and Fugro is: has any of this type of analysis or re-analysis been done, now that you have measured the increased stiffnesses, to see if any of these analytical methods would have been a good thing to use had they been available at the start of the project?

M. HORSNELL, Fugro McClelland Ltd

Certainly for the case of lateral loading a lot of work has been done in looking at non-linearity. Certainly more work needs to be done, I would agree, on the axial case to take that into account. And it was certainly of interest to me to look at the load distribution that the bottom 30 m of those piles in fact know nothing about this mighty structure that is sitting there in the environmental loading. I had to conclude that this probably has to affect the interaction - that in fact the load distribution between the piles is probably affecting more the distribution of load down the piles without necessarily giving rise to any significant additional displacement, over and above that resulting from compression of the steel itself.

Professor M. RANDOLPH, University of Western Australia

The apparent high axial stiffness of the Magnus foundations may be explained in terms of relatively simple non-linear models for the soil, and an

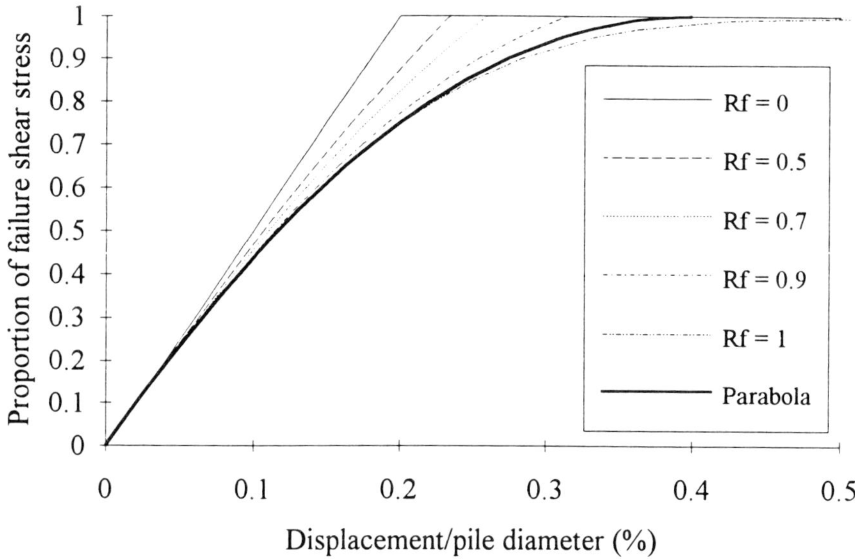

Fig. 2. Load transfer curves derived from hyperbolic stress–strain model for soil

appropriate load transfer approach. If the stress-strain response of the soil is modelled as hyperbolic, then, for values of R_f in the region of 0.9, the secant modulus near failure will be a factor of 10 lower than the initial (tangent) modulus at small strain. This is consistent with the data from the paper by Jardine and Potts. Load transfer curves for a single pile may then be developed by integrating radially the stress-strain response of the soil, which reduces significantly the degree of non-linearity. Fig. 2 shows resulting load transfer curves for an initial tangent shear modulus of 1000 times the failure shaft friction. At failure, the secant modulus is now a factor of only 2–3 lower than the initial value, and the non-linear response may be modelled reasonably by a parabolic load transfer curve.

In order to apply such a load transfer curve to the analysis of a pile in a group, it is important not to abuse the use of interaction factors. Interaction effects apply only to the elastic part of the curve, not the non-linear component. The approach is illustrated in Fig. 3, for an overall group settlement ration, R_s. For the Magnus case, the 'elastic' group settlement ratio would be about 4. However, making simple allowance for non-linearity of the soil response using this modified load transfer approach, the net interaction effect turns out to be only 30–40%, which ties in very well with the non-linear finite element (FE) analyses that Dr Jardine described. Thus, it is possible to use interaction factors to estimate the pile group stiffness at Magnus with reasonable accuracy, but they must be applied in an appropriate way.

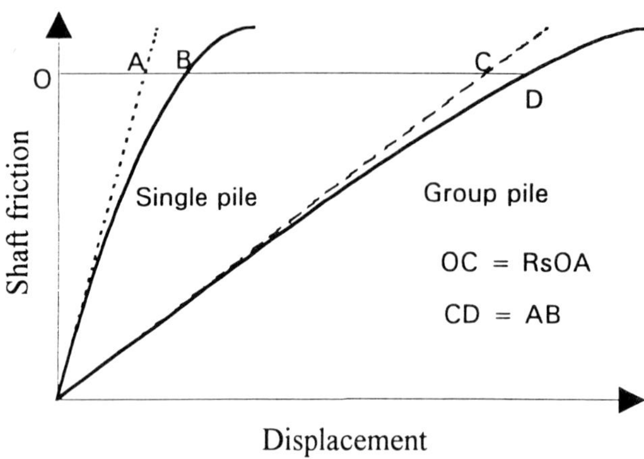

Fig. 3. Modelling interaction with non-linear load transfer curves

Dr R. JARDINE, Imperial College of Science, Technology & Medicine

When David Potts and I performed non-linear finite analyses of the Magnus and Hutton TLP pile groups back in the 1980s, we were surprised by the results concerning the distributions of shear stresses with depth and the radial distribution of ground surface movements.

We soon realised that both features were critically dependent on the combined action of local shaft failure and non-linearity in the small-strain stiffness characteristics of the foundation soils. I was, therefore, very interested in the field measurements made at Magnus of the distribution of total pile load, *Q*, with depth. The data reported by Sharp and Kenley were interpreted as giving a straight-line distribution of *Q* with depth (see Fig. 13, Paper 1). Whilst all of us engineers love linearity, I wonder if this is the most reasonable interpretation to make.

As shown on Fig. 4, a linear load profile implies a rather unlikely block diagram of local shear stress τ with depth: why should τ be constant and then suddenly stop dead? Standard elastic (constant stiffness) pile group analyses give a quite different shape, as indicated on the figure. But such a rapid decay of τ with depth does not agree at all with the field trends for *Q*.

However, predictions for the distribution of τ with depth can be modified by allowing for local shaft failure, variations of stiffness with strain and changes in stiffness with depth. Fig. 13 of Paper 4 shows the trends expected from 'Class A' predictive analysis which incorporated all three of these additional factors. The profiles of τ have markedly different shapes to those predicted by elastic analyses: the trend applying to the typical Magnus

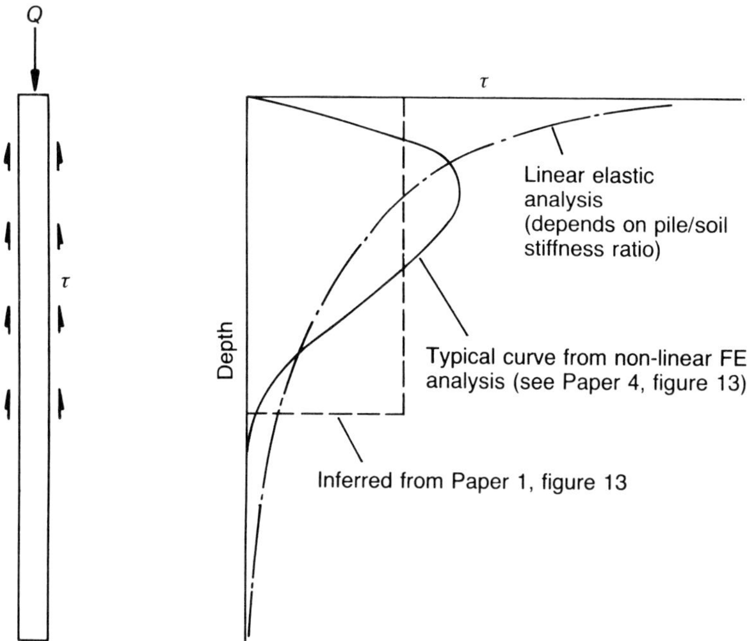

Fig. 4. Alternative predictions for the distribution of shaft friction with depth at Magnus

loading case is shown on Fig. 4 of this discussion. It would be interesting to see if the integrals obtained from the non-linear FE predictions for τ gave a worse fit to the field measurements than the 'engineering' straight line drawn through the data spots.

The shape of the τ profile has an important effect on pile group interaction. Paper 4 shows that local failure combined with stress-strain non-linearity is the main reason for the divergence between the load-displacement response expected from conventional calculations and that measured on site. It is now completely feasible to perform fully non-linear analyses using realistic soil models. Such analyses do not cost very much money and give much more realistic predictions of field behaviour.

Dr R. M. SEMPLE, Dames & Moore, USA

I should like to ask Dr Jardine how do tests on residual fabrics relate to conditions at the wall of the driven pile? Secondly, in selecting interface slip element characteristics for making analytical predictions what criteria does he use?

Dr R. JARDINE, Imperial College of Science, Technology and Medicine

The experiments used to model the pile–soil interface failure were ring-shear interface tests. The soil was remoulded into the apparatus, consolidated, and then subjected to a series of ten fast shearing 'pulses'. Each pulse involved a 1000 mm shear displacement applied over 2 minutes, followed by a 10 minute pause period. The aim was to 'model' pile driving. The samples were then reconsolidated to an appropriate effective stress, before being resheared to failure at about 0.002 mm/min. The slow shearing stage, which was completely drained, was performed to obtain the effective stress parameter that would apply during slow pile loading to failure. The interfaces were made of steel that had been air abraded to give a CLA roughness of 8 microns. Measurements had been made in the field against real offshore piles, using plastic moulds and later laboratory analysis with a Tallysurf device. These indicated that the 8 microns roughness would be the most appropriate.

The results were programmed into the FE analysis using a special interface criterion that related the maximum possible ratio of vertical shear stress to radial effective stress via a directionally specified Coulomb law. One additional feature that was modelled was the mild post-peak reduction in δ with relative displacement seen in the laboratory. Relative displacement was seen as the key parameter (rather than shear strain) because the process is controlled by the behaviour of the shear surface discontinuity.

Dr R. M. SEMPLE, Dames & Moore, USA

Do you have any evidence that the types of fabrics that you have used in the laboratory are comparable with actual?

Dr R. JARDINE, Imperial College of Science, Technology & Medicine

As part of Andrew Bond's PhD research he performed pile tests in London Clay at the Canons Park test site. There he used highly reliable instrumentation to measure the precise δ angles developed when the piles were loaded to failure. We also excavated around the piles and used X-rays, thin sections and photographs to analyse the effects on soil fabric of pile installation and load testing (further details are given by Bond and Jardine (1991) in *Géotechnique* **43**, No. 1).

Both the effective stress measurements made in the field and the laboratory fabric studies showed that displacement pile installation sets up polished shear surfaces with a reduced strength residual soil fabric close to the pile. The degree of particle re-orientation and shear surface development depends on the rate of installation, with driven and fast jacked piles showing higher peak δ than slow-jacked piles. All pile types tested at Canons Park tended towards a low fully-residual angle after a few mm of post-peak displacement. PhD work by Lemos (1986) and others with the ring-shear apparatus at Imperial College has shown that pile-soil friction can also depend on interface hardness and roughness, especially with low plasticity clays. In general there has been very good agreement between the laboratory data and field (on-pile) measurements. The PhD study by Martins (1982) at Imperial College predicted these features through a series of model pile tests in kaolin. But we have since seen direct proof, including visual observation of shiny shear surfaces, of residual fabric controlling pile capacity through field tests at Cowden and Canons Park.

D. E. SHARP, BP Exploration

I would just like to try and sum up some of the overall conclusions from the foundation monitoring project. First of all, regarding the system performance, I think overall we can say that the objectives were fully achieved, we certainly measured the variation of static load distribution with time, we measured the environmental load acting on the group, the proportion of this load taken on the mudmat and the pile group, its distribution within the group and the effective stiffness of the pile group. Unfortunately, because of the loss of the originally fully instrumented pile, we did not measure the variation of pore and total pressure, and because of the difficulties we had with settlement gauges we were unable to measure the absolute vertical movement of the leg. I think, although Mr Kenley was only able to show a few of the actual results, the overall conclusion is that the data quality from the programme was exceptionally high. Looking at the static loading, the contribution of the mudmat was relatively small — 5–10% of the long-term load — and generally changes measured in the static loads compare well with the design predictions. There was, however, a lack of evidence of a reduction in the static axial load over the upper 25 m of the pile.

Dynamically, certainly the environmental loading is less than predicted and this is mainly due to environmental and structural issues, although the higher foundation stiffnesses will contribute to an overprediction of the dynamic amplification factors. The platform and the foundation response were seen to be quasi-static. There was a significant contribution from the mudmat dynamically 15–20%. In general, while the results compare well with predictions, there are some significant differences in numerical values because of reasons that were presented earlier, particularly the high foundation stiffnesses both vertical and laterally. We believe this is due both to soil modulus and particularly to interaction effects. I think, in terms of the difference in numerical value, certainly the different pile head conditions encountered in situ, compared to those assumed in design, contributed significantly to the deviations between predicted and design values. The foundation stiffness response is linear during any individual storm, although during high energy environments it was slightly softer. I think the reason for this may be due potentially to pore pressure accumulation or a small loading rate effect, although, as shown in the papers, the differences were particularly small.

7. Large diameter pile test project — overview

C. R. MULLIS, BP Engineering

Project description

The project consisted of two large-scale pile load tests performed at onshore locations in clays similar to those encountered by hydrocarbon installations in the North Sea. One test was carried out in very strong, heavily overconsolidated clay (OC) at Tilbrook in Cambridgeshire, and the other in essentially normally consolidated clay (NC) at Pentre in Shropshire.

The test piles were 762 mm in diameter, open-ended pipe piles 33.5 m and 58.5 m long, fabricated in RQT 701 and RQT 501 steels. The piles and surrounding soils were instrumented to measure response to pile installation and subsequent static load testing. The piles were driven to pre-determined depths 30 m and 55 m. Following a suitable period to allow for soil reconsolidation (set-up), the piles were loaded to failure in compression. The load was applied via hydraulic jacks reacting against an anchored frame.

The load tests and associated soil investigations were intended to provide information on:

(*a*) The validity of existing design criteria for predicting skin friction in clays similar to those found in the North Sea, and
(*b*) Effective stress changes in soil around a pile during pile installation, clay reconsolidation and loading of the pile to failure.

In particular, it was intended to compare the actual failure load of the piles to that calculated by conventional design methods. Data from the tests were to be used to review the interpretation of the existing database with the objective of recommending a more fundamental and rigorous approach to the computation of the ultimate axial capacity of piled foundations in clay.

The scope of work was extended during the project and included driveability tests at Tilbrook and liaising with Norwegian Geological Institute to allow separate tests to be carried out by NGI at both Pentre and Tilbrook sites. On completion of the main project, additional tests were carried out at Tilbrook to test a pile in uplift and in a lateral direction utilising existing instrumental piles and test rig.

Table 1. Summary of soil investigations and pile instrument sensors at the two sites

Pentre Site (NC)	Tilbrook Grange (OC)
Soil Information 5 borehole location 2 cone test location Pneumatic, Electrical & Standpipe piezometers	**Soil Information** 5 borehole location Pneumatic, Electrical, Standpipe piezometers
Pile Instrument Sensors (55m length) 10 accelerometers 68 strain measures 16 soil pressure cells (total and pore) 2 temperature	**Pile Instrument Sensors (30m length)** 10 accelerometers 40 strain measuring 16 soil pressure cells (total and pore) 2 temperature

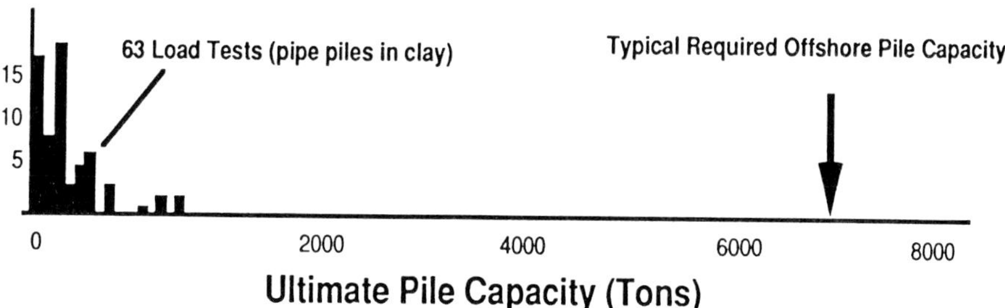

Fig. 1. Summary of database of load tests for pipe piles in clay up to 1984

Participating organisations

The main project was funded by an EEC loan, the Department of Energy, seven of the Oil Company Majors, Lloyd's and a part share by the Managing Consultant namely:

- Commission of European Communities
- Department of Energy (Petroleum Engineering Division)
- Amoco (UK) Exploration Company
- BP Petroleum Development Limited
- Britoil Plc
- Esso Exploration and Production UK Limited
- Marathon Oil (UK) Limited
- Shell UK

Table 2. Principal observations

- ❑ Pile deflections during driving
- ❑ Static pile head load - displacement
- ❑ Static load transferred (T-Z)
- ❑ Stress changes at end and near pile wall

- Statoil Netherlands BV
- Lloyd's Register of Shipping
- McClelland Limited

The support of the participating companies is acknowledged, not only for their financial backing, but also for their technical contributions and permissions to publish the papers at this conference.

Contracts

The project was managed on behalf of the participants by BP International. The managing consultant was McClelland who also awarded the majority of the sub-contracts for the work.

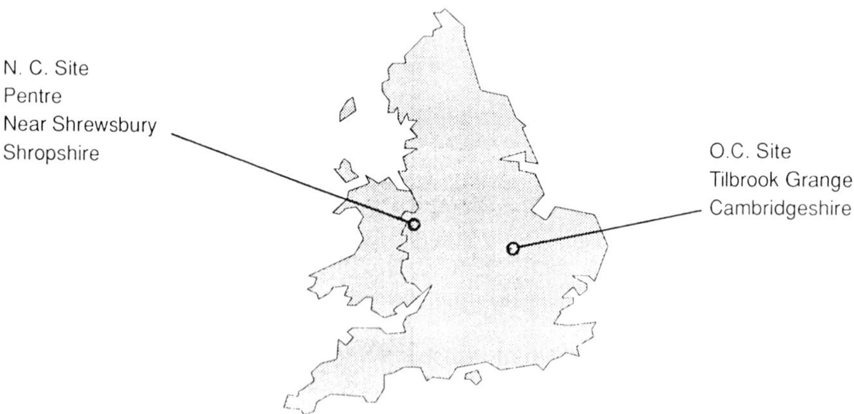

Fig. 2. Location of test piles

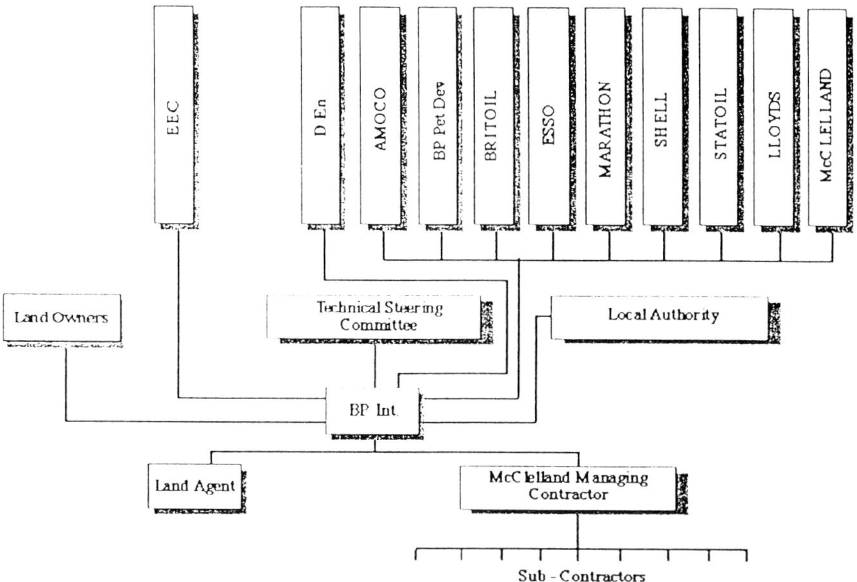

Fig. 3. Contractual relations

Costs

The total costs of the project, including management, engineering, sub-contracts construction, testing and analysis was £3 988 000 plus £337 000 for additional tension and lateral tests. Significant sub-contract elements were:

- Test Piles – £635 000
- Instrumentation – £294 000
- Load Frame – £216 000
- Site Preparation – £201 000
- Pile Driving – £161 000
- Site Acquisition – £138 000

Programme

The key dates for the project were:

- December 1985 – Contract starts
- February 1986 – Start testing instruments
- March 1986 – Pre-order materials
- (Feb) June 1986 – Site investigation start

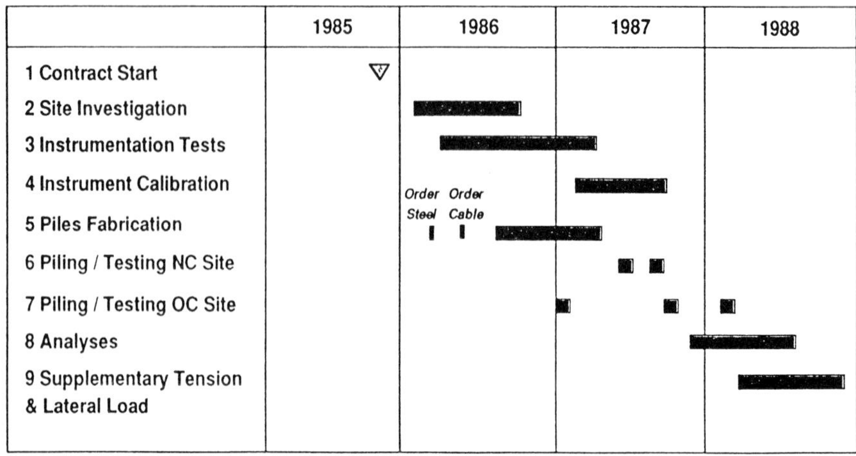

Fig. 4. Large Diameter Pile Test Programme with supplementary tests

- August 1986 – Site preparation
- January 1987 – Commence installation of instruments and calibration
- February 1987 – Driveability tests at Tilbrook
- June 1987 – Drive test pile Pentre
- September 1987 – Test pile Pentre
- October 1987 – Drive test pile Tilbrook
- February 1988 – Test pile Tilbrook
- June 1988 – Complete analysis
- October 1988 – Complete supplementary tests Tilbrook

Lessons learned: a personal view

This high-cost joint industry programme took over a year for all the participants to sign and commit to a contract. This long and costly lead time was risk money for BP International Limited.

It was difficult for the public and local authorities to accept that this oil industry-related test programme was purely a research project. Our ability to overcome this suspicion was the biggest single threat to the test programme continuing.

The participants agreed to a schedule of payments and the money held in an interest-bearing account. The delays at the beginning of the project, delayed expenditure and the interest assisted greatly in funding additional costs associated with the delay.

BP Shipping were able to secure insurance cover against total loss of results. Up to that time we had not anticipated the test results to be insurable.

We were confident in predicting static load capacity of test piles. However, there was far more uncertainty in predicting drivability and the rate of dissipation of pore pressures after driving (pile set-up).

Consideration should have been given to a separate instrument-developing programme. With hindsight instrument development was time-consuming and added significantly to the project costs.

There was a claim by a sub-contractor for unforeseen ground conditions despite the high concentration of boreholes, and research levels of sampling and testing. The claim was rejected.

8. Investigation and interpretation of Pentre and Tilbrook Grange soil conditions

M. D. LAMBSON, BP Engineering, BP International Limited, D. G. CLARE, Ove Arup & Partners, D. W. F. SENNER, Consultant, and R. M. SEMPLE, Dames & Moore

1. Introduction

This paper describes the ground conditions at the two test sites selected for the Large Diameter Pile Test Programme. This followed an initial comprehensive study of the ultimate capacity of offshore piled foundations in cohesive soils. The study recognised that differences in pile size, the magnitude and type of loading, and the soil conditions that can be encountered offshore required significant extrapolation of empirical design criteria beyond the existing database. Two important soils conditions not included in the API RP 2A database are overconsolidated soils with shear strengths greater than 400 kPa and stiff, essentially normally consolidated soils. Two UK sites, Pentre in Shropshire and Tilbrook Grange in Cambridgeshire, were selected as providing the normally consolidated (NC) and the overconsolidated (OC) clay soil conditions, respectively. They represented soil conditions typical of those found beneath the North Sea.

The object of this paper is to provide a comprehensive data set for the ground conditions at each site which gives a sound basis for engineering parameter selection. Site selection is discussed first. The investigation methods and laboratory testing techniques are then outlined before the ground conditions at the Pentre and Tilbrook Grange sites are described. For each site the geological setting is given, after which the stratification and groundwater conditions are presented. Finally the engineering properties are summarised together with the design profiles used for the subsequent pile axial capacity analyses.

2. Site selection

For the purposes of resolving uncertainties in North Sea piled foundation design practice, two types of soil profile were identified as being of particular interest. Firstly, a profile termed NC and comprising a uniform soft becoming

 Large-scale pile tests in clay. Thomas Telford, London, 1993

stiff, essentially normally consolidated silty clay profile up to 60 m in thickness. Secondly, a profile termed OC and consisting of uniform hard, heavily overconsolidated silty clay, up to 40 m in thickness. A thorough literature review of UK Quaternary deposits was undertaken to identify potential onshore sites similar to North Sea sediments. This identified eight regions where thick sequences of cohesive Quaternary sediments are found. Twelve borings were undertaken to complement existing borehole information in the regions of interest. Thomas (1990) summarises the literature review and reports on the ground conditions at sites within the regions of interest. The Quaternary deposits in the areas reported divide into two categories, NC and OC clays. The first of these consists of predominantly glacial lake deposits or glacio-estuarine deposits which comprise about 60 m of essentially normally consolidated clays and silty clays. The sediments of the Forth and Devon valleys, Team valley, Aire valley, Shrewsbury and north-west Norfolk were considered. The second category contains the sediments of Edinburgh, Tyneside, and Bedford and Huntingdon, which comprise predominantly lodgement tills being some 35 m of heavily over-consolidated silty clays with

Table 1: Sites considered for pile test

	REGION	SITE	PI %	Su/σ_v'	DEPTH (m)	COMMENTS
NC	Forth Valley	Alva	30	0.2 - 0.1	70	Insufficient mean strength
		Grangemouth	50-20	0.2 - 0.1	73	
	Team Valley	Lady Park	25-50	0.7 - 0.3	30+	Deposit too thin, plasticity generally too high
		Ouston	25-50	0.7 - 0.3	26+	
		Chester-le-Street	25-40	0.9 - 0.3	15+	
	Aire Valley	Silsden	35-55	0.14	57	Too Plastic
	Shrewsbury	Pentre	10-30	0.16 - 0.32	60	Chosen
	North West Norfolk	Watlington	20-40	0.6 - 0.8	48	Insufficient mean strength
OC	Edinburgh	Leith	10-20	5 - 1.5	25	Insufficient depth and gravel/ cobbles more predominant
	Tyneside	North Shields	20	4 - 1.5	28	
	Bedford and Huntingdon	Tilbrook	20-30	10 - 1	41	Chosen
		Chelveston	15-30	10 - 1	34	Insufficient depth

Fig. 1. Site location

occasional gravel. Table 1 summarises the sites considered. Reference to this shows that of the five regions reviewed for the NC site, one was considered to have too thin a deposit combined with plasticity which was generally too high, two sites had an insufficient mean strength and at another, the clay was thought too plastic to be truly representative of most north Sea soil conditions. Hence, the site at Pentre was selected as being the best compromise. For the OC site, three regions were considered. The borings at Edinburgh and Tyneside revealed insufficient depth of strong clays with gravel and cobble inclusions being more predominant than investigations disclosed in the Bedford and Huntingdon region soils. The site at Tilbrook Grange was chosen from this region. Fig. 1 shows the location of the Pentre and Tilbrook Grange sites with National Grid references of SJ 359 177 and TL 009 710, respectively.

3. Site investigation and laboratory testing techniques

3.1 Fieldwork

Detailed site investigations were carried out at both sites during the period June to October 1986. The soil conditions were investigated by a combination of soil sampling and in-situ testing. The drilling and sampling methods were chosen to be of a similar high standard to those employed offshore. Hence, drilling operations were performed using a skid mounted Failing 1500 drilling rig. The drill string was made up of 127 mm OD drill pipe, with a latch-in sub for downhole testing and push sampling, and a 230 mm open centre drag bit. The borings were advanced by rotary drilling methods with an external flush and open hole. A prepared polymer drilling mud was used with some Barytes weighting additive at Pentre. Sampling and in-situ test equipment was wireline based.

The layout of the fieldwork with respect to the Pentre test pile is presented in Fig. 2. Two surface cone penetration tests, designated C-1 and C-2, were

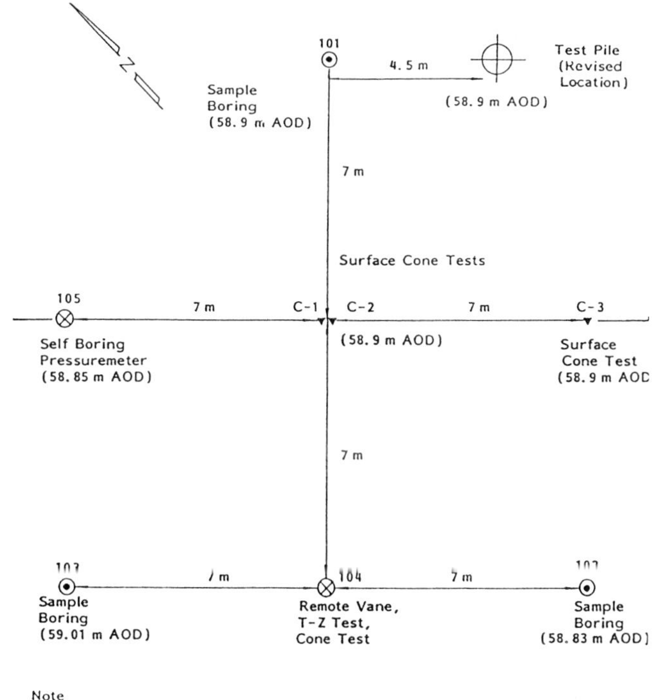

Fig. 2. Layout of the Pentre site investigation

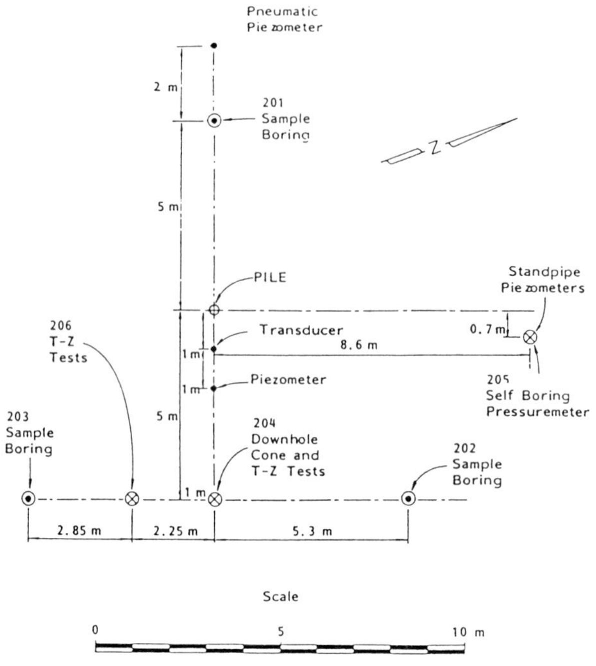

Fig. 3. Plan of borings and cone tests at Tilbrook Grange site

completed to a maximum depth of 41 m using a piezocone and a third test, designated C-3, to a depth of 50.7 m using an advanced multi-piezometer cone, designated Type A+B and described by Bayne and Tjelta (1987). Borings 101 to 104 inclusive were drilled to nominal depths of 60 m. Borings 101 to 103 were made to recover samples for laboratory testing. The samples were taken at nominal 1 m intervals in 76 mm OD thin wall tubes. Sampling was performed using a hydraulic piston sampler in the soft soils and by pushing the tubes open ended using the drill string in the stronger. A few hammered samples of 64 mm OD and 600 mm in length, were also taken in Boring 103. Boring 104 was made to perform in-situ vane tests, axial load-displacement (*T-Z*) probe tests and in-situ cone tests, including dissipation testing. The cone tests were performed using the multi-piezometer cone between 50 and 60 m penetration to supplement the surface single element piezocone tests, which were undertaken from a truck mounted rig. Natural gamma logging was undertaken in Boring 10 m through the drill pipe prior to backfilling the hole with cement bentonite grout. A 100 mm diameter PVC pipe was grouted into Borings 102, 103 and 104 upon reaching target penetration, to facilitate cross-hole and surface to down-hole seismic survey. Boring 105 was drilled to facilitate self boring pressuremeter testing. On completion of this boring, two standpipe piezometers were installed at nominal tip depths of 10.4 and 39.5 m below ground level.

The location of the soil borings at Tilbrook Grange is given in Fig. 3. Eight soil borings were drilled at the site. Borings 201, 202 and 203, drilled to depths of about 40 m, were made to recover soil samples for laboratory testing. The samples were taken at nominal 0.5 m intervals using 75 mm OD thin and thick wall tubes. The thin and thick walled tubes had area ratios of 0.90 and 0.71, respectively. Generally the samples were recovered by hydraulic jacking using the drilling rig which pushed the tubes into the soil. Two samples in Boring 202 were obtained by hammer sampling. Borings 204 and 206 were drilled to perform in-situ *T-Z* probe and downhole piezocone penetrometer testing. The downhole cone tests were performed between ground level and 40 m penetration in Boring 204 using a multi-piezometer cone penetrometer. Seven cone dissipation tests were conducted at intervals between 8 and 37 m

Table 2. Summary of laboratory testing

Test Description	Numbers of Tests	
	Pentre Site	Tilbrook Site
• CLASSIFICATION TESTS		
Atterburg Limits	204	103
Particle Size Distribution	59	22
Specific Gravity	25	11
Carbonate Content	32	32
Organic Content	29	22
• STRENGTH TESTS		
Hand Tool Estimates	789	274
Miniture Vane	24	-
Triaxial Tests		
Total stress (UU)	150	125
Effective stress (CIU, CAU, CK_0U)	72	46
Direct Shear	83	35
Ring Shear	13	12
Simple Shear	19	4
• RESONANT COLUMN TESTS	3	
• CONSOLIDATION TESTS	46	25
• PERMEABILITY TESTS	19	4

depth in Boring 204. These involved stopping the cone penetrometer and monitoring the decay of pore pressures generated. In Boring 206, some six monotonic and cyclic *T-Z* probe tests were undertaken between depths of 4 and 30 m. Boring 205 was drilled for self boring pressuremeter testing. Owing to difficulties of installing the pressuremeter in the upper soils, seven tests were conducted between depths of about 16 and 35 m. On completion of the pressuremeter testing, two standpipe piezometers were installed at nominal tip depths of 12.1 and 35.8 m below ground level. Additional in-situ testing included natural gamma logging of Boring 202, performed through the casing, and crosshole and surface to downhole seismic testing in Borings 202, 203 and 204. In September 1988 the UK Building Research Establishment (BRE) undertook Marchetti dilatometer testing at the site. Two further shallow sample boreholes were drilled at Tilbrook Grange following the pile lateral load test which was later added to the programme. Results from these were unavailable for the pile axial capacity analyses and are reported by Long *et al.* (1993).

3.2 *Laboratory testing*

A comprehensive laboratory testing programme was undertaken to evaluate pertinent physical properties of the foundation soils. In the field unconsolidated-undrained (UU) triaxial compression tests and water content and density determinations were performed on selected samples, together with estimates of undrained shear strength made using a hand-operated penetrometer. Additionally at Pentre, mini-vane testing was also undertaken in the field. All other tests, including additional UU triaxial tests, were conducted by fully equipped soil mechanics laboratories. Where applicable, test procedures were in accordance with BS1377 (1975). The effective stress testing followed the guidelines given by Head (1986). For both sites the testing philosophy was to characterise the soil properties to provide sufficient information as a basis for any pile behaviour analyses to be undertaken, both now and in the future. The laboratory test programme also evaluated a number of the soil properties measured by in-situ testing to enable correlation and confirmation of the soil behaviour. Likewise, some of the soil classification and index test results were related to the geological testing. The scale of the laboratory test programme may be judged by reference to Table 2, which summarises the type and number of tests. Most emphasis was given to soil strength measurement, both in terms of total and effective stress. Undrained shear strength was assessed from triaxial tests using both undisturbed and remoulded samples. Soil strength and behaviour in terms of effective stress was assessed by triaxial compression testing of isotopically consolidated (CIU), anisotropically consolidated (CAU) and K_0 consolidated (CK_0U) undisturbed specimens. Precision strain measurements were made on some CAU tests. Additionally, for some Pentre samples, triaxial extension testing was also performed. The effective stress testing programme was planned once the basic interpretation of the in-situ field data and the majority of the classification plus UU triaxial testing was complete. This enabled proper estimation of the test pressures to be representative of in-situ conditions. Direct shear tests using a shear-box were performed to measure the undrained strength of undisturbed and remoulded specimens, and to measure the shear stress at

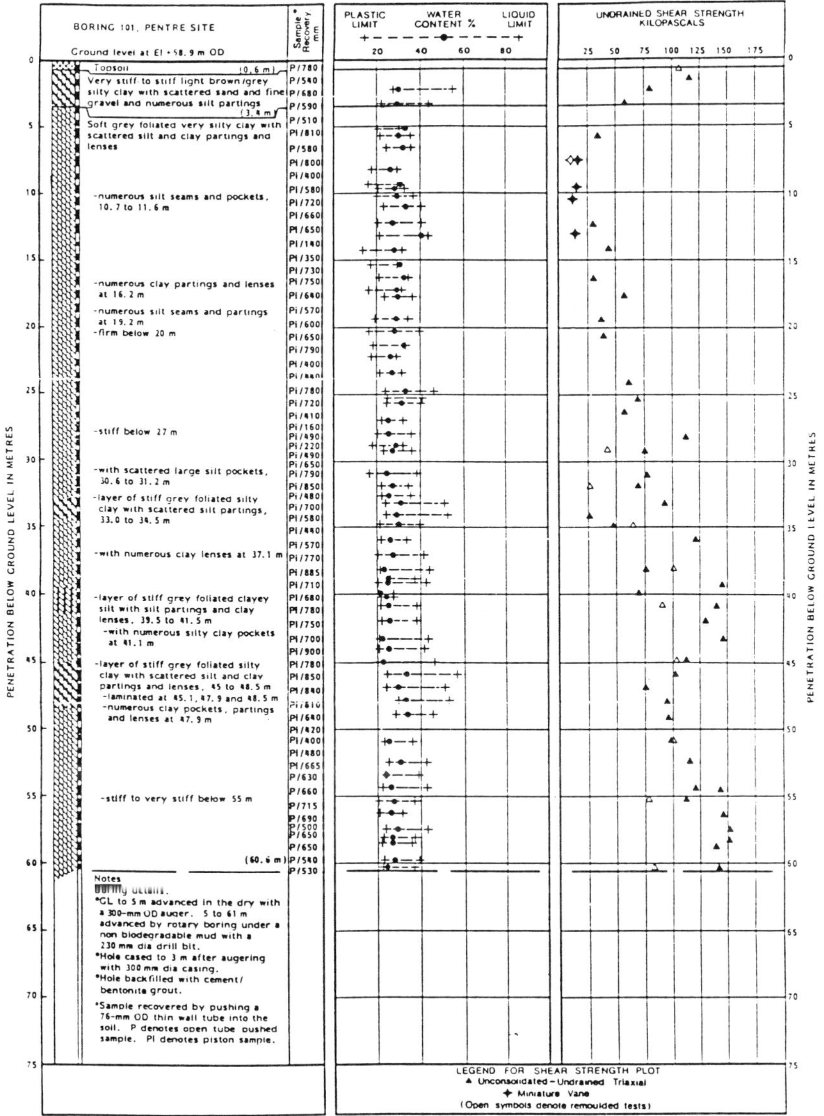

Fig. 4. Log of boring and test results

the interface of undisturbed and remoulded soil and the pile steel material. Ring shear tests were carried out on remoulded soil specimens to determine the residual soil to soil and soil to steel interface shear strengths. The soil strength in simple shear was also assessed from vertically and horizontally orientated specimens which were consolidated to estimated in-situ stresses.

For the Pentre soils results from resonant column tests were used to augment soil stiffness measurements obtained from the triaxial precision strain and simple shear tests. This enabled stiffness to be assessed over a range of strains of practical interest.

The compressibility characteristics of the Pentre and Tilbrook Grange soils were investigated by one-dimensional consolidation tests on undisturbed specimens. Soil permeability was measured in the laboratory by the constant head permeability test, using undisturbed specimens which had been consolidated isotopically to the estimated in-situ vertical effective stress within a triaxial cell.

3.3 Geological testing

At Pentre a geological testing programme was undertaken to examine the microfabric and minerals of the sediments between about 17 and 56 m penetration. A comparison was made with soils from the North Sea Cyrus and Troll Fields. A geological testing programme was performed to define and compare the macrofabric, microfabric, clay minerals and age of the two main strata at the Tilbrook Grange site. In addition, a similar testing programme was performed on clay samples of the same age and type from three other sites in the UK. These locations were selected to compare the microfabric and clay mineralogy of this clay at sites with different stress histories.

For both studies the macrofabric was defined using a binocular microscope on hand specimens while the microfabric was defined using a scanning electron microscope. The age of each sample was identified using palaeontological techniques on hand specimens viewed through a binocular microscope. The clay minerals was defined for selected samples using X-ray diffraction techniques and interpretation of the clay species was after the criteria of Brindly and Brown (1980) and semi-quantification by the method of Weir *et al.* (1975).

4. Pentre ground conditions

4.1 Geology

Pentre is located in the Severn Valley. The bedrock formation, known as the Severn Trench, comprises sandstones and marls of Carboniferous and Triassic age.

The bedrock is overlain by Quaternary deposits which reach a maximum thickness of 80 m in the vicinity of Pentre. These sediments were derived from

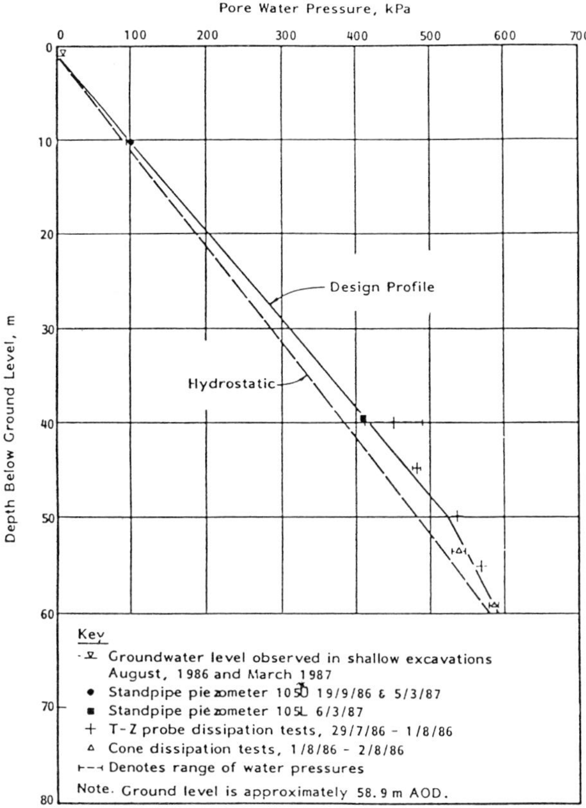

Fig. 5. Pore water pressure profile at Pentre

chlorite and illite-mica rich Palaeozoic metasediments present in the Upper Severn Valley.

The Quaternary sediments comprise a 3 to 4 m thick surface layer of Holocene or Recent alluvium overlying essentially normally consolidated very silty clays considered to have been deposited in an ice proximal glacial lake environment some time after 18 000 years before present. Gravel is occasionally present in the glacial lake clays due to dropstone activity from floating ice.

4.2 Strata encountered

A detailed log of Boring 101 is presented in Fig. 4. A generalised summary of the soil strata encountered in the three borings is as follows:

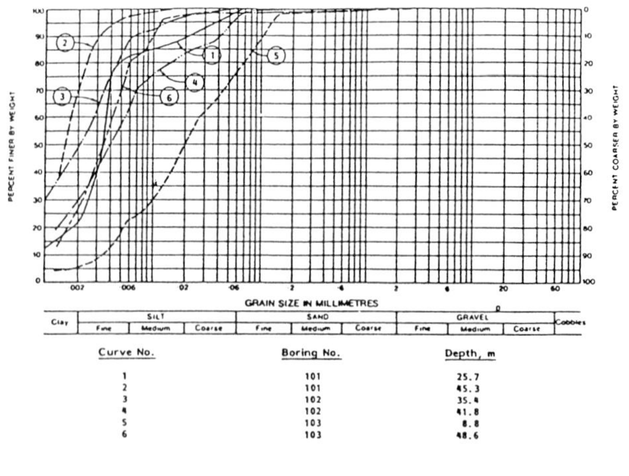

Curve No.	Boring No.	Depth, m
1	101	25.7
2	101	45.3
3	102	35.4
4	102	41.8
5	103	8.8
6	103	48.6

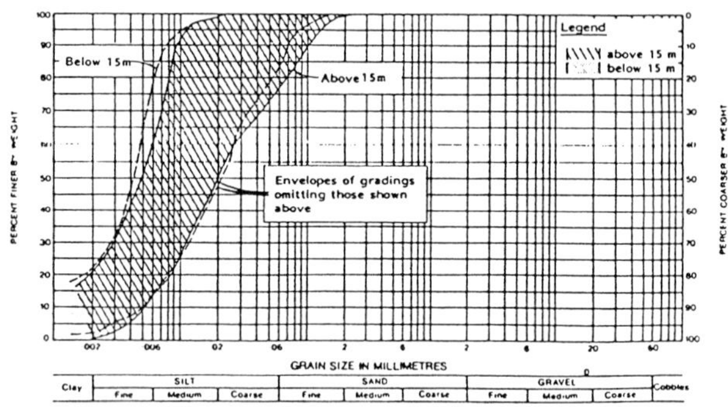

Fig. 6. Summary of grain size curves, stratum II, Pentre

Stratum	Depth - GL (m) From	To	Description
Ia	0	0.6	Topsoil
Ib	0.6	3.5	Stiff to very stiff yellowish brown and grey mottled SILTY CLAY (Recent alluvium)
II	3.5	60.9+	Soft to very stiff grey very SILTY CLAY (glacial lake deposit)

The silt and clay content of Stratum II varies with depth resulting in occasional layers of clayey silt of low to intermediate plasticity and silty clay of intermediate to high plasticity. The clayey silt layers are between 0.5 and

3.0 m thick and generally occur between depths of 36 and 41.5 m. The layers of silty clay are between 2 and 8 m thick and generally occur between depths of 40.5 and 48.5 m. They appear to thicken north-west to south-east across the site. Bands of siltstone gravel, up to 300 mm thick, were encountered in Borings 102 and 103 between depths of 24 and 30 m.

4.2.1 Microfabric and mineralology

The sediments of Stratum II, between 3 and 15 m depth, have a relatively massive fabric of very low anisotropy generally comprising mixed clay and silt sized minerals in a well developed aggregate structure with relatively large voids. This structure is believed to be the result of a continuous rather than a pulsed sediment supply in the later stages of deposition in a glacial lake. The sediments below 15 m, however, appear to have been deposited following the input of seasonal pulses of sediment into the glacial lake, and may be described as laminated to very finely laminated. The degree of

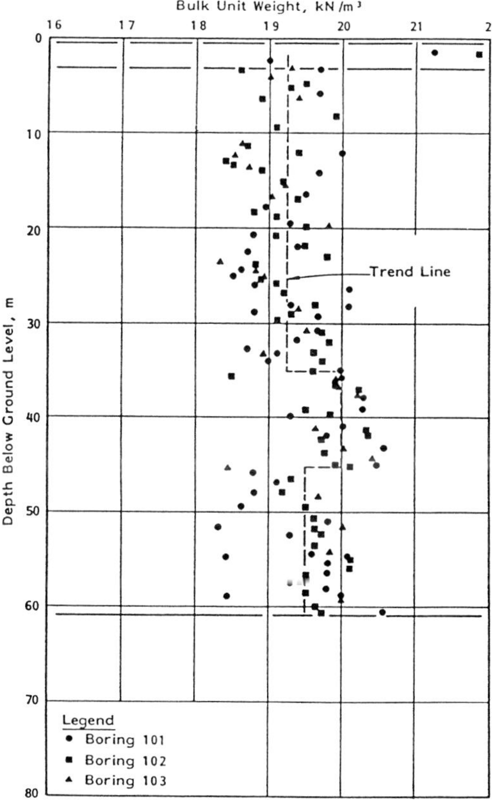

Fig. 7. Bulk unit weight, Pentre

sub-horizontal fabric anisotropy tends in increase with depth from low to medium at 25 m, to medium to high at 55 m. The clay minerals occur generally as simple aggregates. The microfabrics examined are generally not flocculated and do not show any evidence of glacial over-riding of the sediments.

The sediments comprise typically between 35 and 60 per cent clay minerals, illite-mica and chlorite, 35 and 60 per cent quartz, and about 5 per cent plagioclase feldspar. A summary of the composition is given below.

Fraction	Mineral	Proportion
Clay	Illite-mica and chlorite	99%
	Quartz	1%
Fine silt	Illite-mica and chlorite	60-80%
	Quartz	20-40%

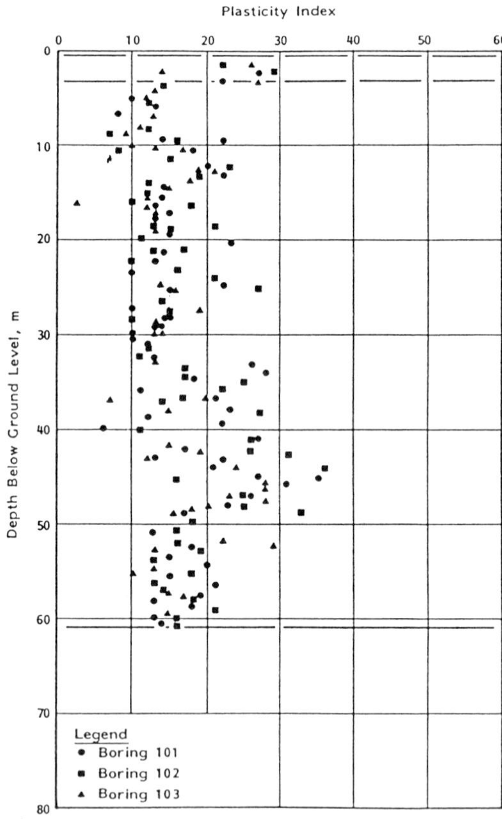

Fig. 8. Plasticity index, Pentre

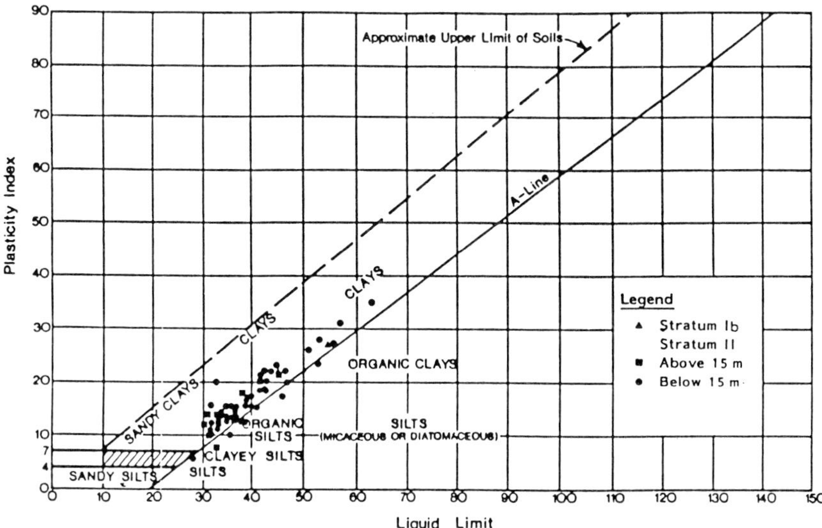

Fig. 9. Plasticity chart for classification of fine-grained soils, boring 101, Pentre

Fraction	**Mineral**	**Proportion**
Medium silt	Illite-mica and chlorite	30-70%
	Plagioclase Feldspar	5%
	Quartz	25-65%
Coarse silt	Illite-mica and chlorite	15-30%
	Quartz	70-85%
Fine sand	Quartz	100%

The Pentre sediments have been compared with normally consolidated North Sea soils from the Cyrus Field (Block 16/28, UK Sector) and the Troll Field (Block 31/6, Norwegian Sector). These sediments are typical of large areas of essentially normally consolidated soils beneath the North Sea. Specimens were examined from 13 and 22 m depth in the Cyrus Field. At 13 m the fabric is uniformly laminated and of medium anisotropy. At 22 m the fabric is massive and of the dependent type, i.e. the fines are wrapped around larger silt and sand grains. Fabric anisotropy was found to be low. The specimen from 13 m depth comprised mainly quartz, plagioclase and orthoclase feldspars, dolomite and calcite, with minor amounts of mica, chlorite and disordered kaolinite.

Photomicrographs of specimens taken from depths of 16, 18 and 78 m in the Troll field show the sediments to be silty clays with an essentially massive, dependent fabric of very low anisotropy, with some areas being essentially isotropic. The clay minerals appear to be predominantly illite, possibly interlayered with smectite, a little kaolinite and chlorite.

The soils examined from the Cyrus and Troll fields appear to be rather finer grained than the Pentre deposits and have a more compact structure of generally lower anisotropy. The mineralogical composition of both the offshore sites and the Pentre site are different due to the different source rocks from which they were derived.

4.3 Groundwater conditions

4.3.1 Background

Water supply wells historically have been drilled in the Pentre area to tap artesian pressures within the sandstone aquifer which underlies the glacial lake deposits of the test stratum. An artesian head up to 5 m above ground level, is reportedly present within the aquifer which is estimated to lie at a depth of about 80 m below ground level at the test site location. Upward flow would be expected to result in greater than hydrostatic pore pressure within the test stratum.

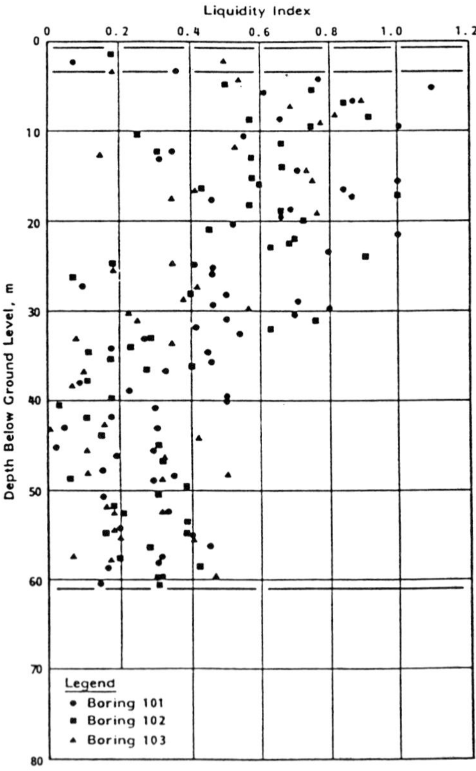

Fig. 10. Liquidity index, Pentre

4.3.2 Measurements

Figure 5 summarises information obtained on groundwater pressures over the period June 1986 to March 1987. The pore pressure profile is broadly as described above and does not appear to vary significantly from summer to winter above a depth of 40 m. Only summer readings are available below 40 m, however, and tend to indicate that fluctuations in aquifer water pressure influence the pore pressures within the glacial lake deposits below a depth of 50 m. Pressure within the aquifer would be expected to fall in the summer and rise in the winter months in response to variations in rainfall.

4.4 Summary and interpretation of soil data

4.4.1 Classification and index tests

Summaries of the grain size curves are presented in Fig. 6. The variation of bulk unit weight with depth is summarised in Fig. 7.

The variation of plasticity index with depth is summarised in Fig. 8. The plasticity index in Stratum II ranges typically between 10 to 30, the higher

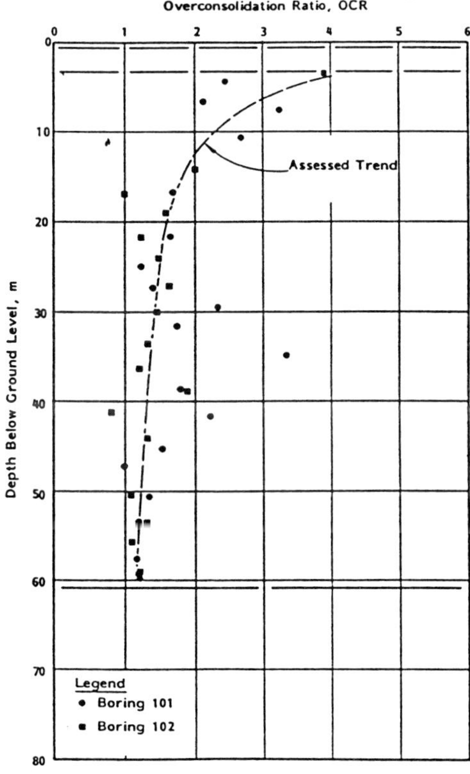

Fig. 11. OCR estimated from oedometer tests using Casagrande's construction, Pentre

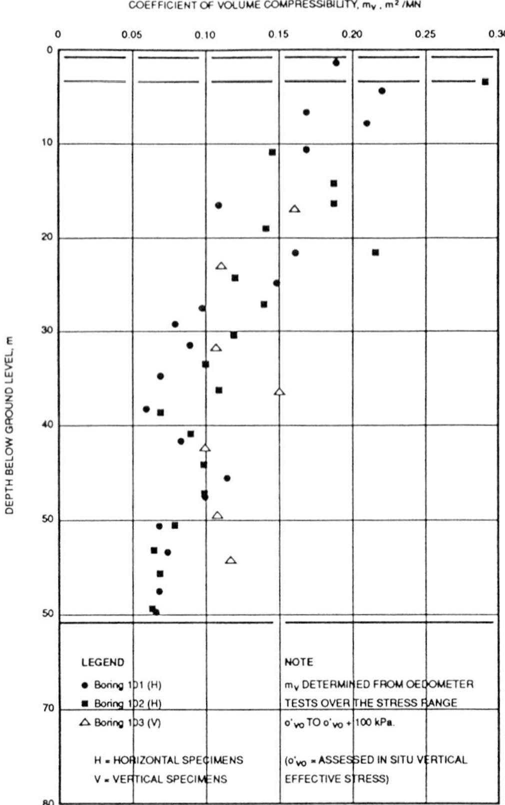

Fig. 12. Coefficient of volume compressibility, Pentre

plasticities occurring in the more clayey zones generally between 35 and 50 m depth.

The plasticity chart for Boring 101 is given in Fig. 9. The majority of results from Fig. 9, Stratum II plot above the 'A' line and may be classified as silty clays of low to medium plasticity. The relatively high plasticity index compared with the relatively low clay size fraction (less than 2 microns) is reconciled by the mineralogical study which revealed significant amounts of the clay minerals chlorite and illite-mica in the fine to coarse silt size fraction. These clay minerals predominantly comprise the clay size fraction.

Liquidity indices are summarised in Fig. 10. The liquidity indices show a trend to decrease with depth in Stratum II.

4.4.2 Overconsolidation ratio

The overconsolidation ratio, OCR, may be estimated from the unconsolidated-undrained triaxial shear strength, vertical effective stress and plasticity data, and also from the oedometer tests using the construction proposed by

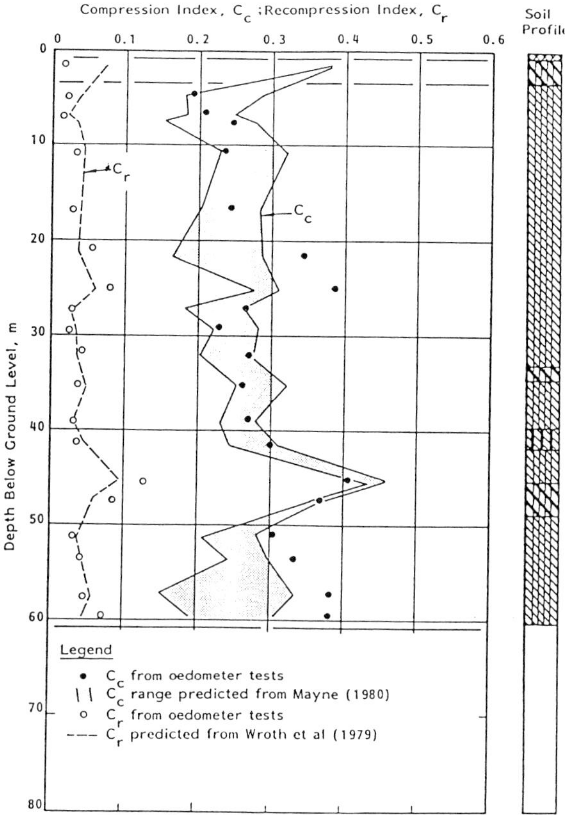

Fig. 13. Compression and recompression indices, horizontal specimens, boring 101, Pentre

Casagrande (1936). Greater reliance is placed on the Casagrande construction and the results of all OCRs obtained by this method are presented in Fig. 11, together with a suggested trend line. The OCR decreases from 1.8 at 15 m to 1.2 at 60 m depth and thus the test stratum is essentially normally consolidated.

It is understood from the geological history that significant erosion has not taken place, and from the soil microfabric studies that the deposits have not experienced glacial overriding. Thus the observed OCR profile would be expected to be either a result of fluctuation in groundwater pressures or secondary compression. The OCR profile obtained for the site is considered to be largely due to a former mechanism and is consistent with a past lowest water table around 5 to 10 m below ground level.

4.4.3 Compressibility

Profiles of the coefficients of volume compressibility, m_v, obtained from

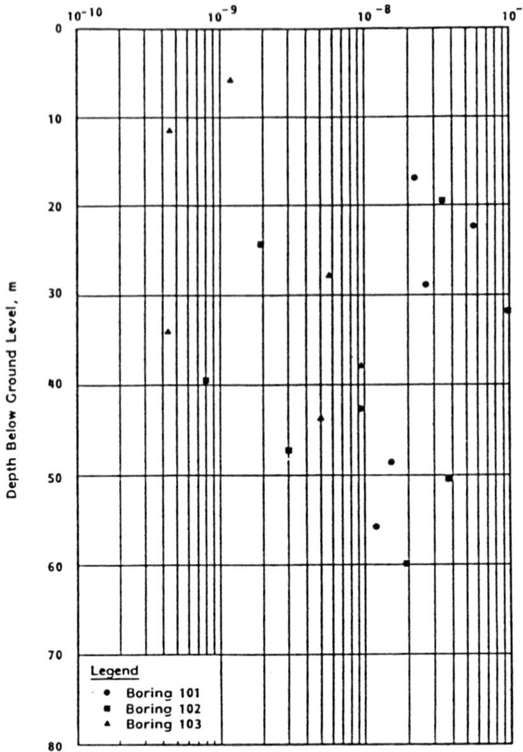

Fig. 14. Vertical permeability, isotropically consolidated triaxial specimens, Pentre

one dimensional oedometer consolidation tests for horizontal and vertical specimens are presented in Fig. 12. The terms horizontal and vertical refer to the orientation of the flat plane of the specimens. The values for vertical specimens are similar to those for horizontal specimens which decrease with depth from about $0.2m^2/MN$ near surface to about $0.065m^2/MN$ at 60 m. The deposits may be classified as being of medium to low compressibility (Tomlinson, 1980).

Compression, C_c, and recompression, C_r, indices determined from oedometer tests are presented for Boring 101 in Fig. 13. The measured values of C_c are compared with predictions based on correlations with index properties proposed by Mayne (1980). The measured values for both horizontal and vertical specimens are in reasonable agreement with the predictions and tend to range typically between about 0.2 and 0.35. The measured values of C_r are compared with correlations proposed by Wroth *et al.* (1979) based on index

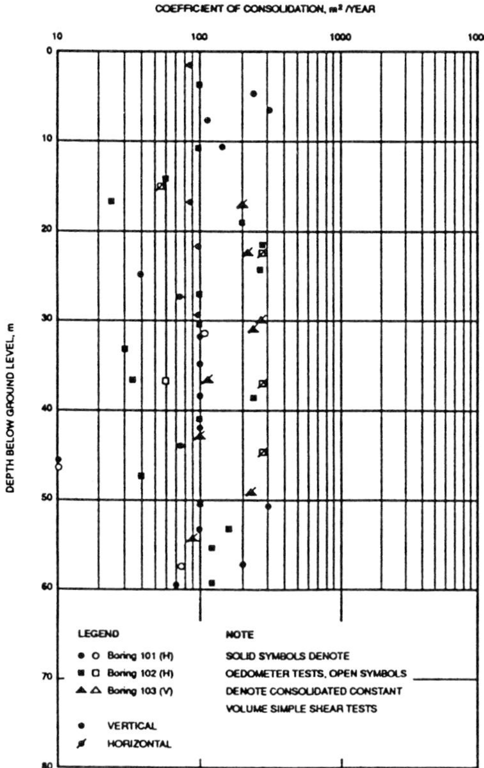

Fig. 15. Coefficient of consolidation from laboratory tests on undisturbed specimens, Pentre

properties and show generally good agreement with values typically in the range 0.02 to 0.07.

Ratios of recompression and compression indices, C_r/C_c for horizontal specimens are similar to those for vertical specimens and are typically between 0.08 and 0.25.

4.4.4 Laboratory permeability tests

Measurements of the vertical permeability of the soils were made in the laboratory on isotopically consolidated triaxial specimens. The tests were carried out on samples from Stratum II and the results are presented in Fig. 14. There is a wide range in permeability from 4×10^{-10} m/sec to 1×10^{-7} m/sec.

4.4.5 Coefficient of consolidation

Laboratory tests

Coefficients of consolidation at estimated in-situ stresses were determined from laboratory oedometer and simple shear tests on horizontal and vertical specimens and are presented in Fig. 15.

Coefficient of consolidation, m²/year

Parameter	Range	Mean	Comments
C_v (undisturbed)	10-300	10 100	Silty clay, 40-50m Very silty clay
C_h (undisturbed)	90-300	100 250	Two groupings evident over profile
C_r (remoulded)	10-100	90	

In-situ dissipation tests

Excess pore pressures are generated during cone penetrometer and *T-Z*

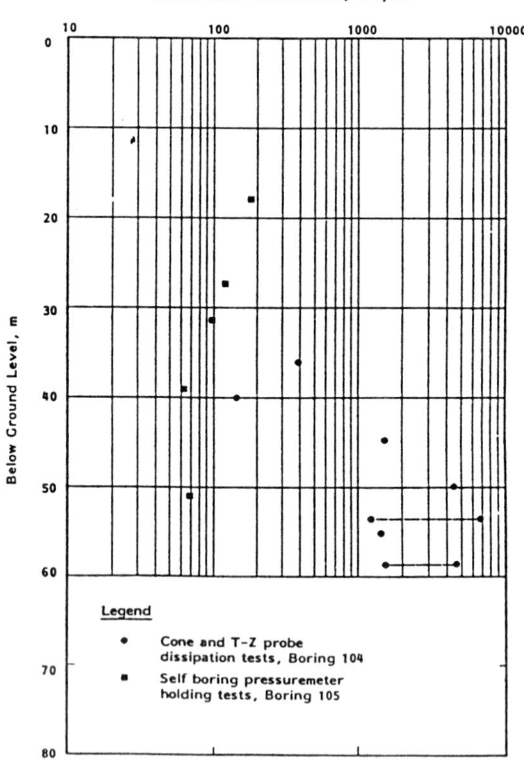

Fig. 16. Coefficient of consolidation from in-situ tests, Pentre

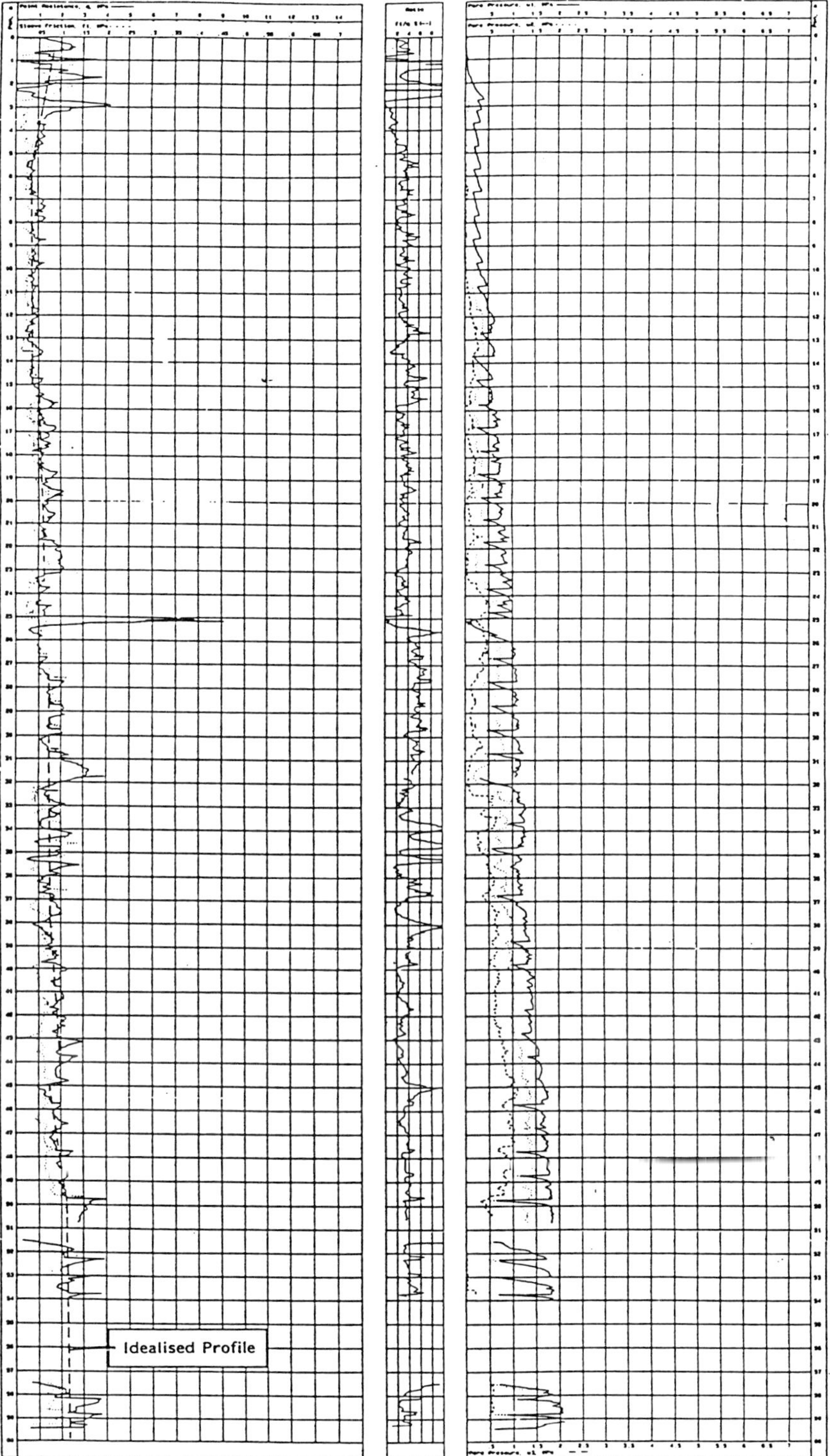

Fig. 17. Cone penetration test results, measured resistances and pore pressures, Pentre

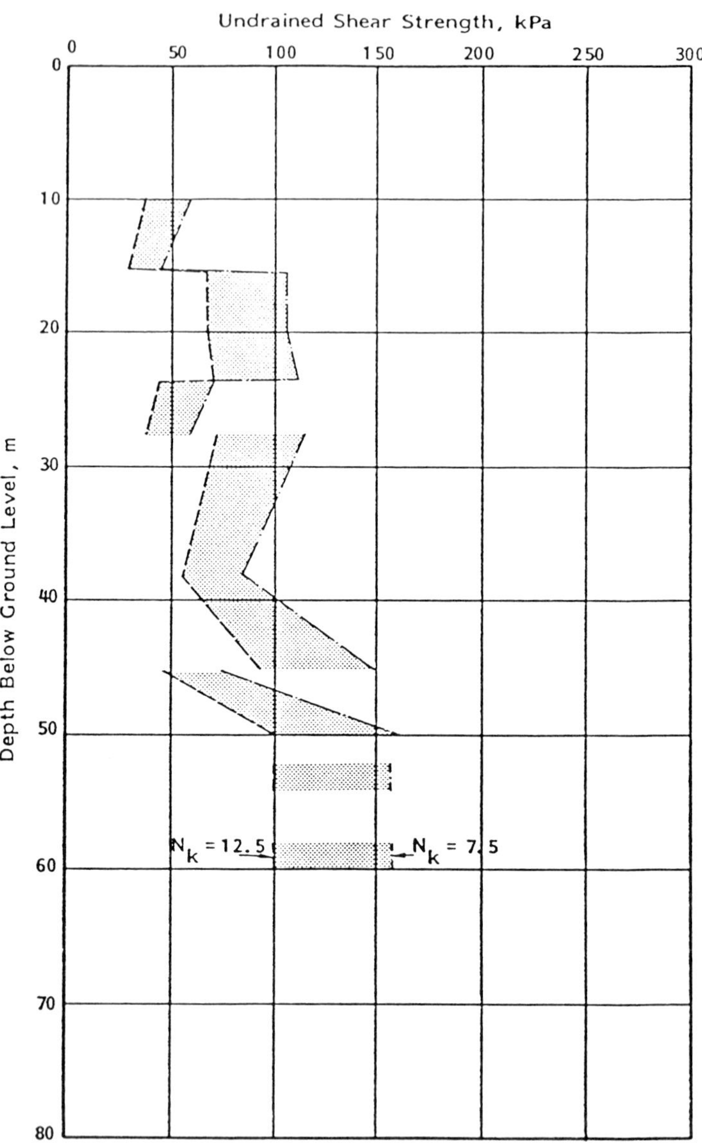

Fig. 18. Undrained shear strength derived from idealised cone point resistance data, Pentre

probe penetration as the advancing cone tip shears and displaces the soil. A dissipation test is the measurement of these excess pore pressures with time after interruption of penetration. The results can be used to estimate the in-situ consolidation characteristics of the soil (Campanella *et al.*, 1982; Lacasse and Lunne, 1982, and Tumay and Acar, 1984).

There are a number of analytical methods available for interpreting drainage characteristics from dissipation tests, all of which depend on the assumed boundary conditions. Interpreted coefficients of horizontal consolidation C_h measured along the *T-Z* Probe and cone penetrometer shafts lie in the range 140-700 m^2/year (Fig. 16).

4.4.6 Field shear strength data

4.4.6.1 Cone penetration tests

Results of the surface and downhole cone penetration tests performed using the multi-piezometer cone are presented in Fig. 17, which shows measured values of point resistance, q, sleeve friction, f_1, and pore pressure at the cone tip u_1 and at various locations on the shaft, u_2, u_3, and u_4.

Data obtained in silty deposits offshore California indicate that partially drained conditions may prevail during a cone penetration test performed at the standard rate (2 cm/sec) in soils with permeabilities between 1×10^{-6} and 1×10^{-8} m /sec (McNeilen and Bugno, 1984). Villet and Darragh (1985) proposed that undrained conditions may be assumed in overconsolidated silts where the ratio of tip pore pressure to tip resistance (u_1/q_c) is greater than about 50 per cent, and that drained conditions may be assumed when the pore pressure ratio is less than about 20 per cent. Work by Lacasse *et al.* (1981) shows u_1/q_c ratios between 0.5 and 0.7 in clays with over consolidation ratios between 1.15 to 2.5.

The tip pore pressure ratios at the site below a depth of 10 m are generally greater than 50 per cent and thus it is considered reasonable to interpret the cone tests assuming undrained behaviour.

The shear strength of cohesive soils can be interpreted from cone penetration tests using bearing capacity theory. Net cone point resistance and undrained shear strength is related by a point resistance and undrained shear strength is related by a cone factor N_k. The N_k values depend on many variables (Schmertmann, 1975) such as the type of soil stiffness in-situ, lateral stress, overconsolidation ratio, rate of cone penetration, shape of penetrometer tip, method of penetration and soil structure. The effects of these factors are discussed by Marsland and Quartermann (1982). Comparing idealised profiles of cone resistance and unconsolidated undrained shear strength at Pentre a mean range of N_k values between 7.5 and 12.5 have been obtained and the resulting variation in undrained shear strength is presented in Fig. 18.

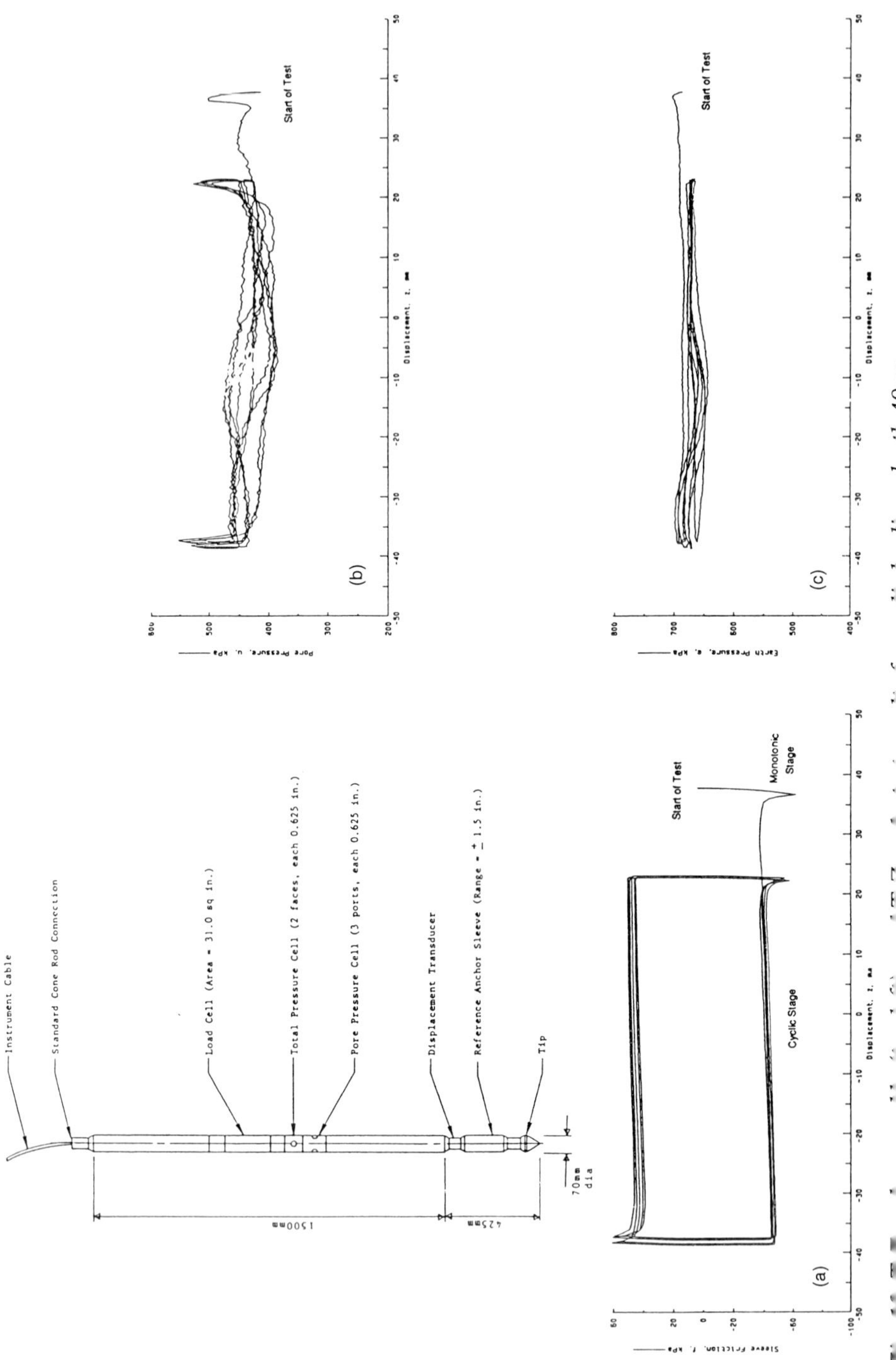
Instrument Cable
Standard Cone Rod Connection
Load Cell (Area = 31.0 sq in.)
Total Pressure Cell (2 faces, each 0.625 in.)
Pore Pressure Cell (3 ports, each 0.625 in.)
Displacement Transducer
Reference Anchor Sleeve (Range = ± 1.5 in.)
Tip
1500mm
425mm
70mm dia
(a)
Start of Test
Monotonic Stage
Cyclic Stage
(b)
Start of Test
(c)
Start of Test

Following the approach of Lunne *et al.* (1985) the cone tests may also be interpreted in terms of total point resistance, with the constant of proportionality being N_{kT}.

The derived N_{kT} values show a similar scatter to the N_k values, although the absolute values are higher as would be expected due to the pore pressure correction. The trend line indicates N_{kT} in the range 12 to 22 near surface reducing to 8 to 12 at depth.

4.4.6.2 Self boring pressuremeter tests

The use of the self boring pressuremeter is described by Gibson and Anderson (1961) and by Windle and Wroth (1977). The variation of shear strength determined in the self boring pressuremeter tests is presented in Fig. 28. These strengths were evaluated assuming undrained behaviour. However, the pore pressure responses observed indicated partial drainage during the expansion phases of the tests, which were performed at 1 per cent per minute cavity strain. The derived shear strengths should therefore not be regarded as undrained shear strengths.

4.4.6.3 Field vane tests

Shear strengths determined from in-situ remote vane tests are summarised in Fig. 28. The vane strengths are similar to the shear strengths obtained with the self boring pressuremeter, both of which are somewhat higher than unconsolidated-undrained triaxial shear strengths.

The tests were undertaken with vanes of 32 to 55 mm diameter rotating at rates of 12 and 18 degrees per minute. Times to reach peak shear stress ranged between 40 and 80 seconds. Based upon an analysis of excess pore pressure dissipation around a cylindrical body it is estimated that partial dissipation of excess pore pressure generated during shear may occur when the horizontal coefficient of consolidation exceeds about $50 m^2/year$. Laboratory and in-situ tests indicate coefficients of horizontal consolidation between 100 and $7000 m^2/year$, and it is therefore probable that partial drainage has influenced the field vane shear strengths.

4.4.6.4 T-Z probe tests

The *T-Z* probe (Fig. 19) provides direct measurement of axial local adhesion and displacement characteristics of the soil, together with the total lateral stress and pore pressure. The testing procedure includes insertion of the tool into the ground, dissipation of excess pore pressure, monotonic loading to determine peak and residual shear strengths, and cyclic loading. The results may be used to predict the local *T-Z* response for a pile and by integration of the response at different depths, obtain the global pile/soil behaviour.

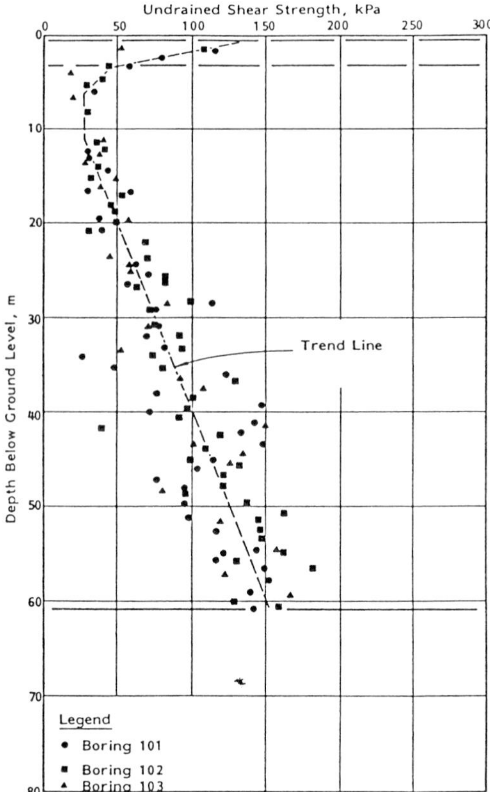

Fig. 20. Undrained shear strength, unconsolidated-undrained triaxial tests, Pentre

a. Testing procedure

The borehole was advanced to approximately 1 m above the required test depth. The tool was then inserted into the soil at a rate 1200 mm/min until the sleeve section of the tool was at the required depth. The tool was then immediately pulled back by about 75 mm at rates varying between 60 and 600 mm/min.

This set the position of the displacement reference anchor for a monotonic compression test following dissipation of excess pore pressures. Typically the interval between insertion of the probe and a monotonic test varied between 20 min and 2 h. A number of tests were left to set up overnight. The monotonic load test was carried out at a rate of 1 mm/min. The probe was advanced approximately 60 mm into the soil. Sleeve friction, pore pressure, total lateral stress and probe displacement measurements were recorded twice a second. Cyclic loading of the probe was then performed. Typically the probe was cycled approximately +30 mm at a rate of 20 mm/min. Typical results are presented in Figs 19a to 19c.

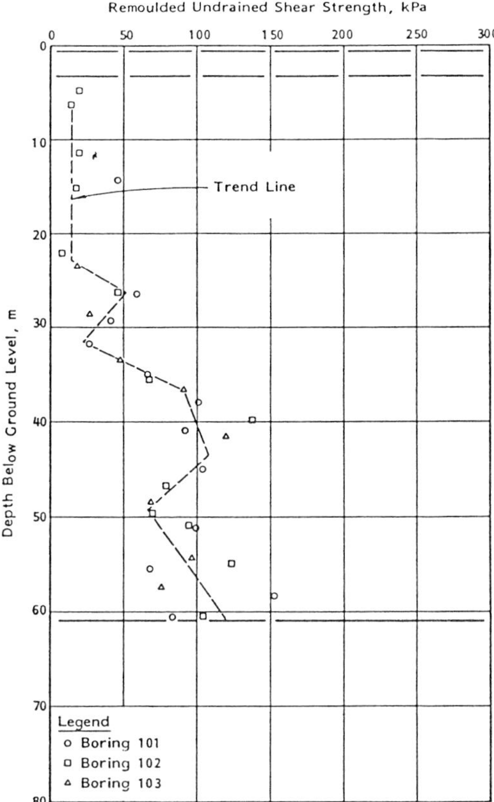

Fig. 21. Remoulded undrained shear strength, unconsolidated-undrained triaxial tests, Pentre

b. Monotonic load tests

Typically the sleeve friction/displacement plot shows the following features.

(a) Initial residual friction acting on the sleeve due to installation.
(b) A peak within 4 rim displacement generally followed by a sharp drop of 10 to 20 kPa.
(c) A constant or slightly rising skin friction after the peak

Parameter	Initial peak	Large displacement
Adhesion factor, α	0.5 to 0.8	0.3 to 0.8
Friction angle	14° (40 m) 18° (45 and 55.2 m)	10 to 16°

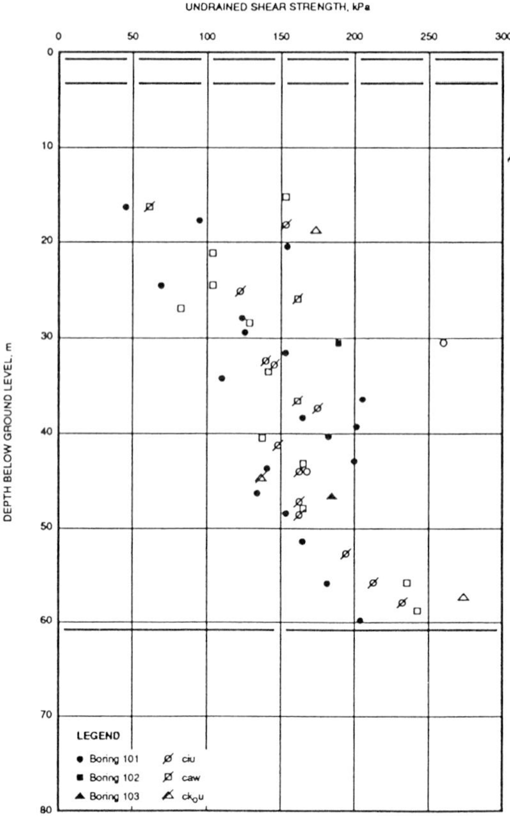

Fig. 22. Undrained shear strength, isotropically consolidated-undrained triaxial compression tests, Pentre

The large displacement friction angles are similar to the friction angles measured at large displacements in the stainless steel interface tests performed in the direct shear box.

c. Cyclic load tests

The plots of sleeve friction against displacement are similar in shape to the monotonic tests, showing an initial peak within a few millimetres displacement, a sudden drop followed by a gradual increase in skin friction with increasing displacement. The plots are nearly symmetrical showing similar behaviour in tension and compression. Little degradation of skin friction with number of cycles, typically up to five, was observed. The residual cyclic skin friction is similar to the residual skin friction measured in the monotonic tests.

The pore pressure/displacement plots show characteristic peak pore press-

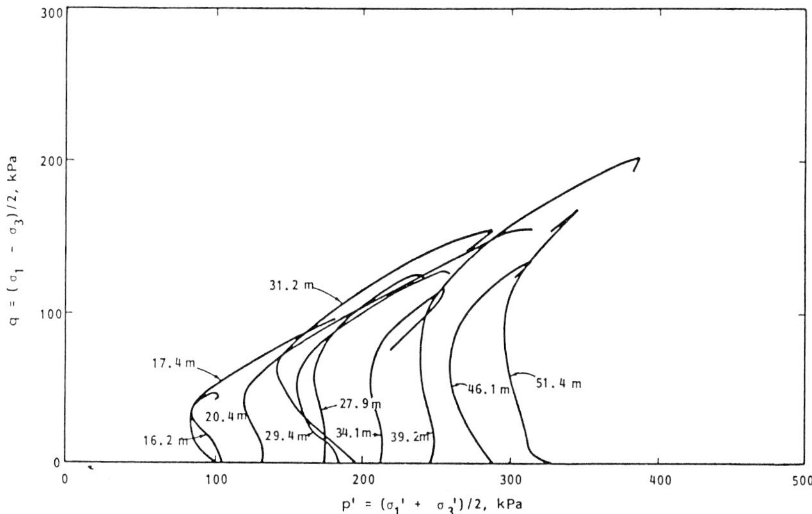

Fig. 23. Stress paths, isotropically consolidated-undrained triaxial tests, boring 101, Pentre

ure increases of about 100 kPa on reversal of shearing and hysteresis of up to 60 kPa in pore pressure during cycling. The earth pressure/displacement cycles are generally repeatable for each test typically showing hysteresis between compression and tension loading of up to about 40 kPa. The lower earth pressure generally occurs during tension loading which may be due to

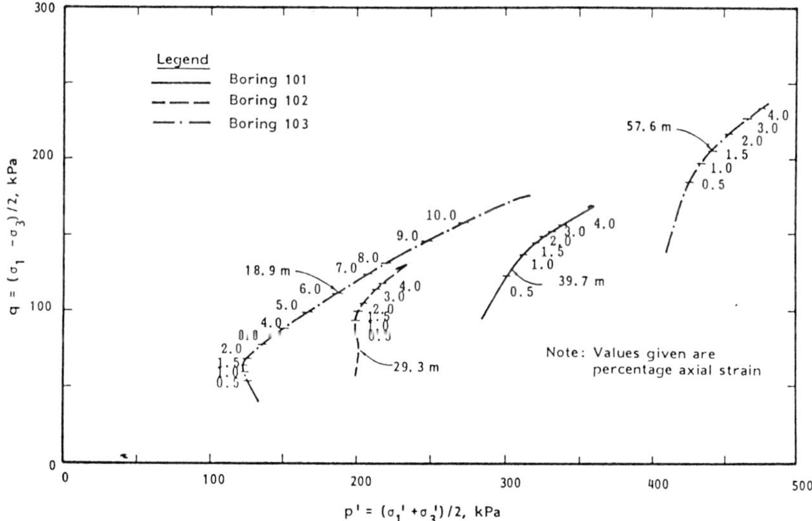

Fig. 24. Stress paths, anisotropically consolidated-undrained triaxial tests with precision strain measurements

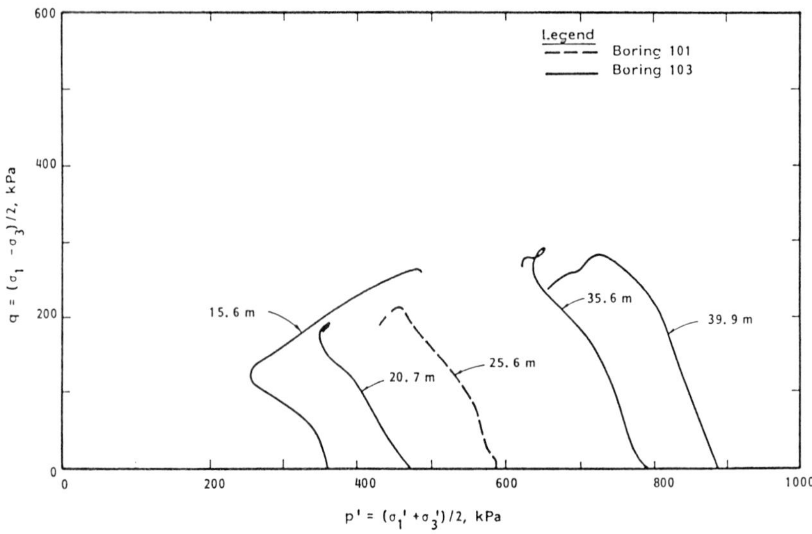

Fig. 25. Stress paths, isotropically consolidated-undrained (SHANSEP) triaxial tests

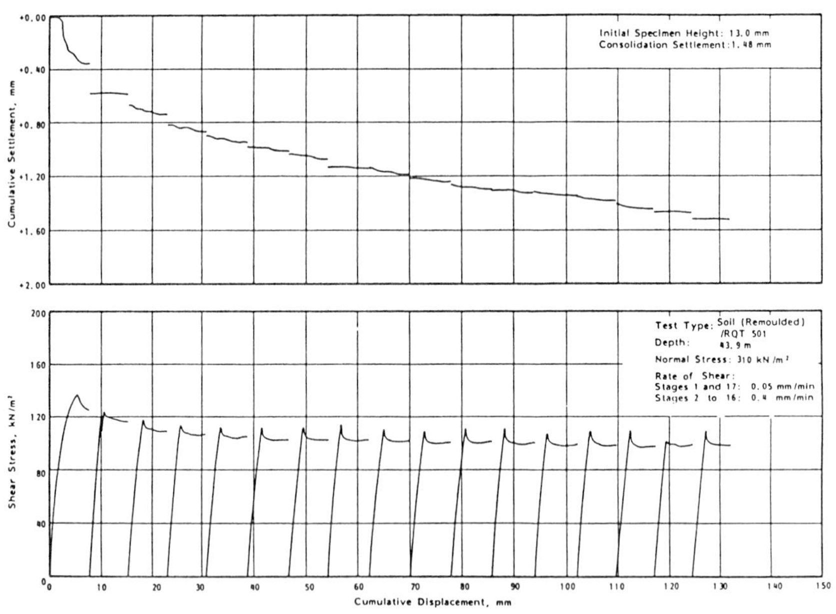

Fig. 26. Direct shear test results, boring 101, Pentre

the change in direction of the thrust causing loading and unloading around the bottom of the borehole and the probe.

d. Discussion

Skempton (1985) indicates qualitative changes in clay shearing behaviour in ring shear tests when the rate of shear exceeds 100 mm/min. Samples which at slower rates displayed sliding shear appeared to exhibit turbulent shear behaviour above 100 mm/min. The rates at which the *T-Z* probe was pulled back to set the anchor were expected to result in turbulent shear and lead to soil fabric similar to that resulting from pile installation. However, it would appear from the results of the monotonic and cyclic tests that an orientated structure was created which exhibits little degradation. Furthermore, the shear stress–displacement relationship generally observed of a sharp peak followed by a post-peak rise in shear stress, is considered to be associated with reversal of shearing direction. Skempton (1985) reports similar behaviour in reversal shear box tests when the normal stress is maintained on the specimens. It would appear, therefore, that in this instance the monotonic test results may not be directly applicable to a *T-Z* analysis of axial pile response, due to the reversal of direction of shear.

4.4.7 Laboratory shear strength data

4.4.7.1 Unconsolidated-undrained triaxial tests

Shear strength measured on undisturbed samples is summarised in Fig. 20. Specimens were tested at the size they were sampled. There is no discernible difference in strength between samples tested on site during the investigation and those tested up to a month later. Above 11 m depth the shear strengths appear to be influenced by movements in the groundwater table and desiccation. Below 11 m there is some scatter in the data, however, a strength profile increasing linearly with depth below ground level is suggested. Remoulded shear strengths are summarised in Fig. 21 for all borings. The ratio of undisturbed to remoulded shear strength typically range between 1 and 2 indicating low sensitivity.

4.4.7.2 Consolidated-undrained triaxial tests

a. Tests at estimated in-situ stresses

The results of isotopically consolidated (CIU) compression, anisotropically consolidated (CAU) compression, and K_0 consolidated (CK_0U) compression tests with pore pressure measurement are presented in Fig. 22 in terms of total stress. The K_0 consolidation stage was strain path driven. Typical stress paths followed during the shearing stages of CIU and CAU tests are presented in Figs 23 and 24. The specimens initially developed positive pore pressures.

Test Type	Friction Angle, degrees ϕ δ'	Range	Mean	Plasticity Index, percent	Peak/Residual	Remarks
UNDISTURBED CONSOLIDATED UNDRAINED TRIAXIAL						
CIU Compression	ϕ'	24-31	28-31 25	6-22 22-31	Peak	
CAU Compression	ϕ'	26-34	32 27	10-21 26-36	Peak	
CK_oU Compression	ϕ'	24-35	28-31 24	10-21 24-25	Peak	
CK_oU Extension	ϕ'	28-46	35	10-25	Peak	
CONSOLIDATED CONSTANT VOLUME SIMPLE SHEAR *						
Undisturbed Vertical Specimens	ϕ'	25-31	28-31 25	10-15 16	Peak	
Undisturbed Horizontal Specimens	ϕ'	21-29*	25 21	17-21 12-26	Peak	
Remoulded Specimens	ϕ'	24-29*	29 24	10-13 14-16	Peak	
DIRECT SHEAR BOX soil-soil						
Undisturbed Vertical Specimens	ϕ'	20-38	25-31	18-28	Peak	
	ϕ'	13-26	22-25 13	8-18 28		At 130 mm displacement
Remoulded Specimens	ϕ'	17-31	24-29	13-28	Peak	
	ϕ'	11-24	17 11	15-17 28		At 130 mm displacement
Stainless Steel Interface						
Against undisturbed Specimens	δ'	12-24	19 12	15-19 28	Peak	
	δ'	8-14	14 8	12-19 28		At 130 mm displacement
Against remoulded Specimens	δ'	13-20	19 13	8-21 28	Peak	
	δ'	7-16	15 8	8-21 28		At 130 mm displacement
RQT 501 Interface						
Against undisturbed Specimens	δ'	22-28	25	12-28	Peak	
	δ'	15-25	19 15	8-17 28		At 130 mm displacement
Against Remoulded Specimens	δ'	19-26	24 19	8-24 28	Peak	
	δ'	14-20	18 14	8-19 28		At 130 mm displacement
RING SHEAR soil-soil						
Remoulded Specimens	ϕ'	13-30	24-30 13	8-18 28	Residual	Shear between plattens and soil
RQT 501 Interface						
Against Remoulded Specimens	δ'	9-18	14-17 11	12-24 28	Residual	

Fig. 27. Summary of friction angles measured in laboratory tests

With increasing strain a different type of behaviour occurs with the pore pressures generally decreasing and at large strains negative pore pressures were occasionally measured in the shallow specimens. This type of behaviour is not normally associated with essentially normally consolidated clays and may be due to sample disturbance or clayey silt type behaviour (Wang and Vivatrat, 1982).

Values of Skempton's pore pressure coefficient 'A' at failure, defined at the maximum deviator stress, increased with depth from around -0.2 at 10 m to +0.5 at 60 m. These values are somewhat less than would be anticipated using the OCR v A correlation of Bishop and Henkel (1962) which suggest values increasing from 0.4 to 0.8 with depth.

b. Normalised shear strength

A series of isotopically consolidated compression tests was carried out on

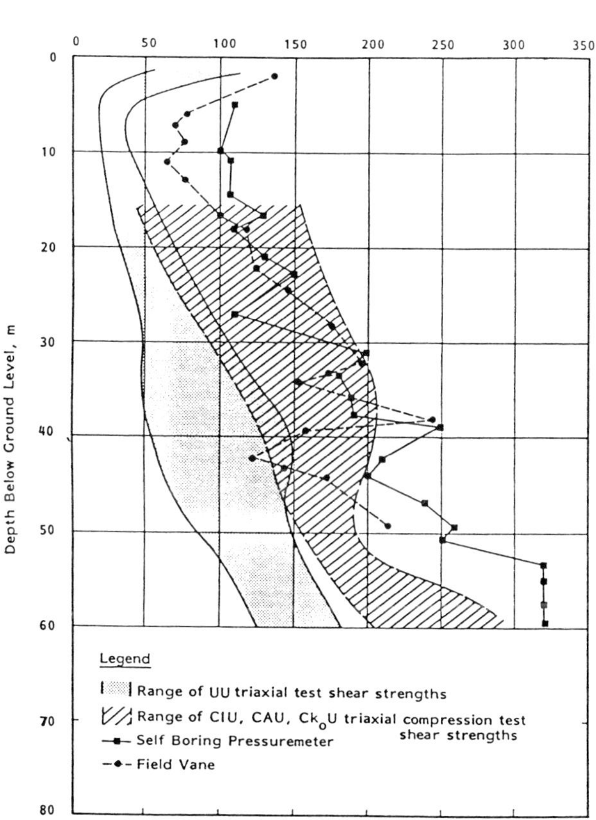

Fig. 28. Comparison of laboratory and in-situ shear strengths

undisturbed samples following the SHANSEP approach (Ladd and Foott, 1974) whereby the specimens are consolidated to stresses greater than the previous maximum stress, generally 2.5 times the estimated in-situ vertical effective stress. This procedure is intended to result in a normally consolidated condition in which previous stress history is over-ridden and the effects of sampling disturbance are minimised.

Stress paths followed during the shearing stages of the tests are presented in Fig. 25. With the exception of the test at 15.6 m the stress paths are generally as would be expected for essentially normally consolidated clays.

The 'A' coefficient at failure ranged between 0.6 and 0.8 and compares reasonably with values obtained from the Bishop and Henkel (1962) relationship for normally consolidated clays.

4.4.7.3 Consolidated constant volume simple shear tests

Simple shear tests were carried out on specimens with horizontal and

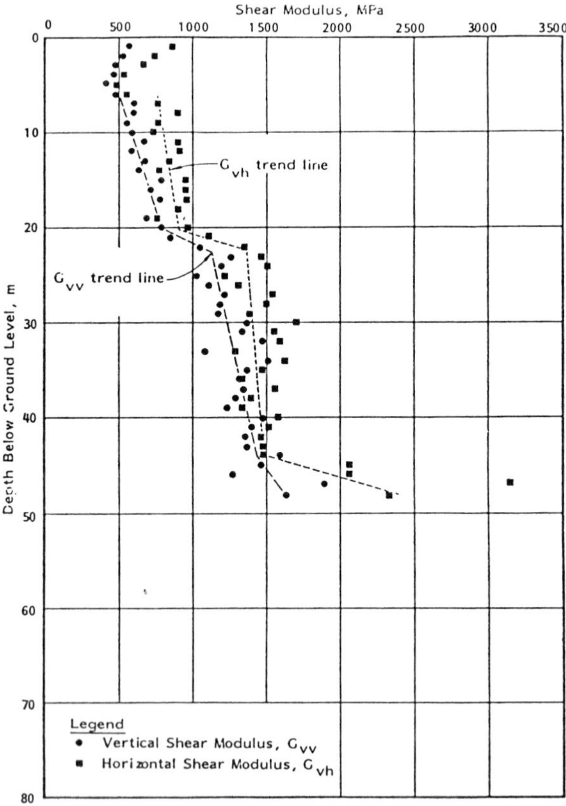

Fig. 29. In-situ shear modulus, cross-hole seismic survey, borings 103–104, Pentre

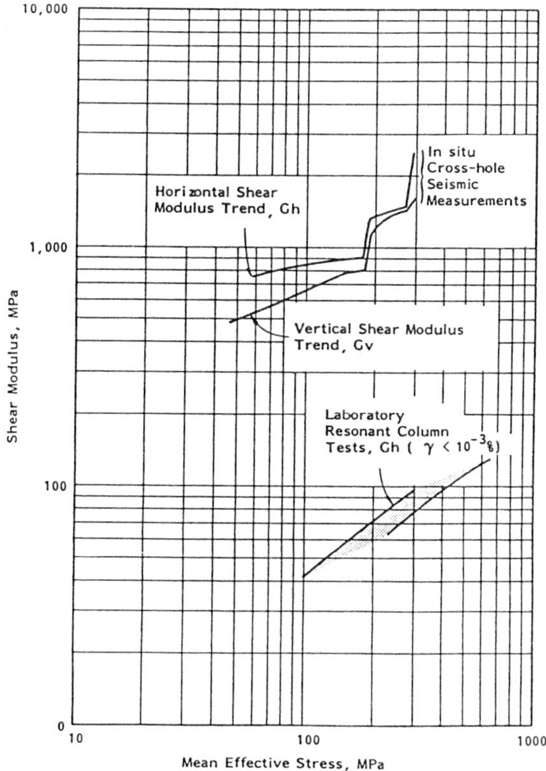

Fig. 30. Comparison of laboratory and in-situ measurements of shear modulus, Pentre

vertical orientations, consolidated to estimated in-situ vertical and horizontal effective stresses respectively and on remoulded specimens consolidated to the estimated horizontal effective stress. Undrained shear strengths obtained from tests on horizontal specimens are similar to the unconsolidated-undrained triaxial test results.

4.4.7.4 Consolidated-drained direct shear tests

Shear box tests were performed on vertically orientated specimens generally at the estimated in-situ horizontal effective stress. Undisturbed and remoulded specimens were tested to determine soil to soil shear strength and soil to steel interface shear strength. Stainless and RQT 501 steels were used in the interface tests to model the soil to steel interface for the *T-Z* probe and test pile respectively. The tests were taken to displacements of up to 130 mm to investigate possible degradation of shear strength. Typical test results are shown in Fig. 26. The residual strengths are typically 70 per cent of peak shear strength. Interface tests performed with the lower plasticity very silty clays sheared at the soil/steel interface whereas the higher plasticity clays ap-

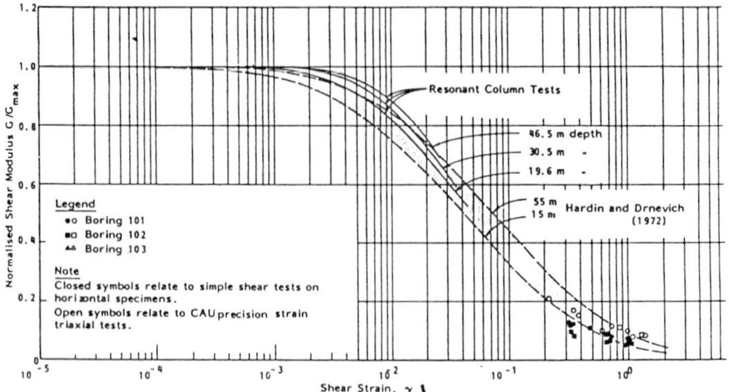

Fig. 31. Normalised shear modulus versus shear strain, Pentre

peared to adhere to the interface with shear taking place through the soil, i.e., the soil failure plane occurred across the top of the roughened platen.

4.4.7.5 Consolidated-drained ring shear tests

Ring shear tests were performed on remoulded specimens to determine the residual soil to soil and soil to RQT 501 steel interface shear strengths. Prior to commencing the tests the specimens were subjected to five complete revolutions to establish a shear plane, resulting in a displacement of about 270 mm. The residual shear strength was then measured after a further displacement of about 210 mm.

A relatively large range of strengths was measured in the soil to soil tests which is considered due to ploughing shear between the platens and the soil.

4.4.7.6 Effective stress shear strength parameters

The effective stress shear strength parameters assessed from the various laboratory tests carried out are summarised in Fig. 27. There is trend of reducing friction angle with increasing plasticity index which is in general agreement with the trends proposed by Kenney (1959) and Navfac DM-7 (1971). In all tests no significant cohesion intercept, c', was measured as is consistent with a normally consolidated cohesive deposit.

4.4.7.7 Discussion of field and laboratory shear strength data

In-situ and laboratory shear strength measurements are compared in Fig. 28. The unconsolidated-undrained shear strengths are generally much lower than the shear strengths obtained from in situ self boring pressuremeters, field vane, and laboratory consolidated-undrained tests, which are in general agreement to 40 m depth.

Shear strengths obtained from the self boring pressuremeter are known to have been influenced by partial dissipation of excess pore pressures during the test. The field vane tests appear to have been similarly affected.

The shear strengths obtained from the consolidated-undrained tests are greater than the unconsolidated-undrained tests. A possible explanation is that lower excess pore pressures were generated in the consolidated-undrained tests which may be the result of an increase in sample density arising from the susceptibility of the very silty specimens to disturbance during sampling and sample preparation.

4.4.8 Shear modulus

a. In-situ measurements

In-situ measurements of shear modulus were made using crosshole seismic techniques and the self boring pressuremeter. Shear moduli measured

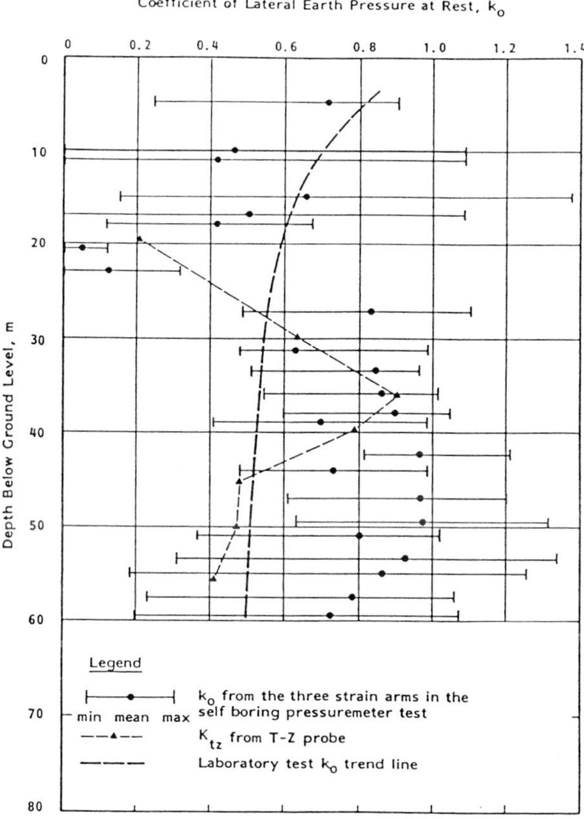

Fig. 32. Coefficient of lateral earth pressure at rest; comparison of in-situ and laboratory measurements, Pentre

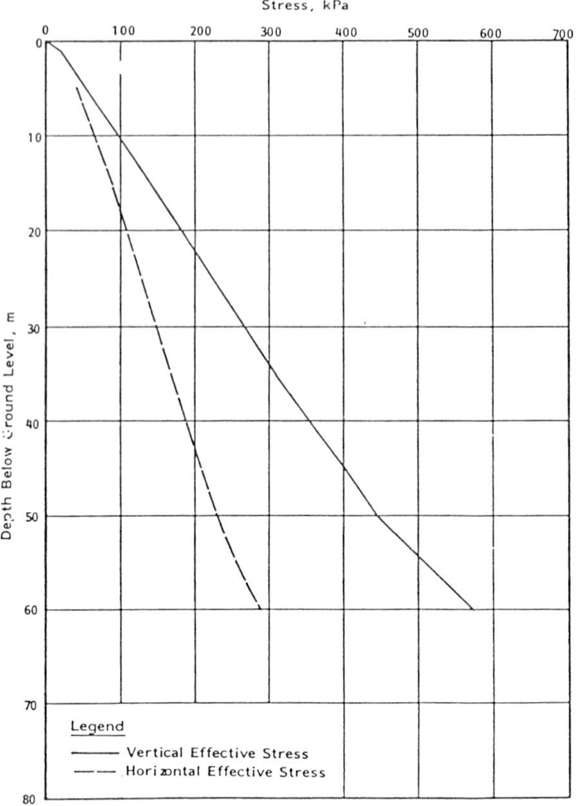

Fig. 33. Estimated in-situ vertical and horizontal effective stresses, Pentre

using the surface to downhole seismic technique were considered unreliable due to waves travelling down a grouted section.

Profiles of shear modulus derived from the cross-hole survey are presented in Fig. 29. The shear moduli were obtained from vertically polarised shear waves. Values of shear modulus obtained from the vertical set of shear waves are denoted G_{vv} and from the horizontal set are denoted G_{vh}. Due to the rotation of the shear wave the horizontal component should only be regarded as an estimate, albeit a reasonably good one.

The horizontal shear modulus is greater than the vertical shear modulus which is an indication of anisotropy. The ratio of horizontal to vertical shear modulus decreases from 1.5 near surface to about 1.1 at 45 m, where there is a marked increase to about 1.6 to 48 m, the lowest test depth.

Shear moduli measured in the self boring pressuremeter tests increase with depth from about 20 MPa near surface to 65 MPa at 60 m depth. These values are much lower than those measured in the seismic work as the mean shear strain level is much higher.

Both the seismic and the self boring pressuremeter moduli show similar trends at 45 m depth. However, the sharp increase in shear modulus evident between 22 and 23 m in the seismic results is not reflected in the self boring pressuremeter data.

b. Laboratory measurements

The variation of shear modulus with mean effective confining pressure measured in resonant column tests at small strain (less than 10^{-3} per cent) is presented in Fig. 30. The maximum shear modulus, G_{max}, was measured about one log cycle of time after the end of primary consolidation and varies between 42 and 130 MPa for confining stresses between 100 and 650 kPa. Shear modulus decreases with increasing shear strain. No single laboratory test is capable of measuring shear modulus over the complete range of shear strains of practical interest. The range of interest can, however, be investigated by combining the results of resonant column, triaxial precision strain and simple shear tests. Precision strain testing used sample mounted instrumentation as described by Jardine *et al.* (1984).

Figure 31 presents a plot of normalized shear modulus versus average shear strain. On this illustration, shear moduli from resonant column tests have been normalised with respect to G_{max} measured for the individual test. For the precision strain and simple shear tests, the shear moduli have been normalized with respect to G_{max} obtained from the shear modulus versus confining pressure data shown in Fig. 30. In obtaining G_{max}, the difference in the state of stress in the three types of tests is accounted for by using the standard method relating mean consolidation stress to vertical and horizontal stresses. The measured values compare favourably with the curves computed using the expressions given by Hardin and Drnevich (1972).

c. Comparison of in-situ and laboratory measurements

In-situ and laboratory measurements of shear modulus are compared in Fig. 30. Although no precise value is available it is believed that the in-situ seismic shear moduli were measured at a shear strain of about 10^{-4} per cent (Baligh and Levadoux, 1980) and therefore correspond to the shear strain level at which the maximum shear moduli were measured in the resonant column test, however the in-situ measurements are about an order of magnitude larger. This is probably largely due to sample disturbance and the different stress conditions in the resonant column test. The seismic shear moduli therefore give in-situ maximum values, and the laboratory tests give an indication of the variation with shear strain level.

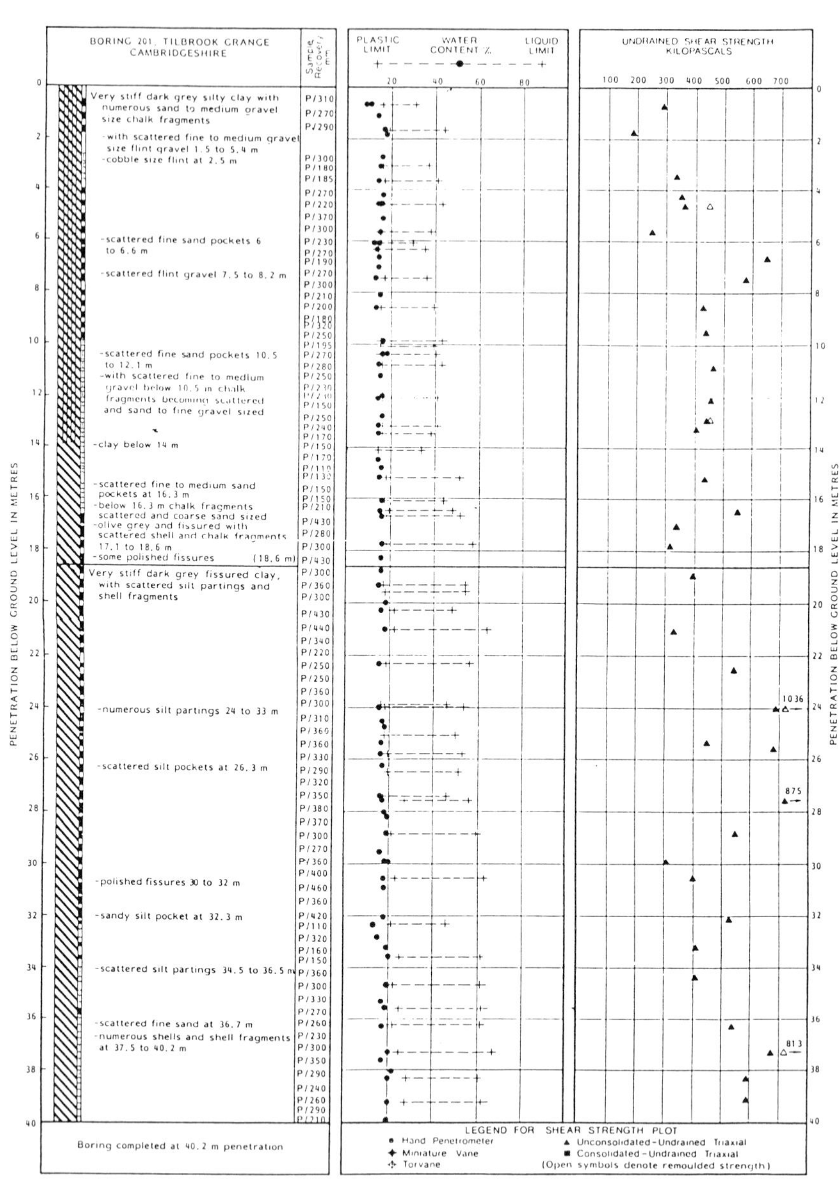

Fig. 34. Log of boring and test results, boring 201, Tilbrook Grange

4.4.9 Coefficient of lateral earth pressure at rest

The coefficient of lateral earth pressure at rest, K_0, was determined from laboratory and in-situ tests and values are summarised in Fig. 32. K_0 has also been derived from oedometer tests using the expression after Brooker and Ireland (1965) and was measured in the consolidation stages of K_0 consolidated-undrained triaxial tests.

Values of K_0 have also been determined from the lift off pressures in self boring pressuremeter tests. The range of values is large and arises from the different lift-off pressures measured in each of the three strain arms. The change in the trend of K_0 values at 25 m is coincident with a change of pressure cell and is not believed to be due to any geological feature. The mean K_0 values above 25 m of 0.4 to 0.7 imply overconsolidation ratios in the range 1 to 2.5, whereas below 25 m the mean K_0 values are generally between 0.7 to 1.0 which indicate overconsolidation ratios between 2.5 and 5.5 which are not consistent with the understanding of the site geological history or micro-fabric.

The ratio of the horizontal to vertical effective stress, K_{tz}, measured following insertion and set up of the *T-Z* probe are shown for comparison. As the tool is pushed into and displaces the soil it would be expected to yield K_{tz} values equal to or slightly greater than K_0. Different tools were used above and below 40 m depth. Above 40 m the K_{tz} values are similar to the self boring pressuremeter, whereas below they are similar to the laboratory derived values.

On balance it is believed that the trend of K_0 versus depth determined from the laboratory tests presents a best estimate that is consistent with the geological evidence and with essentially normally consolidated soils. This trend is generally within the range of values determined in the self boring pressuremeter tests.

4.4.10 In-situ effective stresses

Assessed profiles of vertical and horizontal effective stress are presented in Fig. 33. The vertical effective stress profile is based on the design unit weight profile shown in Fig. 7 and the groundwater pressure profile presented in Fig. 5. The horizontal effective stresses are based upon the K_0 trend profile presented in Fig. 32.

4.4.11 Organic content and carbonate content

The organic content varies typically between 0 and 1 per cent. The carbonate content is generally between 5 and 10 per cent.

5. Tilbrook Grange ground conditions

5.1 Geology

The Tilbrook Grange test site is located on an eroded lodgement till plateau that overlies a low bedrock escarpment. Regionally, the bedrock comprises Jurassic rocks with small outcrops of Cretaceous rocks. These bedrock strata dip gently to the east and south east.

The upper bedrock deposit is the Oxford Clay, which was deposited in a shallow shelf sea with restricted conditions giving rise to reducing conditions at the seafloor (Hallam, 1975). Oxford Clay is very hard fissile clay with limestone bands and concretionary nodules (Hains and Horton, 1969). In the Bedford and Huntingdon area, the Oxford Clay has been overlain by over-

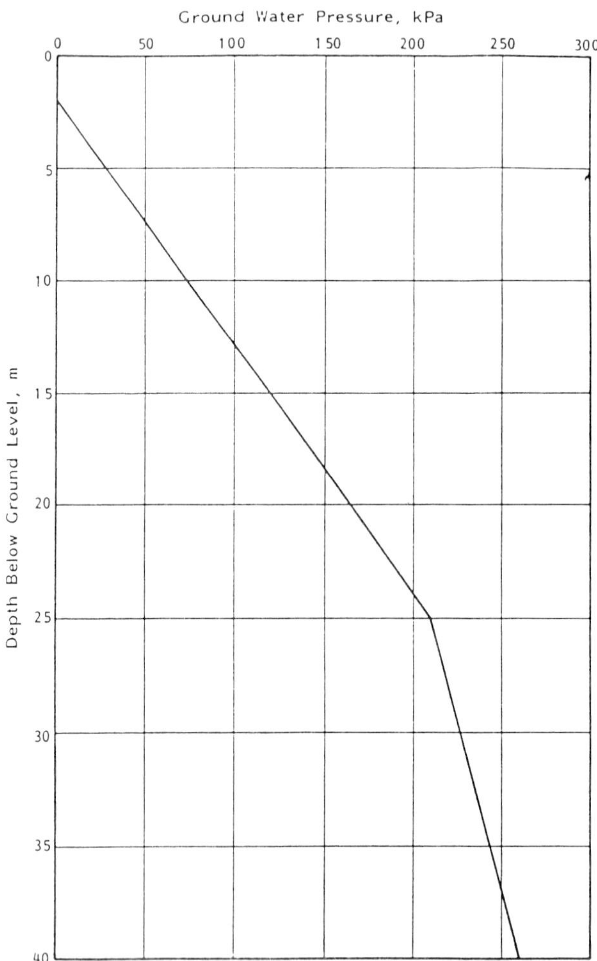

Fig. 35. Groundwater pressure profile, Tilbrook Grange

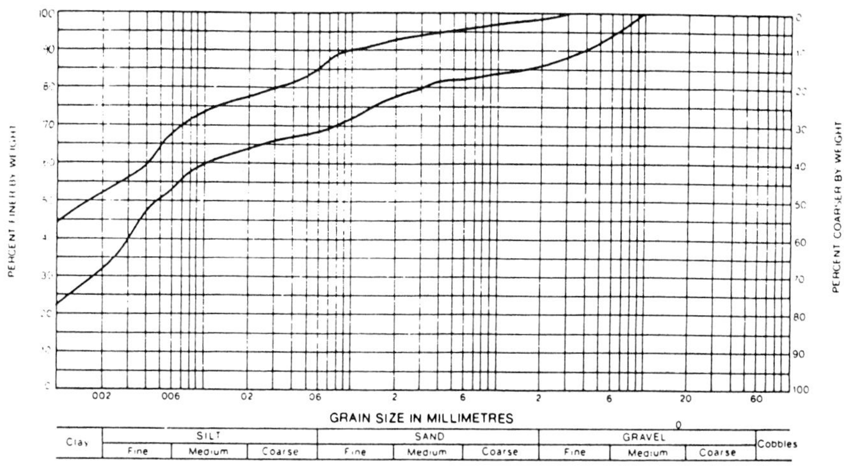

Stratum I

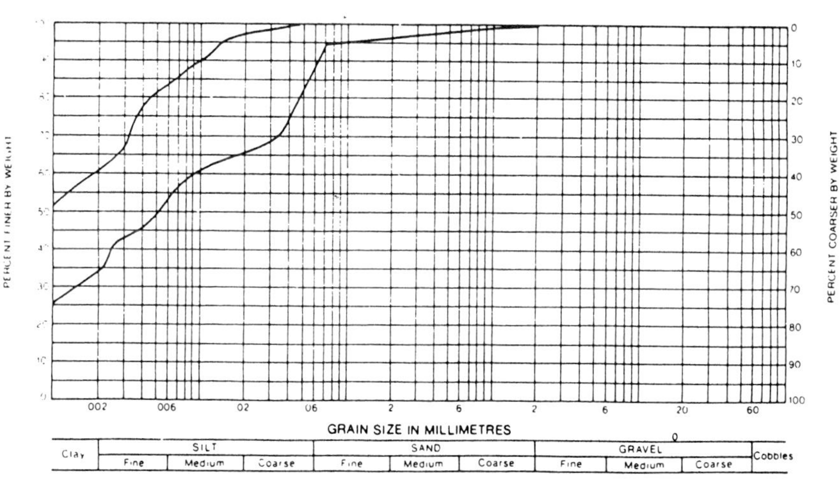

Stratum II

Fig. 36. Summary of grain size curves, Tilbrook Grange

burden to a thickness of about 260 m to 460 m, and consequently the deposit is now heavily overconsolidated. Measured shear strengths show strong anisotropy (Burland, Longworth and Moore, 1977).

The bedrock is overlain by Lowestoft Till that was deposited in the Anglian glacial stage. This till has been subjected to two glacial and two interglacial stages since deposition. Lowestoft Till is a dark bluish grey, silty clay with chalk and flint clasts gravel erratics (Funnell and Wilkes, 1976).

5.2 Strata encountered

Figure 34 is a detailed log from Boring 201. The stratigraphy disclosed by the three sample borings is very uniform, as given below:

Stratum	**Depth - GL (m)** **From**	**To**	**Description**
Lowestoft Till	0	17.1-18.6	Hard dark grey silty CLAY (Lowestoft Till)
Oxford Clay	17.1-18.6	40.2+	Hard dark grey fissured CLAY (Oxford Clay)

5.2.1 Microfabric and mineralogy

5.2.1.1 Lowestoft Till

The clay mineralogy is relatively uniform, comprising illite (40%), mixed-layer species (37%) and kaolinite (23%). Chalk clasts in the fine sand to fine gravel size range give the relatively high carbonate content of between about 30 to 50%.

The Lowestoft Till has a variable degree of anisotropy both in orientation of platy clay minerals and in the distribution of silt sized quartz and other mineral grains. Anisotropy increases with depth, and can be high in close proximity to coarser mineral grains in the matrix. The following table shows the degree of anisotropy determined from microfabric examination.

Depth - GL (m) **From**	**To**	**Degree of anisotropy**	**Comments**
0	7	Low	Occasional micro-features
7	12	Moderate to high	Crenulate
12+		High	Platy clay and micas

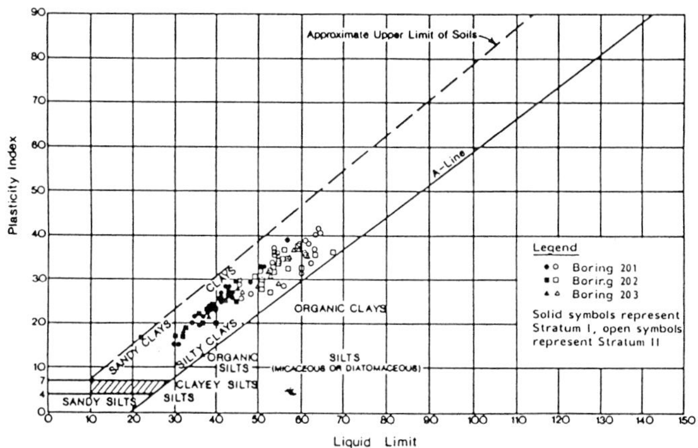

Fig. 37. Plasticity chart for classification of fine-grained soils

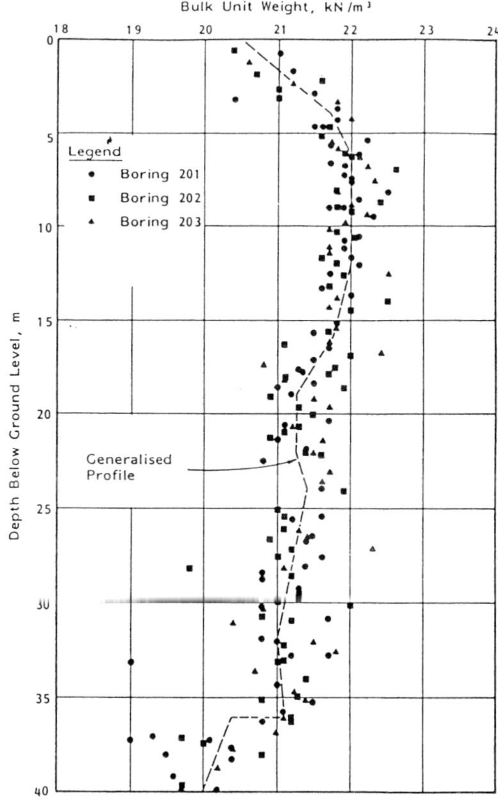

Fig. 38. Bulk unit weight: borings 201,202 and 203, Tilbrook Grange

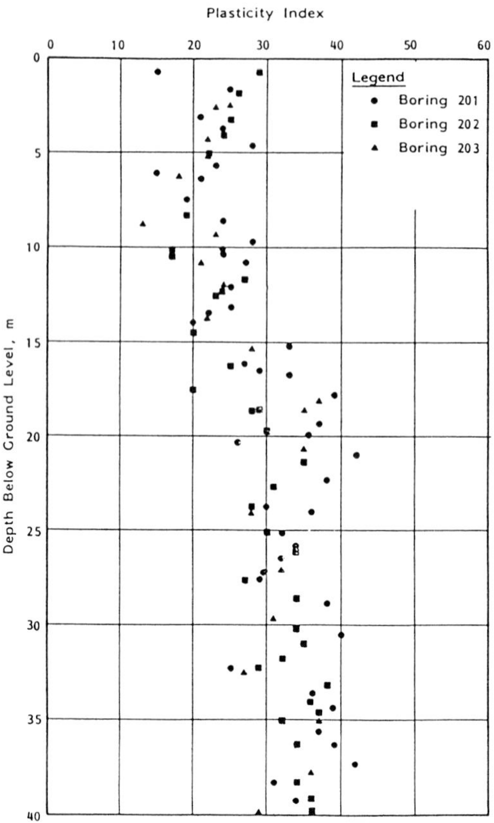

Fig. 39. Plasticity index: borings 201, 202 and 203, Tilbrook Grange

5.2.1.2 Oxford Clay

Mineralogy of the Oxford Clay is relatively uniform comprising: mixed-layer illite-smectite species (45%), illite (36%) and kaolinite (18%). Shell material is present in variable quantities. In general, the former aragonitic fossils have all been replaced by pyrite non-organic fossil particles and small quantities of gypsum are present. Organic matter occurs throughout, tending to increase with depth.

Anisotropy of the Oxford Clay microfabric tends to become more pronounced with increasing depth, as shown in the following table:

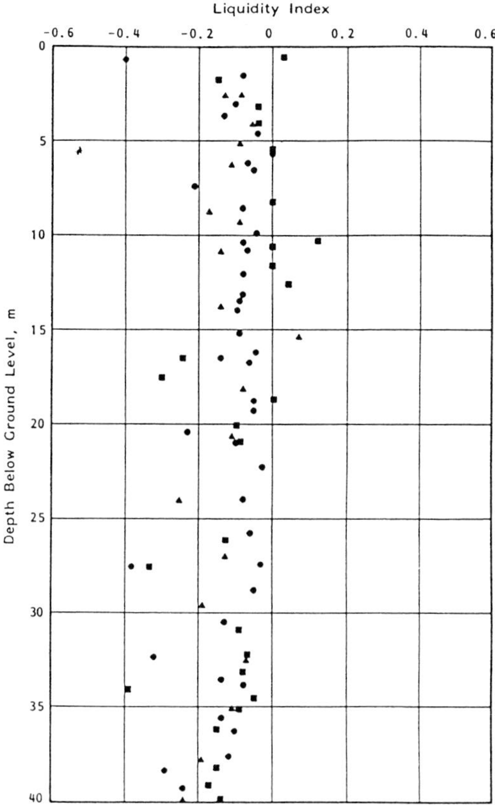

Fig. 40. Liquidity index: borings 201, 202 and 203, Tilbrook Grange

Depth - GL (m) From	To	Degree of anisotropy	Comments
17	30	Moderate	Platy clay, sub-horizontal (some subordinate steep plates)
30	37	High	Platy clay and micas (sub-horizontal)

5.3 Groundwater conditions

Groundwater pressures were monitored from October 1986 until testing in February 1988. Two standpipe piezometers were installed at 12.1 m and 35.8 m below ground level. Equalisation took around five months in the

Lowestoft Till (-12.1 m) and some 12 months in the Oxford Clay (-35.8 m). Fig. 35 presents the groundwater pressure profile. The reduction from the hydrostatic below 25 m -GL reflects probable underdraining of the Oxford Clay associated with water extraction from the deeper Cornbrash strata. Measurements indicated little seasonal variation.

5.4 Summary and interpretation of soil data

5.4.1 Classification tests

Figure 36 shows the range of particle size distributions for the Lowestoft Till and Oxford Clay. The relatively large proportion of sand and gravel erratics in the till is consistent with that observed at many North Sea sites. The plasticity chart presented in Fig. 37 shows that data from both strata are generally parallel to the A-line, indicating the soils are of the same geographical origin within each stratum.

5.4.2 Index properties

Profiles of bulk unit weight, plasticity index and liquidity index are shown in Figs 38, 39 and 40, respectively. For each index property, similar trends were observed for each borehole. The Lowestoft Till is low to moderately plastic, and the Oxford Clay is of moderate to high plasticity. Liquidity index indicates that both strata are heavily overconsolidated, as expected from geology.

5.4.3 Carbonate and organic contents

The relatively high carbonate content from the Lowestoft Till is due to the presence of weathered chalk erratics. In the Oxford Clay, carbonate content reduces with depth in the range 35 to 15%, which is greater than typical North Sea clay values.

5.4.4 Coefficient of consolidation

5.4.4.1 Laboratory tests

Horizontal coefficient of consolidation, C_h, at the existing in-situ lateral effective stress was estimated from oedometer tests in horizontally orientated samples. Measured values were:

Stratum	**C_h, m²/year**
Lowestoft Till	12 to 20
Oxford Clay	1 to 5

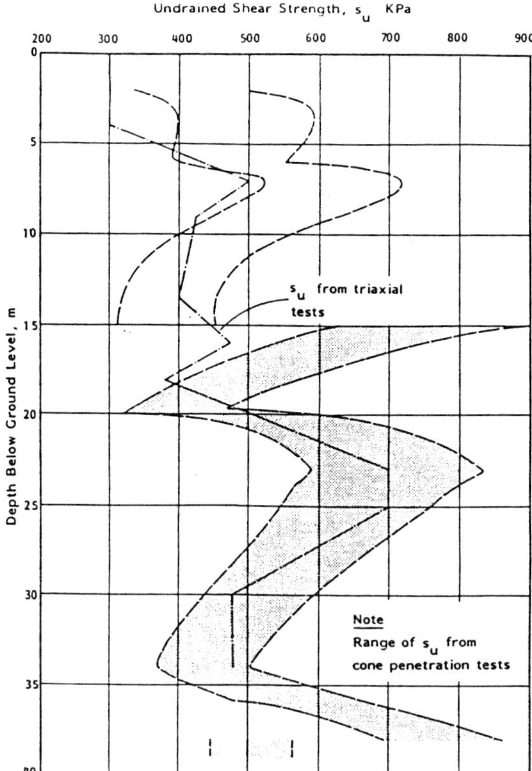

Fig. 41. Undrained shear strength from cone test results, Tilbrook Grange

5.4.4.2 Probe dissipations tests

The analytical method used to estimate coefficient of consolidation from probe tests has been described in the Pentre Site section of this paper. At Tilbrook Grange the pore pressures measured by the *T-Z* probe (see section 4.4.6.4) were considered unreliable, and therefore have not been used. In the piezocone tests small negative pore pressures were generated in the near surface soils, and positive pore pressures up to four times undrained shear strength (consistent with cavity expansion in an elastic material) were measured at depths greater than about 30 m.

For interpretation purposes, a theoretical excess pore pressure distribution decreasing linearly to zero across a cylindrical cavity with a radius 10 times that of the probe shaft was assumed. Interpreted coefficients of consolidation are in the range 2 to 100 m^2/year, showing the highly variable excess pore pressure dissipation response.

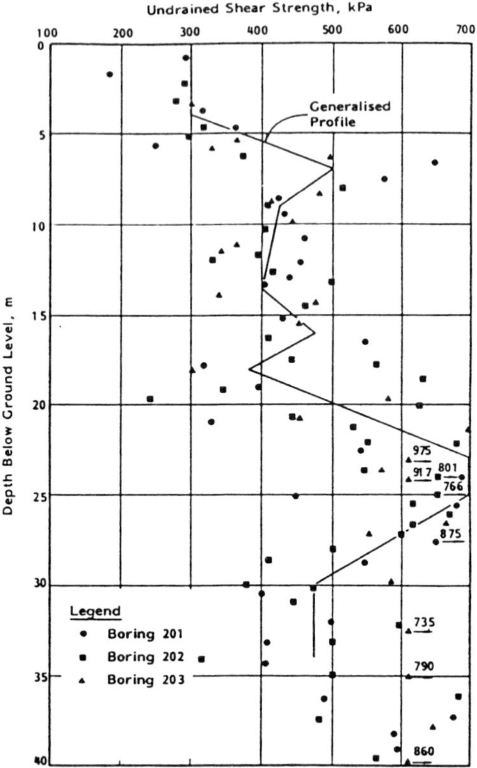

Fig. 42. Undrained shear strength, unconsolidated-undrained triaxial tests, Tilbrook Grange

5.4.4.3 Self boring pressuremeter holding tests

Three holding tests were performed following the expansion phase. At an average cavity strain of about 10 per cent, the strain was held constant and the lateral pressure decay monitored. The coefficients of consolidation are 4 m^2/year in the Lowestoft Till and 5.5 to 19.7 m^2/year in the Oxford Clay. These values compare reasonably well with the oedometer test results.

5.4.4.4 Laboratory permeability tests

Vertical permeability was measured in isotropically consolidated triaxial specimens. Permeability ranges from 6×10^{-12} to 3×10^{-11} m/s, with the Oxford Clay being less permeable than the Lowestoft Till.

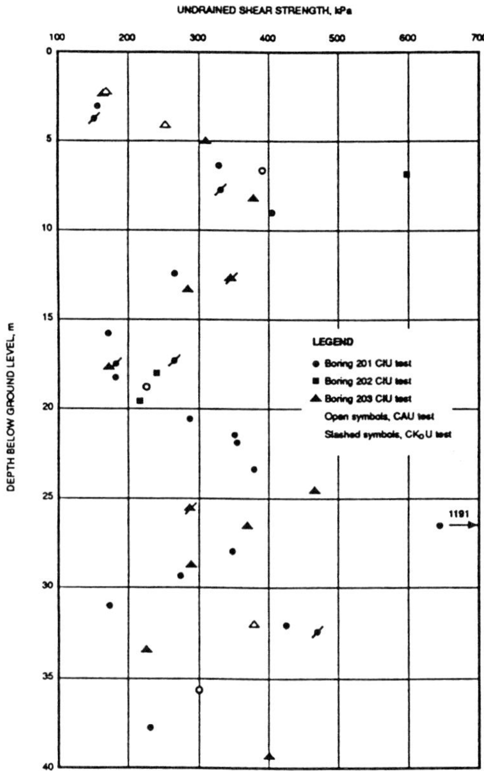

Fig. 43. Undrained shear strength, consolidated-undrained triaxial compression test results, Tilbrook Grange

5.4.4.5 Compressibility

Estimates for the coefficient of volume compressibility, m_v, obtained from one-dimensional consolidation tests on vertical specimens range from about 0.01 to 0.04 m^2/MN, with no apparent trend to the data. The soils are therefore relatively incompressible (Tomlinson, 1980).

5.4.5 Field shear strength

5.4.5.1 Cone penetration tests

Correlations between undrained shear strength (from unconsolidated-undrained triaxial compression tests) and cone point resistance are described in section 4.4.6.1 of this paper. Mean values of cone factors N_k and N_{kt} are shown in the following tabulation:

Stratum	**Depth interval**	N_k	N_{kt}
Lowestoft Till	0 - 5 m	12.2	15.5
Lowestoft Till	5 - 10 m	14.8	15.7
Lowestoft Till	10 - 15 m	13.3	14.0
Lowestoft Till	<15	21.7	23.5
Oxford Clay	<20	21.6	24.6
Oxford Clay	20 -25 m	14.7	14.7
Oxford Clay	25 - 30 m	16.6	14.9
Oxford Clay	30 - 35 m	16.5	16.3
Oxford Clay	35 - 40 m	22.5	21.5

The wide scatter in N_k values is reasonably consistent with the findings of O'Riordan, *et al.* (1982) for clays with plasticity index in the range 15 to 40. Undrained shear strength was determined using the correlations suggested by Schmertmann (1975) and Lunne *et al.* (1985) and results are shown in Fig. 41.

5.4.5.2 T-Z probe: monotonic and cyclic load tests

In general the sleeve friction versus displacement data have different characteristics in the two soil types. For tests in the Lowestoft Till, peak friction was mobilised at displacements of less than 1.5 mm, and after a small drop in skin friction gradually increased with displacement to more than 130 per cent of the initial peak value. In the Oxford Clay the initial peak displacement was about 2 to 4 mm, and after slight post peak reduction, skin friction returned to approximately peak value at large displacements. Adhesion factors, α, were:

	Adhesion factor, α	
Stratum	**Initial peak**	**Large displacement**
Lowestoft Till	0.1 to 0.3	0.2 to 0.4
Oxford Clay	0.5 to 0.85	0.5 to 0.65

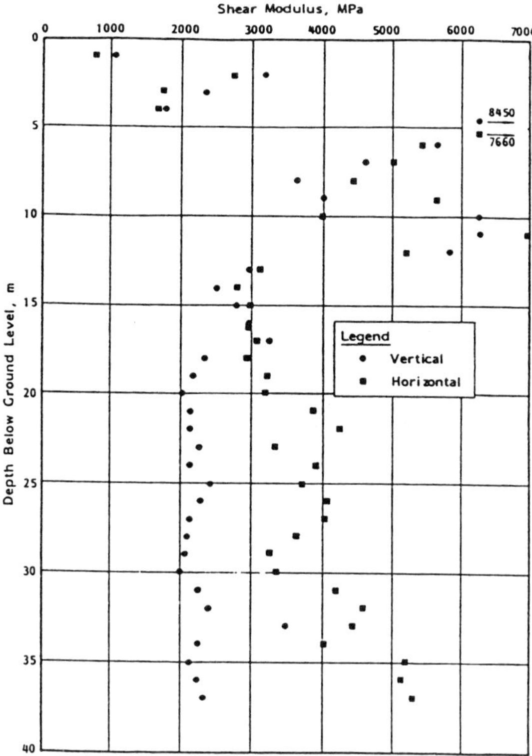

Fig. 44. In-situ shear modulus, cross-hole seismic survey, borings 202–204, Tilbrook Grange

Displacement controlled cyclic load tests were performed after monotonic loading. With five cycles of 60 mm displacement up to about 30 to 40 per cent reduction of skin friction occurred.

5.4.6 Laboratory shear strength

5.4.6.1 Unconsolidated-undrained triaxial compression test

Undrained shear strengths measured in unconsolidated undrained triaxial compression (UU) are shown in Fig. 42 for the three borings at Tilbrook Grange. Full sized samples were tested both on site and soon after in the laboratory, with no apparent strength differences. The generalised profile was developed for axial pile capacity, and is consistent with interpretation of North Sea sites. Similar test samples remoulded at natural water content and density indicate the soils have little to no sensitivity (Skempton and Northey, 1952), as is typical of stiff, overconsolidated North Sea clays.

5.4.6.2 Consolidated-undrained triaxial compression test

A series of triaxial compression tests were undertaken on samples that were consolidated: isotropically (CIU), anisotropically (CAU) and at K_0 (CK_0U). Undrained shear strength tends to be less than that obtained from UU tests, although there is scatter in the data. Values of pore pressure coefficient 'A' at failure (Skempton, 1954) tend to increase with depth in the ranges -0.4 to

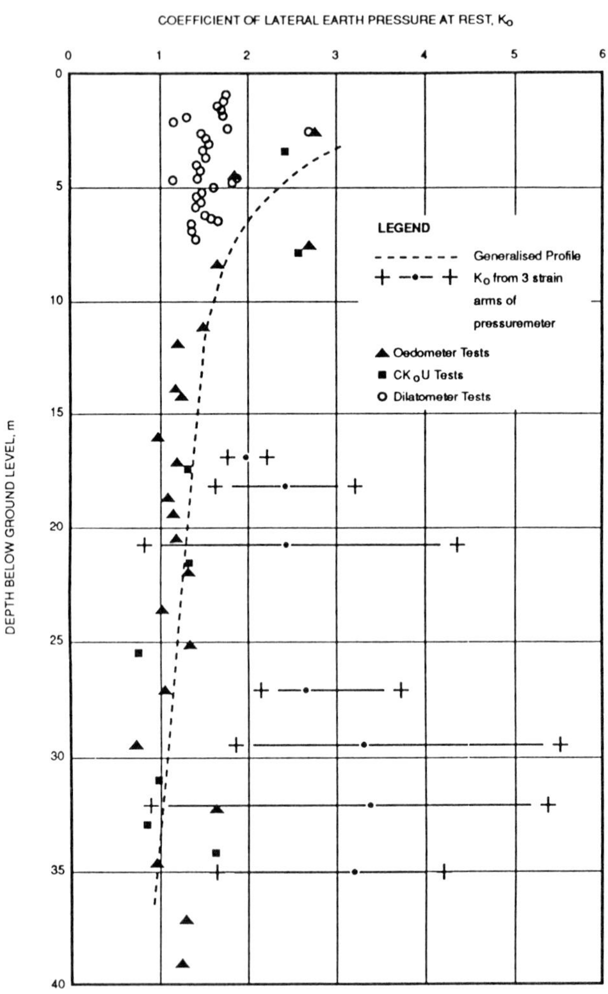

Fig. 45. Coefficient of lateral earth pressure at rest; comparison of in-situ and laboratory measurements, Tilbrook Grange

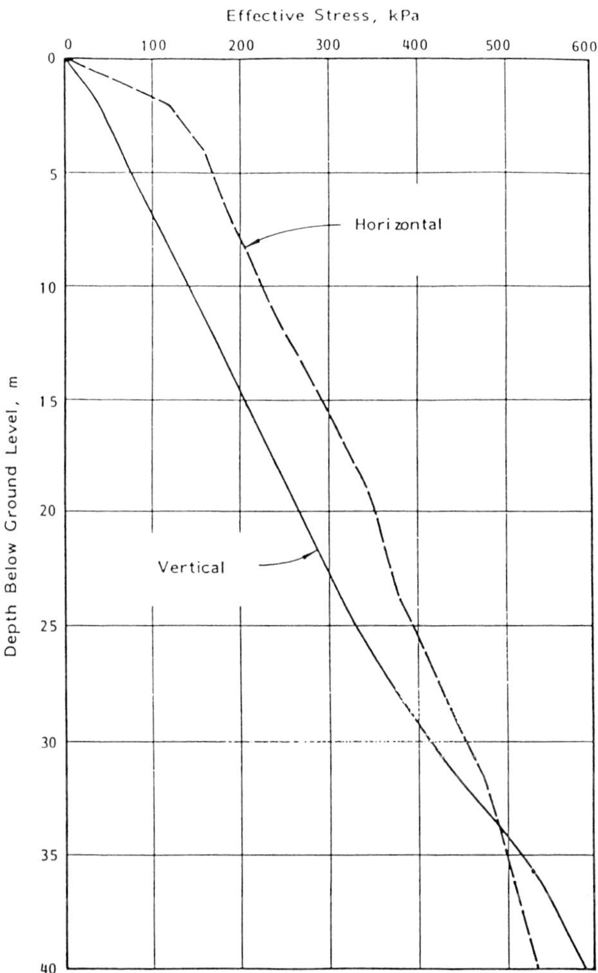

Fig. 46. Estimated in-situ effective stresses, Tilbrook Grange

-0.2 near the surface, -0.2 to 0 at the base of the Lowestoft Till, and from 0.2 to 0.4 over the depth of Oxford Clay investigated.

Results from CIU, CAU and CK_0U tests are given in Fig. 43 in terms of total stress. Effective friction angles are shown in the following table:

Test	**Specimen type**	**ϕ' Lowestoft Till**	**ϕ' Oxford Clay**
CIU	Undisturbed	26	28
CAU	Undisturbed	26	24
CK_0U	Undisturbed	27	30

5.4.6.3 Consolidated constant volume simple shear tests

One specimen from each stratum was tested, with undisturbed samples cut vertically and horizontally, and a sample remoulded at natural water content. Undrained shear strength is about 10 to 20 per cent lower than that from unconsolidated-undrained triaxial compression tests. Effective angle of friction is about 28°.

5.4.6.4 Consolidated-drained direct shear tests

Tests were performed to measure undrained shear strength of undisturbed samples and samples remoulded at natural water content, and to measure the shear stress at the interface of soil and RQT701 steel. Specimens were vertically orientated and consolidated to the estimated horizontal effective stress. Test rates allowed full dissipation of excess pore water pressures, and displacements were taken to about 100 mm to investigate shear stress degradation.

The results are summarised in the following table in terms of effective friction angle ϕ' (soil to soil) and δ' (soil to steel):

Specimen type	**Shear stress**	**Lowestoft Till**		**Oxford Clay**	
		ϕ'	δ'	ϕ'	δ'
Undisturbed	Peak	27.5	26	25.5	25.5
Undisturbed	Residual	23.5	17.5	12.5	11.5
Remoulded	Peak	25	22	22	20.5
Remoulded	Residual	23.5	17	13.5	11.5

Consolidated-drained ring shear tests

These tests were performed on samples remoulded at natural water content to determine residual soil to soil (ϕ'_r) and soil to RQT701 steel δ'_r interface shear strengths. The results are consistent with data published by Lupini *et al.* (1981) and are summarised below:

Specimen type	Lowestoft Till ϕ'	Lowestoft Till δ'	Oxford Clay ϕ'	Oxford Clay δ'
Remoulded	27.5	16	13.5	11

5.4.6.5 Effective stress shear parameters

The Lowestoft Till and Oxford Clay have plasticity indices of about 20 to 30 and 30 to 40, respectively. From published data (Kenny, 1959; Navfac DM-7, 1971), friction angles are expected to be about 29° for the Lowestoft Till and 27.5° for the Oxford Clay.

Consolidated-undrained triaxial compression tests (CIU, CAU, CK_0U) gave similar effective angles of internal friction for each stratum. As would be expected, ϕ' is unlikely to change significantly for plasticity index in the range 20 to 40.

Measured peak and residual friction angles from direct shear and ring shear tests are summarised in the tables shown above. The main conclusion from the tests is that the reduction in friction angle, due either to large displacements or a soil/steel interface, is more pronounced in the Oxford Clay. This is the result of several factors, including:

(*a*) The Lowestoft Till (Stratum I) is of lower plasticity and contains granular inclusions, and
(*b*) The Oxford Clay (Stratum II) is fissured and has a relatively high carbonate content.

5.4.7 Shear modulus

Values of shear modulus derived from a cross-hole seismic survey between Borings 202 and 204 are given in Fig. 44. Where the horizontal shear modulus exceeds the vertical value is an indication of strong anisotropy. This occurs in the Lowestoft Till.

Pressuremeter tests performed in the Oxford Clay show a similar trend in the shear modulus from the seismic survey. However, as would be expected the pressuremeter values, which are measured at relatively large deflections, are only about one-thirtieth the magnitude obtained from the small strain seismic values.

5.4.8 Coefficient of lateral earth pressure at rest (K_0)

K_0 has been estimated from both laboratory and in-situ tests. Index properties can be related to K_0 using published correlations by Brooker and Ireland (1965) and Alpan (1967), as described for the Pentre Site. K_0 was also determined from the consolidation stage of triaxial compression tests, and from oedometer tests.

Figure 45 shows values of K_0 derived by the various methods, and indicates how consistent values from laboratory data are. Near ground level K_0 approaches the value of the passive earth pressure coefficient, K_0. As would be expected, K_0 reduces rapidly from near the passive value at the ground surface to about 1.5 at 10 m depth, and thereafter reduces to near unity at 40 m below ground level.

Values of K_0 were also determined from the lift-off pressures in self boring pressuremeter tests. The range of values determined is relatively large due to different lift-off pressures measured by the three strain arms. Mean values of K_0 from the pressuremeter approach the passive value throughout the Oxford Clay stratum, clearly overestimating K_0.

The K_0 values derived from interpretation of the dilatometer tests are also shown in Fig. 45. Tests were undertaken in the upper 7 m of the Oxford Clay, with interpretation following the method of Powell and Uglow (1988). These in-situ K_0 estimates are slightly lower bound to those measured in the laboratory.

5.4.9 In-situ effective stresses

Figure 46 shows the estimated vertical and horizontal effective stress profiles for Tilbrook Grange. These profiles were derived using generalised groundwater pressure, bulk density and K_0 profiles.

6. Conclusions

Suitable normally consolidated and over consolidated clay soil sites were selected for the pile load test programme following a literature review of UK Quaternary deposits and outline investigations at twelve locations. The Pentre site was chosen as having primary plasticity, stress history and strength characteristics similar to those under the North Sea. Further investigation revealed the following characteristics:

- Essentially normally consolidated soft to stiff silty clay of low to intermediate plasticity as required.
- Lower clay fraction/higher silt content than expected for the plasticity. The test stratum is generally a very silty clay becoming a clayey silt or silty clay in places.
- Consolidation and permeability characteristics are generally similar

to clayey silt. Stress paths during shearing at insitu stresses reflect clayey silt type behaviour or sample disturbance.

The Tilbrook Grange Site was selected as having OC clay deposits similar to those under the North Sea. Investigations here showed the soil to be:

- Heavily over-consolidated stiff to hard clay of intermediate plasticity as required.
- The Lowestoft Till is typical of many North Sea clay deposits. The carbonate content of the Oxford Clay, although higher than typical North Sea clays, is derived from discrete inclusions and therefore not believed to influence the engineering properties.
- Laboratory testing revealed that the soil properties and effective stress behaviour were typical of many North Sea sites.

Acknowledgements

The authors are particularly grateful to the participants in the Large Diameter Pile Test Programme for permission to publish this paper. Other organisations that undertook work for the site investigation were: BRE (dilatometer testing), Delft Geotechnics Limited (surface CPT), Geotechnical Engineering Research Centre at City University (ring shear tests), Soil Mechanics Limited (pressuremeter testing) and Wimpey Laboratories (cross-hole seismic survey). Geological testing was undertaken by Professor E. Derbyshire (University of Leicester), and Dr A. Parker and Dr B. W. Sellwood (University of Reading). A project of this size and quality cannot be successfully completed without the diligent efforts of many people. Recognising this, the authors appreciate the assistance of their former colleagues at McClelland Limited, who undertook the majority of the site investigation and laboratory test programme. Finally, gratitude is also due to BP Engineering for their active support.

References

ALPAN, I. (1967), The Empirical Evaluation of Coefficients K_o and K_{or}. *Proc. Conf. on Soil Mech. and Fndn Engng*, Tokyo, Vol. 7, No. 1, pp. 31-40.

BALIGH, M. M. and LEVADOUX, J. N. (1980), *Pore Pressure Dissipation After Cone Penetration*. Massachusetts Institute of Technology, Report No MITSG 80-13.

BAYNE, J. M. and TJELTA, T. I. (1987), Advanced Cone Penetrometer Development for In-situ Testing at Gullfaks C. *Proc. 19th Offshore Technology Conf.*, Houston, Paper No 5420.

BISHOP, A. W. and HENKEL, D. J. (1962), *The Measurement of Soil Properties in the Triaxial Test*. Edward Arnold, 2nd Edition.

BROOKER, E. W. and IRELAND, H. O. (1965), Earth Pressures at Rest Related to Stress History. *Canad. Geotech. J.*, Vol. 2, No. 1, pp. 1-15.

BRINDLEY, G. W. and BROWN, G. (1980), *Crystal Structures of Clay Minerals and their Identification*. Mineralogical Society London, p. 495.

BS 1377 (1975), *Methods of Test for Soil for Civil Engineering Purposes.* British Standards Institution, London.

BURLAND, J. B., LONGWORTH, J. N. and MOORE, J. F. A. (1977), A Study of Ground Movement and Progressive Failure Caused by a Deep Excavation in Oxford Clay. *Géotechnique*, Vol. 27, No. 4, pp. 557-591.

CAMPANELLA, R. G., GILLESPIE, D. and ROBERTSON, P. K. (1982), Pore Pressures During Cone Penetration. *Proc. of the Second European Symp. on Penetration Testing*, Amsterdam, pp. 507–512.

CASAGRANDE, A. (1936),The Determination of the Preconsolidation Load and its Practical Significance. *Proc. First Int. Conf. on Soil Mech. and Fndn Engng*, Cambridge, Vol. 3, p. 60.

FUNNELL, B. M. and WILKES, P. F. (1976), Engineering Characteristics of East Anglian Quarternary Deposits. *Q. J. Engng Geol.* Vol. 9, pp. 145-147.

GIBSON, R. E. and ANDERSON, W. F. (1961), In-situ Measurement of Soil Properties with the Pressuremeter. *Civ. Engng and Public Works Review*, Vol. 56, No. 658.

HAINS, B. A. and HORTON, A. (1969), *Central England (Third Edition)* , British Regional Geology Series, HMS0, p. 142.

HALLAM, A. (1975), *Jurassic Environments*. Cambridge University Press.

HARDIN, B. O. and DRNEVICH, V. P. (1972), Shear Modulus and Damping in Soils: Design Equations and Curves. *J. Soil Mech and Fndn Div., Am. Soc. Civ. Engrs*, Vol. 98, No. SM6, pp. 603-624.

HEAD, K. H. (1986), *Manual of Soil Laboratory Testing. Volume 3: Effective Stress Tests*, Pentech Press, London.

JARDINE, R. J., SYMES, M. J. and BURLAND, J. B. (1984), The Measurement of Soil Stiffness in the Triaxial Apparatus. *Géotechnique*, Vol. 34, No. 3.

KENNEY, T. (1959), Discussion. *Proc. Am. Soc. Civ. Engrs*, Vol. 85, No. SM3, pp. 67-79.

LACASSE, S., JAMIOLKOWSKI, M., LANCELLOTTA, R. and LUNNE, T. (1981), In-situ Characteristics of Two Norwegian Clays. *Proc. 10th Int. Conf. on Soil Mech and Fndn Engng*, Stockholm, Vol. 2, pp. 507-511.

LACASSE, S. and LUNNE, T. (1982), Penetration Testing in Two Norwegian Clays. *Proc. 2nd European Symp. on Penetration Testing*, Amsterdam, pp. 661-669.

LADD, C. C. and FOOTT, R. (1974), New Design Procedure for Stability of Soft Clays. *J. Geotech. Engng Div., Am. Soc. Civ. Engrs*, Vol. 99, No. GT7, pp. 763-786.

LONG, M. M., LAMBSON, M. D., CLARKE, J. and HAMILTON, J. (1993), Cyclic

Lateral Loading of an Instrumented Pile in Overconsolidated Clay at Tilbrook Grange. This volume

LUNNE, T., CHRISTOFFERSEN, J. P. and TJELTA, J. N. (1985), Engineering Use of Piezocone Data in North Sea Clays. *Proc. 11th Int. Conf. on Soil Mech. and Fndn Engng*, San Francisco.

LUPINI, J. F., SKINNER, A. E. and VAUGHAN, P. R. (1981), The Drained Residual Strength of Cohesive Soils. *Géotechnique,* Vol. 31, No. 2, pp. 181-213.

MARSLAND, A. and QUARTERMANN, R. S. T. (1982), Factors Affecting the Measurements and Interpretation of Quasi Static Penetration Tests in Clay. *Proc. 2nd European Symp. on Penetration Testing*, Amsterdam, Vol. 2, pp. 697-702.

MAYNE, P. W. (1980), Cam-Clay Predictions for Undrained Strength. *J. Geotech. Engng Div., Am. Soc. Civ. Engrs*, Vol. 106, No. GT11, pp. 1219- 1242.

MCNEILAN, T. W. and BUGNO, W. T. (1985), Cone Penetrometer Tests in Offshore California Silts. *Strength Testing of Marine Sediments: Laboratory and In-situ Measurements*, ASTM STP 883, 1985, pp. 55-71.

NAVFAC DM-7 (1971), *Design Manual – Soil Mechanics, Foundations and Earth Structures.* Department of the Navy Facilities Engineering Command, Alexandria, Virginia.

O'RIORDAN, N. J., DAVIES, J. A. and DAUNCEY, P. C. (1982), The Interpretation of Static Cone Penetrometer Tests in Soft Clays of Low Plasticity. *Proc. 2nd European Symp. on Penetration Testing*, Amsterdam, pp. 755-760.

POWELL, J. J. M. and UGLOW, I. M. (1988), The Interpretation of the Marchetti Dilatometer Test in UK Clays. *Penetration Testing in the UK*, Thomas Telford, London, pp. 269-273.

SCHMERTMANN, J. H. (1975), Measurement of In-situ Shear Strength. State-of-the-art Report. *Proc. Am. Soc. Civ. Engrs Speciality Conf. on In-Situ Measurements of Soil Properties*, Raleigh, North Carolina, Vol. 2, pp. 57-138.

SKEMPTON, A. W. (1954), The pore pressure coefficient A and B. *Géotechnique*, Vol. 4, pp. 143-147.

SKEMPTON, A. W. and NORTHEY, R. D. (1952), The sensitivity of clays. *Géotechnique*, Vol. 3, No. 1, pp. 30-53.

SKEMPTON, A. W. (1985), Residual Strength of Clays in Landslides, Folded Strata and the Laboratory. *Géotechnique*, Vol. 35, No. 1, pp. 3-18.

THOMAS, S. (1990), *Geotechnical Investigation of UK Test Sites for the Foundations of Offshore Structures.* Department of Energy Offshore Technology Report, OTH 89 294, HMS0, London.

TOMLINSON, M. J. (1980), *Foundation Design and Construction.* Pitman Publishing Limited, Fourth Edition, p. 793.

TUMAY, M. T. and ACAR, Y. B. (1985), *Piezocone Penetration Testing in Soft Cohesive Soils, Strength Testing of Marine Sediments, Laboratory and In-situ Measurements*, ASTM STP 883, pp 72-82.

VILLET, W. C. B. and DARRAGH, R. D. (1985), The Interpretation of Piezometric

Cone Tests in Highly Overconsolidated Offshore Silts. *Proc. 17th Offshore Technology Conf.*, Houston, OTC Paper 4916.

WEIR, A. H., ORMEROD, E. C. and EL-MANSEY, I. M. I. (1975), Clay Mineralogy of Sediments of the Western Nile Delta. *Clay Minerals*, Vol.10, pp. 369-386.

WANG, J. L. and VIVATRAT, V. (1982), Geotechnical Properties of Alaska OCS Silts. *Proc. 14th Offshore Technology Conf.*, Houston, Paper 4412.

WINDLE, D. and WROTH, C. P. (1977), The Use of the Self Boring Pressuremeter to Determine the Undrained Properties of Clays. *Ground Engineering*, Vol. 10, No. 6.

WROTH, C. P., CARTER, J. P. and RANDOLPH, M. F. (1979), Stress Changes around a Pile Driven into Cohesive Soil. *Proc. Conf. on Recent Developments in the Design and Construction of Piles*. Institution of Civil Engineers, London, pp. 345-354.

9. The development of instrumentation for drivability testing and load testing of large diameter piles

I. J. SOLOMON, Structural Monitoring, Fugro-McClelland Ltd, Glasgow, W. R. COX, Fugro-McClelland Marine Geosciences Inc, Houston, J. CLARKE, BP International Ltd, and T. J. POSKITT, Queen Mary and Westfield College, London

The Large Diameter Pile Tests Programme took place between 1985 and 1988. A pair of comprehensively instrumented piles were driven and subsequently load tested; one at a normally consolidated clay site in Shropshire (the NC site), and the other at an overconsolidated clay site in Cambridgeshire (the OC site). This paper is one of a series of papers on the planning execution and results of the project. The paper discusses the procedures undertaken to obtain a system of instruments that would withstand the rigours of being driven with a pile, and yet be stable and sensitive enough to record the long-term static behaviour of the pile. The paper describes the test techniques developed to simulate the effects of a piling hammer on the instruments being tested. The types of sensors initially chosen for testing are discussed, together with the modifications shown to be necessary as a result of the testing.

Parameters to be measured

The analysis to be performed after driving and load testing the piles required that the following parameters to be measured, using instruments installed on the pile before driving

(a) Stresses in the pile wall during driving and load testing
(b) Axial and lateral acceleration of the pile during driving
(c) Soil pressure on the pile wall during driving and load testing
(d) Pore water pressures at the pile-soil interface during driving and load testing.

It was decided at the commencement of the project that wherever possible, each parameter of interest would be measured using two different types of instrument in the interests of redundancy.

It was also required to measure and record other parameters, for example pore water pressures at various distances from the pile and pile head displacement during load test. However, as the sensors used for these parameters were not required to withstand driving, their selection and testing shall not be considered in this paper.

Difficulties of instrumenting piles

A driven pile is a very harsh environment for instrumentation. Pile-mounted instruments have repeatedly to withstand very high strain and acceleration levels during driving. Wave equation studies based on the use of the Menck MHU 1700 underwater hammer carried out at the commencement of the project predicted that instruments mounted on the OC pile would have to withstand a strain range of ±3500με and accelerations of up to 850*g*. Under the most severe driving conditions, it was estimated that a total of approximately 2000 hammer blows would be required to install the OC pile.

Under these conditions, the fatigue performance of instruments is just as important as the measurement range. Fatigue of instruments and instrument installation mountings can cause offsets to develop in the instrument outputs during driving, which can lead to difficulties in ascertaining zero load levels if the likely range of these offsets is not known.

Test techniques used

The large number of instruments that it was planned to install, and the harsh environment envisaged meant that for each instrument to be used, it was essential to thoroughly test all aspects of installation, reliability and accuracy. This was done in order that there would be a high level of confidence in the ability of each type of instrument to operate throughout the drive and remain operational through the set-up and load-testing phases of the project.

Selection and testing of a system of instrumentation for the project was carried out in conjunction with the Civil Engineering Department at Queen Mary College (now Queen Mary and Westfield College), part of London University. Queen Mary College had been involved in the development of pile-mounted instrumentation since the late 1970s (Refs 1-2), and many of the initial instrumentation concepts utilised in the project came from this experience.

The objectives of the testing carried out at QMC were to select suitable instruments that would be able to survive driving, and to attempt to quantify the levels of zero error introduced in the instrument outputs due to fatigue during driving and due to long-term drift.

The most numerous instruments to be installed on the piles were strain gauges used to monitor the pile wall stresses during driving. The strain gauges were also the instruments with the largest number of variable factors affecting their successful installation, so probably the largest amount of development effort was devoted to investigating a reliable strain gauge system.

Preliminary small scale tests

A series of preliminary small-scale tests were carried out at QMC using a standard Proctor machine to impart shock loads to a test piece instrumented with foil gauges, and two tests were carried out using the Bashit machine (see below) to investigate whether it was possible to achieve the high strains and accelerations required to test the instruments.

Flexure test rig

A simple test rig was constructed in order to perform laboratory tests on the endurance of the strain gauges under simulated driving conditions, and the high sustained strain levels expected during the load tests.

The rig consisted of a strip of 15mm thick RQT701 steel plate 1500mm long by 79mm wide supported at one end, and with the other end as a cantilever. The RQT701 steel used was the same as that used for the OC pile, as this was the most critical in terms of metallurgy. The strip of steel would be instrumented on one or both sides with the strain sensors under test.

To simulate the sustained levels of stress expected while load testing, the free end of the cantilever was fitted with an extension to which weights could be added. The level of strain induced was varied by altering the amount of weight attached to the extension.

To simulate driving conditions, the extension added for the static tests was removed and the free end of the cantilever was loaded with a jack until strain levels of 3500με were attained, then suddenly released via remote control and allowed to oscillate freely. It was estimated that each release was equivalent to approximately three blows under the most severe driving conditions.

Bashit tests

Following these preliminary tests, an extensive programme of tests using the QMC 'Bashit' machine was carried out. The Bashit machine consists of a frame to hold a 1.5m by 0.5m steel plate, a winch, and a 565kg drop hammer. Instruments to be tested are mounted on the plate, and the hammer is repeatedly dropped on the plate from heights of up to 1.75m. The stress-time and acceleration-time signals generated have characteristics similar initially to those induced in a pile under hard driving conditions. The levels of stress and acceleration induced may be controlled by using cushioning material between the test plate and the hammer, adjusting the drop height of the hammer, or changing the thickness of the test plate used.

Fig. 1. The 'Bashit' machine

Tests using the Bashit machine took place from the end of May 1986 to the beginning of June 1987. All instruments selected for use on the piles were tested until full confidence was attained in the ability of each instrument to withstand the expected environment. In order to accelerate the test programme, several instruments of different types would be tested at the same time. Figure 2 shows a typical instrumented plate mounted in the Bashit machine.

Other tests

In addition to the test programmes described above, which were intended to confirm that the instruments would withstand driving, other tests took place which were more concerned with the long term accuracy and stability of the instruments.

(a) Sustained loading tests were carried out to assess likely levels of load related creep.
(b) Sustained zero loading tests were carried out to assess likely levels of long term drift.

1, 5, 8: FOIL GAUGES
2, 3, 4: WELDABLE GAUGES
6, 7: VIBRATING WIRE GAUGES
9: ACCELEROMETER (metal block)
10, 12, 13: ACCELEROMETERS (Tufnol blocks)
11: WELD PAD FOR STRAIN MODULE
14: PORE PRESSURE CELL (and foil gauges)
15, 16, 17: FOIL GAUGES (to check strain module)
18: STRAIN MODULE
19: TOTAL PRESSURE CELL (and foil gauges)

Fig. 2. Instruments mounted on test plate

(c) Incremental load testing tests of instruments and instrument installations were carried out to check sensitivity, linearity and hysteresis.
(d) Hydrostatic tests were carried out to ensure that the instruments and instrument installations would withstand the hydrostatic pressure of the groundwater head.

Dynamic strain sensors

Four different types of strain transducer were installed in the piles: bondable foil resistance strain gauges ('foil gauges') and weldable wire resistance strain gauges ('weldable gauges') were selected to measure stresses during driving (dynamic measurements), while weldable vibrating wire strain gauges and resistance strain modules were used for long-term measurements of load in the piles where stability was important (static measurements). The instruments used for the static measurements are discussed in a later part of this paper.

The long-term accuracy of a bondable strain gauge system is related to the adhesive used to install the gauges. Over long periods creep in the bonding material can produce offset readings. Hot-cured adhesives exhibit little or no creep under load, but are more difficult to apply than cold cured adhesives.

Foil gauges were used as the main dynamic strain measurement sensor, with the weldable gauges installed as the backup sensor. Foil gauges have a higher accuracy than weldable gauges, although more work is required to mount them on the pile wall.

Initial selection procedure

Little information was available on the fatigue behaviour of strain gauges at the high strain levels expected. It was required that offsets in the strain gauge outputs were minimised so as not to lose the zero load readings. Manufacturers were approached with an initial specification of a range of ±3500, and a maximum offset of 50με after 10^5 cycles with the result that the following gauges were selected for testing

(a) HBM bondable foil type 6/350 LY11
(b) Ailtech SG159 350Ω weldable gauge (standard flange)
(c) Ailtech SG159 350Ω weldable gauge (broad flange)

Both the HBM gauges and the Standard Ailtech gauges had successfully been used before on piles by QMC.

A programme of preliminary tests was then carried out on the bondable gauges at QMC using a Proctor machine and the Bashit machine. The objective of these tests was to check survival of a strain gauge installation and to enable a test plate to be designed for the Bashit machine suitable for the desired strain and acceleration levels.

A conclusion of this preliminary test programme was that additional development work was required on the technique for attaching the strain gauges to the leadout wires, as several gauges became detached in this area.

Flexure tests

The next stage in the development of the dynamic strain gauge system was a series of tests on the flexure test rig. Two foil gauges were bonded to the test piece using cyanoacrylate adhesive. The RQT701 test piece was then given six cycles of static loading in steps to 3500με.

At this time, the RQT701 test piece had not been heat-treated after flame cutting. This meant that the test piece was distorted and contained considerable residual stresses. As the test piece was loaded, these stresses were relieved, leading to substantial changes in the outputs from the strain gauges. The effects of flame cutting were confirmed by testing a small sample of the test piece in a Rockwell hardness testing machine, which showed significant variations in surface hardness.

As a result of the experience gained with the first test, the test piece was stress relieved by heat-treating it. The test piece was then reinstrumented, again with two foil gauges, but this time one was bonded using a cold-cure epoxy adhesive. A narrow-flange weldable gauge was also installed. The test cycle performed previously was then repeated. This time the offsets caused by residual stresses were less than 15% of those observed previously.

The test piece was then subjected to a dynamic endurance test consisting of 160 sudden releases from 3500με, equivalent to 480 hammer blows under the most severe driving conditions. The instrumentation installed remained the same as for the post heat treatment static flexure test. All gauges performed satisfactorily.

The conclusion of the initial test programme was that it was probable that both foil gauges bonded using cyanoacrylate and weldable wire strain gauges would be suitable for use in the project, but that further development work was needed to ensure a consistent and reliable installation. The tests also emphasised the importance of stress relieving the piles if gauges were to provide reliable estimates of pile load.

Bashit tests

Following these preliminary tests, extensive tests using the Bashit machine were carried out. The objectives of the Bashit tests on the dynamic strain sensors were as follows:

(a) to define a procedure for gauge installation such that a gauge would withstand the expected driving conditions

(b) to investigate the magnitude of offsets introduced due to creep during driving
(c) to establish whether the weldable gauges could be successfully welded to the RQT701 steel, as the steel suppliers were uncertain if this could be achieved adequately
(d) to establish which of the two weldable gauges selected for further testing was the more suitable

A total of 59 foil gauges and 30 weldable gauges were tested using the Bashit machine. It proved to be impossible to reach 3500με strain levels without also incurring accelerations of the order of 3000*g*, three times greater than the acceleration levels expected.

These high accelerations caused severe damage both to the test pieces and to the machine itself, which caused considerable delay to the test programme. The solution adopted was to use 30mm test pieces instead of the 15mm used previously.

This limited the accelerations to the desired levels, but meant that peak strains were reduced to the order of 1000με. Additional strain gauges were tested using the flexure test rig to simulate over 2000 blows at up to 2000με.

It was difficult to distinguish between offsets introduced by creep during driving and those caused by stress relieving of the test piece during testing. Gauges installed on previously used test pieces showed significantly smaller offsets than those mounted on new test pieces. Typical offsets measured were less than 30με for both the foil and weldable gauges, less than 1% of full scale.

No problems were encountered in attaching the weldable gauges to the RQT701 steel, and there were no gauge failures caused by welds failing.

Final selection - foil gauges

The foil gauge system selected for use in the piles was HBM 6/350 LY11 gauges bonded with HBM Z70 cyanoacrylate adhesive. The installed gauges were coated with Micromeasurements M-Coat A polyurethane varnish to protect the measurement grid of the gauge. The gauge was then coated with a 1/8 inch layer of Dow Corning 3145 RTV silicone rubber compound to afford mechanical protection and moisture ingress protection. A final layer of HBM ABM75 foil-backed putty was then added to give additional protection against moisture ingress.

The cable attached to the strain gauge was strain relieved by securing it to the pile wall using two bridge-shaped aluminium clamping strips. The strips were manufactured undersize so that they would deform the cable slightly on installation, thus ensuring a tight grip.

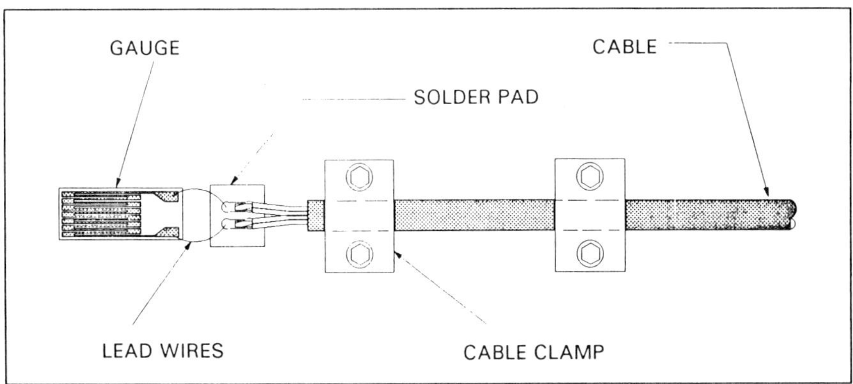

Fig. 3. Foil gauge installation detail

Final selection - weldable gauges

There was no apparent difference in the performance of the two types of gauge tested, so it was decided to use the narrow-flange Ailtech SG159 gauge, as this was a standard item that had been successfully used in previous instrumented pile tests. The gauge and its vulnerable 'gooseneck' area were given mechanical protection using the same silicone rubber and foil-backed putty system that was used for the foil gauges. The gauges were supplied hydrostatically tested by the manufacturer, so it was not necessary to rely on the protective compounds for moisture ingress protection.

The cable from the gauge was secured using the same aluminium clamping strips used with the foil gauges. The positioning of these strips relative to the gauge was found to be critica1 if damage to the gauge was to be avoided as the strips fixing screws were tightened.

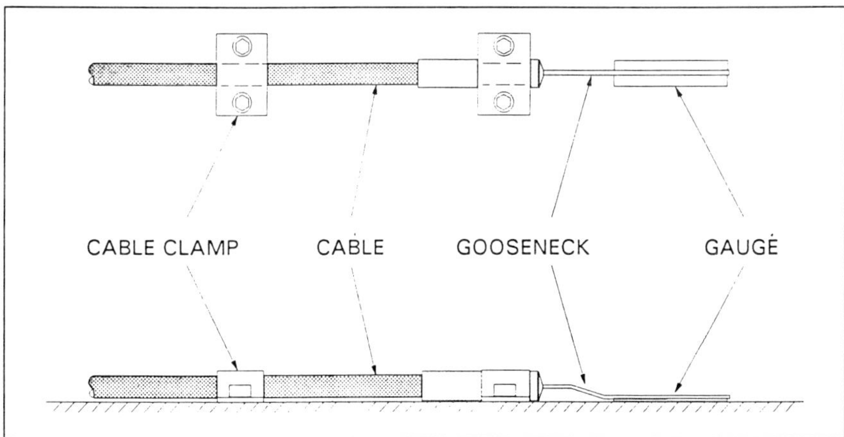

Fig. 4 Weldable gauge installation detail

Static strain sensors

The static strain measurements required stability and accuracy over a period of months, in addition to the ability to withstand driving conditions. Two types of instrument were investigated for the static strain sensors. The primary instrument was to be the McClelland 'strain module' and the backup instrument was a vibrating wire gauge.

Strain module

The McClelland strain module had been used previously in similar applications, but not at such high strain levels. It consisted of a strain-gauged inner tube equal in length to one diameter of the pile. The tube is fitted inside an outer tube and cradle assembly, which are welded to the pile wall. The foil gauges used in the strain module are installed under laboratory conditions using hot-cure adhesives to ensure a highly stable installation.

Vibrating wire strain gauge

Vibrating wire strain gauges consist of a taut wire in a holder that is attached to the test piece and an electromagnetic pickup/exciter unit. The pickup/exciter unit causes the wire to oscillate at its natural frequency. As the test piece is strained, the tension in the wire changes, and so does its natural frequency. The change in natural frequency is proportional to the square root of the applied strain. The oscillation frequency of the wire is detected by the pickup/exciter unit and a pulse train of the same frequency is transmitted as an output. Vibrating wire strain gauges are more limited in strain range and speed of reading than resistance gauges, but have much better long-term stability.

Strain module development

Initially, a single prototype unit was manufactured which was instrumented with three different combinations of strain gauge and adhesive. The strain gauges were Micromeasurements SK and AE series 350Ω foil gauges, and the adhesives were Micromeasurements M-Bond types 610 and AE-15; both adhesives were hot-cure epoxies. A thermocouple was also installed to monitor temperatures inside the instrument during.welding to ensure that the maximum ratings of the gauges and adhesive were not exceeded.

This prototype unit was Bashit tested on three separate test pieces, the securing welds being ground off each time. Despite this harsh treatment it survived 3163 blows producing 600 to 2000με at accelerations of up to 1500*g* before a crack started to form in the outer tube just above a joint in the cradle.

A second prototype was produced with a thicker outer tube and a one-piece cradle. The second prototype was subjected to further extensive Bashit tests. A design flaw eventually caused a failure in the adaptor used to connect the cable penetrator to the strain module body. A stronger adaptor was designed and successfully tested. Typical offsets induced in the output of the strain module prototypes were of the order of 30με per 3000 blows.

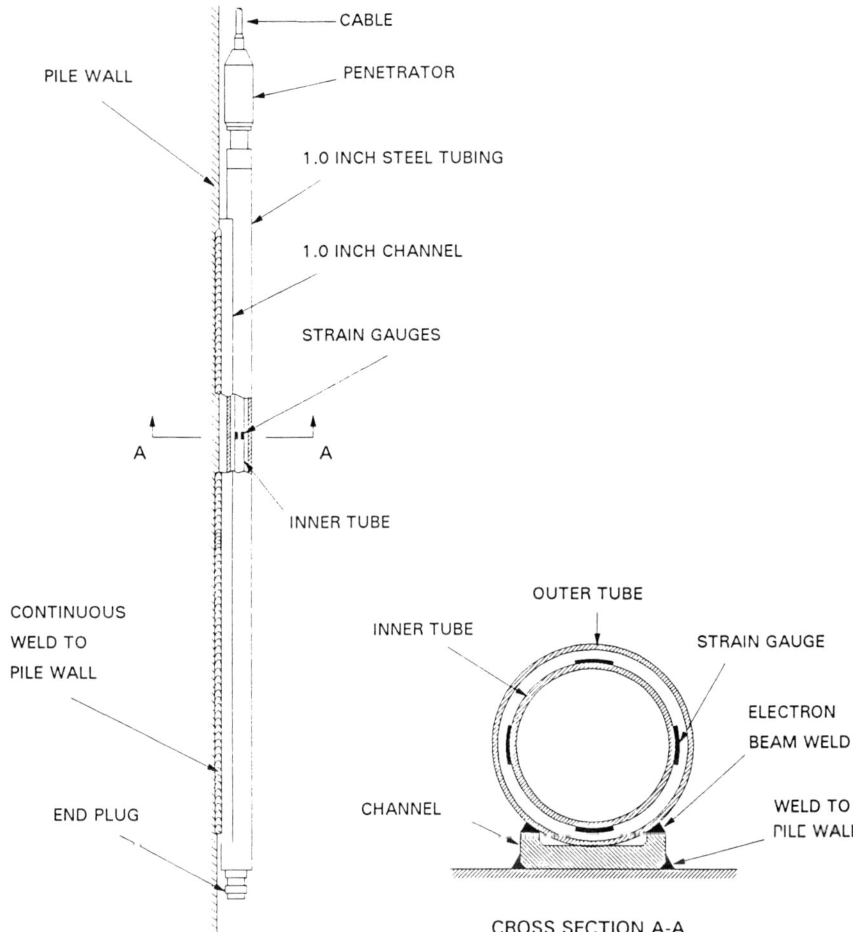

Fig. 5. Strain module

Vibrating wire gauge development

Two types of vibrating wire gauge were selected for evaluation; one manufactured by the Slope Indicator Company (the Sinco gauge), and one manufactured by IRAD (now known as Geokon). Both gauges are continuously excited gauges, which means that the vibrating wire is always oscillating, giving a better dynamic response than the more common type of vibrating wire gauge, where the wire is 'plucked' electronically to make a reading.

The bulkier IRAD gauge disintegrated after only 23 blows during preliminary Bashit testing, and its use was not considered any further. Development work was then concentrated on the Sinco gauges, which werre tested on the flexure test rig and the Bashit machine alongside the dynamic strain gauges.

It was found that the manufacturer's recommended method of holding the pickup to the gauge was not sufficient to withstand driving, so a replacement system was designed and successfully tested which used formed aluminium clamping pieces screwed to the test piece. Fourteen Sinco gauges were Bashit tested in total, and typical offsets of the order of 80με per 3000 blows were observed.

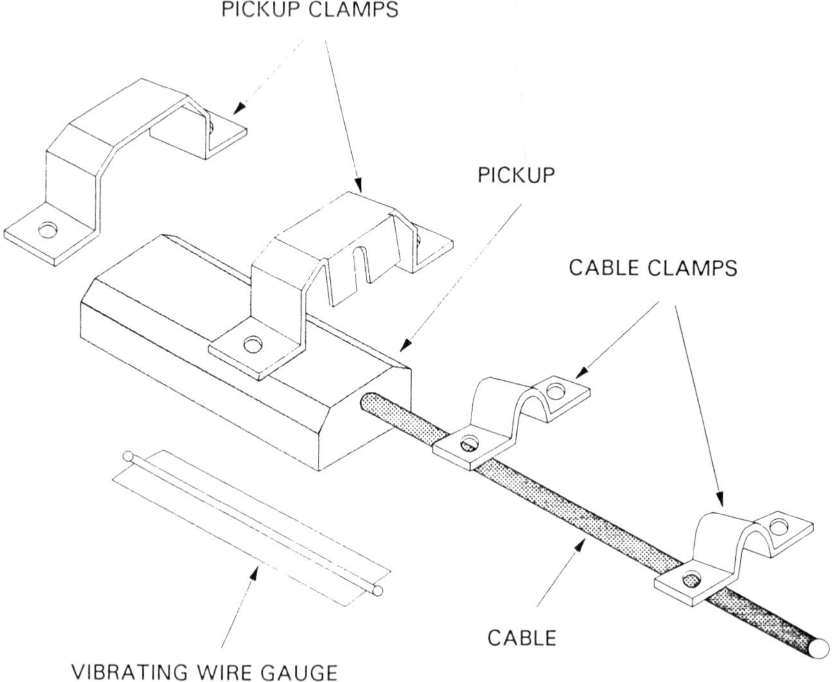

Fig. 6. Sinco vibrating wire strain gauge

The nominal range of operation for the Sinco gauges was ±1500με. Although in practice, the gauges were found to fail at approximately 2000με, it was decided to install the gauges in the test piles only in positions where they were expected to see less than 1500με.

Total pressure cells

Significant changes intotal pressures adjacent to the pile wall were expected during driving and load testing. It was anticipated that these pressures would be in the range of 220kPa to 1700kPa at the NC site, and 110kpa to 3850 kPa at the OC site. It was desired that the cells selected should have a long-term accuracy of better than 10kPa.

Two designs of cell were evaluated, the 'Magnus' cell and the 'McClelland' cell. Both had been used previously in accelerations of up to 1000*g*, but neither had been used in such high-strain environments before.

Pressure on the face plate of the Magnus cell causes a strain gauged tube to go into tension. The measurement range of the instrument is dependent on the wall thickness of the tube.

Pressure on the face plate of the McClelland cell places a load on a central boss which is connected to the body of the cell by four strain gauged shear webs. The measurement range of this cell could be adjusted by altering the dimensions of the shear webs.

The Magnus cell is of a simpler design and easier to strain gauge than the McClelland cell, but is more difficult to manufacture, and gives a smaller electrical output for a given deflection of the pressure plate.

Total pressure cell development

One prototype of each type of cell was manufactured, and each was repeatedly loaded to its rated capacity of 5000kPa using a Budenberg dead weight testing machine. The cells showed good repeatability, with offset changes of less than 0.7% of full range. The cells also underwent tests for cross-sensitivity.

Following the initial static tests, the cells went through extensive Bashit testing. Both types of cells initially suffered failures of the diaphragm connecting the pressure plate to the cell body, and the retaining bolts in the mounting ring sheared. The McClelland cell appeared to be more robust than the Magnus cell, surviving more than twice as many blows before failure occurred.

The cell bodies were redesigned to accommodate more mounting bolts, and to allow fitment of insulation pieces to isolate the cell bodies from the pile wall. This prevents galvanic action between the dissimilar metals of the cell body and the pile from causing corrosion, and from producing hydrogen gas on the face of the cell, which would interfere with the pressure measurement.

The failures of the Magnus cell during Bashit testing had been attributed to a weakness in the front face of the cell. To counteract this, the front face was redesigned to be similar to that of the McClelland cell, with a diaphragm and pressure plate machined from the cell body, instead of being assemled

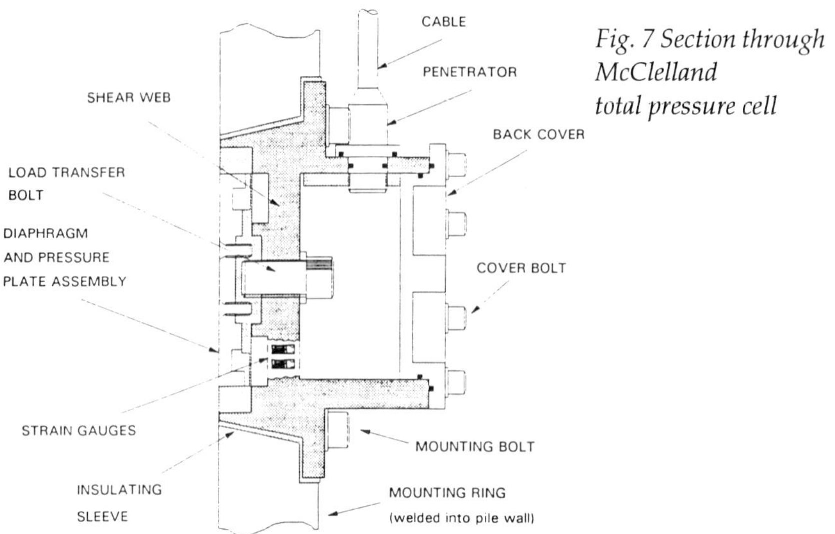

Fig. 7 Section through McClelland total pressure cell

from separate parts and welded together. The diaphragm thickness on both types of cell were increased in thickness and polished to give them a longer fatigue life.

Both types of cell were successfully retested on the Bashit machine with no further problems. Neither the McClelland cell or the modified Magnus design (known as the Hybrid cell) appeared to ahve any grteat advantage over the other, so it was decided to install one of each type at each instrumented level in the piles. Typical offsets induced during testing were of the order of 40kPa per 3000 blows.

Pore pressure cells

Theoretical assessments of the effects of pile insertion show that significant changes in soil pore-water pressures occur adjacent to the pile wall. It was anticipated that the pore pressures would be in the range 150kPa to 1450kPa at the NC site and 40kPa to 3300kPa at the OC site.

Pore pressure cells consist of a measurement chamber which is fitted with a pressure sensor and sealed with a permeable filter. The filter isolates the pressure sensor from the total pressure, but permits the pore water to enter the measurement chamber, thus allowing its pressure to be measured.

Pore pressure cell development

In order to minimise develoment effort, it was decided to take a selection of commercially available pressure transducers and mount them in bodies similar to those used with the total pressure cells. The pore pressure cell body was designed to protect the pressure transducer and external strain. The front

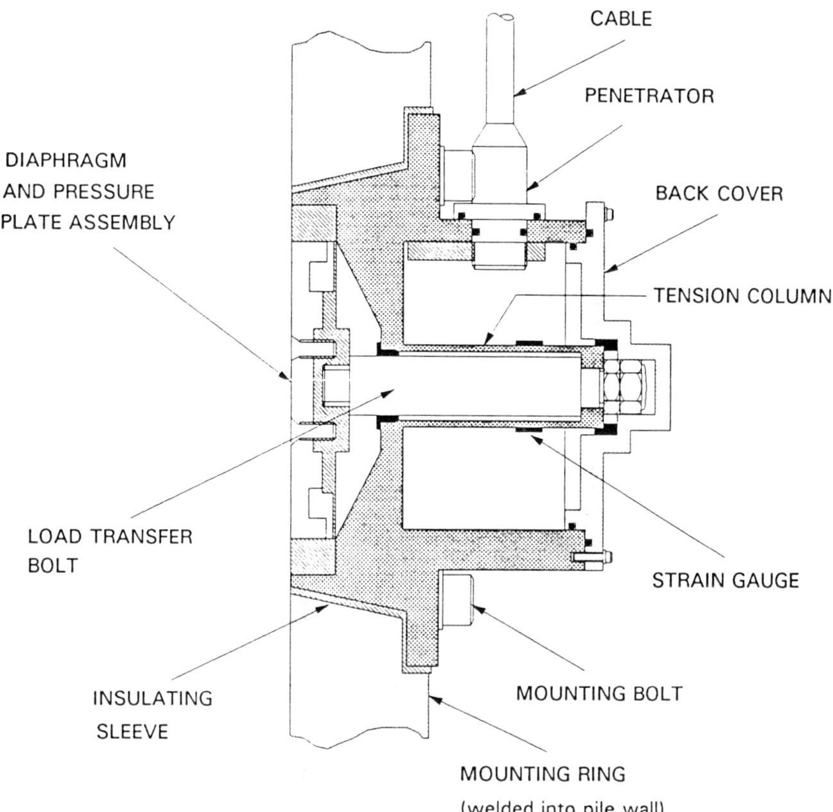

Fig. 8 Section through Hybrid total pressure cell

of the cell body contained a sand/epoxy filter designed to have a similar permeability to the clay at the test sites.

Initial Bashit testing was carried out with a Transamerica (now Transintruments) BHL4700 pressure transducer mounted in a prototype cell body. The transducer proved to be quite rugged once an initial problem with damage to unsupported leadout wires was overcome by potting the back of the cell body with epoxy resin.

Kistler 4140A and Druck PDCR42 transducers were also tested in the cell body, but were not as robust as the Transamerica device, with the Druck transducer failing after only 500 blows.

The prototype cell body design was slightly modified to allow the fitment of insulating pieces similar to thos fitted to the total pressure cells, and further tested.

A total of three prototype cell bodies were manufactured and tested with the different transducers. Typical offsets induced during testing were of the order of 21kPa per 3000 blows.

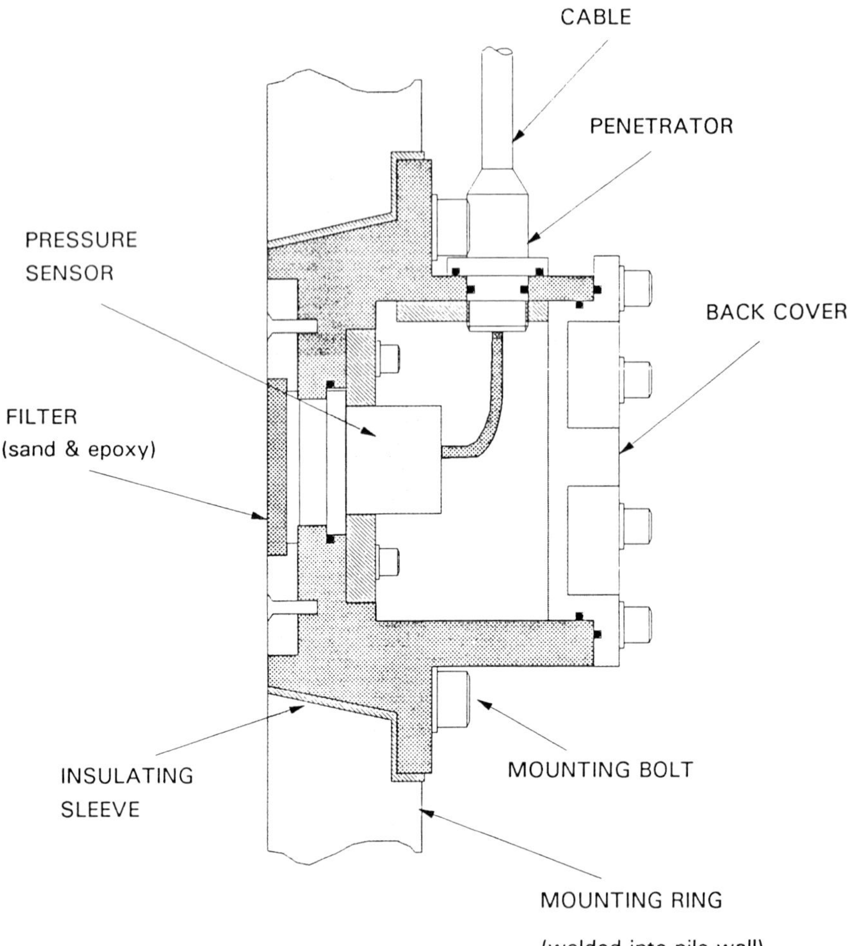

Fig. 9 Section through pore pressure cell

Accelerometers

Accelerometers were required to measure the axial and lateral accelerations in the pile during driving. Accelerometers used in this application are normally of the piezo-electric type, as they are very compact and can withstand the high shock environment.

Piezo-electric accelerometers contain a seismic mass attached to a cantilever made from a piezo-electric material. Accelerations cause the seismic mass to move, flexing the cantilever and causing a change in the electric charge on the cantilever. The amount of charge on the cantilever is proportional to the accerleration experienced by the seismic mass and may be measured.

Accelerometer development

Transducers from three different manufacturers were evaluated for use on the piles. The first to be tested was the Piezotronics PCB305A, as it had been used successfully in the past by QMC, and several examples were held in stock.

Previous practice at QMC had been to screw the accelerometer into a small steel block that would be welded to the pile wall. For this series of tests, the design of the block was modified so that the high strains expected in the pile wall would not be transmitted into the accelerometer, causing spurious signals.

However, initial Bashit tests revealed that this was not sufficient to prevent damage to the accerlerometers due to the high levels of strain and lateral accerlerations present during the tests. In addition, the large high- frequency content of the accelerations imparted to the accerlerometer caused the seismic mass to bo into resonance, causing noisy signals.

Kistler 8618A and Bruel and Kjaer 4371 accelerometers were also Bashit tested using the steel blocks with similar results.

Mechanical filters manufactured by Bruel and Kjaer were tried with the Piezotronics accelerometers. The mechanical filters consist of a disc of resilient material sandwiched between two metal discs. The filters were fitted between the accerlerometers and the steel mounting blocks, and were designed to absorb the high-frequency components of the accerleration signal which were causing the resonance problem.

The mechanical filters gave encouraging results, but were not robust enough for the environment and disintegrated after a few hundred blows.

Alternative methods of mechanical filtering were investigated. The method that gave the best results was to encapsulate the accelerometers in a block of reinforced phenolic resin (Tufnol) using an epoxy potting compound. The Tufnol block containing the accelerometers would then be bolted to the pile wall.

The final configuration adopted was a block with space for two accelerometers, one mounted axially and one mounted laterally. Both the Bruel and Kjaer and the Piezotronics accelerometers were successfully Bashit tested housed in Tufnol blocks. Several blocks may be seen mounted on the test plate shown in Fig. 2.

It was decided to use the Piezotronics accelerometers on the piles because they were considerably cheaper than the Bruel and Kjaer devices, which required special cable and signal conditioning equipment. If no lateral accelerometer was to be installed in a block, a temperature sensor was fitted instead in order to monitor pile wall temperature during driving.

Cable and cable attachment

The type of materials used for the instrument cables were chosen after consultation with QMC and cable manufacturers. The main concerns were that the cables should be waterproof, resistant to abrasion, and with low piezo-electric properties (so that the cables would not generate false signals during driving).

A four-core cable and a seven-core cable were specified and manufactured specifically for the project. The seven-core cable was to be used for the strain modules, pore pressure cells and the total pressure cells, and the four-core cable was to be used for the resistance strain gauges. A total of 7000m of seven-core cable and 10,500m of four-core cable were manufactured for the project. The pickups for vibrating wire gauges came fitted with a suitable cable of the correct length, so it was not necessary to design any special cable for these sensors.

Cable construction - seven core

The individual conductors in the seven-core cable were insulated with polypropylene, which has excellent piezo-electric properties, but relatively poor abrasion resistance and waterproofing. An uninsulated drain wire was laid alongside the seven cores, and an aluminised mylar tape was wrapped around the cores and drain wire to provide electrical screening. An inner sheath of polyethelene (PE) was then extruded over the cores and drain wires, followed by a braided steel-wire armour layer.

Finally, an overall sheath of polyurethane (PU) was extruded over the PE inner sheath and armour. PE has excellent waterproofing qualities, and good piezo-electriic and abrasion resistance properties, but is relatively inflexible and difficlut to mould onto. PU has excellent abrasion resistance and waterproofing properties, and is more flexible than PE, particularly at low temperatures. PU is also much easier to mould onto than PE, making the design of the penetrators into the instruments simpler.

Cable construction - four core

The four-core cable had a central pair of PE insulated conductors twisted together with a drain wire and wrapped in an aluminium foil screen. A second pair of PE insulated conductors and a second drain wire were laid over this central core. The complete cable was then covered in an overall foil screen, and an extruded outer sheath.

The outer sheath of the four-core cable had been specified as PU, but the cable was delivered with a PE sheath. Due to programme constraints, the cable was accepted and not substituted with one of the correct specification. The main implications were that the cable was slightly more prone to abrasion damage, and that the cable was more awkward to work with, particularly at low temperatures, because of the stiffness of the PE.

Cable attachment

Bundles of cables, in combinations typical to those to be found in the piles were Bashit tested to evaluate techniques for securing the cables to the pile walls. It was desired to tie the cables off to upstands welded to the pile wall, rather than mould the cables into a channel because previous experience at QMC had indicated that moulding compounds used in bulk were a significant source of piezo-electric noise.

Initially, high-strength nylon cable ties were used to secure the cable bundles to the upstands, but these proved not to be strong enough, and either broke or let the cable bundle slip. Stainless steel banding and buckles were found to offer an apparently more satisfactory solution.

It was found during driving at the NC site, however, that the Bashit tests had neglected to account for the weight of the typical lengths of cable to be found in the pile, as the bundles tested were only of the order of a metre long. This extra weight meant that the banding was not adequate to prevent the cables inside the piles from slipping, causing the loss of some instruments. More detail on these failures and the remedial action taken for the cables in the OC pile are given in another paper in this series (Ref. 3).

Performance of system

The programme of tests described in this paper showed that long-term errors caused by drift and creep could be expected to be less than 0.1% of measurement range.

Provided that all instruments were installed using the procedures developed during testing, it was confidently expected that very few instruments would fail during driving, as the predicted worst-case strain and acceleration levels had been significantly reduced since the commencement of the instrumentation development programme. This reduction meant that the Bashit testing carried out had exposed the instruments to an environment considerably more severe than that actually expected. The strain and acceleration levels actually measured during driving were of the order of 1000με and 500*g*, compared with the 3500με and 850*g* originally predicted.

When the NC pile was driven, over 75% of the instruments survived. Although this was an acceptable proportion, given the amount of redundancy built into the system, it had been anticipated that a greater number would survive. The majority of the failed instruments were foil gauges, and an investigation revealed that they had been disconnected following slippage of the cable bundles during driving.

Remedial action was taken on the cable bundles in the OC pile (Ref. 3).

This action was successful, and only one instrument failed during driving on the OC pile, a McClelland total pressure cell that had been giving spurious readings before the drive.

Conclusions and recommendations

It is essential that instruments to be mounted on a pile wall are carefully evaluated before selection to ensure that they will withstand the arduous conditions encountered during driving. Cable bundling and securing techniques must also be thoroughly tested.

It is desirable that, wherever possible, system redundancy is enhanced by developing more than one type of instrument to measure each parameter of interest. For example, static strains were measured using both vibrating wire strain gauges and strain modules.

It is essential to stress relieve piles prior to instrumentation if unpredictable instrument offsets are to be avoided. Stress relief may be carried out by any practical means, either thermal, mechanical or a combination of the two.

Acknowledgements

The authors wish to thank the sponsors of the Large Diameter Pile Tests Programme for permission to publish this paper.

References

1. SUTTON V. J., RIGDEN, W. J., JAMES, E. L., St JOHN, H. D. and POSKITT, T. J. A full scale instrumented pile test in the North Sea. *Proceedings of 11th Annual OTC*, Houston, April 1979
2. CUTHBERT, L. G. and POSKITT, T. J. Development of instruments for offshore piles. *Ground Engineering*, January 1983.
3. COX, W. R., SOLOMON, I. J. and CAMERON, K. A. Instrumentation and calibration of two 762 mm diameter pipe piles in clays. This volume.

10. Instrumentation and calibration of two 762 mm diameter pipe piles for axial load tests in clays

W. R. COX, Fugro-McClelland Marine Geosciences, Inc., Houston I. J. SOLOMON, Fugro-McClelland Ltd, Glasgow and K. CAMERON, Precision Monitoring & Control, Ltd, Stockton

In the Large Diameter Pile Tests (LDPT) programme two instrumented pipe piles of 762-mm diameter were driven and tested under axial compression loads at sites in England. This paper is one in a series of papers on the planning, execution, and results of the LDPT project. Discussed in this paper are the instrumentation and calibration of the two test piles. Information is also given on the fabrication of the pile sections prior to instrumentation and the joining of the instrumented pile sections into the completed lengths of 58.5m for the pile at the normally consolidated-clay site (NC pile) and 33.5 m for the pile at the overconsolidated clay site (OC pile). A discussion on the development and testing of instruments is given in a companion paper, Solomon et al *(1992).*

Fabrication of test piles

Fabrication of the two test piles began in September 1986 and was completed in August 1987. Steels for the piles were supplied by the British Steel Corporation (BSC). The two test piles were fabricated from roller-quenched and tempered (RQT) steels (BSC 1982). RQT 501 steel with a minimum yield strength of 470 N/mm was used for the NC pile. RQT 70 1/2 steel with a minimum enhanced yield strength of 770 N/mm^2 was used for the OC pile.

Reasons for selecting steel types

The pile configurations used for the tests represented a compromise between the conflicting requirements regarding instrumentation, pile strength, and compressibility. An important feature of the test was that the piles should be relatively compressible. The relative soil-pile stiffness ratio, was selected as a measure of compressibility of the pile (Murff 1980). For a given skin friction, depends on pile cross section and length.

The length and wall thickness for each test pile were selected to have similitude with piles used in North Sea offshore platforms. The selected wall thicknesses required the piles to be fabricated from high-strength steels.

Specifications for both test piles were agreed early in the contract period when the MHU 1700 pile hammer was being considered for the OC site installation and the detailed site investigations had not yet been carried out.

The NC pile was selected to have a diameter of 762 mm (30 in) and a wall thickness of 15 mm. The OC pile was also selected to have a diameter of 762 mm, but with a wall thickness of 30 mm. The internal protective channels were assumed to add 7550 mm^2 to the cross sectional area of the NC pile and 10,300 mm to the cross- sectional area of the OC pile. This would give effective cross sectional areas of 42,750 mm and 79,300 mm for the NC and the OC piles, respectively.

Estimates of driving resistance and upper bound load capacity based on data from the preliminary 1983/4 soil investigations yielded the following results:

	Maximum predicted soil resistance: MN		Maximum steel stress: N/mm^2	
Test phase	OC	NC	OC	NC
Installation	21.0	4.6	480	180
Static test	40.6	12.1	512	283

The maximum steel stresses during installation by pile driving were dynamic stresses estimated from the results of wave equation analyses.

By restricting the maximum static-load steel stress to about two-thirds of yield and dynamic installation stresses to 80 per cent of yield, the required yield strengths were 768 N/mm^2 for the OC site and 425 N/mm^2 for the NC site. Enhanced RQT 701 steel with a characteristic yield strength of 770 N/mm^2 and RQT 501 steel with characteristic yield strength of 470 N/mm^2 were chosen for the OC and NC sites, respectively.

Fabrication of pile cans

Contract for fabrication of the piles was awarded in September 1986. The work for rolling cans for forming the piles began in October 1986. The cans were rolled from steel plate with lengths of 2 and 3 m. There were no problems in rolling cans from the RQT 501 steel plate.

Several attempts were made by the fabrication contractor to roll the RQT 701 plate as supplied by BSC. Whilst the rolls used by the fabricator had sufficient strength to close the longitudinal seams of the cans during rolling, on releasing the rolls the can seams sprung apart by up to 600 mm.

The problem of spring back was resolved by allowing the fabricator to heat the cans to 600 deg. C followed by hot rolling and pressing the cans. A temperature of 600 deg. C was selected from a series of strength tests performed on RQT 701 heated specimens, which showed that the resultant yield strength would not be less than 600 N/mm^2. A reduced yield strength from 770 to 600 N/mm^2 for the OC pile was considered acceptable. This was based on a revision of the anticipated maximum test load from 45 to 30 MN. It was also based on pile-stress measurements made during the trial-pile drive at the OC site.

Fabrication of pile sections

Pile sections with lengths of 9, 11, and 12 m were formed from the 2- and 3-m length cans. Internal steel work and instrumentation were installed inside the sections before the sections were joined for the final pile lengths.

Preheat for welding was required for the RQT 701 steel but not for the RQT 501 steel. Both RQT 701 and 501 welds were post-weld heat treated (PWHT). PWHT was required for RQT 701 steel. The principal reason for PWHT of RQT 501 was to relieve locked in strains that could cause problems for interpreting strain readings from instruments.

Non-destructive testing was carried out on all welds. Butt welds were tested by magnetic particle inspection, ultrasonic, and radiography. Fillet welds were only tested by magnetic particle inspection.

Layout of instruments in OC and NC piles

The layouts of instruments selected for the NC and OC test piles are shown in Figs. 1 and 2, respectively. Instruments were installed around the inside circumference of the pile at 90-degree spacings. These instrumented quadrants are known as A, B, C, and D. The majority of instruments were installed on the A and B axes. The C and D axes contain only HBM foil gauges and lateral accelerometers.

Before installing internal steel work, the locations for the instruments were marked and the pile surface smoothed by grinding at each spot where strain gauges were to be installed. The area of grinding, coarseness of grit pads, and amount of grinding were specified in procedures for instrument installation.

Internal steel work

Channels were installed inside the piles to protect instruments and cables. These channels were fabricated from steel plate instead of using commercially available channel sections. This was required to avoid welding the pile wall in close proximity to instruments and also to keep the top of the channel open so that instrument cables from lower pile sections could be extended through upper sections after the girth welds were made to join pile sections.

The two upstands for each fabricated channel were welded in place before PWHT, as shown in Fig. 3. The channels were no closer than 45 degrees from a longitudinal seam weld in the can. The channels were fabricated from RQT

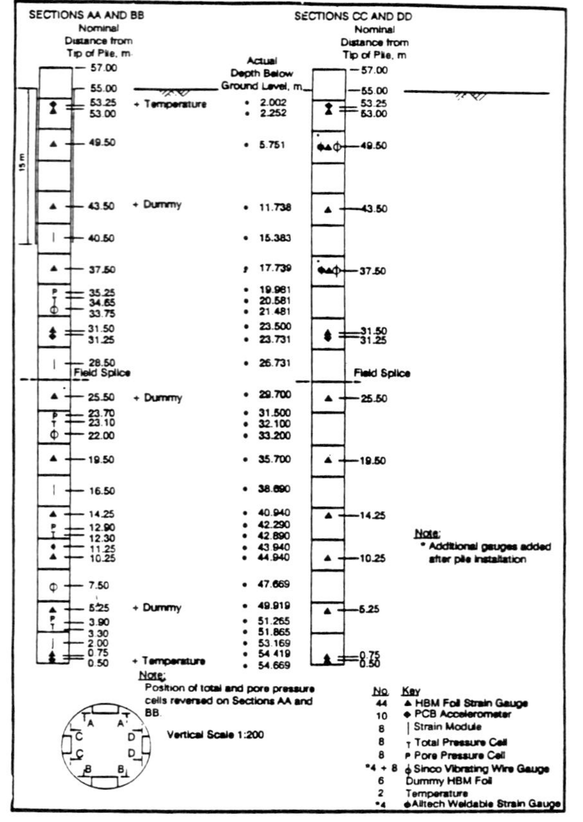

Fig. 1. Layout of instruments on NC pile

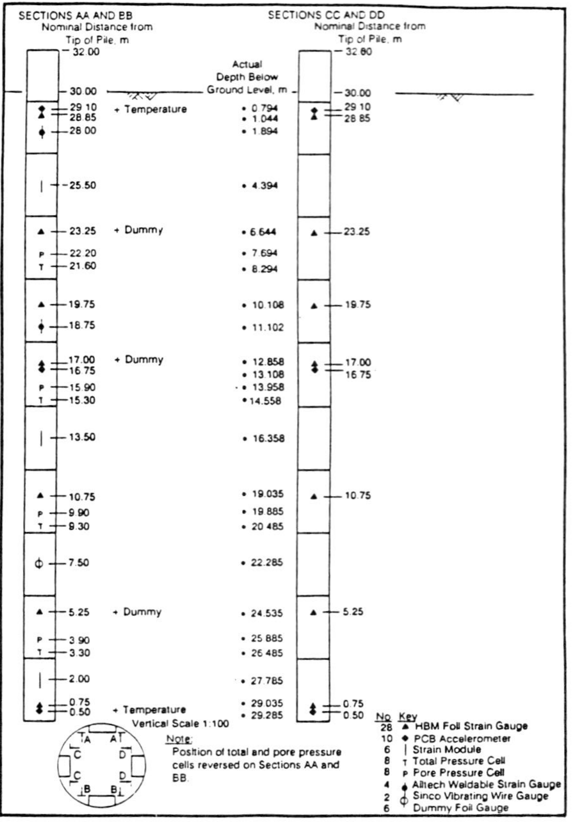

Fig. 2. Layout of instruments on OC pile

Fig. 3. Channel legs and cable brackets

501 and RQT 701 steels for the NC and OC piles, respectively. Girth welds were ground smooth where cables passed over them.

All instrumentation was installed after PWHT and the covers for the channels were then welded. No PWHT could be permitted on the fillet welds for the covers because of the close proximity of instruments and cables. Short upstands, shown in Fig. 3, were welded within the instrument channel to form brackets for clamping cables. Weld pads were placed in the RQT 701 sections for installation of the strain modules, and ground flat prior to PWHT.

Square tubing for the inclinometer was welded to the inside of the pile sections. Holes were flame cut and ground out of the pile wall to accommodate mounting rings for the pressure cells. The mounting rings with dummy cells bolted to them were welded in position before PWHT.

The sections of both piles, with instrument protection upstands, mounting rings, dummy cells, weld-pads, and various brackets, were given PWHT in a furnace.

Installation of instruments

Quality control

An installation procedure was written for each type of instrument to be installed in the piles. This was to ensure a consistent, high quality installation. Additionally, an external Quality Control (QC) Inspector was present during all stages of instrument installation.

The role of the external QC inspector was to ensure that the instrument installation procedures were being adhered to by the personnel installing the instruments. This was done by means of witness points. Once a witness point was reached, instrument installation work was interrupted to allow the QC inspector to satisfy himself that installation work was in accordance with approved procedures.

Fig. 4. Foil gauge under dome

General installation procedure

After a pile section had been released for instrumentation by the fabricator and placed on rollers, the general procedure for instrumenting that section was as follows:

(*a*) All loose material in the section was cleaned out, and the positions of the instruments marked.
(*b*) Fixing holes for cable clamps, Sinco pickups, and accelerometer modules were drilled and tapped. Rough surface preparation took place.
(*c*) Strain modules were installed. This was done first so that other instruments in the pile would not be damaged while the strain modules were being welded.
(*d*) HBM foil gauges were then installed because they were the most common instrument and the most time consuming to install.
(*e*) Next, all other instruments were installed.
(*f*) Instrument cables were tied down in the cable channels, and final resistance checks made.
(*g*) Following completion of all installation records, instrument cables were stowed on drums inside the section (with all free ends accessible), and the section was released for calibration.

Installation of HBM foil gauges

The first step in installation for the HBM foil gauges was the initial marking out, and the drilling and tapping of holes for cable-securing clamps. The area of the pile to be gauged was then cleaned with degreaser, abraded with emery paper until smooth, and then cleaned with conditioner and neutraliser. This first step was common for most other instruments.

The adhesive used to bond the HBM foil gauges to the pile wall was a cyanoacrylate type, which cures by absorbing moisture from the air. For this reason, the temperature and humidity inside the pile section had to be controlled if shock-setting of the strain gauge adhesive was to be avoided.

Electric heating pads were attached to the outside wall of the pile beneath the location for the gauge installation and the steel area at the gauge location was warmed to around 20 deg. C. Relative humidity was controlled by using a dry-air blower to blow warm, dry air through the pile section. Temperatures and relative humidities were monitored using a thermohygrometer. Gauges were installed when the relative humidities were between 30 and 65 per cent.

Following installation of a gauge and connection of a cable, the gauge resistance and insulation from the pile wall would be measured and recorded. The gauge was also 'tap-tested' to ensure a good dynamic response by striking with a sledge hammer the outside of the pile wall near the location of the gauge on the inside wall and observing the output of the gauge on a storage oscilloscope.

The first protective coating was Dow Corning 3145 RTV silicone rubber compound. A layer was spread over the gauge and cable end and left to cure for a minimum of 24 hours. When it was cured, a piece of HBM ABM 75 protective putty was placed over the 3145 RTV.

At this stage, the installation underwent its final test. Resistances were rechecked, and the installation was tap-tested a second time. Following final testing, a dome of 0.010-in.-thick steel shim, as shown in Fig. 4, was spot-welded over the installation to protect it from chafing by cables passing over it.

Installation of Ailtech weldable gauges

The gauge and cable assembly was fixed to the pile wall using the cable securing clamp nearest the end of the cable. A gooseneck was then formed in the stem of the gauge so that the gauge lay flat on the surface of the pile wall. The gauge was now ready for spot-welding.

Prior to welding the gauge, the welding unit would be set up to give optimum results by making a series of trial welds on a piece of 0.005-in.-thick shim (the same thickness as the flange on the Ailtech gauge). The piece of shim would then be peeled away from the pile wall. If the welds were torn out and remained on the pile wall, the welder was set correctly, and the gauge could be welded.

The remainder of the installation procedure for Ailtech gauges was identical to that for HBM gauges: resistance check, tap-test, 3145 RTV, ABM 75, final resistance check and tap test, and protective shim cover.

Installation of Sinco vibrating wire gauges

A welding trial was performed in the same manner as for an Ailtech gauge, but using a 0.007-in.-thick shim to simulate the thicker flange of the Sinco gauge. A Sinco gauge would then be taped in position and welded in place.

The pickup sensor would then be clamped in place over the gauge, as shown in Fig. 5. The gauge was tested by connecting a Sinco Model 52609 Indicator box to the pickup and measuring and recording the indicated strain and the current taken by the pickup sensor. If the readings fell outside those permitted, the gauge would be removed and the procedure repeated.

Installation of strain modules

The strain modules used in the OC pile were welded onto built-up weld-metal pads to match the strain module to the RQT 701 steel of the pile wall. The strain modules used in the NC pile were welded directly to the pile wall.

Prior to welding the strain module, the installation area would be preheated to 70 deg. C using electric heater pads: During welding, the thermocouple in the strain module would be monitored. If the thermocouple indicated more than 100 deg. C, the welding would be interrupted in order to protect the strain gauges inside the strain module.

The strain module is shown before installation in Fig. 6 and after installation in Fig. 7. Following welding, the preheat temperature would be maintained for two hours. All welds for the strain module to pile connection were tested 100 per cent both visually and by magnetic particle inspection. Finally, the thermocouple wire was cut off, and the strain module was pumped full of silicone grease and plugged.

Strain modules on the NC pile were fitted with a support block under the penetrator adaptor. The adaptor had proved to be weak during Bashit testing at Queen Mary College. Strain modules on the OC pile were manufactured with a stronger adaptor.

Installation of pressure cells

The first step for installation of the pressure cells was to remove the dummy pressure cells and clean the taper seat of the mounting ring with emery cloth and then with degreaser. The Tufnol insulator was placed in the mounting ring, followed by the pressure cell itself. The pressure cell was bolted down using a torque wrench. Figs. 8 and 9 show the two pressure cells as seen from the inside and outside of the pile, respectively.

Installation of accelerometer modules

The back of the accelerometer module to be installed was coated with quick setting epoxy resin and bolted into position finger tight so that the epoxy resin began to extrude from underneath the module. After the epoxy resin had cured, the mounting bolts would be removed, cleaned and reinserted. This time they would be torqued to 50 ft-lb in order to compress the epoxy resin.

The accelerometer module was functionally tested by gently tapping it on its active axis, and observing the oscillating signal produced on a storage oscilloscope. The oscilloscope trace was photographed and used to calculate the resonant frequency of the accelerometer module.

Fig. 5 (left). Vibrating-wire gauge

Instrument calibration

Pile section calibration had a number of purposes:

(*a*) to stress-relieve the pile prior to driving and load testing in order to minimise stress relief that could cause offset changes in instruments during driving and load testing.
(*b*) to verify correct operation of strain sensors.
(*c*) to calibrate strain sensors against applied axial load, in order to obtain a sensitivity value in volts per kN that could be used during load testing.
(*d*) to measure the cross-sensitivity of the total pressure cells and pore pressure cells to axial load, i.e. to measure the spurious output caused by compression of the cell body, in order that a correction factor could be applied in the field.

Calibration equipment

As shown in Fig. 10, the calibration frame consisted of two anchorage assemblies, one fixed to a bed and the other free to move backwards and forwards. The two anchorage assemblies were tethered together with 16 high tensile Macalloy bars of 40-mm diameter (BSC 1972). Each bar had an allowable tensile load of 780 kN (79.5 tonnes). The pile section to be calibrated was placed between the two anchorage assemblies, touching the free anchorage assembly.

As shown in Fig. 11, a bearing head, 1000-tonne load cell, and 1000-tonne hydraulic jack were then placed between the fixed anchorage assembly and the free end of the pile section. When the jack was operated, the pile section would be pushed against the free anchorage assembly.

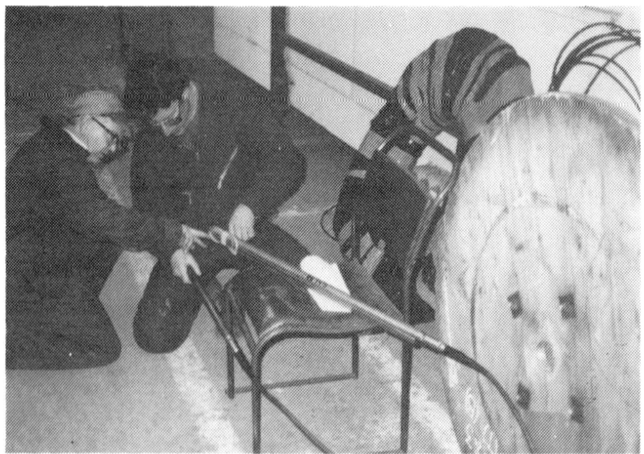

Fig. 6(above). Inspecting the strain module

Fig. 7 (right). The strain module in place

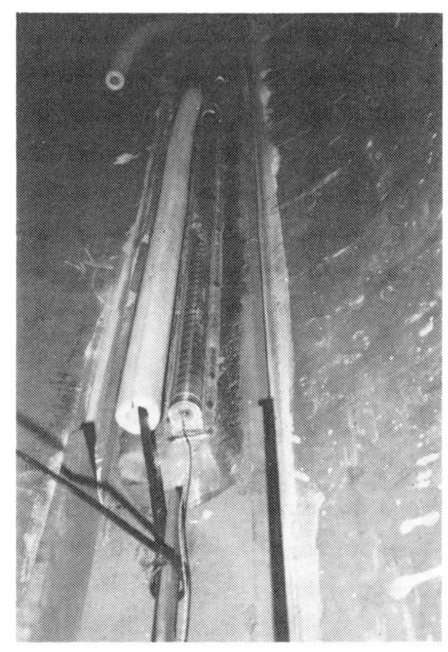

Calibration method

The pile section to be calibrated was exercised in order to stress relieve it prior to calibration. This was done by loading the section to 870 tonnes, and then loading six times to 837 tonnes (these figures apply to the NC pile only; the equivalent figures were 1018 and 1005 tonnes for the OC pile, respectively).

After a pile section had been sufficiently exercised, calibration runs could begin. A calibration run was performed by loading the pile section from zero to 800 tonnes in nominal 100-tonne steps, and then unloading back to zero, also in 100-tonne steps (the load figures given apply to NC pile sections only; OC pile sections were calibrated to a peak load of 970 tonnes).

Fig. 8 (left). Pore and total pressure cells

Fig. 9 (below). Total and covered pore pressure cells

Fig. 10 (bottom). Calibration frame and pile sections

Fig. 11 (above). Jack and load cell for calibration system

Fig. 12 (right). Clamping cables

Fig. 13 (below). Insulating with mineral wool

The lack of repeatability of data for the first section that was calibrated was believed to be caused by drag of the pile on the top and side constraints of the calibration frame. These constraints were provided to prevent the section from buckling. Monitoring of lateral movements of the pile during calibration indicated that no buckling would occur. The side constraints were then relaxed and all sections demonstrated better repeatability of calibration data.

For the sections above the toe sections, it was not possible to calibrate with the channel covers welded into place. These upper sections had their channels left open to receive cables from lower sections.

Calibration of toe sections

The toe sections of the OC and NC piles were calibrated twice, once with the channel covers off, and once with the instrument cables secured and the channel covers welded in place. This permitted correction factors for sensitivities to be calculated to compensate for the lack of stiffness caused by the missing channel covers. These correction factors were used for sections in which the channel covers could not be installed until after the sections had been joined.

Before the channel covers were welded in place on the toe sections, the cables were banded to the brackets that were welded at approximately 1m intervals along the pile, as shown in Fig. 1. The cables were positioned around the brackets and three plastic-covered stainless-steel bands were inserted through each bracket, around the cable bundle, and then tightened with a commercial strapping device. This operation is shown in Fig. 12. Next a protective layer of heat-resistant mineral wool was installed around and over the instruments and cabling. This operation is shown in Fig. 13.

Fig. 14 shows the top of a toe section after the channel covers have been installed. It can be seen that the covers are not extended to the top end of the section. This allows for the cables to be doubled back into the section to avoid heat produced when the section is welded to the next section. Fig. 15 shows the bottom of a toe section. Also shown in Fig. 15 is the box section that served as a conduit for the inclinometer.

Calibration results

A set of three calibration runs could be achieved in 15 minutes, and good linearities and repeatabilities were obtained. Bending in the pile sections was low, typically less than 1 mm.

The total pressure cells were found to have significant cross sensitivity. In the NC pile cross sensitivity ranged from –4 kPa/MN to +7 kPa/MN prior to channel covers being welded on. Welding on channel covers increased the values dramatically by a factor of 9.2 for McClelland cells and 5.2 for Magnus cells. This gave cross sensitivities ranging from –27 kPa/MN to +64 kPa/MN. In the OC pile cross sensitivities of the total pressure cells were not signifi-

Fig. 14. Top of toe section

cantly affected by the addition of cover plates. Neither was there any notable difference between McClelland and Magnus cells in the OC pile. Cross-sensitivity values ranged from –11 kPa/MN to +6 kPa/MN for the OC pile cells. Typical cross-sensitivities of the pore pressure cells varied between 0.1 kPa/MN and 2 kPa/MN, or less than 0.3% of full scale.

The cross-sensitivities measured during calibration may have been greater than those occurring in the field as there was no pressure on the active face of the sensors during calibration. Application of a hydrostatic pressure to the active faces of the sensors would have corrected some of the distortion to the sensor bodies during calibration, thus giving a smaller spurious output, and hence a lower figure for cross sensitivity.

Fig. 15. Bottom of toe section

Assembly of sections

After completion of instrument calibration, the pile sections were circumferentially welded together to form two lengths of 27 and 30 m for the NC pile and one length of 32 m for the OC pile. On the inside surface of the piles the welds were ground level with the pile wall in areas where instrument cables were to pass. The channel upstands were then spliced across the circumferential welds. The circumferential welds and channel extensions were then stress relieved using localised PWHT.

During the joining of the pile sections, the cables were stored on reels placed inside the piles to keep them away from the area for welding and heat treatment. The reels with the wound cables are shown in Fig. 16.

Cans of 1.5-m length were required at the top of each pile to keep the hammer shroud clear of the cable exits during driving. The driving can for the NC pile was welded to the 30-m long pile section in the fabrication yard. In the case of the OC pile, transport difficulties for a pile longer than 32 m required the driving can to be welded to the OC pile at the test site to form the total length of 33.5 m.

After joining the pile sections, the cables were banded into position inside the 32-m long OC length and inside the 27-m long NC length (Fig. 12). Protective heat-resistant mineral wool was installed around instruments and cabling (Fig. 13). Cover plates were welded across the channel upstands and over the pressure cells. The welding sequence was controlled to minimise heat input.

The 31.5-m and 27-m long sections for the NC pile were joined on the test site (Fig. 17). Cables were then pulled from the bottom 27-m length into the

Fig. 16. Storage reels for cables

31.5-m length and out the cable exits located below the driving can. The cable exits for the OC pile were identical to those shown for the OC pile in Fig. 18. Cables were banded, mineral wool was placed, and the channel tops were welded into place. Temperatures in the pile at instrument locations were monitored during welding of the cover plates to ensure minimum heat input to the instruments.

Final pile make-up

The completed NC pile had an overall length of 58.5 m. The bottom 54 m was 15-mm thick and the top 4.5 m was 20mm thick. The increase in wall thickness at the top was specified to minimise damage during driving. Fig. 19 shows the completed NC pile ready for pitching. It was reported (Department of Energy 1987) that the NC pile with a length of 58.5 m was the longest onshore pile to be continuously driven in one length in the UK.

The completed OC pile had an overall length of 33.5 m. The bottom 27 m was 30 mm thick. The next 5 m had a thickness of 40 mm. The top 1.5 m length was also 40 mm thick but was fabricated from Grade 50 material.

The overall lengths reported above are the design lengths. The actual lengths as measured on the sites prior to driving were 58.416 m for the NC pile and 33.243 m for the OC pile.

Remedial work on the OC pile

Failure of 36 of the 90 instruments in the NC pile occurred during driving and in the ten days following driving. An investigation was made of reasons for the failures so that remedial work might be considered for the OC pile before driving.

The NC pile was driven through a 15 m deep casing in which the soil had been removed. After the NC pile was driven, the top of the soil plug in the pile was found to be 1.73 m above the bottom of the casing. There was no soil in the top 13.3 m of the pile and the water in the pile was pumped out to prepare for access. Ladders were hung inside and personnel went into the pile to remove sections of the cover plates in the vicinity of the top three foil gauges on the C axis. An examination showed that the clamping system for the cable bundle had not been adequate and the bundle had moved upward approximately 10 mm during driving causing the foil gauges to be broken.

As discussed by Solomon *et al* (1992), dynamic tests had been made of the cable clamping system on the Bashit machine at Queen Mary College and the clamping system was indicated to be satisfactory. It is clear now that the dynamic environment for cable clamping was not accurately modelled on the Bashit machine.

For the OC pile the decision was made to investigate bonding compounds

Fig. 17 (left). Splicing NC pile on site

Fig. 18 (below). Cable exits on OC pile

that would supplement the clamping action from the three bands. Full-size mock-ups of the bracket and cables were bench tested under static loads. Seven epoxy type bonding compounds were investigated. The decision was made to use a urethane material with the trade name of Devcon Flexane 94. The static tests with the Devcon material showed an increase in cable clamping strength of 10 to 12 times greater than tests with only the three stainless steel bands and no bonding compound.

A 300-mm length of cover plate at the first cable clamp above each instrument on the A and B axis of the OC pile was removed, the cable clamps were removed, and Devcon material was applied to the cables and bracket.

Fig. 19. NC pile, 58.5 m long

The cables were then again clamped to the bracket with three stainless steel bands.

Two threaded plugs were installed in each replacement cover plate before welding the plate to the channel. After welding, the plugs were removed and polyurethane foam was injected in each hole to give additional fixing of the cables to the pile and to the channel upstands.

Because of time and cost constraints, sections of cover plates on the C and D axes were not removed and hence no Devcon material was applied to cable brackets on those two axes. Two holes were drilled and tapped in the cover plate at the locations of the first cable clamping bracket above each instrument on the C and D axes. Polyurethane foam was injected through these holes to improve the cable fixing as was done on the A and B axes. Plugs were then screwed into the holes to prevent water from entering the channels.

The remedial measures on the OC pile were successful since no gauges were lost during driving.

Observations and recommendations

(*a*) To model the axial stiffness of an offshore pile, it was necessary to use high-strength steels for the NC and OC piles. Fabrication of the OC pile was delayed for four months because of the large spring back of some 600 mm that occurred after rolling the 3 m long cans from the RQT 701 steel with an enhanced yield strength of 770 N/mm^2. No problems were encountered in rolling the cans for the NC pile from the RQT steel with a yield strength of 470 N/mm^2. In future pile tests, caution should be exercised in using steels of ultra-high yield strength, say greater than 500 N/mm^2.

(*b*) Steel tubular piles have large locked-in strains that result from the rolling and welding processes. Heat treatment in a furnace to relieve most of these locked-in strains before installing instruments should become a standard procedure for instrumented pile tests.

(*c*) Considerably more time was required for installation of foil gauges than was anticipated. This was primarily caused by the need for humidity control during laying of gauges and the long wait time required for curing of coating materials. In retrospect, the Ailtech weldable gauge would probably have been a more economical gauge to install than the foil gauge and would have performed satisfactorily.

(*d*) Potting of the channels was considered as a means for restricting cable movements during driving. This was not accepted since there was evidence from a previous test that during driving piezo-electric effects are generated in cables that are fixed to the pile by potting. For this reason the cables were clamped to steel brackets that had beeh welded along the pile at approximately 1 m intervals. During driving of the NC pile, there were instrument failures caused by cables pulling away from some of the instruments. As a precaution, before the OC pile was driven, the cables were bonded with a urethane material to the cable brackets.

There was no pulling away of cables from instruments during driving of the OC pile.

(*e*) No significant problems occurred with the general procedures for installation of the instruments.

(*f*) The lack of repeatability of data for the first section that was calibrated was believed caused by drag of the pile on the top and side constraints. These constraints were provided to prevent the section from buckling. Monitoring of lateral movements of the pile during calibration indicated that no buckling would occur. The side constraints were then relaxed and all sections demonstrated better repeatability of calibration data. All other features of the calibration system performed satisfactorily.

(*g*) Although there was considerable development work for LDPT instruments, it can be expected that future programmes with instrumented piles will have different testing regimes than LDPT and additional development work will be required.

Experience with LDPT would indicate that, in a future instrumented-pile project, instrument development should first be completed before proceeding with other major contracts.

(*h*) Redundancy of instruments continues to be a wise choice for pile tests. In relationship to the overall costs of a programme, a 50 per cent increase in number of instruments above the minimum required would perhaps not increase costs significantly. There was a sufficient redundancy of instruments in the LDPT programme.

References

BSC (1972), *Macalloy Prestressing Ground Anchors*. Publication No. 44, Second Edition, April 1972, British Steel Corporation, Motherwell, Lanarkshire, UK.

BSC (1982), *The RQT Series, BSC Plates*. British Steel Corporation, Motherwell, Lanarkshire, UK.

DEPARTMENT OF ENERGY (1987), *Offshore Research Focus*, No. 61, October 1987

MURFF, J. D. (1980), Pile Capacity in a Softening Soil. *International Journal of Numerical and Analytical Methods in Geomechanics*, Vol. 4, pp. 185-189.

SOLOMON, I. J., COX, W. R., CLARKE, J., and POSKITT, T. J. (1992) The development of instrumentation for drivability testing and load testing of large diameter piles. This volume.

11. Large diameter pile tests data acquisition systems

C. CURRIE, W.S. Atkins Consultants Ltd, and I. SOLOMON, Fugro-McClelland Ltd.

The Large Diameter Pile Tests Programme took place between 1985 and 1988. A pair of comprehensively instrumented piles were driven and subsequently load tested; one at a normally consolidated clay site in Shropshire (the NC site); and the other at an overconsolidated clay site in Cambridgeshire (the OC site).

This paper is one of a series of papers on the planning, execution and results of the project. The paper discusses the factors influencing the design of a large multi-channel data acquisition system, and goes on to describe the solutions adopted for the project, and the performance of the system during the test programme.

The data acquisition system was used at two sites, to collect data from instrumented piles during driving, to record pressures and stresses during the set-up period, and to monitor short term trends during load tests. The system consisted of signal conditioning units, patch units, tape recorders and a computer system. It operated successfully and proved to be reliable and adaptable. Problems with vibration and earthing were anticipated and measures were taken to reduce these.

Function

The Data Acquisition Systen (DAS) was designed to collect data from the instrumented piles during driving during the set-up period, and during the load tests. In addition the DAS was used to interface with other users of the data:

(*a*) During driving, signals were passed to the on-site drivability analysis equipment and analogue signals were recorded on tape recorders,
(*b*) during load tests, signals were passed to plotters used to control the rate of loading,
(*c*) after testing, data was given to participants wishing to carry out further analysis.

Requirements

Flexibility

The DAS had to be capable of being used at both the NC and OC sites without major reconfiguration being necessary for the differing instrumentation regimes at the two sites.

Signal conditioning

The DAS had to interface with several different sensor types and the first setting was one of converting, filtering ,offsetting and amplifying the signals to provide peak voltages of about ±5 volts. This is a suitable voltage range for computer acquisition of data, and for tape recorders, plotters etc.

Interfacing

The amplified signals had to be output to a number of devices, each of which required a different selection and order of channels. The devices receiving signals from the DAS included:

- the DAS computer
- tape recorders
- drivability analysis equipment
- plotters

Logging

Approximately 100 channels of data had to be logged but the requirements for each stage of the test were different:

(*a*) during driving, dynamic data was required to record the response for a few milliseconds during each hammer blow;
(*b*) during set-up, long term trends were required over a period of several months,
(*c*) during load testing at lest one reading every minute was required from each channel.

Reliability

During the critical phases of the tests, the DAS had to provide a back-up method of data storage where practicable to guard against one method failing. For example, during load testing, analyogue recordings were made of all signals at the same time as they were being recorded digitally by the DAS. In addition, possible defects of the signal conditioning during the set-up phase had to be catered for, particularly the re-establishment of bridge balancing resistances to maintain the correct offset.

Outline description

The DAS consisted of the following major subsections:

- Signal conditioning units - used to amplify and filter the signals from the sensors,
- Patch units and tape controls - to monitor and distribute selected sensor signals and to provide remote control of up to eight analogue tape recorders
- Analogue tape recorders - used as the primary means of data acquisition during pile driving and as a backup system of data acquisition during the load testing,
- Computer system and peripherals - used to control operation of the DAS to digitise the sensor signals and to store the digitised data for subsequent analysis.

The system was housed in a specially strengthened air conditioned cabin that was transported between sites. Most of the DAS was housed in two cabinets as shown on the following page.

The cabinets which were approximately 1.9m high, also contained fan units for cooling ad a ferro-resonant ultra isolation transformer to smooth any transients in the power supply. Individual subsections making up the DAS are described in more detail in subsequent sections.

Signal conditioning

The signal conditioning system consisted of seven subracks each with a capacity for 16 channels. The output from the signal conditioning was fed to a 128 channel data highway (the record data bus) contained in the record patch unit.

The majority of subracks contained general purpose amplifier circuit boards – with each board carrying two channels. Low-pass filters were set at 7.2 kHz to cater for dynamic data during the pile driving and these were not altered subsequently. Otherwise, the boards were configured to suit the sensor type as shown in Fig. 1.

Strain gauges, pressure and load cells

These sensors were all resistive sensors configured as one or more arms of a Wheatstone bridge. The signal conditioning boards for these sensors provided a bridge excitation voltage, and had precision resistors installed as necessary to complete the bridge.

Balancing of the bridge was achieved by switching in a series of standard resistors rather than using a variable resistor (potentiometer) – this was to enable possible replacement of any board during the set-up period without

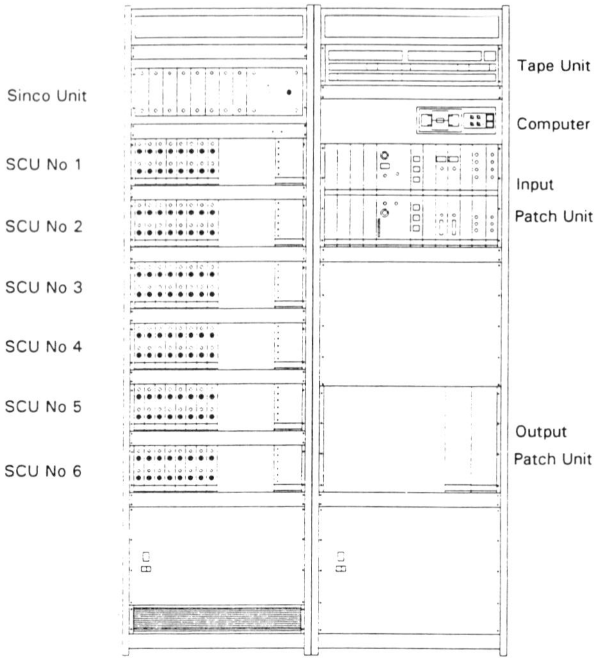

Fig. 1. Front view of cabinets

losing the offset reading. Gains of up to 2000 were used to amplify the signals to nominal full-scale ranges of the order of ±5 volts.

Accelerometers

Previous experience with the accelerometers used had indicated that noisy signals were possible due to the sensors being electrically connected to each other via the pile wall, leading to the formation of earth loops. Special floating input circuitry was designed that would electrically isolate the sensors from one another, preventing this type of problem from occurring. The isolated signals were then amplified in the normal way to the same levels as the other signals.

This floating input circuitry was in fact later found to be not strictly necessary as it had been decided, during the instrument development phase, to install the accelerometers inside protective housings of electrically insulating material thus ensuring that there was no electrical connection between sensors.

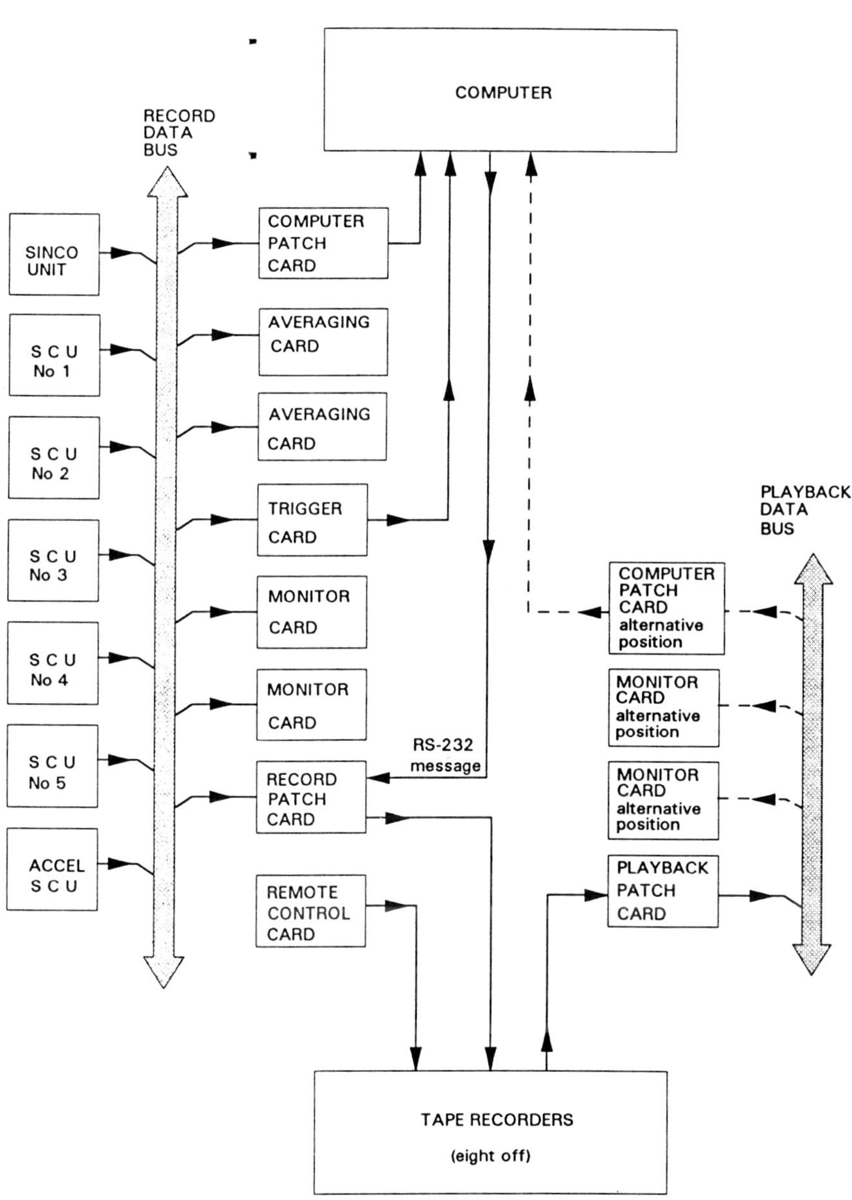

Fig. 2. Block diagram of the Data Acquisition System

Vibrating wire strain gauges

The gauge manufacturer's standard signal conditioning circuits were used which provided a variable frequency output proportional to the square of the strain. These outputs were converted to variable voltage signals by specially designed converter circuits to enable a standard logging system to be used for all sensors. The square root conversion of the signal to make it proportional to strain was carried out later by the software during analysis.

Patch units

The two patch units were housed in the cabinet adjacent to the signal conditioning units. The record patch unit had three main functions

Selecting channel order

Eight 14-channel tape recorders were used to record data. A record patch board was used to select signals in groups of 13 from the record data bus and direct these along with the 14th channel (which was used for identification marks and synchronisation signals) to the appropriate tape recorder. Similarly a computer patch board was used to select signals for logging on the computer.

Providing signals for other users

Two averaging boards were used to pick up four signals from the record data bus and provide these as four individual outputs, one output as the average of all four signals, and two outputs as the average of each pair of signals.

During the pile driving in particular, this provided a significant degree of redundancy so that the drivability analysis would not be affected by any sensor failures. Two monitor boards were also available. These were used to dial up any channel on the data bus for diagnostic purposes.

Tape recorder control and synchronisation

A remote control board was designed to provide remote control (stop, play, rewind, etc.) to the eight tape recorders. It was also necessary to devise a system to enable synchronisation and unique identification marks to be recorded on the tapes so that the each tape could be replayed subsequently – it would be impossible to load the tapes to exactly the same start position each time.

Trigger. A trigger board was developed to issue a pulse to the computer. The trigger board monitored a signal level and issued the pulse if the signal exceeded a threshold value for longer than a specified time (thus avoiding spurious triggers caused by noise). Multiple triggers caused by the same hammer blow were avoided by providing a lock-out period immediately following a valid trigger pulse such that no further trigger inputs were accepted for a selected period.

Blowcount. The trigger pulse was interpreted by the computer which kept a tally of the blow number and issued an R5232 message consisting of the blow number and the date and time. This message was recorded on channel 14 of all the tape recorders and printed out on a printer during the test.

The playback patch unit provided a data bus similar to the record data bus but carried the output signals from the tape recorders (either during playback or during recording for monitoring purposes). Some of the above boards were used in either bus depending on the requirements of the stage of testing.

Tape recorders

Eight Racal Store 14 tape recorders were used to record data. During pile driving, these were the primary method of dynamic data collection (although the drivability analysis equipment did provide a back-up by collecting data from a limited number of blows and channels).

During load tests the tape recorders were used purely as a back-up storage device. The tape recorders were mounted vertically on specially designed racks and were all controlled centrally via the remote control system.

The tape recorders stored data on open reel magnetic tape. The bandwidth or frequency input range that it was possible to record varied with the tape speed.

During driving, the high frequency content of the signals meant that the tapes had to be run at a high speed, and a 2400 foot tape would only last for about 30 minutes. During load testing, the signals were not changing so fast, and the requirement to record high frequencies was relaxed. This meant that the tapes could be run at a lower speed, with a 2400 foot tape lasting for about 5 hours.

Computer hardware

The computer system was based on a Digital (DEC) micro PDP- 11/23. At the time this was considered to be a very powerful and compact computer. The system was housed adjacent to the signal conditioning.

Computer

The computer consisted of:

(*a*) 11/23 processor
(*b*) 512 KByte memory
(*c*) 10 MByte hard disc drive
(*d*) two 5 $^1/_4$" floppy disc drives
(*e*) serial port interfaces
(*f*) a calendar/clock to maintain date and time during a power failure

(*g*) two 64-channel analogue to digital (a/d) converters
(*h*) a real-time clock board to control the data acquisition sampling rate.

Data conversion

The continuous analogue input signals had to be converted to a stream of discrete digital values to make them suitable for computer storage. Two 64-channel analogue to digital (a/d) converters along with a real-time clock (to provide the sampling frequency) were used. The a/d converters had a maximum throughput of 250 kHz; this implied an absolute maximum sampling speed of 4000 kHz for 64 channels. The required sampling speeds (and intervals between samples) were

(*a*) 0.03 Hz (30 secs) during set-up
(*b*) 0.5 Hz (2 secs) during load tests
(*c*) 15 kHz (0.07 millisec) for dynamic data during driving.

The requirements for acquisition of dynamic data from all relevant channels were too onerous for the budget and timescale of the project. It was therefore agreed to use the computer to monitor static values in the gaps between hammer blows at a sampling speed of 4 Hz (0.25 sec interval between samples).

Peripherals

Peripherals included a terminal for operator interface, a printer, and a cartridge tape drive for back-up of data from the hard disc. The DC300 tape cartridges held 10 Mbytes of data; this capacity (which is equal to the size of the hard disc) provided storage for about:

(*a*) a week's data at the slow acquisition rate during the set-up period
(*b*) 10 hours' data at the medium rate during load testing
(*c*) 2 seconds of data if it had been possible to acquire dynamic data for all channels during pile driving

The onerous storage requirements for acquiring dynamic data were another reason why it was decided to acquire static data only during the pile driving.

Software

One of the key requirements of the data acquisition software was to minimise the data handling overhead of the computer. Reading to and from disc had to be kept to a minimum and the data had to be stored in a very compact and efficient form if the computer was to be able to keep up with the volume of data coming in.

System software

The standard DEC RSX-11M operating system was used. It is a multi-user, multi-tasking system and this enabled a number of tasks to be run simultaneously, for example

(*a*) collection of data
(*b*) dealing with trigger signals
(*c*) backing-up data
(*d*) displaying data

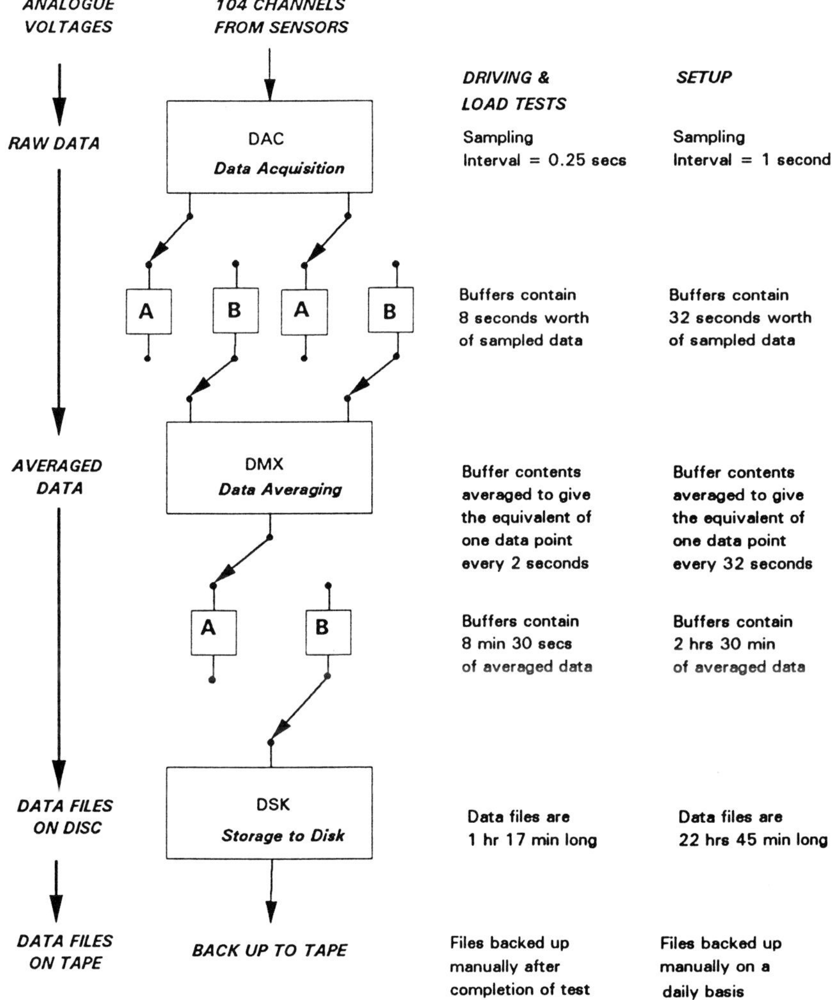

Fig. 3. Data acquisition and initial processing

In addition, the standard manufacturer's routines were used to drive the a/d converters. One of the features of these routines was the use of direct memory access (DMA). DMA enables the a/d converters to write data directly into memory. Normal data transfer would require the a/d converters to send data to the processor which would then store the data in memory. DMA removes the bottleneck caused by the processor and greatly improves the throughput and efficiency of the system.

Initial data acquisition

The a/d converters collected 12-bit data, i.e. for a full scale range of ±10 volts the corresponding digital values were 2048. The resolution (1 bit) was therefore 5 millivolts. The 12-bit data was split into two 8-bit bytes and the data remained in this compact form throughout the subsequent processing.

A disadvantage of DMA is that areas of memory being filled with data cannot at the same time be read. To get round this, small packets of data were collected on a double buffer system. This can be compared to a two drawer filing cabinet – while the top drawer is open the second must remain closed. The memory was defined to provide two front end storage areas (buffers); while one buffer was being filled the other was processed to a different area of memory not governed by the DMA rules. The time taken to fill these front end buffers was

(*a*) 8 secs during driving and load tests;
(*b*) 32 secs during set-up.

These time lags between the data being collected and being available for inspection were found to be acceptable for this project. For future systems it would be desirable, if technology permits, to be able to use the system as a multi-channel oscilloscope, displaying graphs of signals in real time without this time-lag.

Initial processing

The raw data released from the front end buffers was processed and stored in a further set of double buffers (averaged data buffers) as follows:

Spike detection. Spikes were defined as being up to three data points which varied by more than a threshold value from the mean. A record was kept of spikes detected for each channel and this proved to be a useful method of diagnosing noisy signals. In addition, the spikes could be removed and this was used to remove the data point corresponding to a hammer blow from the static data during driving.

Averaging. It was decided that, for the slow sampling speed (1 sample every 32 secs), better quality data would be obtained by taking 1 sample per second and then taking the average of every 32 points.

This system was also used during the faster acquisition with initial sampling at 4 Hz and the average of 8 points being taken to give an averaged sample every 2 seconds. This meant that only two sampling speeds were used as the requirements for the pile driving and load testing could both be fulfilled by the faster acquisition rate.

Demultiplexing. Data acquisition systems collect data in timestep order (a series of channels for each timestep) and it is normally required in channel order (a series of timesteps for each channel). The transformation (which can be likened to swopping rows and columns in a table) is called demultiplexing and was carried out at this stage.

Storage

The data from the averaged data buffers was transferred to data files on the hard disc. The data was stored in its raw format with a file header being created for each file containing the calibration information to convert the data to engineering units. The data was written to file in 256-point frames and the files were restricted to 10 frames in length.

Monitor

The data in both the raw data buffers and the averaged data buffers could be inspected in real time using the monitor programme. The data could be displayed in either bits, volts, or engineering units. In fact, the displays in bits and volts proved to be most useful for diagnostic purposes during the driving and testing as a full scale reading of either 2048 or 10 volts was easily detected and indicated a failed channel.

SPiDAR

The software package SPiDAR was used to give analysis and graphical display of data in the data files. As the programme displayed the data, it was calibrated into engineering units. The programme automatically displayed each channel in turn unless commands were received from the operator to change channel or frame or to carry out zoom functions.

A large number of routines were available within SPiDAR apart from the plot routines described above. These extended to frequency analysis and manipulation of the signals (e.g. the square root conversion of the vibrating wire strain gauge signals).

Operation

During driving and load tests, the DAS was manned by three engineers. Two engineers carried out a routine of monitoring the status of all sensors, signal conditioning, tape recorders and the computer. The other engineer was responsible for communicating key information on the progress of the test to other parties. During the set-up period the DAS ran unattended apart from a daily routine check.

Power supply

The power supply for the instrumentation cabin and DAS was from a 30 kVA generator. A back-up generator was also provided; this was connected so that it would be utilised immediately in the event of failure of the main generator thus minimising any potential loss of data.

As the generator was generally lightly loaded it tended to run fast causing the standard clock in the computer to gain time as it depended on the supply frequency. The calendar clock board was therefore used to update the computer standard clock at regular intervals.

Earthing

Various configurations of earthing were available to avoid earth loops and the minimise the effects of mains noise. Noise levels were generally reduced to about 5 mV although noise levels were as high as 50 mV on a few channels. The averaging system used during data acquisition, however, compensated for even the highest levels of noise.

Vibration

To isolate vibration effects on the DAS, the instrumentation portacabin was mounted on sleepers and a 60 mm layer of vibration absorbing composite (Ticopad).

Reliability

The system operated satisfactorily after commissioning with only a daily back-up of the data file being carried out and a weekly change of tape cartridge during the set-up period. The DAS computer broke down on one occasion during the set-up period and manual readings were taken until it was repaired.

Vibration caused some minor problems with certain channels on some tape recorders while driving the OC pile. This was due to the proximity of the cabin to the pile and the fact that the pile was not sleeved over the initial length as the NC pile was. However, the fault was detected during the routine monitoring of the recorded data and the signals were recorded on spare channels of the tape recorders without significant loss of data.

Conclusions

The DAS performed very well throughout the testing period and proved to be very reliable and adaptable to the requirements of both sites. It was designed and built during the first half of 1987 to a tight timescale and budget and many developments have taken place in technology since then. However, any similar project to the Large Diameter Pile Tests would require the development of a similar Data Acquisition System. Whilst the hardware and software would undoubtedly be different, the general principles would be applicable to any such system.

12. Measurements of axial strain and pile wall displacement, and the determination of skin friction during the driving of instrumented piles

T. POSKITT, Queen Mary and Westfield College,
K. YIP-WONG, Maunsell & Partners and W. COX,
Fugro-McClelland Marine Geosciences Inc.

Data is presented on axial strains and accelerations, and on radial accelerations obtained during the driving of three instrumented test piles. The load transfer characteristics mobilised during driving are obtained and these are shown to vary as driving progresses. The cause is remoulding of the soil at the interface between the pile wall and the soil, and, reductions in the horizontal pressure between the soil and pile wall caused by flexure of the pile. Finally, a comparison is made between the quakes during driving and values obtained from static load tests.

Introduction

The method of installing a pile has an important effect on its skin frictional resistance. When piles are driven into the ground, soil in contact with the pile wall is remoulded, and, lateral movement of the pile wall causes changes in the soil pressure on the pile wall. The result is that skin friction will not be strictly related to the shear strength.

There is evidence to suggest that for some cohesive soils the skin friction reduces as driving progresses; a phenomenon sometimes called friction fatigue, Hereema (1979). It should not be confused with the reduction in effective stress caused by the generation of pore pressures as the pile penetrates the ground. It depends upon the strength of the soil and Tomlinson (1957, 1971) has noted a marked tendency to a decrease in adhesion factor with increasing shear strength. He goes on to say that 'This is no doubt due to the failure of the stiffer clays to close the gap caused by flexure of the pile during driving, and probably to a greater tendency to cracking of the soil and infiltration of water around the upper part of the shaft in clays of high shear strength.'

From this it appears that three mechanisms affect the skin friction, viz

(*a*) remoulding with possible changes in moisture content,

Large-scale pile tests in clay. Thomas Telford, London, 1993

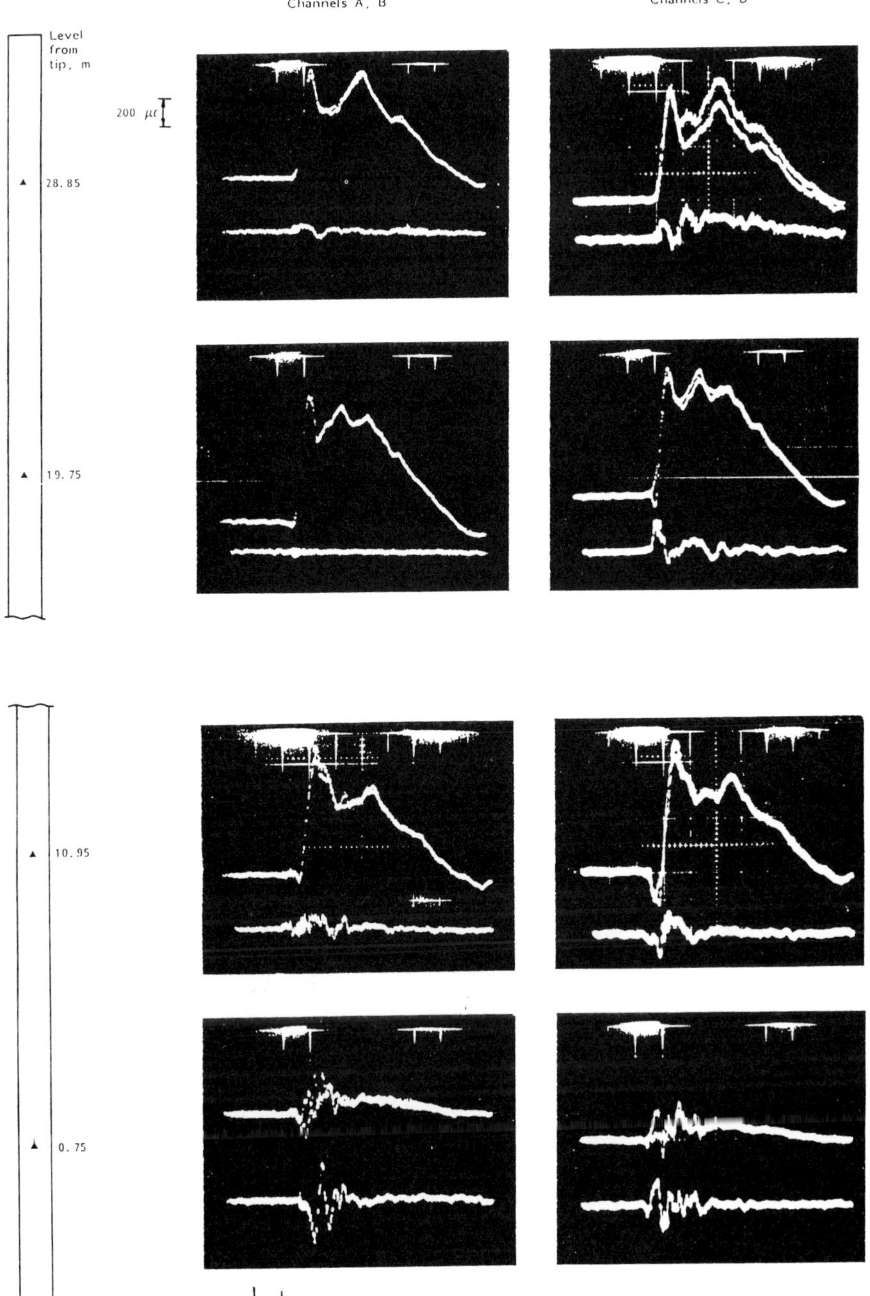

Fig. 1. Bending strains 30 m penetration, OC pile (top trace – individual channels; bottom trace – difference between channels

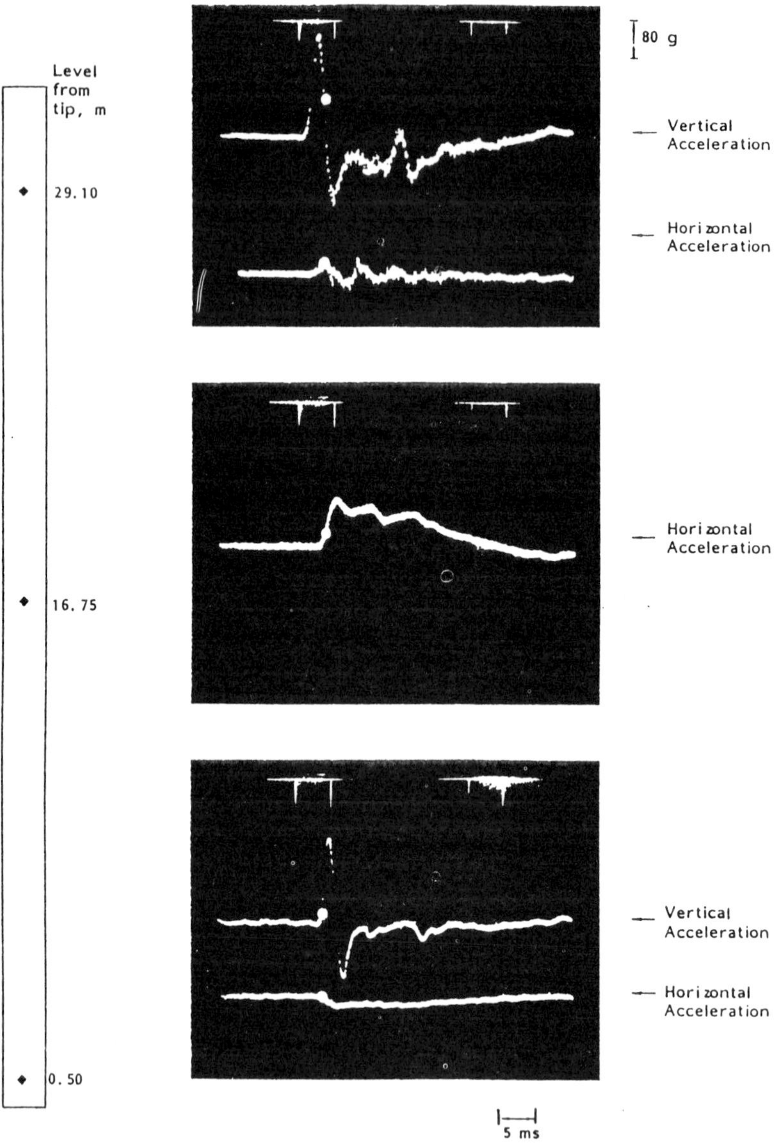

Fig. 2. Accelerometer signals, 30 m penetration, OC pile

(*b*) changes in pore pressures, and
(*c*) flexural deflexions of the pile with potential loss of contact between pile and soil.

In order that the installation of the test piles should be representative of conditions offshore, it was necessary that the driving procedures should simulate those offshore. To ensure this, elaborate instrumentation was included on the test piles in order that all the factors affecting the driving could be recorded.

During driving a high level of surveillance was maintained using the instrumentation. This enabled aspects such as hammer energy, penetration resistance, axiality of the blow etc. to be frequently examined.

Instrumentation

Field testing was done at two test sites. The first at Tilbrook Grange in Cambridgeshire was heavily overconsolidated Till overlying Oxford Clay with an average undrained shear strength of 470 kPa. This is referred to as the OC site. The second at Pentre in Shropshire was normally consolidated silty clay with layers of clayey silt of average undrained shear strength 87.5 kPa, and is referred to as the NC site.

One partially instrumented and one fully instrumented pile were driven at Tilbrook. These are referred to as the TP and OC piles respectively. At Pentre only one fully instrumented pile, referred to as the NC pile, was driven.

The TP pile was a tube 762 mmOD, 35 mmWT, and 33 m long. It was driven in two sections the first of which was 15 m long and carried no instruments. The second section was 18 m long and was spliced to the first section in the field. It carried instruments 3.5 m from its top.

The TP pile offered a valuable opportunity to field test the prototype strain gauge and acceleration transducers. These were attached in the factory on the outside wall of the pile. A taped record of the 2231 blows required to advance the pile from a penetration of 12.25 m to 28 m was obtained. Because of the similarity of the TP and OC piles it also provided valuable corroboration of results which were subsequently obtained from the OC pile.

The OC pile was a tube 762 mmOD, 30 mmWT, and 32 m long. It carried 72 transducers attached to the inside wall of the tube. Signal cables were carried in channels. These channels and their protective shoes increase the blockage at the pile toe from 0.0690 m^2 to 0.1620 m^2. A taped record of the 2858 blows required for its installation was taken.

The NC pile was also tubular with an outside diameter of 762 mm, wall thickness of 15 mm, and 55 m long. It carried 92 transducers and again had

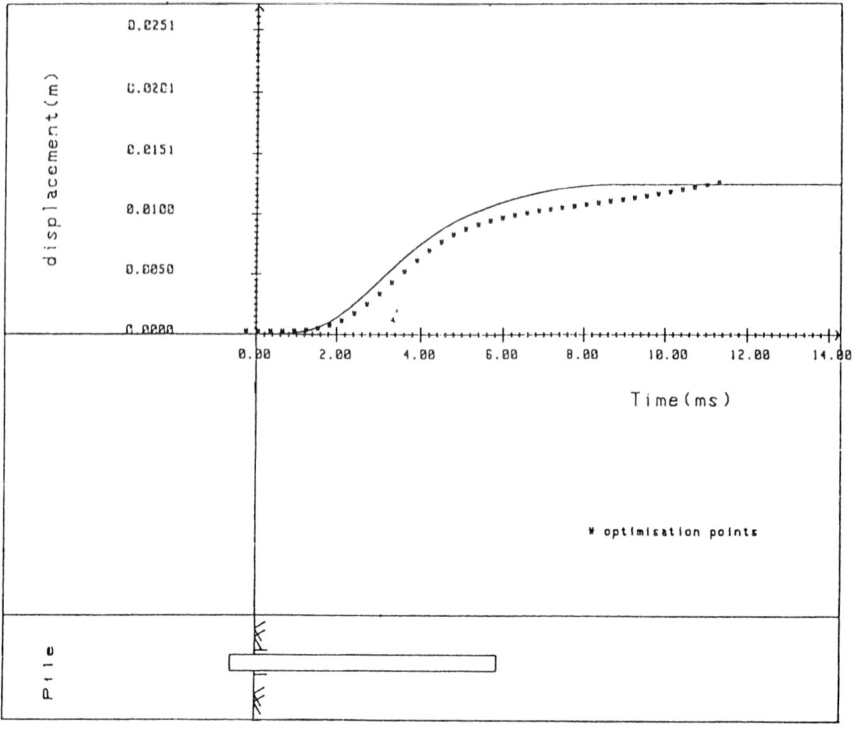

Fig. 3. Displacement–time curve, OC pile, 30 m penetration

internal channels for the signal cables. A taped record was taken of all 2243 blows required to install it.

The hammer used at Tilbrook was the BSP HA40, rated energy 470 kJ and ram weight 40 t. At Pentre the BSP HH7, rated energy 84 kJ and ram weight 7 t was used. In both cases the driving was only moderately difficult.

The sensors consisted of strain gauges, both bonded and spot-welded, accelerometers, pore pressure and total pressure cells, and temperature sensors. Of these, only the strain gauges and accelerometers are suitable for studying the rapid transients caused by ram impact. Thus for the dynamic work only strain and acceleration signals are discussed. This is not to imply that the other sensors do not give valuable data between blows.

There is very little experimental data on pile flexure (sometimes called whip) during driving. Since this affects transducers such as pressure cells, it was decided to mount accelerometers with their axes in the radial planes of the pile in order to study horizontal movements of the pile wall. The acceler-

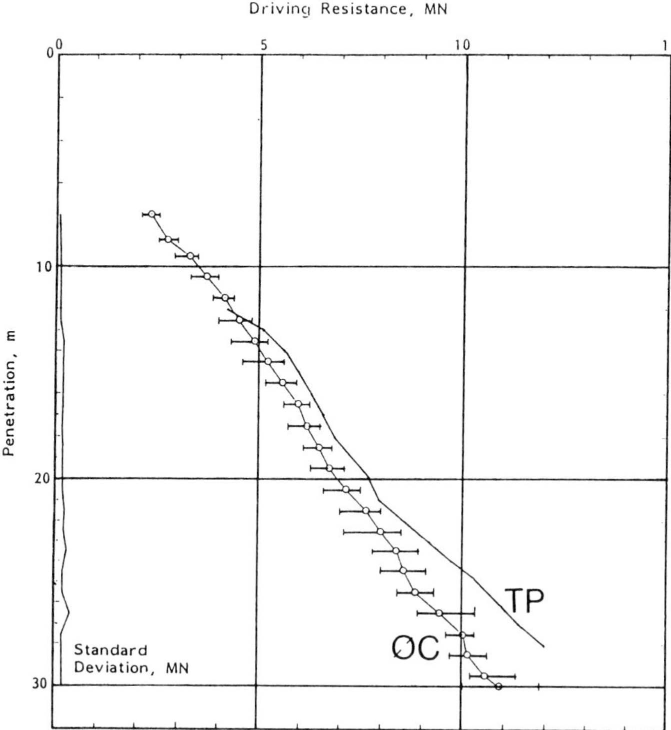

Fig. 4. Driving resistance, OC pile, 2858 blows; TP pile, 2231 blows

ations which were recorded are thus axial and radial. On the TP pile they were at the pile head only, while on the OC and NC piles sensors were mounted at the head, centre, and toe of the piles.

Corresponding to the acceleration measurements strain records at these locations are also considered. To summarise therefore, the data analysed is for axial strains, and for axial and radial accelerations at the top centre and toe of the two fully instrumented test piles. In the case of the TP pile results are for the top only in both the main drive and the re-tap test. Sensors were mounted in pairs diametrically opposite each other. This allowed signals to be averaged or differenced electronically both during and after driving.

Measured signals

During the installation of the three piles over 7600 blows were recorded. All of these blows have been studied, although it is not possible to give more than a few representative signals here.

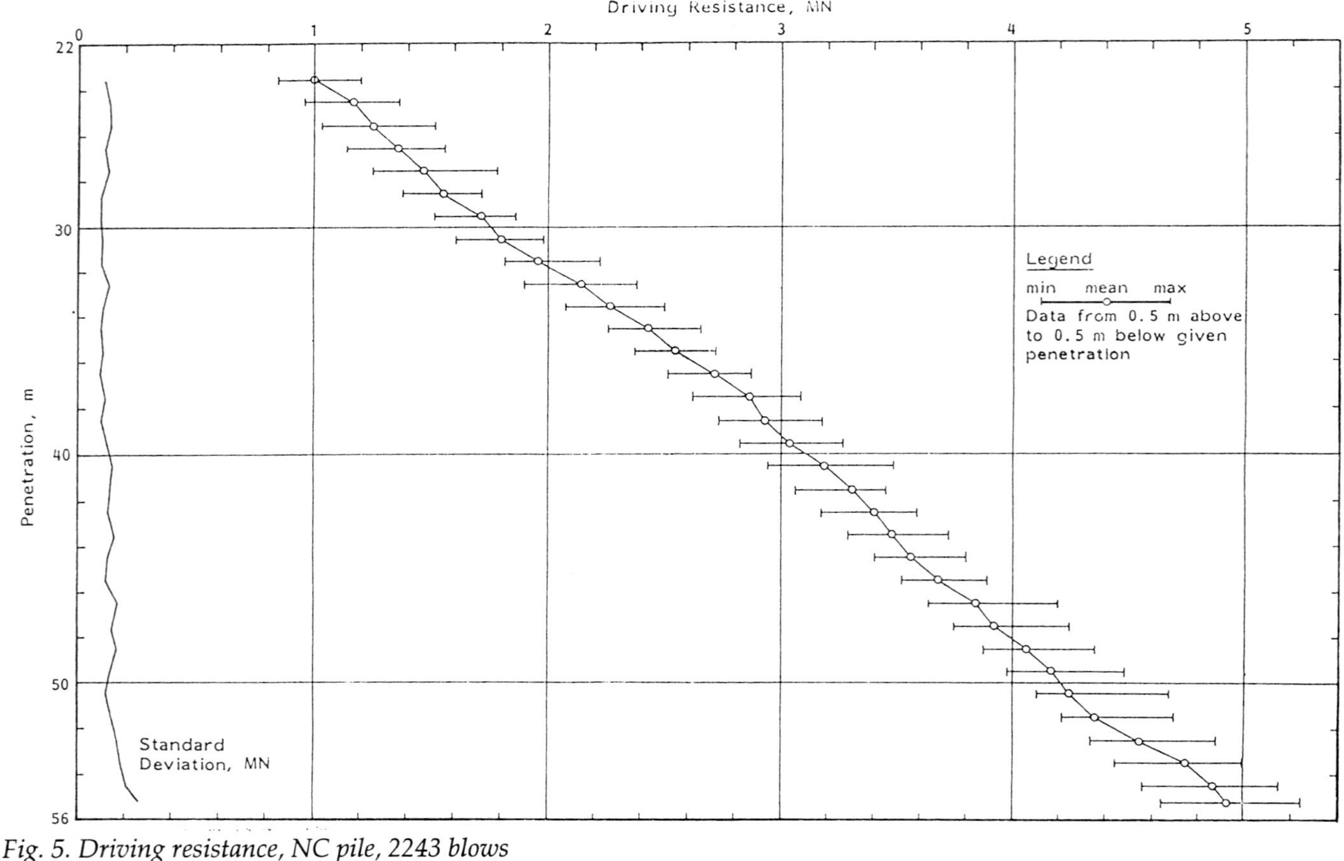

Fig. 5. Driving resistance, NC pile, 2243 blows

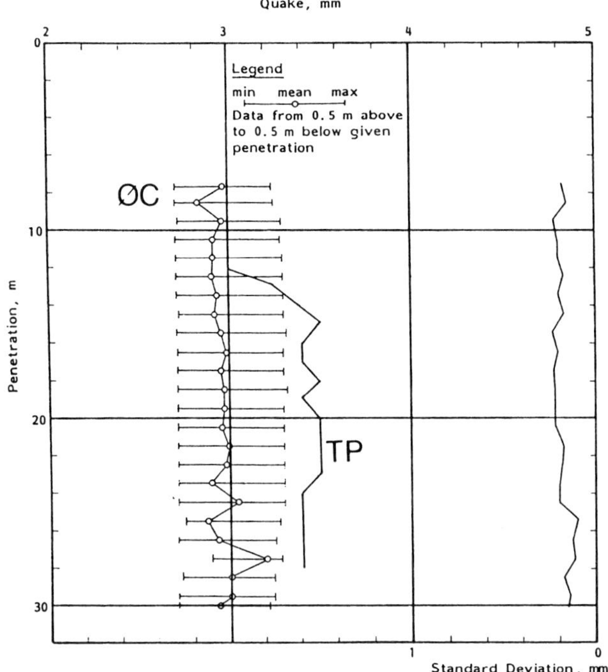

Fig. 6. Quake, OC and TP piles

It was found that the bonded HBM foil gauges performed well under all driving conditions. Typical results for the OC pile are shown in Fig. 1. Similar signals were obtained for the TP and NC piles. The Ailtech spot-welded gauges showed a tendency to spikiness, which under hard driving conditions was judged to be quite serious.

The acceleration signals showed that the problem of resonance under moderately hard driving conditions had not been solved entirely satisfactorily. Of special interest is the unexpectedly high horizontal acceleration at the pile centre, Fig. 2.

Signals such as those shown in Fig 1 were sampled and processed during driving. The maximum sampling rate which could be achieved was about 1 in 15–20 blows. This corresponds to a processing time of about 30 s/blow. It provided useful information during driving of hammer energy, pile penetration resistance, blow eccentricity etc.

Generally the ram eccentricity was not large, however, it was sufficient to impart a significant flexural impulse into the piles. The result was a low-frequency whipping of the piles.

The OC pile shows a surprisingly high acceleration at the centre with smaller values at the top and toe, Fig. 2. The corresponding values for the NC

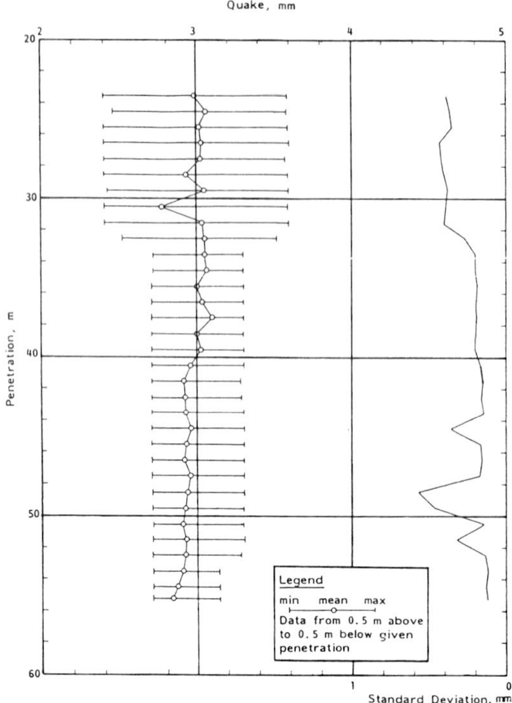

Fig. 7. Quake NC pile

pile were much smaller. Because the NC pile is longer and more flexible than the OC pile it nevertheless showed as much tendency to whip as the OC pile.

It was concluded that the mechanism which generates lateral motion of a pile is in two parts. The first is a non-uniform stress wave which travels down the pile at the normal wave speed. As it passes any particular cross-section of the pile, it generates a flexural impulse at that section. This flexural impulse then causes the pile to set off on the second stage. This is a low frequency flexural oscillation which can readily be detected by eye, Poskitt (1990).

Nine months after installation a re-tap test was carried out on the TP pile. This was performed with the BSP HA40 hammer operating at maximum energy. The first blow caused massive resonance in the accelerometers and there was no detectable set. During the next few blows as the pile began to break free resonance effects diminished. A total of 274 blows were applied and these moved the pile 0.5 m.

Wave equation parameters

In order to determine the skin friction mobilised during driving a dynamic analysis for the hammer/pile/soil system is required. The most widely

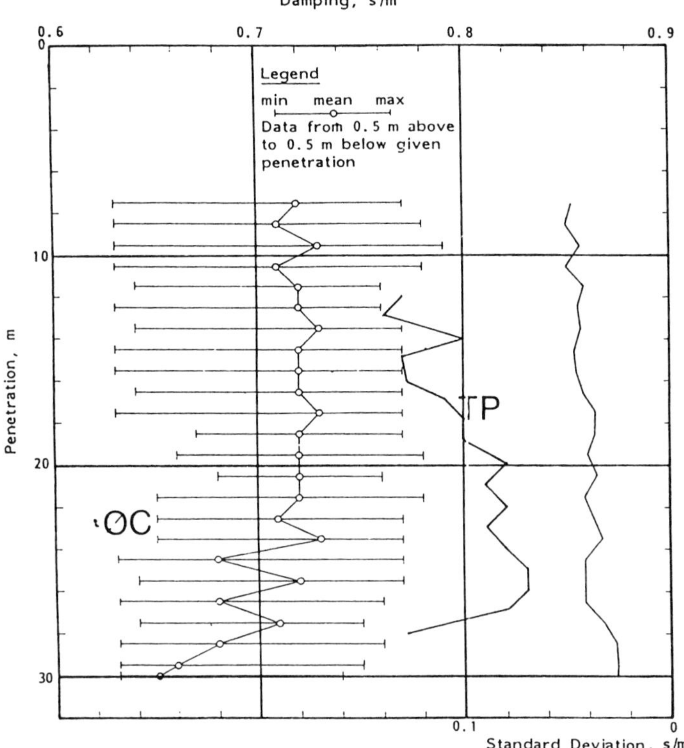

Fig. 8. Damping, OC and TP piles

accepted theory for this is the numerical wave equation analysis of E. A. L. Smith, and this has been adopted here. In Smith's theory the relation between skin friction and pile displacement is assumed to be elastic/plastic at slow rates of penetration. This defines the basic *t-z* curve at any specific depth in the soil. Rate effects, which are important for the Tilbrook and Pentre soils, are accounted for by a damping parameter *J*.

The load transfer characteristics of the soil under dynamic conditions are thus divided into three groups governed by three parameters

(a) an elastic range which extends until the pile slips relative to the soil. The displacement at which slipping occurs is called the quake, *Q*
(b) a plastic range during which slipping occurs at constant shear stress *Ty*, and,
(c) a viscous range.

This depends only on the rate of slipping and leads to an increase in the mobilised frictional resistance. It is related to the pile penetration velocity through the Smith damping parameter *J*. Rate effects are poorly understood

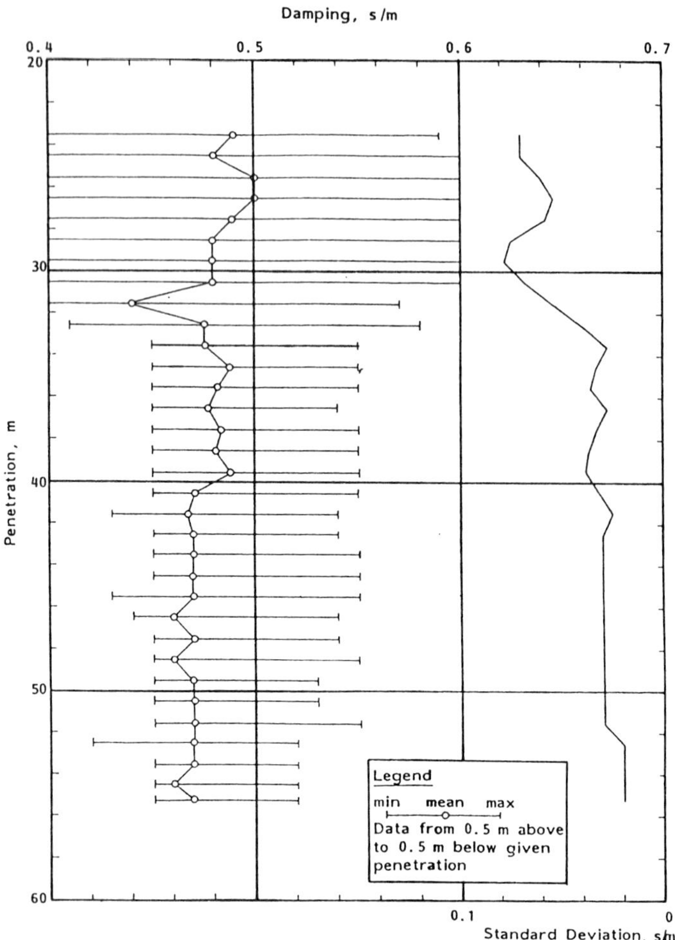

Fig. 9. Damping pile, NC pile

and only simple expressions are justified to allow for its inclusion. In the present it is assumed that rate effects lead to an increase of $J\ V\ T_y$ in the mobilised skin friction, where V is the pile wall velocity.

In pile design it is generally T_y which is of most interest, and Q and J are only significant as a part of the process used for its determination.

The use of strain gauge and acceleration signals for determining T_y, Q and J can be done in a number of different ways. In the method used here, the pile is imagined cut at the transducer location. Continuity is then restored by applying the measured force-time curve to the cut section. The wave equation analysis is then used to study the motion of that part of the pile beneath the

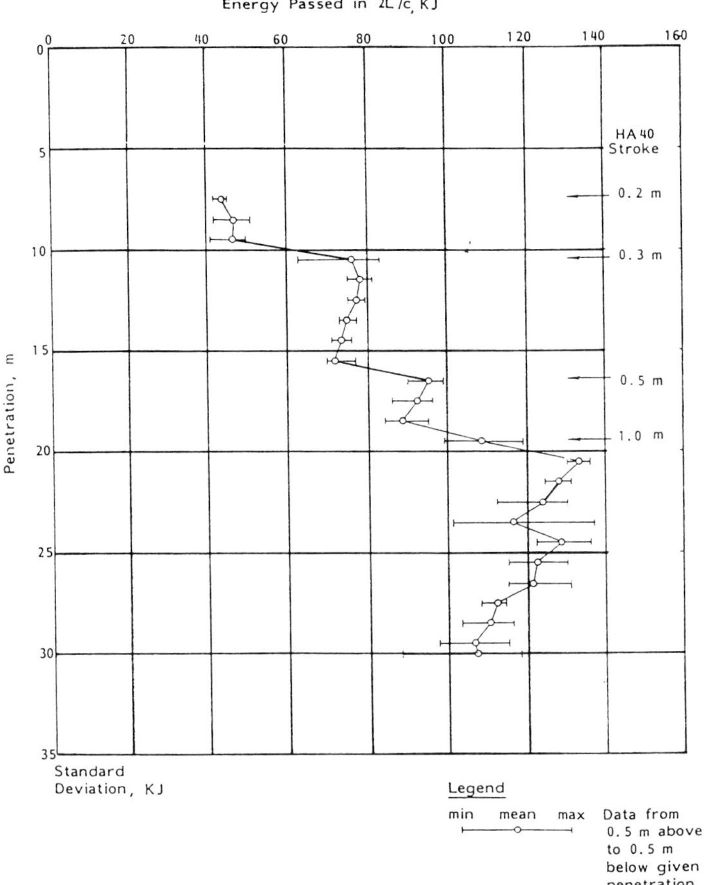

Fig. 10. Hammer performance, OC pile, 2858 blows (BSP HH40)

cut. In particular the analysis is used to calculate the displacement-time curve. This is then compared with the measured curve. The method of nonlinear least sum of squares, Nash (1979), is then used to find the load transfer parameters which give the best fit between the calculated and measured signals.

There are several advantages and disadvantages in this procedure. First, it is fairly quick and on site gives an analysis rate of about one blow in fifteen. This is equivalent to a computation time of about 30 s per analysis. A second advantage is that least squares enables a compromise to be made between errors inherent in the assumptions of the theoretical one-dimensional wave propagation model, and errors in measurements, especially those associated with acceleration transducers and their resonance problems.

A major disadvantage in the method is that it necessitates integration of

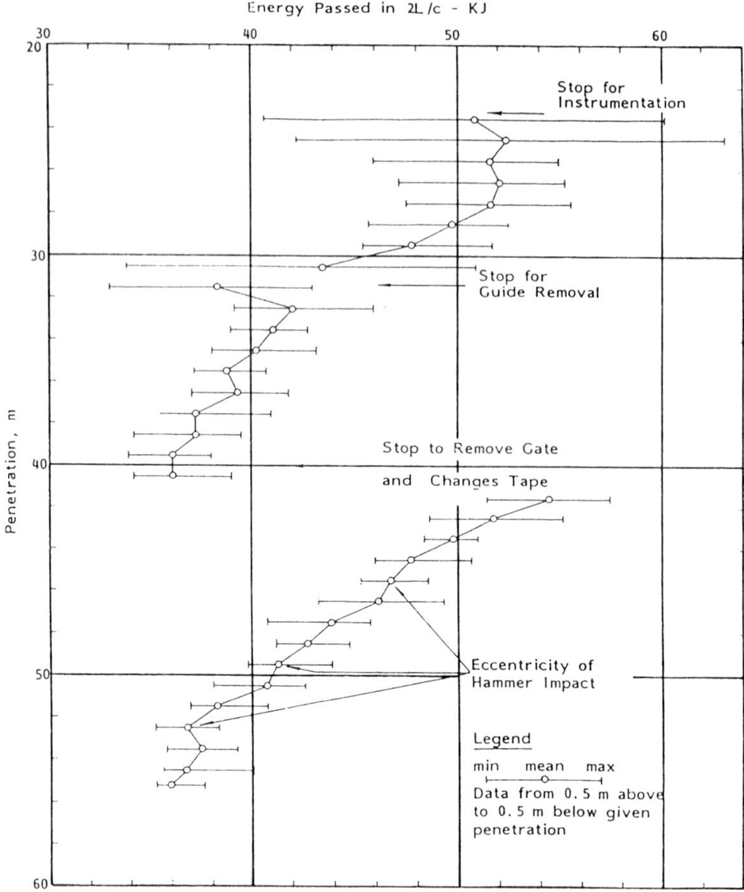

Fig. 11. Hammer performance, NC pile, 2243 blows (BSP HH7)

the acceleration signals, Yip(1988). Spurious components in the measured signals lead to a build-up in errors with time. However, experience suggests that if the integration is restricted to time intervals of the order of that required by the stress wave to travel down to the toe and back again then build-up in errors is not unduly great. Since the results are being used to study changes in the load transfer characteristics (i.e. friction fatigue) as driving progresses, errors in absolute values are not a major concern. This has been confirmed by two independent checks which are given below. A typical integrated signal for the OC pile is shown in Fig. 3. Similar curves were obtained for the TP and NC piles.

Using the above analysis, all the recorded signals for the pile-head sensors on the the TP, OC, and NC piles have been analysed. When doing this, variations of Ty in the different strata have been assumed to be similar to the

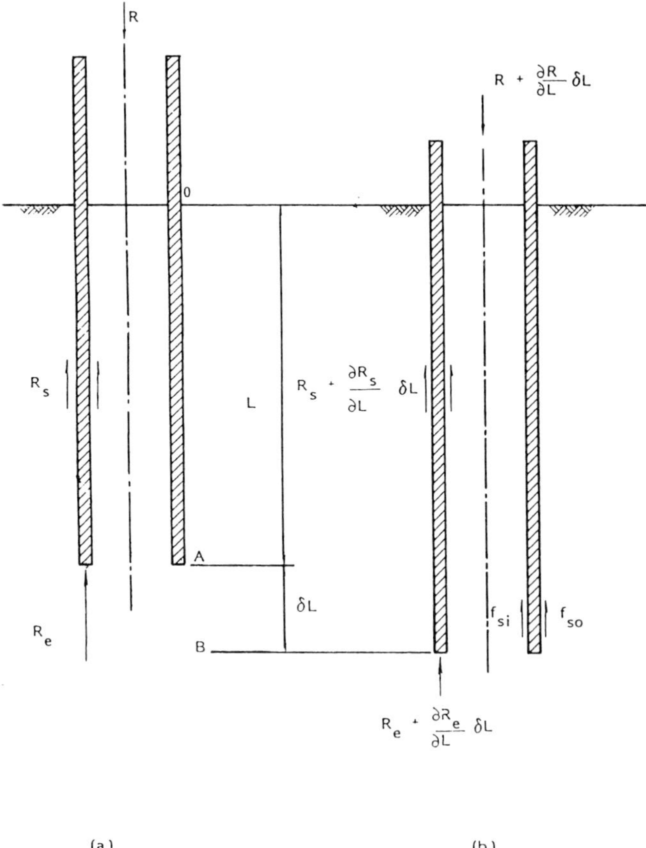

Fig. 12. Failure mechanism during driving

variations in undisturbed shear strength. Any errors introduced by this are likely to be fairly small since it is known that the total driving resistance is relatively insensitive to the precise distribution of resistance along the pile shaft. Frictional resistance due to movement of the soil plug relative to the inside pile wall has been taken as one-half that of the corresponding friction on the outside wall.

Quake and damping have been assumed to have the same value for all the soil strata. This was made necessary by the absence of any information relating to these parameters in the engineering report on the site soils. The driving resistance, quake and damping which have been calculated are shown in Figs 4 to 11.

The most obvious characteristic of the driving resistance penetration curves for the three piles is their linearity. [This confirms the earlier observation that

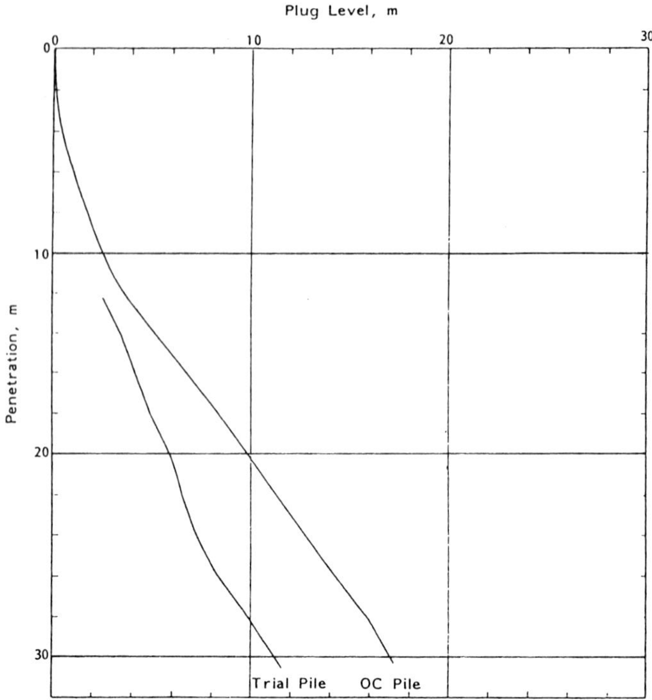

Fig. 13. Plug movements

mobilised skin friction is not strongly related to undisturbed shear strength.] If the curves are projected backwards they cut the vertical axes at 2.5, -0.6 and -14.6 m.In other words they do not deviate much from passing through the origin. The deviation of the TP pile is the greatest, but it should be remembered that this is being extrapolated over the first 12 m of penetration.

The quake values of the clays at the two sites are remarkably similar and in the region of 3 mm. This is a widely used value in wave equation studies and agrees well with 0.1 in. used in classical pile driving formula. The load tests which were carried out oh the OC and NC piles provided load displacement curves of high accuracy. Poskitt and Ward(1988) have described a procedure whereby load test data can be used to make estimates of quake. The procedure uses the initial slope of the loading curve and the rebound from the unloading curve. When this is applied to the OC and NC piles quakes of 3.4 and 3.5 mm respectively are obtained. These agree remarkably well with the values obtained by optimisation.

The damping values for the Tilbrook and Pentre soils are in the range which would be expected from the literature. The Pentre soil is a silty clay and has a noticeably lower damping parameter than the Tilbrook clay.

Figs 10 and 11 show the energy delivered to the OC and NC piles by the HA40 and HH7 hammers in the time it takes for the stress wave to travel to the toe and be reflected back. Some variation in hammer output can be seen as the hammers get hot.

The general conclusion is that the method for analysing the dynamic signals is consistent and provides reasonably good estimates of the dynamic load transfer parameters.

Results

Referring to Fig. 4 it can be seen that the OC pile drove more easily than the TP pile. This is due to the shoes at the toe of the OC pile which protect the cable channels. They reduce by 42.5% the area of the inside wall which contributes to friction from the plug. The net effect is a reduction in the driving resistance of the OC pile.

The instruments on the TP pile were removed when the penetration reached 28 m and the remaining 2 m were driven without monitoring. By extrapolating the driving resistance curve in Fig. 4 to final penetration a driving resistance of 13 MN is obtained. The re-drive oh this pile was carried out 9 months later, and the driving resistance was then found to be 22 MN. The set-up was therefore 9 MN.

The most striking feature of Figs 4 and 5 is the linearity of the curves. There is clearly little direct relationship between the mobilised skin friction and the distribution of the undisturbed strength with depth. The mechanism which causes this is remoulding of the soil at the pile/soil interface, and, reduction in the horizontal pressure caused by flexure of the pile during ram impact. As a result there is a loss of friction on the pile shaft which offsets the gain in resistance which comes from the pile penetrating further into stronger soil; i.e. friction fatigue.

Referring to Fig. 12(a) consider the pile when it has been driven to a penetration L. The total driving resistance R is made up of a shaft component R_s and an end component R_e.

Following the application of a few blows the pile will penetrate a small amount dL and R, R_s and R_e will increase as shown in Fig. 12(b). In addition, due to the shaft having penetrated a further dL there will be an increase in shaft resistance due to the friction on the extra length AB:

$$f_{so}\, p_o\, \mathrm{d}L + f_{si}\, p_i\, \mathrm{d}L$$

where $p_{o,i}$ are the perimeters of the outside and inside of the pile For equilibrium

$$\frac{\partial R}{\partial L} = \frac{\partial R_s}{\partial L} + f_{so}\, p_o + f_{si}\, p_i + \frac{\partial R_e}{\partial L} \qquad (1)$$

With cohesive soils the end resistance depends upon the undrained shear strength and is not dependent upon the depth. Thus to a good approximation $\partial R_t/\partial L$ is zero. Hence equation (1) becomes

$$\frac{\partial R}{\partial L} = \frac{\partial R_s}{\partial L} + f_{so}\, p_o + f_{si}\, p_i \tag{2}$$

If the soil plug does not move then $\partial R_t/\partial L$ will be due entirely to friction fatigue. This was the case with the NC pile. With the TP and OC piles the soil plugs did move, Fig. 13, and so $\partial R_t/\partial L$ will be affected by this. However, it thought that this is not a major factor.

Before equation (2) can be used it is necessary to know f_{so} and f_{si}. An upper limit for these is c_u the undrained strength of the soil at the pile tip. In practice, a better estimate of f_{so} would be the cone sleeve friction, however, this would still leave uncertainty about the value of f_{si}. Using c_u gives an upper estimate for the friction fatigue. With these assumptions equation(2) can be written

$$\frac{1}{p_o + p_i}\frac{\partial R_s}{\partial L} = \frac{1}{p_o + p_i}\frac{\partial R}{\partial L} - c_u \tag{3}$$

This gives the rate of loss of shaft capacity per driven metre per unit length of perimeter in contact with the soil.

Equation (3) applies at any penetration. For the present conditions at final penetration are examined. Using average values of $\partial R_t/\partial L$ taken from Figs 4 and 5 the values given in Table 1 were obtained:

Table 1. Loss in shaft capacity.

Pile	$\partial R/\partial L$ kN/m	c_u: kPa	p: m	p: m	$\frac{1}{p_0 + p_i}\frac{\partial R_s}{\partial L}$ kN/m/m
TP	457	470	2.394	2.174	-370
OC	356	470	2.394	1.267	-373
NC	122	138	2.394	1.363	-105

The final column in Table 1 shows significant losses in shaft capacity due to driving, with the stronger Tilbrook soil giving a greater rate of loss than the weaker Pentre soil.

Conclusions

The tests provided a large volume of good quality data which has enabled reliable estimates to be made of the load transfer characteristics during driving. It is shown that these characteristics change as driving progresses due to intensive remoulding of the soil at the interface between the pile and soil, and, due to major changes in the horizontal pressure on the pile wall brought about by flexural deflections of the pile.

References

HEEREMA E. P. Relationships between wall friction displacement velocity and horizontal stress in clay and in sand for pile driveability analysis. *Ground Engineering*, Jan. 1979.

NASH J. C. *Compact numerical methods for computers*. 1979, A Hilger Ltd, Bristol.

POSKITT T. and WARD G. The evaluation of pile load test data in calcareous soils for use in pile driving calculations. *Engineering For Calcareous Sediments*, Vol. 2, pp. 449-460.

POSKITT T. Lateral motion of piles during driving. *The Structural Engineer*, Vol. 68, No 22, 1990, pp. 441-443.

SMITH E. A. L. Pile driving analysis by the wave equation. *J. Soil Mech. & Fndn Engng Div., Am. Soc. Civ. Engrs*, Vol. 86, SM4, 1960, pp. 35-61.

TOMLINSON M. The adhesion of piles driven in clay soils. *Proc 4th Int. Conf. Soil Mech. & Fndn Engng*. 1957, Vol. 2, pp. 66-71.

TOMLINSON M. Some effects of pile driving on skin friction. *Behaviour of piles*, pp. 107-114, 1971, ICE, London.

YIP K. L. *Dynamic analysis of pile driving*. PhD thesis, University of London, 1988.

13. Static and cyclic axial load tests on two 762 mm diameter pipe piles in clays

W. R. COX, Fugro-McClelland Marine Geosciences Inc., Houston, K. CAMERON, Precision Monitoring & Control Ltd, Stockton, and J. CLARKE, BP International Ltd, Uxbridge

In the Large Diameter Pile Tests (LDPT) programme two instrumented pipe piles of 762 mm diameter were driven and tested under axial compression loads at sites in England. One pile was driven to a penetration of 55 m at a normally consolidated clay site (NC site) and the second pile was driven to a penetration of 30 m at an overconsolidated clay site (OC site).

This paper is one in a series of papers on the planning, execution, and results of the LDPT project. Discussed in this paper are preparation of the sites, installation of anchorages for the reaction system, installation of piezometers, descriptions of the reaction and loading systems, procedures for the testing, dissipation of pore pressures after driving and before testing, and results of the static, sustained, and cyclic loadings on the two piles. Analyses of the test results are discussed in a companion paper, Gibbs et al. *(1993).*

Site preparations

The works for the NC site at Pentre in Shropshire and for the OC site at Tilbrook Grange in Cambridgeshire were carried out during September and October 1986. This timing provided optimum weather conditions to minimise subsoil damage and to provide a reasonable working surface for heavy plants.

At the NC Site 1200 m^3 of topsoil was excavated and stockpiled and 2300 m^3 of hardcore was placed. At the OC site 1350 m of topsoil was excavated and stockpiled and 2000 m^3 of hardcore material was placed. At both sites the hardcore was placed on a Terram 1000 geotextile fabric to form a working surface. At Pentre, site preparation works included road widening to allow delivery of the pile. In addition, at both sites a temporary access road was laid.

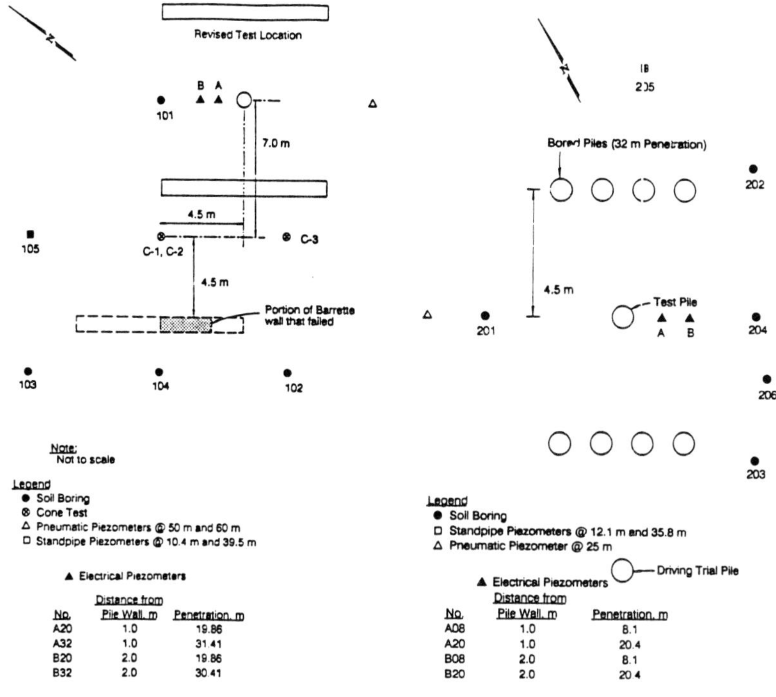

Fig. 1. Layout at NC site

Fig. 2. Layout at OC site

Anchorages for the reaction system

Design considerations

The maximum design test loads at the NC and OC sites were 14 MN and 30 MN, respectively. These loads were computed for the 762 mm diameter test piles using skin frictions based on soil data recovered during the 1986 site investigations (Lambson *et al.* 1993). The design test loads were based on α values of 1.2 for the NC site and 0.75 at the OC site. The α values were chosen to give a high probability that the reaction system would have sufficient capacity to fail the piles.

Horizontal clearance from test piles

The positioning of the anchorages for the reaction frame, as shown in Figs 1 and 2, were designed to give a horizontal clearance between the anchorages and the test piles of five test-pile diameters. The actual clearances were 3.72 m at the NC site and 3.74 m at the OC site, which correspond to 4.9 test-pile diameters.

Barrette walls

For the NC site, post-tensioned barrettes (diaphragm walls) were chosen as anchorages for the pile-load reaction frame. These were installed in February and March 1987. Four barrette panels, each 0.8 m wide, 4.5 m long by

Fig. 3. Reinforcement cage for barrette wall at NC site

24 m deep, were constructed using diaphragm-walling techniques. Soil was excavated under bentonite using a rope-operated grab mounted on an Ajax crane.

The contractor commenced excavation of the first 4.5 m long panel using unweighted bentonite mud to support the excavation. A delay by the contractor in fabricating the steel cages precluded immediately concreting the panel and the excavation remained partially complete and open for three days. The panel walls collapsed when the excavation was recommenced. There were concerns about the safety and pull out capacity of the collapsed panel, so the panel was abandoned and backfilled with mass concrete. It was decided to relocate both barrette walls and the test pile to the new positions shown on Fig. 1.

The Contractor changed to a different procedure to excavate for the walls. The new method required the use of weighted mud, and raising the level of the mud to 1 m above ground level. After the excavation, a 24 m long reinforcement cage was raised and placed in each of the four panels, as shown in Fig. 3. The cage for each panel contained eight fully-debonded Macalloy bars of 40 mm diameter. These bars were later post tensioned to anchor the post bases for the reaction frame. About 90 m of concrete was placed in each panel using a tremmie pipe and the concrete was cast to ground level. All

Fig. 4. Reinforcement cage on strong back for bored pile at OC site

operations for each panel, including excavation, placement of the reinforcing cage, and pouring of concrete, were carried out in a continuous sequence lasting about 12 hours.

Bored piles

The anchorages for the reaction frame at the OC site comprised two groups of four 0.75 m diameter by 30 m long bored cast-insitu concrete piles. These were installed in March and April 1987. Positions of the bored piles relative to the test pile are shown in Fig. 2.

The piles were augered without the use of a drilling fluid immediately concreted to minimise stress relief of the walls. Six debonded Macalloy bars of 40 mm diameter were placed in each pile for the full length. The Macalloy bars were later post tensioned during erection of the posts for the reaction frame. Fig. 4 shows the lifting of the reinforcing cage on a strong back for one of the eight bored piles.

Piezometers

Pore pressure in the soil must be known at several stages during the test programme to predict the in situ stress state. The stages of interest are before, during, and after both driving and load testing the pile. Initial in situ pore pressures were monitored over a period of several months using both standpipe and pneumatic piezometers. Changes in pore pressure in the yield zone adjacent to the pile were monitored using electrical pressure transducers.

At both the NC and OC sites, the piezometers were installed after completion of the anchorage systems. The NC piezometers were installed in April 1987 and the OC piezometers were installed in May 1987. The positions and depths of the piezometers at the NC and OC sites are shown in Figs 1 and 2, respectively.

Standpipe piezometers

This type of piezometer was installed in a predrilled borehole and consisted of a 25 mm diameter standpipe connected to a porous tip. The pressuremeter borehole was used at both sites for standpipe installation. The porous tip was surrounded by a 1 m-deep well-graded sand cell and sealed both above and below the sand cell with a bentonite seal. The remainder of the borehole was backfilled with a cement grout. Water pressure was measured with a water level probe. A Bourdon gauge was used for measuring artesian pressure. The ceramic tips were placed at penetrations of 10.5 and 39.5 m at the NC site and penetrations of 12.35 and 34.65 m at the OC site.

Pneumatic piezometers

The pneumatic piezometer was operated by water pressure acting on a diaphragm. This pressure is balanced by an externally applied nitrogen pressure. When the externally applied pressure equals the ground water pressure acting on the reverse side of the diaphragm, a valve opens and allows flow along a return line to a detector in the readout unit. Water pressure is then displayed on a digital readout unit. The piezometers were installed at penetrations of 50 and 60 m at the NC site and at 25 m penetration at the OC site.

Electrical piezometers

This type of piezometer was installed in the yield zone adjacent to each pile. The piezometer consists of a standpipe installed in a predrilled borehole with a Druck PCDR 830 transducer coupled to a pneumatic packer lowered into the standpipe.

All boreholes were cased to final penetration to ensure stability of the borehole walls. A directional survey of the boreholes was undertaken at intervals of approximately 5 m to check that the required verticality tolerance of 1 in 150 was maintained and to determine the position of the piezometer relative to the pile.

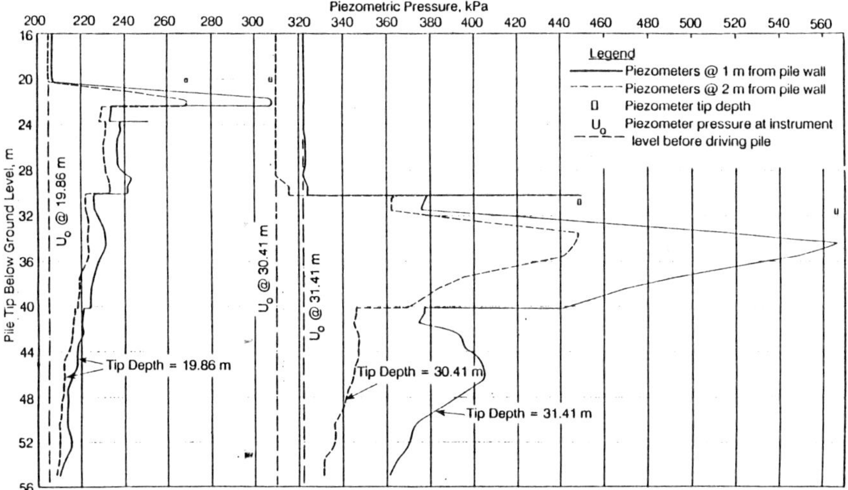

Fig. 5. Piezometric pressure versus pile tip depth during driving at NC site

Piezometer response during driving

Piezometers at NC site

The response of piezometers during driving of the NC pile are shown in Fig. 5. The two curves on the left side of the figure are for piezometers installed at 19.86 m depth at 1 and 2 m away from the pile wall. On the right side of the figure are two piezometers installed at depths of 30.41 and 31.41 m at 2 and 1 m away from the pile wall, respectively.

The piezometric pressures existing before the pile was driven are indicated by the dashed lines labelled U_O. The data in Fig. 5 indicate that pore pressure response was unaffected until the pile tip advanced past the elevation of the piezometers.

Piezometers at OC site

The response of piezometers during driving of the OC pile are shown in Fig. 6. The two curves on the left of the figure are for piezometers installed at 8.1 m depth in the Lowestoft Clay that extends from the ground level to a depth of 17.1 to 18.6 m. The two curves on the right of the figure are for piezometers installed at a depth of 20.4 m in the Oxford Clay that underlies the Lowestoft Clay to at least 40 m. The data show the large increase in pore pressure measured in the Oxford Clay in comparison with the overlying Lowestoft Clay.

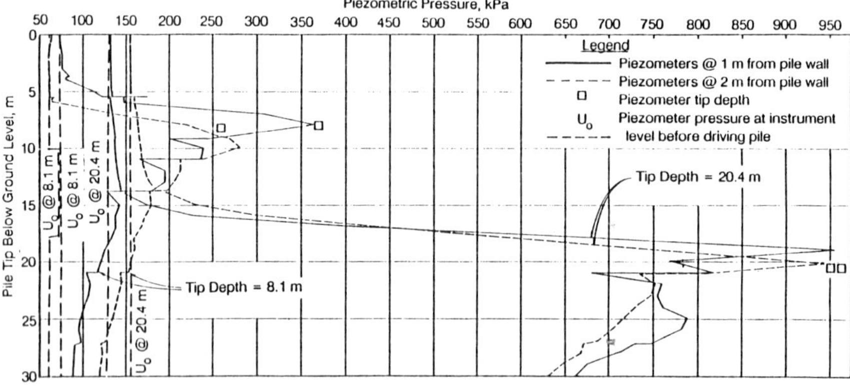

Fig. 6. Piezometric pressure versus pile tip depth during driving at OC site

Reaction frame

Design scheme

Figures 7 and 8 show the general arrangement of the load reaction system used at the NC and OC sites respectively. The reaction system was designed by Ove Arup & Partners. Four post bases were tied to the anchorage system by extending Macalloy bars through the bases and then tensioning the bars. The four posts and the two secondary beams were then connected to the post bases by 50 mm diameter Macalloy bars, six at each end of each secondary beam.

For the NC pile tests, the arrangement of the reaction frame was similar to a frame that Ove Arup & Partners had used in a high-capacity bored-pile test in the London area.

As shown in Fig. 7, three hydraulic jacks were placed on the top of each end of the primary beam. The loading pistons of the jacks reacted upward on the bottom flanges of the secondary beams. This scheme allowed for a minimum height between primary beam and ground level. This decreased the possibility of instability of a column made up of the pile, jacks, and load cells. The scheme also alleviated the problem of fitting six jacks on the pile head.

During unloading from the static test on the NC pile, the three jacks on one side of the primary beam tended to unload more rapidly then the other three. This caused the beam to rotate. Since the maximum load for the static test was only 6.03 MN, it was possible during the cyclic loading tests to use only two of the 750-tonne jacks, one at each end of the primary beam. In addition, each jack was connected to a common manifold in the pumping system and this gave a more equal pressure drop through the hydraulic lines from the two jacks. In this manner it was possible to complete the NC series of tests with one jack at each end of the primary beam.

Fig. 7. 30 MN capacity reaction frame at NC site

It should be noted that the reaction system and pumping system for the load jacks were not designed for cyclic loading. The decision to perform cyclic loading tests was made after fabrication of the reaction system and acquisition of the pumping system.

Modified scheme

At the OC site four jacks were placed directly on the pile and the load cells placed on top of the jacks, as shown in Figs 8 and 9. It was possible to use only four of the six jacks since each jack had 750-tonne capacity that would provide a total load of 3000 tonnes or approximately 30 MN. Levelling heads

Fig. 8. 30 MN capacity reaction frame at OC site

on the load cells remained active whilst the levelling heads on the jacks were disabled. This system proved satisfactory in all respects.

Dimensions

The reusable primary beam, shown in Figs 7 and 8, was made up of three plate girders connected side-by-side. The primary beam had a length of 9.65 m, a centreline height of 3.3 m, an end height of 1.7 m, and a width of 1.8 m.

Structural concrete slab

A reinforced concrete slab was cast around and under the entire reaction system, as shown in Figs 8 and 9. This slab not only provided a clean working area, but also provided lateral support to the pile at the ground line. Before the slab was cast, the pile was wrapped with two layers of PTFE sheeting which minimised friction between the pile and the slab during testing. Slots were formed in the slab at the two cable exits to allow for vertical movement of the pile and cables.

Loading system

The loading system was fabricated and supplied by Arbil Equipment Ltd.

Jacks

Six 750-tonne single acting jacks were supplied. Each jack had a maximum stroke of 500 mm and contained an integral levelling head capable of accommodating up to 5 degrees of rotation. As mentioned earlier, the levelling heads were disabled for the OC tests because the load cells, which were positioned on top of the jacks, had integral levelling heads.

Fig. 9. Four 750 tonne jacks and four 1000 tonne load cells at OC site

Load cells

Four 1000-tonne compression load cells with integral levelling heads were supplied by Arbil Equipment Ltd. The cells were calibrated to the National Physical Laboratory standards in accordance with BS 1610 (1985), both before and after load testing of the piles. Differences in load cell sensitivities for the two sets of calibrations were found to be negligible. The cells were provided with electrical readouts for connection to the data acquisition system.

Pumping system

The pumping system is shown in Fig. 10. Three air-powered hydraulic pumps, each capable of working independently, were used. The system was capable of maintaining a load of at least 600 tonnes on each of the six jacks simultaneously at the full 500 mm stroke. Sufficient output to stroke the jacks at a minimum rate of 2 mm per minute was provided. Control valves in the manifold gave an option of any of the three pumps to be used, or any pair of the three pumps, or the three pumps could be in parallel operation simultaneously.

General procedures for testing

Three different load tests were performed on each of the NC and OC piles. These were a static test, a creep (sustained) load test, and a cyclic load test. Creep testing commenced 24 hours after static testing and cyclic testing commenced 24 hours after completion of the creep testing. The static test was performed at a constant rate of penetration of 1 mm/min.

During the creep tests, the pile was loaded in four increments of 60, 70, 80 and 90 per cent of the ultimate load reached during the static test. Each load level was sustained for five hours and any resulting creep was measured. If

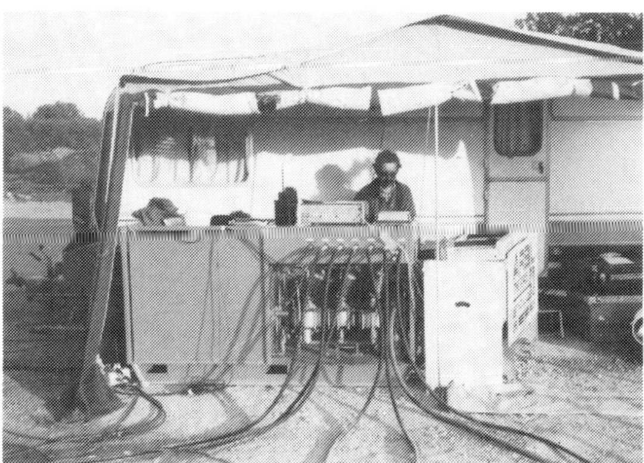

Fig. 10. Pneumatic–hydraulic pump for load jacks

the pile plunged during a sustained load, then the creep load test was concluded after a pile movement of 30 mm.

After the end of the creep tests, a static test was performed to determine a new ultimate pile load. The cyclic hoads applied corresponded to 20 to 80 percent, 20 to 90 percent, and 20 to 100 percent of the new static load. The numbers of cycles applied were 100, 50 and 20 cycles, respectively. If the pile plunged during cyclic testing, then the cyclic testing was concluded after a pile movement of 30 mm. All cyclic testing was one-way, that is, testing in axial compression only.

Consolidation of NC pile

The NC pile was driven on 20 July 1987. The static load test was conducted on 2 September 1987, 44 days after driving. At the time of initiating the static load test, there was complete dissipation of the excess pore pressures generated by pile driving.

The period for 90 per cent dissipation varied between 1.5 and 5.7 hours. At 43 m penetration where the clay is highly plastic, there was a significant difference in pore pressure response and 90 per cent dissipation occurred after some 100 hours. Rapid dissipation of excess pore pressures is attributed to the silt content and the foliated structure of the soil.

Axial compression tests on the NC pile

Static load test

Peak load measured at the pile head during the static test was 6.03 MN at a pile head movement of 36.1 mm. Loading of the pile continued to about 115 mm movement where a post peak load of 5.48 MN was measured. Upon unloading, the pile head rebounded by 21 mm. Pile head load versus deflection for the static test is presented on Fig. 11. At the peak load the end bearing was 0.86 MN or about 14 per cent of the total capacity and was mobilised at a tip movement of 10 mm.

Creep load tests

In creep tests initiated 23.9 hours after the static test, no substantial residual movements of the pile were noted until the load level of 5.4 MN (90 per cent). This load level was held for 3 hours and 58 minutes. In this time the pile plunged 30.6 mm and the creep testing was concluded.

Cyclic load tests

A second static test was initiated 24.7 hours after concluding the creep tests. The second static test was run at 1 mm/min. and a peak load of 5.63 MN was reached 31 minutes after start of the test.

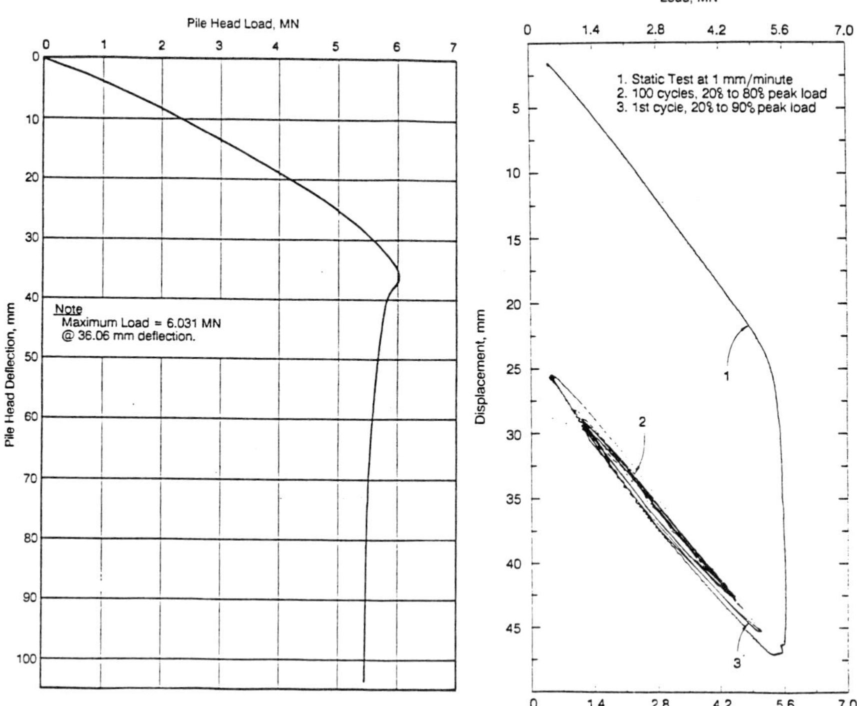

Fig. 11. Load versus deflection for static test on NC pile

Fig. 12. Load–displacement curves for cyclic testing of NC pile, 100 cycles at 20% to 80% peak load.

The first level of cyclic loads was initiated 38 minutes after unloading from the second static test. Two hours and 12 minutes were required to produce 100 cycles at the load level of 20 to 80 per cent (12.12 to 4.48 MN). Average time for each cycle was 1.3 minutes. Load-displacement information on the new static test, the first level of cyclic loading, and the first cycle at 20 to 90 per cent are given on Fig. 12.

Fifty cycles at 20 to 90 per cent (1.12 to 5.04 MN) were completed in 1 hour and 17 minutes. This is an average of 1.5 minutes for each cycle. Load- displacement data on this series of cyclic loadings are given on Fig. 13. Twenty cycles were made at 20 to 100 per cent (1.12 to 5.6 MN). Time required was 38 minutes, or an average of 1.9 minutes per cycle. At the twentieth cycle the pile had plunged about 30 mm. Results of these tests are shown in Fig. 14.

At the conclusion of the cyclic tests, two fast static tests were run at a rate of 14.1 mm/min. Peak loads of 6.2 MN were reached in about 4 minutes of loading. A third static test at the constant rate of penetration of 1 mm/min.

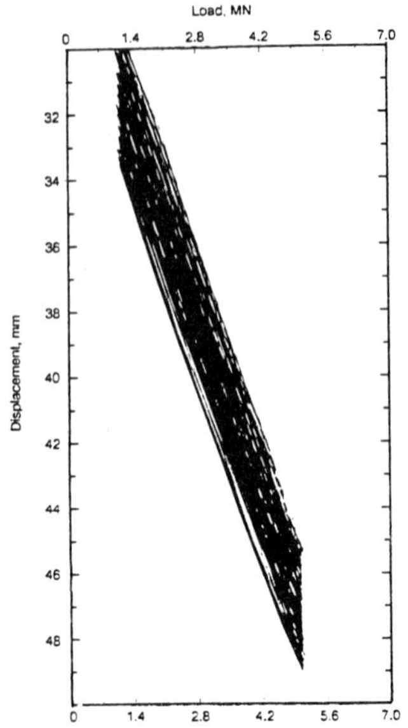

Fig. 13. Load–displacement curves for cyclic testing of NC pile, 50 cycles at 20%–90% peak load

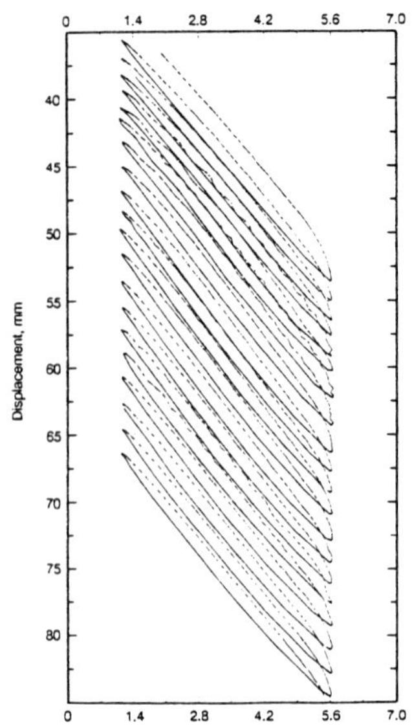

Fig. 14. Load–displacement curves for cyclic testing of NC pile, 20 cycles at 20%–100% peak load

was the final tests on the NC pile. A maximum load of 5.5 MN was reached in 45 minutes.

Consolidation of OC pile

The OC pile was driven 5 October 1987. The static load test was conducted 130 days later on 16 February 1988. Percentage dissipations of excess pore pressure versus log time after driving are given in Fig. 15 for cells at 8, 15 and 27 m. The influence of tensioning the reaction piles is very evident. The percentage dissipations are between 80 and 95 per cent in the Lowestoft Clay (0 to 18.6 m) and between 77 and 84 per cent in the Oxford Clay (18.6 to 30 m).

The pore pressures measured by cells at 21 m are uncertain and for this reason they are shown in Fig. 16, together with the pressures measured by the piezometers at 20.4 m. The influence of tensioning the reaction piles is noted. The ambient pore pressure at 21 m is also shown. Immediately after driving the pore pressures measured by the cells at 21 m are + 40 kPa and - 40 kPa, which are below ambient. The pore pressures then continued to increase reaching a maximum value some 15 days after driving. After this time, the pore pressures started decreasing and were behaving as expected. The percentage dissipation as determined from the piezometer readings are between 66 and 69 percent. It is believed that these percentages probably represent lower bound dissipation rates for the cells at 21 m.

Excluding the data from the 21 m cells, excess pore pressures of 10 percent and 20 per cent are estimated to have remained in the Lowestoft Clay and Oxford Clay, respectively, at the time of load testing. Although consolidation of the soil over nearly the lower half of the pile (Oxford Clay) was not considered complete, the degree of consolidation was believed to be well within the API data base.

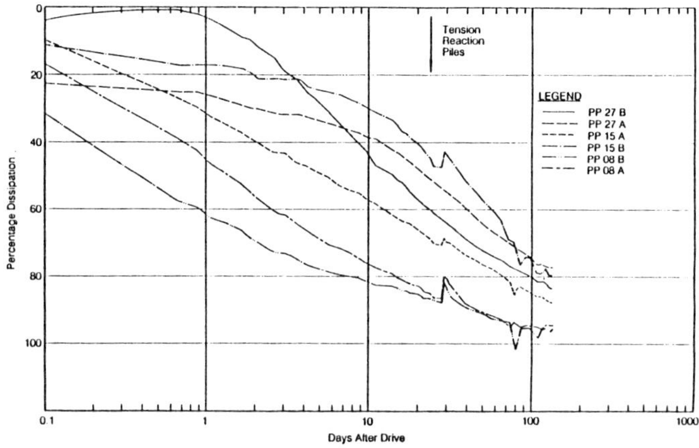

Fig. 15. Pore pressure dissipation on wall of OC pile versus log time of set-up period

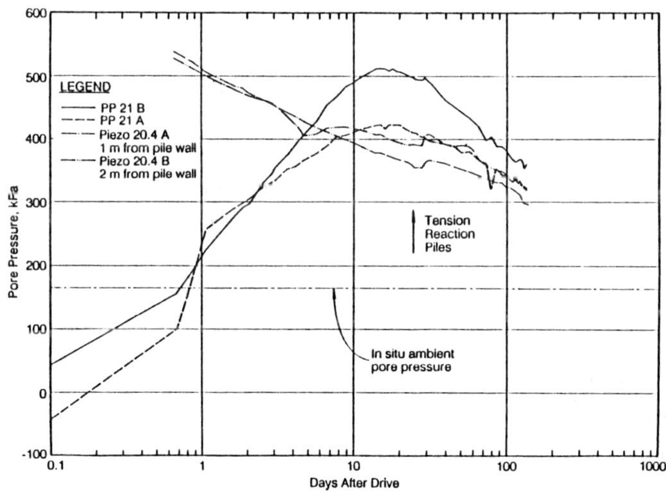

Fig. 16. Pore pressure dissipation on wall of OC pile versus log time of set-up period (21 m penetration)

Axial compression tests on the OC pile

Static load test

Peak load measured at the pile head during the static test was 16.13 MN at a pile head movement of 28.1 mm. Loading of the pile continued to 128.7 mm movement where a post peak load of 14.21 MN was measured. Upon unloading, the pile head rebounded by 17.4 mm. Pile head load versus deflection for the static test is presented on Fig. 17. At the peak load the end bearing was 1.45 MN or about 9 percent of the total capacity and was mobilised at a tip movement of 8.6 mm.

Creep load tests

Creep tests were initiated 24.6 hours after conclusion of the static test. Sustained loads of 60, 70 and 80 per cent of the peak static load were applied for 5 hours. At the 60 and 70 per cent load levels, relatively small movements were observed. A deflection of 2 mm was noted after 5 hours at the 80 per cent load level. An attempt to apply the 90 per cent load level was abandoned when the pile moved down 30 mm before reaching this load level.

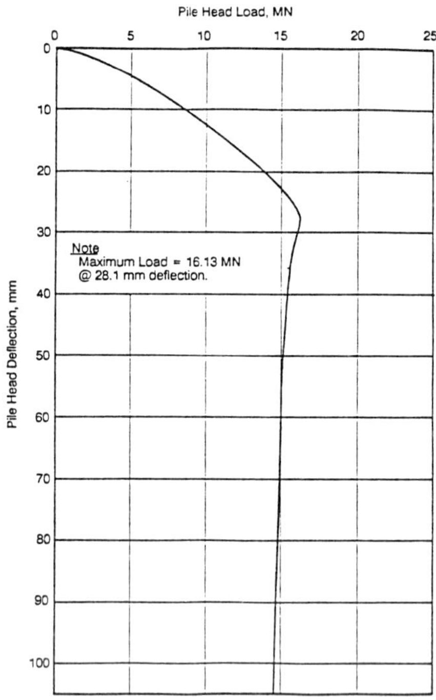

Fig. 17. Load versus deflection for static test on OC pile

Cyclic load tests

A new static test at a constant rate of penetration of 1 mm/min. was performed 24.5 hours after the conclusion of the creep tests. This test developed a pile head load of 14.6 MN at a pile head deflection of 48 mm.

The first level of cyclic loads was initiated 20 minutes after unloading from the second static test. Four hours and 26 minutes were required to produce 100 cycles at the load levels of 20 to 80 per cent (2.92 to 11.7 MN). Average time for each cycle was 2.7 minutes. Results of these tests are shown in Fig. 18.

Fifty cycles at 20 to 90 per cent (2.92 to 13.17 MN) were completed in 2 hours and 48 minutes. This is an average of 3.4 minutes for each cycle. Results of these tests are shown in Fig. 19.

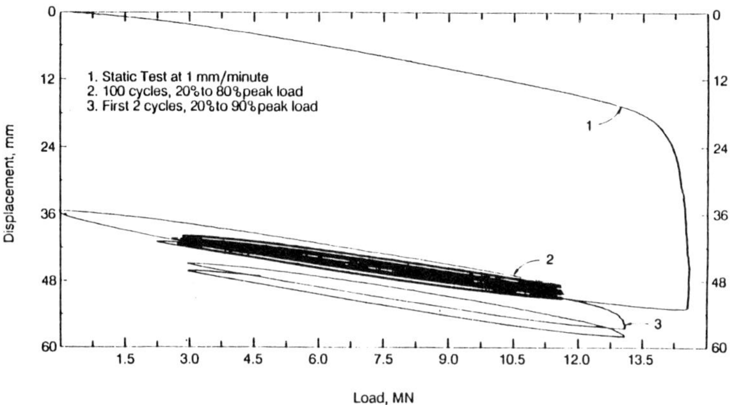

Fig. 18. Load–displacement curves for cyclic testing of OC pile, 100 cycles at 20%–80% peak load

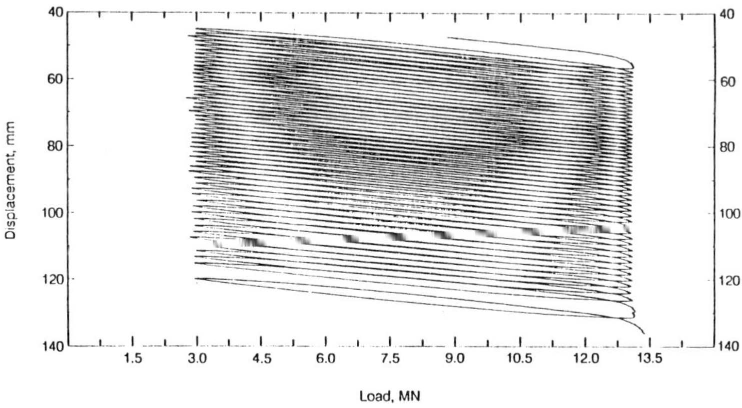

Fig. 19. Load–displacement curves for cyclic testing of OC pile, 50 cycles at 20%–90% peak load

A cyclic load level of 20 to 100 percent was attempted but the pile plunged at a rate of about 6 mm/min as the load was applied. This test was accepted as the rapid load test and a maximum load of 14.1 MN was recorded.

A post-cyclic static test at 1 mm/min gave a capacity of 13.5 MN with a pile top displacement of 93 mm. The precyclic static capacity was 14.6 MN, so the cyclic loading lead to a reduction of about 8 per cent in capacity.

Observations

Structural concrete slab

The piles were wrapped above the ground line with two layers of PTFE material before the structural concrete slab was cast. This proved an excellent means for restraining lateral movement of the piles and causing little to no friction resistance to vertical movement of the piles. There was also the obvious benefit in the slab's providing a clean working surface around the piles.

Performance of the reaction system under loading. The scheme for the reaction frame of placing the load jacks at each end of the primary beam was not designed for cyclic testing and has merit for cyclic testing only if the stroke of all jacks can be maintained the same. This was not possible with the jacks and pumping system that were acquired before the decision to run cyclic tests. For the cyclic tests on the NC pile, the three jacks at each end of the primary beam were replaced by one jack. The cyclic tests were conducted satisfactorily although there continued to be unequal vertical movements at the ends of the primary beam. For the OC tests the load jacks were placed in the more conventional system on top of the piles and the OC tests were completed with excellent service from the reaction system.

At the NC site the vertical deflection of the pile loading frame at the maximum static load of 6.03 MN was less than 1 mm. Deflection of the frame at the OC site was about 2.5 mm at the maximum load of 16.13 MN.

Test results

The test results that are summarised in Figs 11 to 14 and 17 to 19 give evidence to the satisfactory performance of the instrumentation, loading and data acquisition systems. Analyses of the results are discussed in a companion paper, Gibbs *et al.* (1993).

References

GIBBS, C. E., MCAULEY, J., MIRZA, U., and COX, W. R. (1993) Reduction of field data and interpretation of results. This volume.

LAMBSON, M. D., CLARE, D.G., and SEMPLE, R. M. (1993) Investigation and interpretation of Pentre and Tilbrook Grange soil conditions. This volume.

14. Reduction of field data and interpretation of results for axial load tests of two 762 mm diameter pipe piles in clays

C. E. GIBBS, John Brown Engineers & Constructors PLC, London, J. MCAULEY, BP International Ltd, Uxbridge, U. A. MIRZA, Earl and Wright, London, and W. R. COX, Fugro-McClelland Marine Geosciences Inc., Houston

Introduction

1.1. Project and paper description

The Large Diameter Pile Tests (LDPT) programme involved load testing two instrumented 762 mm OD tubular steel piles which were driven vertically at two UK onshore sites. The sites, one at Pentre in Shropshire, and the other at Tilbrook Grange in Cambridgeshire, provided respectively normally consolidated (NC) and over consolidated (OC) clay soil conditions. The aim of the project was to monitor pile and soil load deflection responses during driving and subsequent load testing. This paper is one of a series of papers on the planning, execution and results of the LDPT project.

This paper presents a brief description of the background to the pile tests, details the full test results and interpretation of the data and describes the methods used during data reduction.

1.2. Background

The test piles were fabricated from high strength steels, the OC test pile from RQT 701 steel with a nominal enhanced yield stress of 770 N/mm^2, and the NC test pile from RQT 501 steel with a nominal yield stress of 470 N/m^2 Both test piles were internally instrumented with strain gauges, strain modules, accelerometers, total earth pressure cells and pore pressure cells. The instruments and attached cables were protected during driving by steel sections welded to the inside pile wall. Details of the instrumentation can be found in companion papers, Solomon *et al.* (1993) and Cox *et al.* (1993a).

The NC pile was driven to approximately 55 m penetration below original ground level. The top 15 m section of the pile being sleeved-off from the surrounding soil by a pre-installed 914 mm OD by 10 mm wall thickness steel casing. The OC pile was driven to about 30 m below original ground level without a sleeve.

Each test pile was loaded to failure in axial compression and then sustained loading and cyclic loading tests were performed. The load tests were applied quasi-statically using hydraulic jacks reacting against a purpose-built load reaction frame and anchorage system. The static tests were performed at a constant rate of penetration of 1 mm/min. The free standing section of each pile was supported laterally by a pile head restraint slab to prevent pile buckling and over stress during testing. Full details of the reaction frame, anchor systems and test procedures can be found in a companion paper, Cox *et al.* (1993b).

1.3. Soil conditions

1.3.1. NC Site, Pentre

A detailed site investigation was carried out at the site which included three sample borings, *T-Z* probe and cone penetration testing in situ selfboring pressuremeter testing and gamma logging. Crosshole and downhole seismic surveys were also carried out. Soil conditions revealed at the site comprise a 3 to 3.5 m thick stratum of desiccated recent alluvium of stiff silty clay underlain to at least 60 m penetration by a silty clay stratum formed by glacial lake deposits.The silt and clay content in the stratum varies with depth resulting in occasional layers of clayey silt and silty clay. The measured shear strength profile for the NC site is included in Fig. 12.

1.3.2. OC site, Tilbrook Grange

The site investigation carried out at the OC site was similar to that at the NC site. Soil conditions revealed at the site comprisetwo clay strata, designated respectively, Stratum I and Stratum II referenced from the ground. Geologically the stratigraphic units, strata I and II, are known as Lowestoft and Oxford Clay respectively. Stratum II is a very stiff silty clay of intermediate plasticity which extends between the ground level and 17 to 18.6 m. The clay is heavily over consolidated and the average undrained shear strength over the stratum thickness is about 400 kPa. The underlying Stratum II clay extends to at least 40 m penetration. The clay in this stratum is very stiff and fissured and is of high plasticity. Undrained shear strength obtained from unconsolidated undrained tests vary within a very wide range. The estimated average undrained shear strength between 18 m and the pile tip level at 30 m is 585 kPa.The measured undrained shear strength profile for the OC site is included in Fig. 33.

Full details of the soil conditions at both sites can be found in a companion paper by Lambson *et al.* (1993). Included in Lambson *et al.* are results of additional testing carried out at the OC site after the completion of the work

Table 1. Installation and test data

	NC Pile	OC Pile
Hammer Type	BSP 357	HA40
Installed	20.7.87	5.10.87
Static Test	2.9.87	16.2.88
Creep Test	3-4.9.87	17-18.2.88
Cyclic load Test	5.9.87	19.2.88

described in this paper. This will result in slight differences between the data presented here and that given within Lambson *et al.*.

1.4. Installation and testing

Table 1 presents the information on pile installation and testing for both the NC and OC piles.

2.0 Interpretation of results

The following two sections present the interpretation of the data obtained during installation of the piles, during the set-up periods and during the pile tests. Data for the NC site at Pentre are presented in section 2.1, and for the OC site at Tilbrook in section 2.2.

2.1. N.C. Pile

2.1.1. Pile installation

The top of the soil plug in the pile was continuously monitored during driving. At the end of driving the soil plug was 1.73 m above the tip of the 914 mm diameter casing. All of the soil in the casing had been removed before driving the pile.

Pile driving generates excess pore pressures in the soils. This is because the soil mass surrounding the pile is disturbed and the stress state is altered by the installation process. The effective stresses and pore water pressure distribution immediately after installation have been found to have a significant impact on the long term pile frictional capacity (Esrig and Kirby, 1979, Kraft, 1982). With elapsed time, the excess pore pressures generated during driving gradually dissipate as the soil surrounding the pile consolidates. This is accompanied by some gain in the shear strength of the clay (Randolph *et al.*, 1979).

The displacement ratio of the NC pile is 0.16 after allowing for the protective channels along the pile wall. This value expresses the ratio of net to gross volume of the pile, hence provides a measure of soil disturbance and may be compared to a displacement ratio of 1 for a closed ended pile of which the installation aspects have been thoroughly investigated by Randolph *et al.* (1979). The displacement ratio at the tip becomes about 0.27 if the ratio is based on the areas of the pile wall and ploughs at the end of the protective channels.

During installation the measurements included both total earth pressure and pore water pressure with cells located on the pile at average depths corresponding to 20.3 m, 31.8 m, 42.6 m and 51.6 m below the ground level.

Piezometers, pre-installed to about 20 m and 31 m below the ground level at approximately 1.0 m and 2.0 m away from the pile wall, allowed monitoring of the pore pressure variation at these depths as the pile was driven to the target penetration.

2.1.1.1. *Total earth pressure and pore pressure measurement*

Five of the eight total pressure cells and seven of the eight pore pressure cells survived pile driving. There is some uncertainty about the reliability of the TP52BN (total pressure) cell. TP52BN became unstable during the period between pitching the pile and driving and underwent a large zero shift. During set-up this cell appears to behave normally, however the pressures recorded before the test are lower than would be expected.

During driving, the total earth pressure and the pore pressure cells provided continuous profiles of measurements versus penetration down to 52 m below ground level. The data are presented in summary form in Fig. 1.

The excess pore pressure Δu, generated during installation increased with depth. The average value of $\Delta u/s_u$ over the pile shaft is about 2.7, and compares favourably with theoretical results based on plane strain undrained expansion of a cylindrical cavity. Over the lower half of the pile, the ratio of Δu normalised by the effective overburden pressure, σ'_v, was above unity. Interpretations of data from piezometers pre-installed at 1.0 and 2.0 m from the pile wall indicate that the plastic zone of influence (beyond which zero Δu is generated) is of the order of 10 to 15 times the pile radius.

Immediately after installation, the radial effective stresses along the shaft are extremely small, approaching zero, indicating the severe remoulding of the soil that occurred during installation. It is believed, however, that the radial effective stresses immediately after installation were of the order of 5 to 10 per cent of the effective overburden pressure. This finding of negligible effective radial stresses is in strong contradiction with results of theoretical analyses that are based on the Cam Clay Model using critical state concepts. Such models would indicate that the radial effective stress immediately after installation is much higher, of the order of $3s_u$ for the soil conditions at the test site.

2.1.2. *Consolidation stage*

The pile was tested after interpretation of the pore pressure data indicated complete dissipation of the excess pore pressures generated by pile driving. Figs 2a and 2b show the measured pressures.

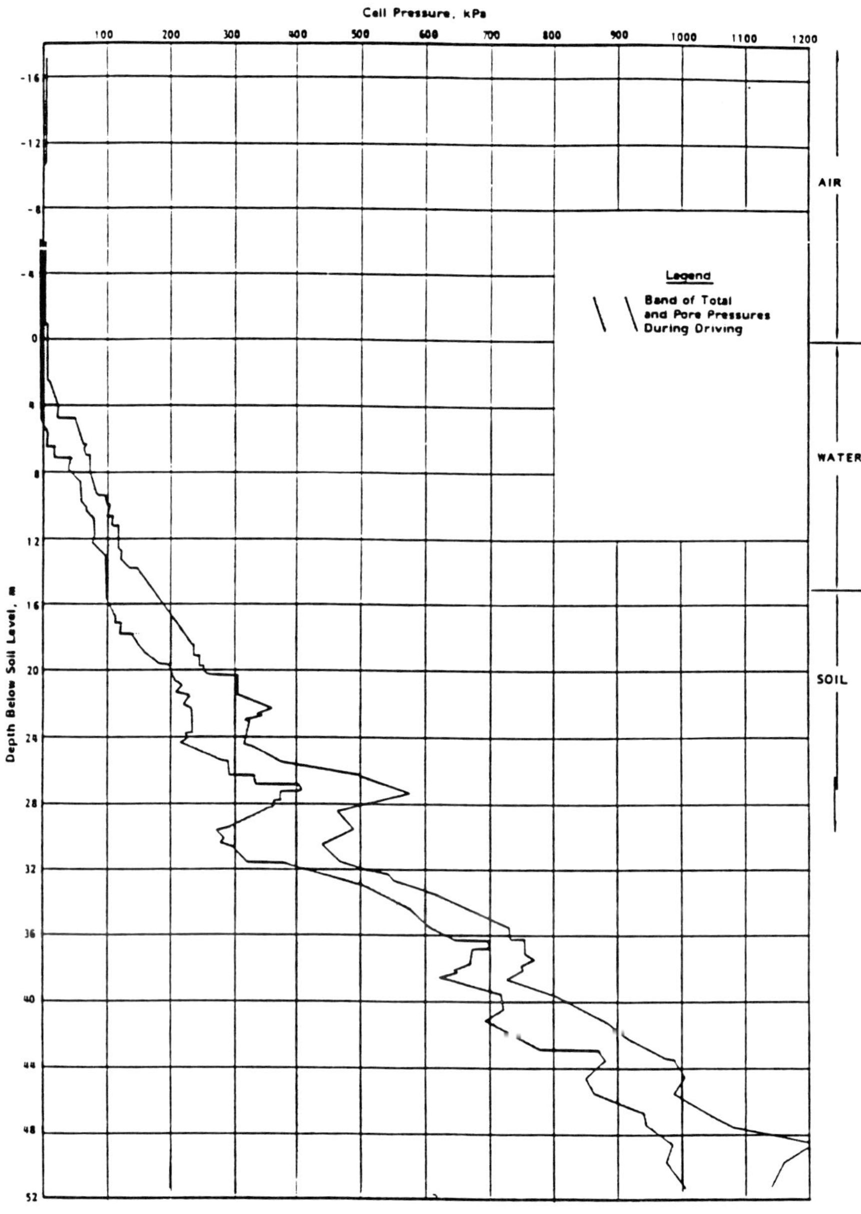

Fig. 1. Cell pressure versus instrument depth during driving – summary plot, NC pile

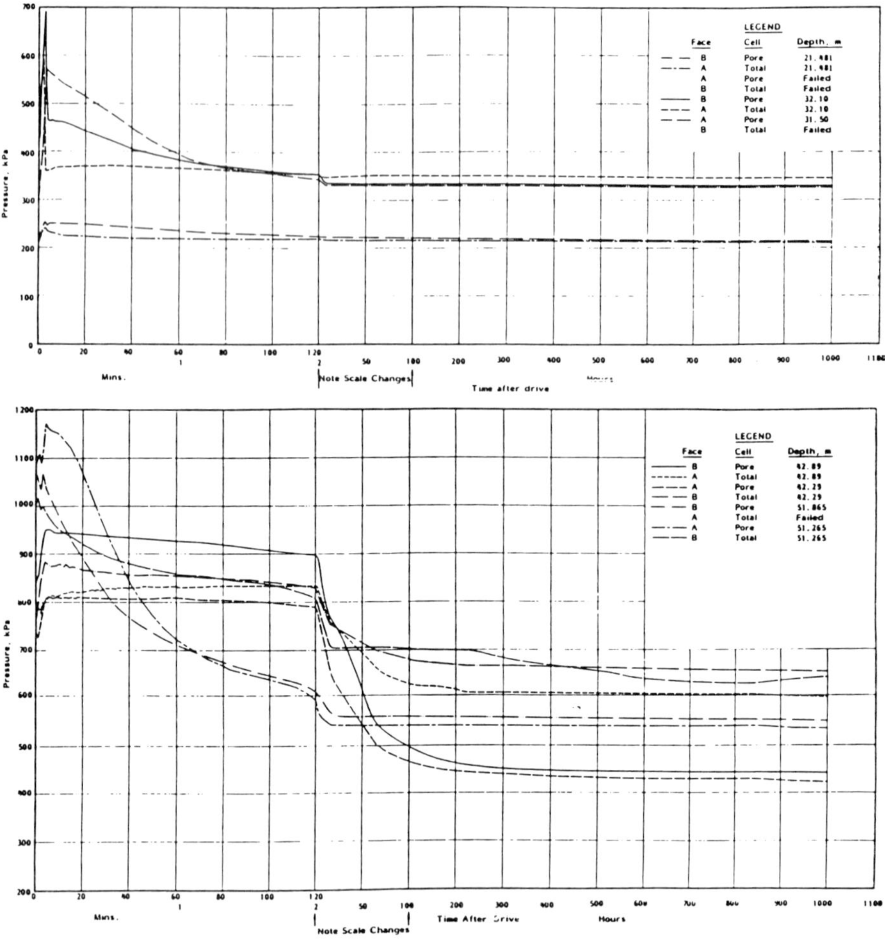

Fig. 2. Total and pore pressures versus time after drive, NC pile: (a) (top) at 20 m and 30 m; and (b) (bottom) at 43 m and 52 m

Except at a penetration of 43 m, the period for 50 per cent dissipation was consistent and varied between 0.45 and 1.5 hours while for 90 per cent dissipation, it varied between 1.5 and 5.7 hours.

At 43 m penetration, there was a significant difference in pore pressure response in that 50 per cent dissipation occurred after 25 and 90 per cent dissipation after some 100 hours, thus reflecting the large difference in permeability between this zone and elsewhere along the shaft. The soils data show that at 43 m the silty clay is of high plasticity. The rapid dissipation of excess pore pressures at the other measurement locations is attributed to the variations in the silt content and the foliated structure of the soil. Utilisation of the procedure by Massarch (1978) indicated that the potential for hydrofracturing by driving is very small and therefore hydrofracturing is not believed to have contributed to the fast pore pressure dissipation.

The predicted consolidation times were higher than the actual. The nearest comparison, however, was provided by the predictions from the in situ cone and *T-Z* probe data which gave lower bound T_{50} and T_{90} estimates of 2 hours and 36 hours, respectively, while the upper bound estimates were 4 days and 10 weeks, respectively.

The total earth pressure measurements indicate that the total radial stress was more or less constant throughout the consolidation stage down to 32 m while below this depth reductions of up to 35 per cent from the value immediately after installation were measured. The reduction in total radial stress is again in contradiction with reported theoretical findings based on the critical state concept. This indicates that the radial effective stress at the end of consolidation is the same as the total radial stress after installation, i.e. total radial stress at the pile wall remains constant as the pore pressures dissipate.

At the end of consolidation, however, the radial effective stresses were much smaller than expected except at the highly plastic silty clay zone at 43 m, where the ratio of the effective radial to the original effective vertical stresses is about 0.54. This is in reasonable agreement with the k_0 value for a normally consolidated state. Down to 32 m, the effective radial stress was nearly zero, and at the other measurement location, at 52 m, the effective stress was smaller than expected.

It is not known why the final effective radial stresses were so small. This finding is not supported by the results of Karlsrud and Haugen (1985) who reported an average effective stress ratio of 1.2 which is slightly greater than the estimated k_0 value at the Haga site, and also by Azzouz and Morrison (1988) from experience with the MIT cell at two soft clay sites.

Several hypotheses may be put forward to try and explain the very low final effective stresses. One possibility is that the experimental findings noted above are for piles or model pile segments much more rigid than the test pile which is long and highly compressible.

It is known from the strain module measurements, and confirmed by the load test data, that the pile exhibited some bending during installation (residual stress) and it is conjectured that arching might have occurred in zones richer in clayey silt than others thus leading to loss of contact between cell and soil. The more sensible data obtained at about 43 m penetration is from a silty clay zone that is highly plastic which could have flowed back on to the pile. Other hypotheses are associated with the stiffness and accuracy of the total pressure cell readings and it is noted that the total earth pressure and the pore pressure cells are some 0.6 m apart along the pile shaft. Thus the measurements were not made at the same location.

Another consideration, though unlikely, might be the effect of installation of the casing down to 15 m on the shallower soils but this cannot be reliably

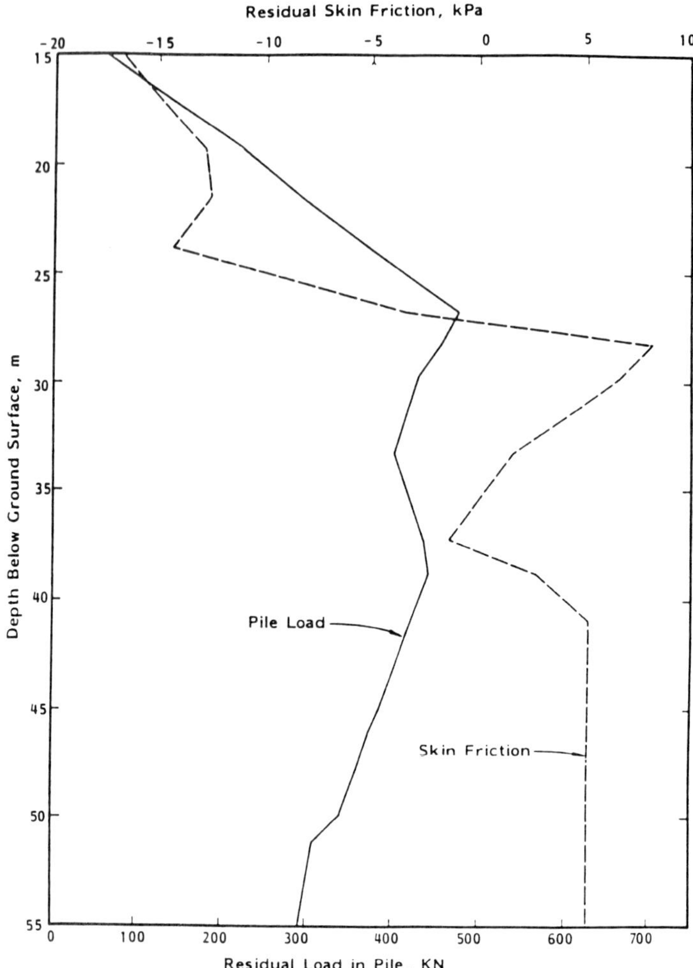

Fig. 3. Residual pile load due to installation, NC pile

quantified since the measurements were obtained using two opposite cells (A and B).

2.1.3. Installation residual stresses

The existence of residual stresses in driven piles and the influence of these stresses on interpretation of pile load tests and axial pile response has been described by several authors, e.g. Mansur and Hunter (1970), Holloway *et al.* (1978), Briaud and Tucker (1984), Poulos (1987), Holloway *et al.* (1978) point out that residual compression stresses cause stiffer pile response than an initially stress-free pile having the same total capacity. Correct formulation of load transfer, *T-Z*, curves must therefore account for residual stresses, though most of the available pile load test data do not consider this effect.

The pile load distribution, residual unit skin friction profile and end bearing before test commencement are shown on Fig. 3. Residual negative skin frictions due to pile installation were of the order of 12 to 16kPa (0.25 to 0.45 s) in the upper 10 m of the pile (about 0.25 of pile length). Below this depth, positive skin friction was developed together with a residual tip force of 290 kN.

The positive skin friction below about 25 m was small, of the order of 5 to 8 kPa. The negative stresses had an influence on the subsequent mobilised values during pile loading.

2.1.4. *Static pile load test data*

The peak capacity achieved was 6.03 MN occurring at a pile head settlement of 36.06 mm. The residual capacity was about 5.5 MN. At the peak load the end bearing was 0.86 MN or about 14 per cent of the total capacity and was mobilised at a tip movement of 10 mm or about 1.5 per cent of pile diameter.

During the residual stage end bearing was continuously increasing albeit at a slow rate probably due to partial drainage and consolidation effects. At the end of the test the tip load was 1.22 MN. The field load versus pile head deflection curve is given in a companion paper.

2.1.4.1. *Results of static load test*

Results of the static load test are given on Figs 4 to 10 as axial load distributions, skin friction distributions along the pile, the load transfer (*T-Z*) curves and the end bearing transfer (*Q-Z*) curves.

The axial load distribution curves on Figs 4 and 5 were developed by applying very small adjustments to the curves resulting from a direct reduction of field data. These small adjustments give a smooth curve that eliminates abrupt, and unrealistic, changes in slope of the load curve at several elevations for data points.

The skin friction curves shown on Figs 6 and 7 are a closed form mathematical solution using the load distribution curves on Figs 4 and 5. Section 3.6 describes the method used. Fig. 6 presents the skin friction distributions for load levels up to the maximum pile head load of 6.03 MN. In addition, the skin friction is shown for the residual load system which existed at the start of the test. Top load on the pile for this residual system was 0.02 MN corresponding to the dead weight of the load cells and bearing plate. Five curves of skin friction distribution at load levels beyond the peak pile load are presented in Fig. 7.

The skin friction below about 40 m penetration for a given pile head load is relatively constant and corresponds to the linear portion of the axial load

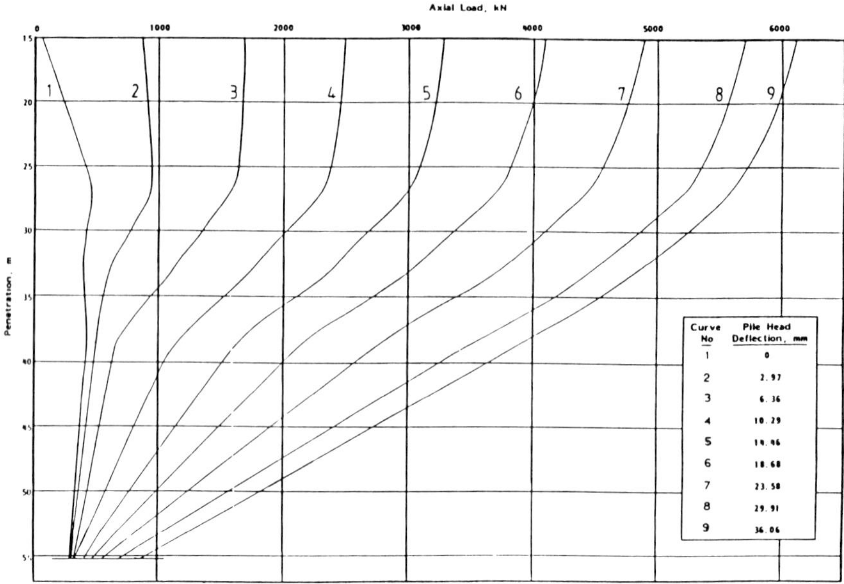

Fig. 4. Axial load distributions:static load test, pre-peak loads (NC pile)

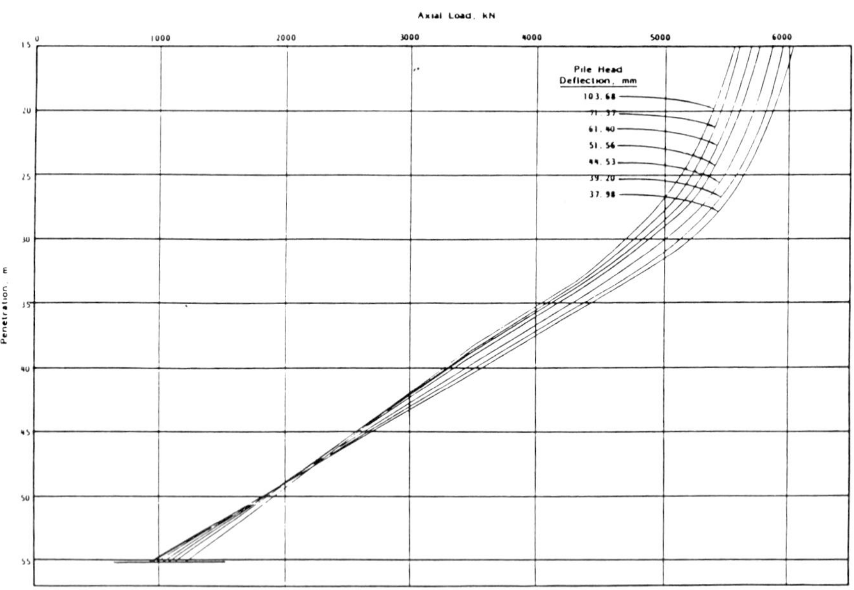

Fig. 5. Axial load distributions: static load test, post-peak loads (NC pile)

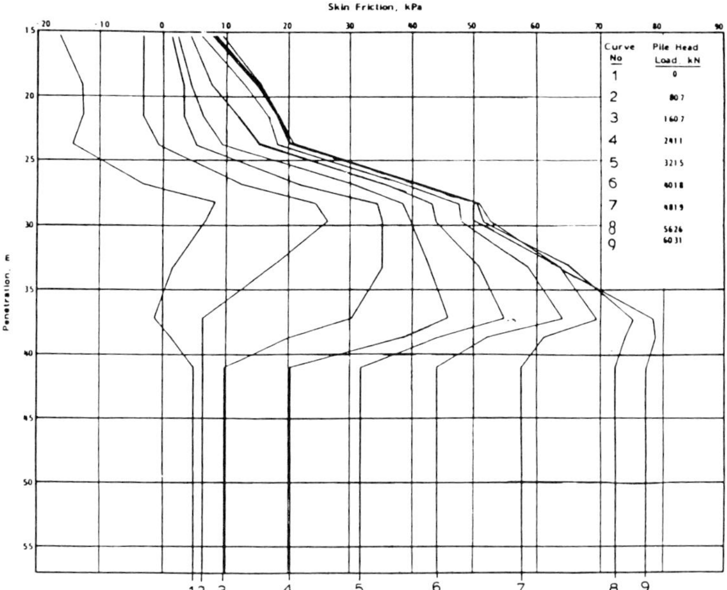

Fig. 6. Skin friction versus penetration: static load test, pre-peakloads (NC pile)

distribution curves below this depth. This is because a linear axial load distribution was assumed below 40 m penetration, where the only functioning instruments were the strain modules at 53.169 m penetration.

Load transfer curves were generated for ten penetrations along the embedded length of the pile using methods described in section 3.4. These curves are presented on Figs 8 and 9. Load at the pile tip versus tip movements is plotted in Fig. 10.

2.1.4.2. *Results of pressure cell measurements*

Total and pore pressures measured during the load test are presented versus time on Fig. 11 for the cells at 43 m. Total pressure readings are shown uncorrected and corrected for cross-sensitivity. Where there are two total pressure cells the average total pressure is also shown.

Corrections due to cross-sensitivity are generally less than 150 kPa, however, for cell TP32A corrections of up to 330 kPa are used. Cross sensitivity corrections have proved to be greater than anticipated for the NC pile, however, it is believed that the data presented here are consistent with the calibration data available.

It should be noted that in the analysis of the pile load test data all total pressures have been corrected for cross-sensitivity.

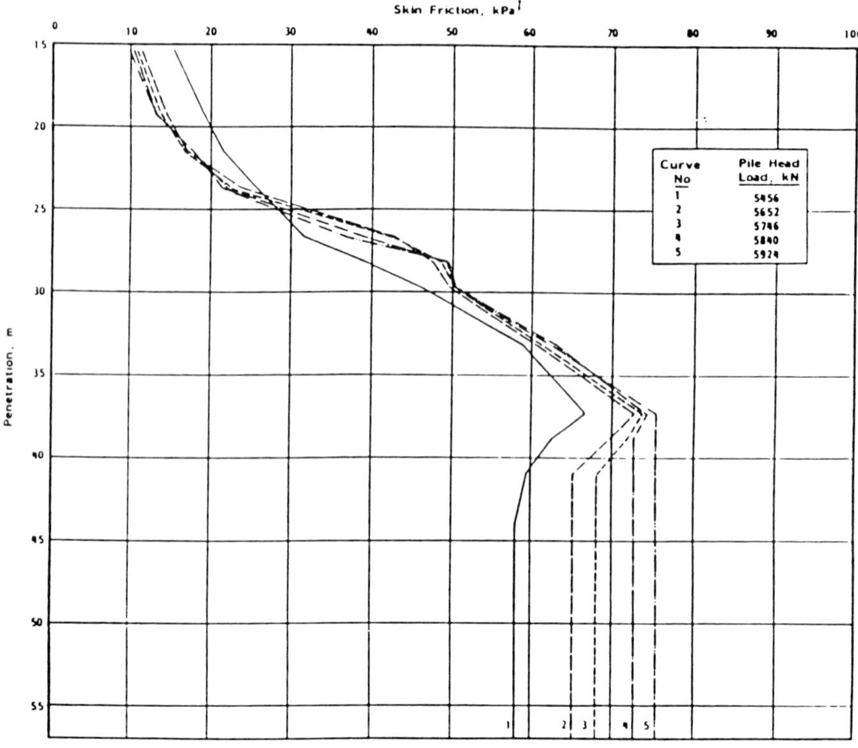

Fig. 7. Skin friction versus penetration : static load test, post-peak loads (NC pile)

2.1.4.3. Skin friction curves

The measured unit skin friction profile at peak load is shown on Fig. 12 together with undrained shear strength profiles from the three sample borings (101, 102 and 103), for comparison purposes. These strength profiles are based only on results of unconsolidated-undrained, UU, triaxial tests. Selection of the UU profiles rather than any other test type is consistent with the recommendations of API RP2A (1987) which also recommends performing UU tests on high quality push samples to be recovered with a 76 mm thin wall sampler, as has been performed in this study. The undrained shear strength profile corresponding to $s_u = 2.5$ times the penetration below ground level is superimposed on the UU profiles. It is generally considered that this idealised profile fits the UU data reasonably well though variations above and below the trend are apparent below about 40 m.

The average unit skin friction at peak load was about 54 kPa, corresponding to a conventional friction/strength a value of 0.62. The skin friction profile at peak load consisted of a linearly increasing portion between 15 and 37 m followed by a constant portion between 37 and 55 m, albeit the undrained

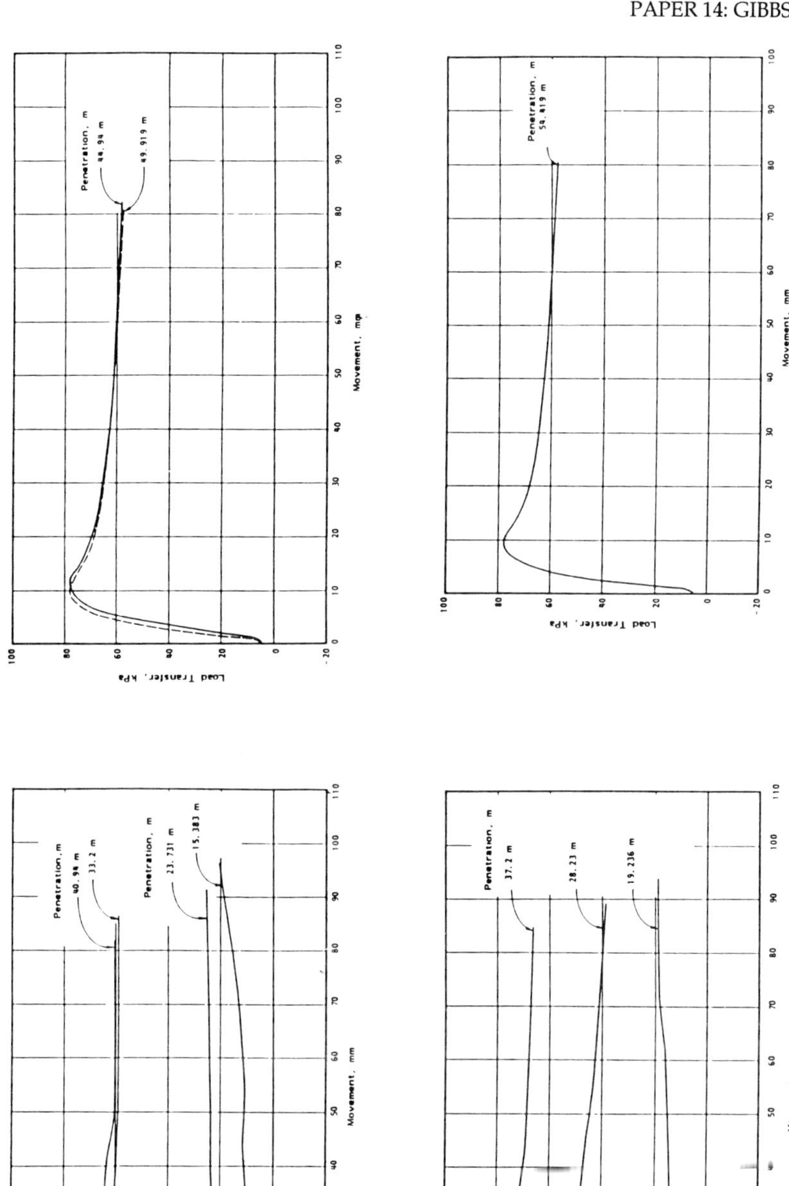

Figs 8 (left) and 9 (right). Load transfer curves: static load test, NC pile

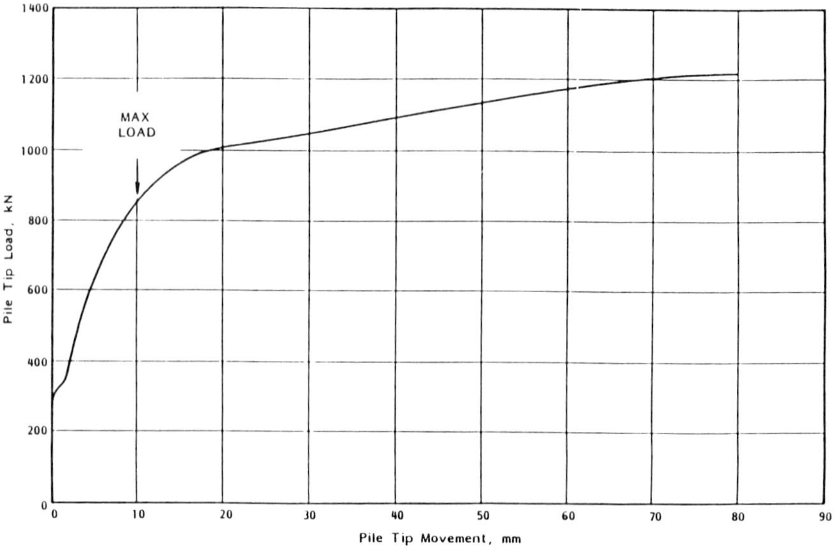

Fig. 10. End load versus pile tip movement: static load test, NC pile

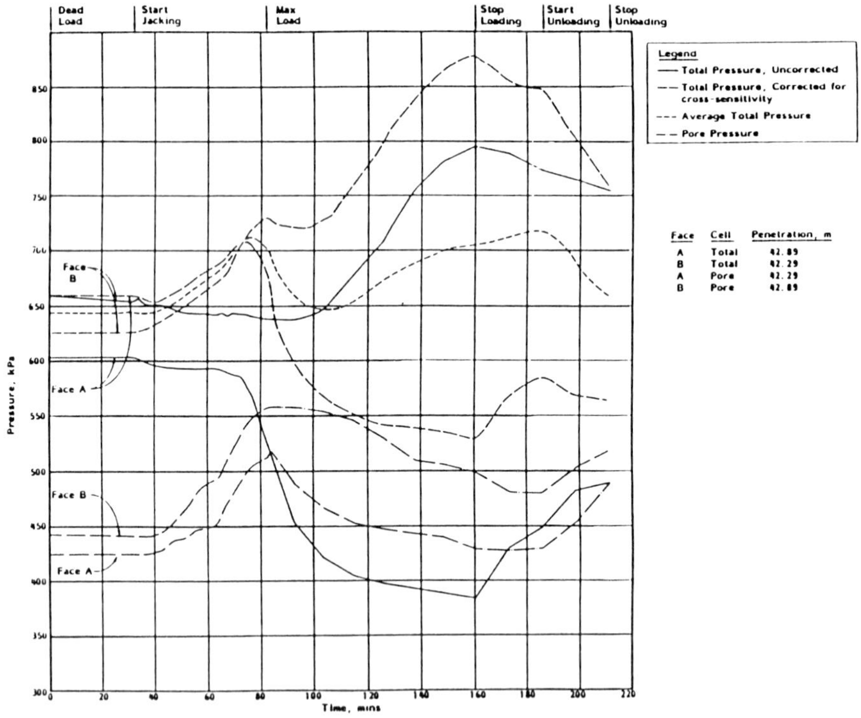

Fig. 11. Load test pressure cells at 43 m penetration, NC pile

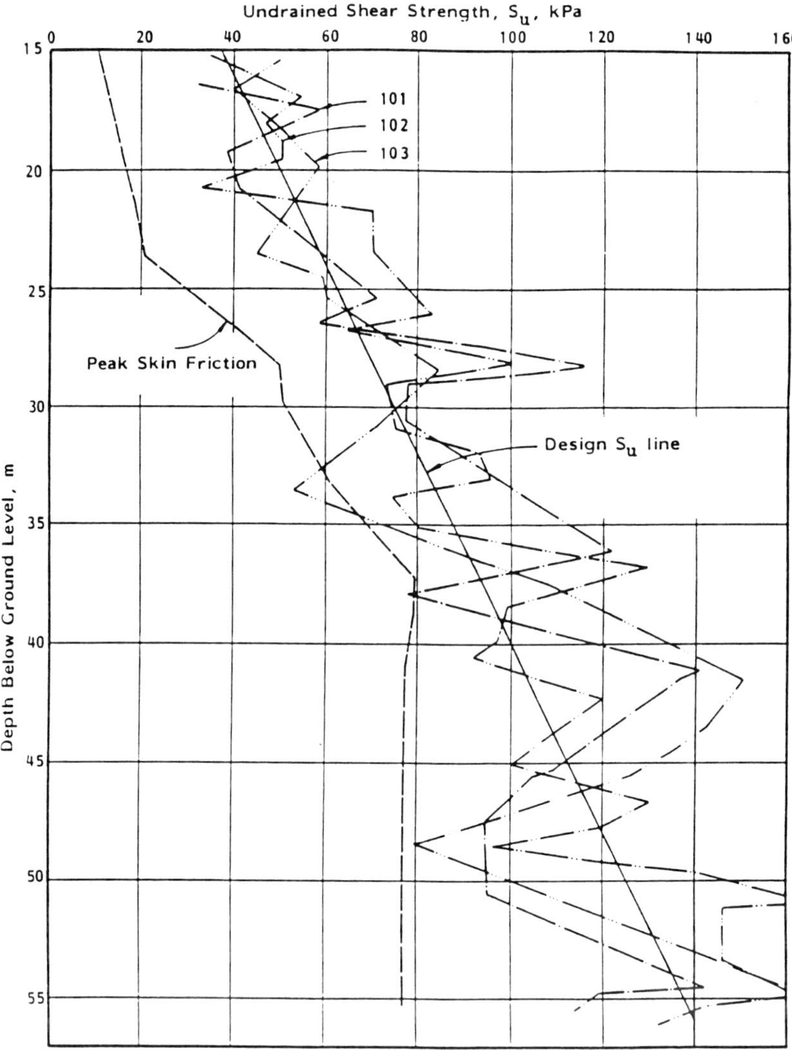

Fig. 12. Comparison of peak skin friction and undrained shear strength, NC pile

shear strength increases linearly with depth. It should also be noted that a linear axial load distribution was assumed below 40 m penetration due to failure of instrumentation. In the uppermost 10 m (12 pile diameters) skin friction values lower than the trend were measured and it is believed that the reduced values could be due to pile whip during driving and the residual negative stresses mobilised during installation. The backfigured α values are 0.3 to 0.35 in the upper 10 m, 0.4 to 0.8 over the middle third of the pile reducing to between 0.6 and 0.7 over the lower third.

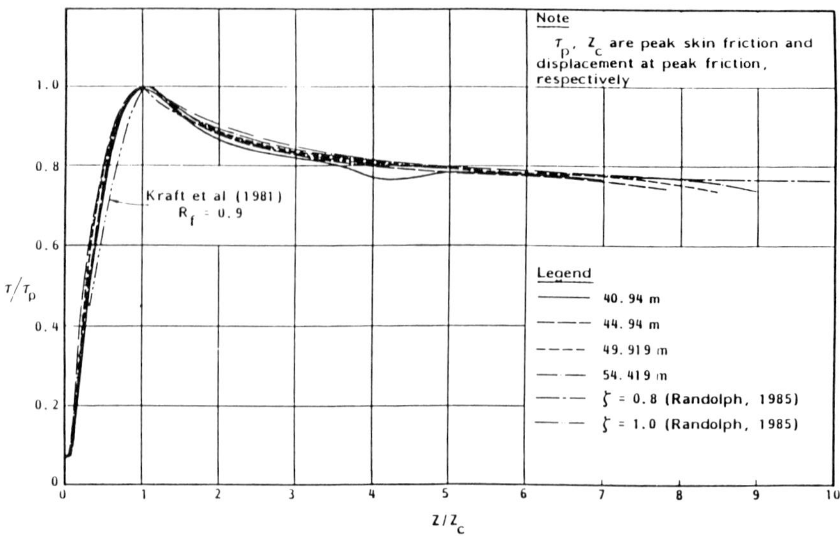

Fig. 13. Normalised T–Z curves: comparison with analytical procedures, NC pile

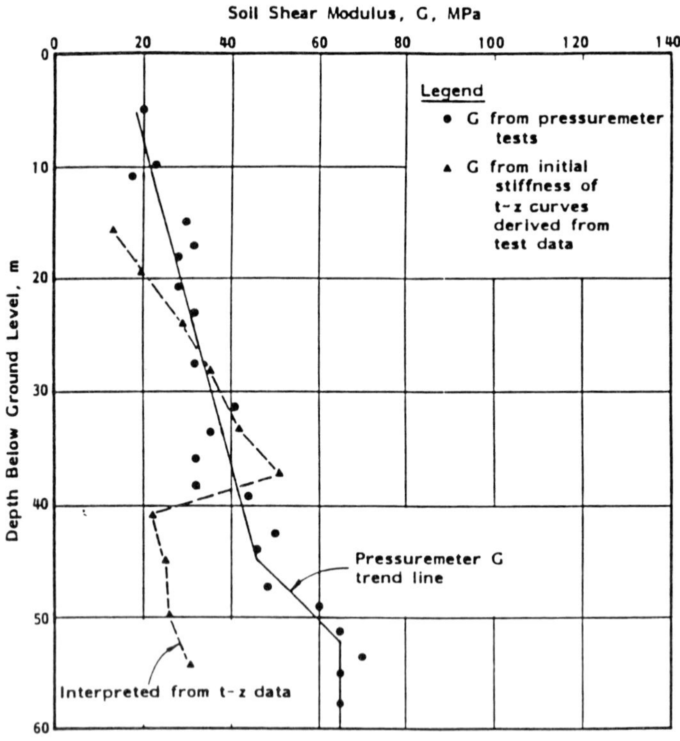

Fig. 14. Measured and backfigured shear moduli; NC pile

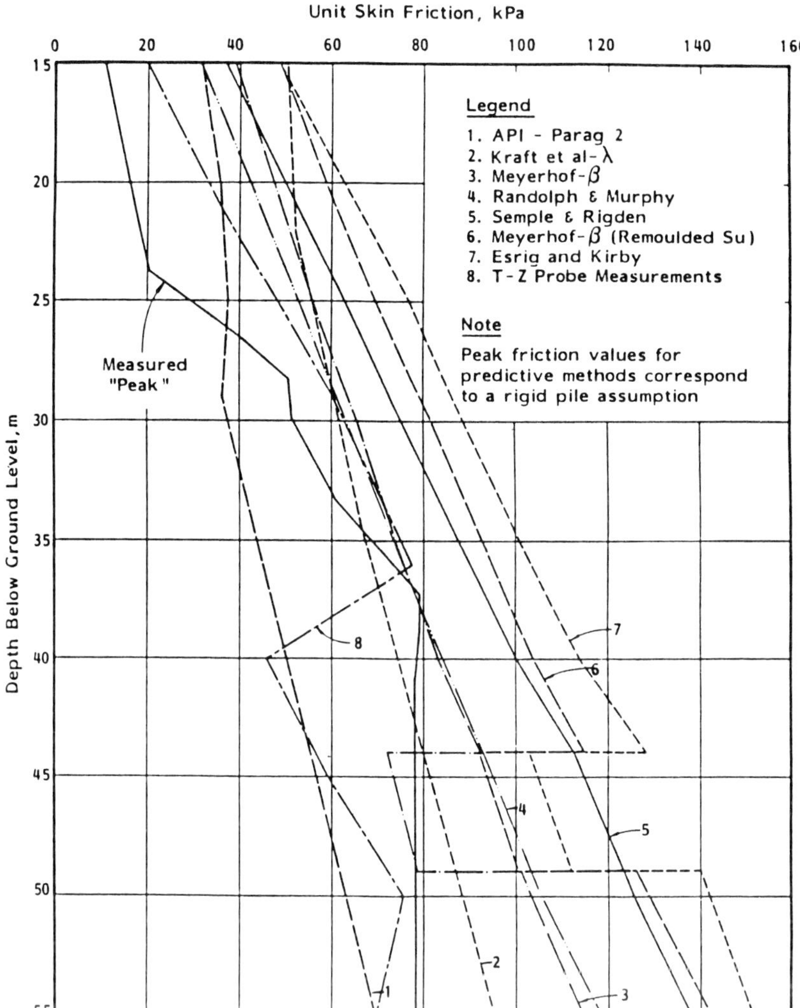

Fig. 15. Comparison of measured and predicted peak skin friction, NC pile

2.1.4.4. *T-Z curves*

Measured *T-Z* curves are presented in Figs 8 and 9. Over the lower third of the pile length, the *T-Z* curves exhibited distinct peak values followed by a strain softening residual stage, and were in accordance with theoretical models for non-linear work hardening behaviour prior to yield followed by strain softening after peak, Randolph (1983, 1985), Kraft *et al.* (1981b) see Fig. 13.

The peak friction appears to be mobilised at a displacement of 10 to 12 mm (1.5 per cent *d*) and the ratio of residual to peak skin friction is about 0.75, with the residual values being reached after an additional displacement of 40 mm. Results of laboratory direct shear tests gave residual to peak values of about 70 per cent and additional displacements to reach residuals of about 20 mm. In contrast to the well defined *T-Z* curves in the deeper soil, the curves in the shallower soil did not exhibit peaks but values comparable in magnitude to residual friction, probably due to continuous exhaustion of friction by the length of pile driven. In this shallower zone, the maximum friction appears mobilised at displacements of 25 to 30 mm. Over the middle third of the pile there is a transition zone where the peak is less well defined than in the deeper zone. Based on elasticity procedures, shear moduli can be back-figured from the initial (elastic) portion of the *T-Z* curve (Fig. 14). The results indicate shear moduli comparable in magnitude to those measured by the pressuremeter, however below about 40 m penetration, where the curves exhibited well defined behaviour, the shear moduli correspond to about half of those measured with the pressuremeter and it is believed that this reduction may be due to the influence of residual stresses induced during installation.

2.1.4.5. Comparison with predictions

Several analytical procedures were employed in comparing the measured peak friction with the predictions. The predictive procedures included total as well as effective and quasi-effective methods. In all, seven predictive methods were used. In addition, peak skin friction values measured with the *T-Z* probe under monotonic loading were compared to the measurements.

Peak skin friction profiles predicted by the various methods are shown on Fig. 15. Superimposed on this figure is the unit skin friction profile measured at peak load. An additional skin friction profile based on the *T-Z* probe measurements under monotonic loading is also presented. Measurements of skin friction with the *T-Z* probe were made at discrete depths below 20 m. These measurements have been joined together by straight lines to provide a profile and the value at 15 m has been estimated based on the trend. Results of the predictions are summarised in Table 2.

It should be noted that pile compressibility and soil strain softening effects on peak capacity are not included in Table 2 but are subsequently discussed. The predicted capacities given in Table 2 are therefore for a rigid pile.

The distribution of skin friction with depth is different from the predictions particularly in the shallow soils, however these methods are not expected to provide good predictions in the upper 10 pile diameters since neither residual stresses nor gapping are explicitly accounted for. Residual stresses however are measured with the *T-Z* probe, and the measured peak friction with the probe is found to be in reasonable agreement with that for the test pile. The *T-Z* probe is a rigid cylindrical model pile, thus the measured *T-Z* response is expected to provide skin friction values for a rigid pile to compute peak capacity for a compressible pile the measured curves are used in conjunction with a subgrade reaction program to account for pile compressibility effects. In the present study however we have found that the probe measurements can be applied directly to the load test measurements.

The peak pile capacity corresponds to the capacity associated with the peak load transfer only when the pile is infinitely rigid. The peak capacity is affected by the shaft resistance at each point along the pile and by the soil-pile interaction as defined by the load transfer (*T-Z*) response for each point. In addition, the amount of softening in the *T-Z* curve (residual skin friction) also affects the capacity, Murff (1980), Kraft *et al.* (1981a), Randolph (1983), as well as the pile-soil stiffness ratio. Therefore for a long compressible pile, peak frictional capacity may not correspond to the integration of the peak load transfer curves over the pile length.

The effects of strain softening of the soil and pile length (compressibility) on ultimate capacity have long been recognised. Randolph (1983) for example, provides normalised charts relating skin friction reduction factors, R_f, to the pile–soil stiffness ratio, *K*. This parameter, *K*, is given by the equation below, and is an adaptation of the parameter π_3 originally proposed by Murff (1980). The reduction factor R_f expresses the reduction in capacity from that of an infinitely rigid pile.

$$K = \frac{\pi\, d\, \tau\, l^2}{(AE)_p\, \Delta w_r}$$

where d is pile diameter, τ is peak skin friction, l is pile length, $(AE)_p$ is net cross-sectional area of pile times pile Young's modulus, and Δw_r is additional displacement necessary to mobilize residual skin friction. Based on Randolph's chart and the average skin friction determined for the various methods, it is

Table 2. Predicted rigid frictional capacity, NC pile

Method	Predicted frictional capacity, MN	Average Peak skin friction, kPa	Predicted/ Measured
API (1984) Para 2	4.47	46.6	0.86
Kraft et al λ (1981)	6.59	68.6	1.27
Meyerhof β (Remoulded Interface Strength) (1976)	6.92	72.3	1.34
Randolph and Murphy (1985)	7.09	74.0	1.37
Semple and Rigden (1984)	8.41	87.8	1.63
Meyerhof β (Remoulded Soil Strength) (1976)	8.66	90.5	1.68
Esrig and Kirby (1979)	9.46	98.8	1.83
T-Z Probe	5.37	56.0	1.04

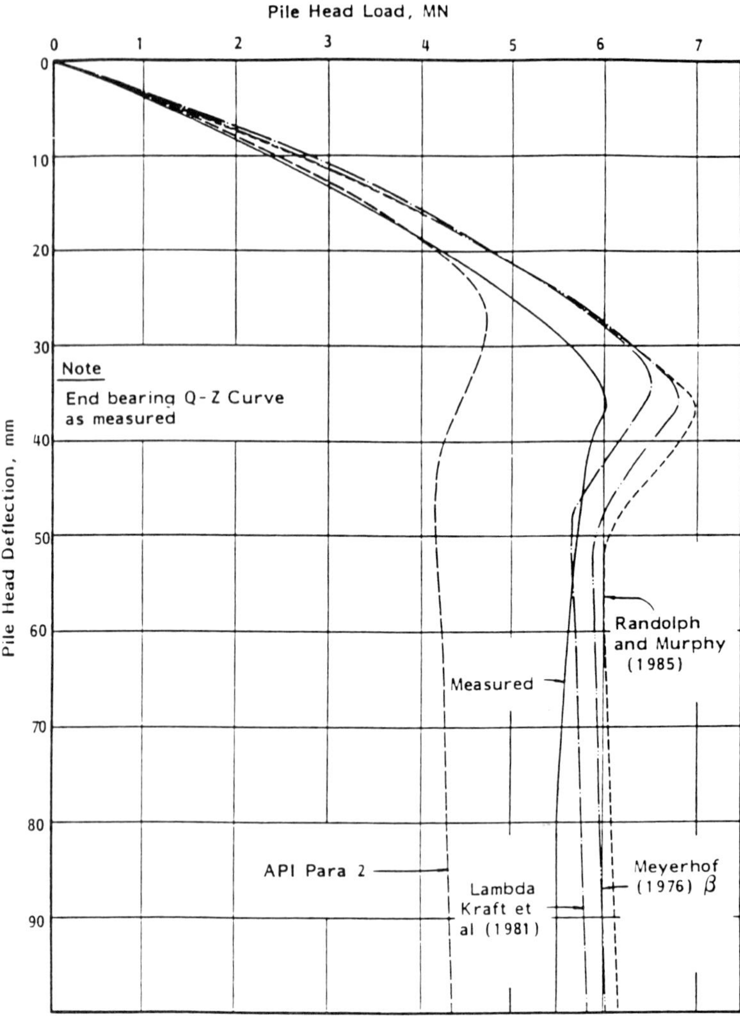

Fig. 16. Comparison of measured and predicted pile head load–deflection response, NC pile

possible to estimate the reduction in skin friction due to the pile compressibility effect and the strain softening of the soil. In the analyses the parameter, ε, expressing the ratio of residual to peak friction, is taken as 0.7 with the value of the additional displacement, Δw_r, taken as 20mm and the pile length as 40 m. The parameter K varies from about 1 for API (1984) Para 2 method to 2.1 for the Esrig and Kirby method thus giving $1/K^{1/2}$ values of between 1 and 0.69. The estimated reduction factors, R_f, from Randolph's chart are between 0.87 and 0.93, and these indicate the magnitude of reduction to be applied to the rigid pile skin friction curves

Table 3. Predicted 'compressible' peak friction capacity, NC pile

Method	Predicted frictional capacity, MN	Average Peak skin friction, kPa	Predicted/ Measured
API (1984) Para 2	4.17	43.5	0.81
Kraft et al λ	5.93	62.0	1.15
Meyerhof β (Remoulded Inter-face Strength)	6.23	65.0	1.20
Randolph and Murphy	6.38	66.6	1.23
Semple and Rigden	7.40	77.3	1.43
Meyerhof β (Remoulded Soil Strength)	7.58	79.2	1.47
Esrig and Kirby	8.28	86.5	1.60

presented in Fig. 15. The computed peak friction capacities for the various methods are given in Table 3.

2.1.4.6. *Comparison of measured and predicted load-deflection response*

Predicted load–deflection curves for the pile head are presented on Fig. 16. The predicted curves are results of computations of a subgrade reaction numerical solution similar to that described by Coyle and Reese (1966). Input to the computer program consists of the *T-Z* curves for the shaft and *Q-Z* curve for the tip, as well as the stiffness of the pile. The method of solution employs a finite difference approximation.

The comparisons given in Table 4 have been made for load deflection curves generated using the measured *Q-Z* response, described in section 2.1.4.10, and hence are comparisons between skin friction predictions. The comparisons indicate that the measured peak capacity was some 10 per cent lower than the revised Lambda method by Kraft *et al.* (1981a). The Meyerhof Beta method, Meyerhof (1976), based on remoulded interface parameters was the next best.

Table 4. Comparison of measured and predicted load–deflection response, NC pile

Method	Peak capacity, MN	Pile Head Deflection, mm	Residual Capacity, MN	Predicted/Measured Peak	Residual
API (1984) Para 2	4.72	26.6	4.1	0.78	0.76
Kraft et al (1981)	6.5	34.3	5.6	1.08	1.03
Meyerhof (1976)	6.8	36.0	5.9	1.13	1.07
Randolph and Murphy (1986)	7.0	37.3	6.0	1.16	1.09
Measured	6.0	36.0	5.5	-	-

Note : The measured end bearing curve has been used in the predictions.

The API (1984) Para 2 method underestimated peak capacity by some 20 per cent. The Randolph and Murphy (1985) method, a version of which is recommended by API (1987), over predicted capacity by about 20 per cent. The Kraft *et al.* (1981a) method also very closely predicted the pile head deflection at peak as well as the residual capacity. The *T-Z* probe measurements gave excellent agreement at both peak and residual friction capacities. A common feature of the analytical methods is that they predicted a friction profile increasing with depth which is not in agreement with the measurements while the *T-Z* probe predicted the measured profile far more closely than any of the analytical methods.

Comparisons of load deflection curves generated using the theoretical end bearing curve, i.e. ultimate value of 9 times s_u, show that the peak capacities predicted by the Lambda (1981), Meyerhof (1976) with remoulded interface parameters, and the Randoph and Murphy (1985) methods were within 10 per cent of the measured peak value. In applying Randolph and Murphy's method, the strength ratio for remoulded normally consolidated clay was taken as 0.2, for both sets of predictions.

2.1.4.7. Interface friction angles

The mobilised soil–pile friction angles at peak load could not be deduced reliably from the measured earth and pore pressure data from the cells at both 32 and 52 m. However interpretation of the measured data at 20 and 43 m gave mobilised angles of 15 and 24 degrees at peak and 20 and 17 degrees at residual capacity. The results are in good agreement with those from the direct shear box test on soil/RQT 501.

2.1.4.8. Comparison with T-Z probe

Generally a good comparison was found between the installation total earth pressures of both the test pile and the *T-Z* probe. However, the agreement was less good for the pore pressures.

The peak and residual skin friction values measured by the probe were in very good agreement with those measured for the test pile. Indeed the distribution of skin friction with depth was very closely predicted by the probe. Since the *T-Z* probe is much more rigid than the test pile, a subgrade reaction approach which caters for pile compressibility effects should be used to develop the load–deflection response (capacity) of the prototype pile.

In this study however it was found that the probe's peak and residual skin friction values can be applied directly to the test pile. The predicted peak friction capacity from the probe data is found to be only 4 per cent higher than the measured capacity.

2.1.4.9. End bearing

The end bearing (*Q-Z*) curve for the tested pile is shown on Fig. 17 designated Curve 1. To account for the weight of the soil displaced by the pile, taken as the product of the effective overburden pressure times the gross area of the tip, Curve 2 has been derived by subtracting this component from the measured curve. The measured curve exhibits a parabolic portion followed by a straight line.

By excluding the component of effective overburden pressure, the tip bearing capacity at peak load is 630 kN, corresponding to a bearing capacity factor, N_c, of 10 compared to the conventional value of 9. End bearing was increasing throughout the load test and upon termination of the test, bearing capacity was 990 kN corresponding to an N of 15.8. However, beyond a tip load of 1000 kN, corr sponding to bearing capacity of 770 kN, there was strong evidence of consolidation resulting from partial drainage. It is suggested that the value of 770 kN be taken as ultimate and this corresponds to a N_c value of about 12, where the reference s_u is that measured by UU tests (ignoring consolidation effects).

In conventional pile design analyses, ultimate end bearing is assumed fully mobilised at a tip displacement of 0.1d where d is pile diameter, and it is not uncommon to use smaller percentages down to 0.06d. The measured response however is much stiffer than predicted and does not correspond to methods used in current practice. The assumed 'ultimate' tip load of 1000 kN, equivalent to a bearing capacity of 770 kN, is mobilised at 20 mm or about 2.6 per cent of pile diameter.

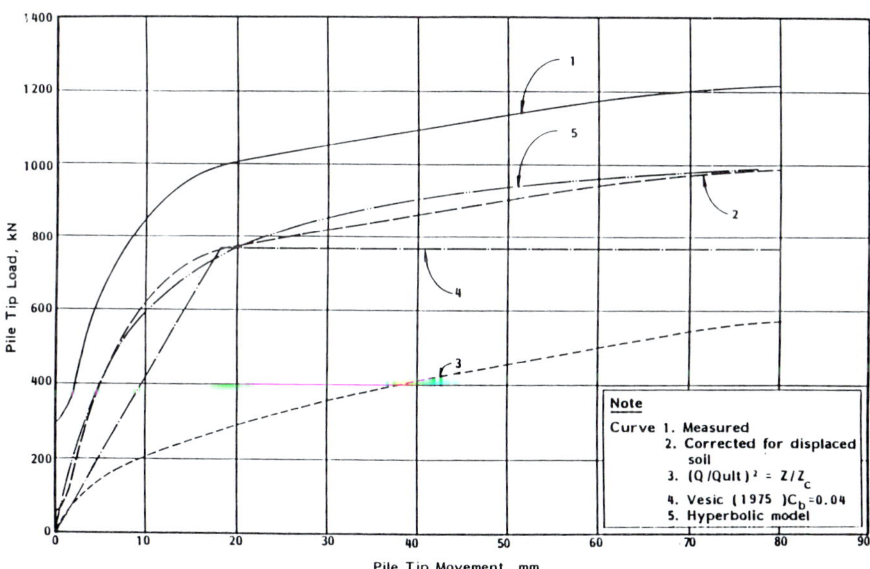

Fig. 17. Tip load versus movement (Q–Z curves), NC pile

TOTAL AND PORE PRESSURE CELLS MOUNTED ON THE PILE

Average Depth (m)	Channel	Static Theoretical Values s_u kPa	u_o kPa	σ_{vo} kPa	σ'_{vo} kPa	k_o	σ'_{ho} kPa	σ_{ho} kPa	Maxima at end of driving u_i kPa	Δu kPa	σ_{hi} kPa	σ'_{hi} kPa	$\Delta\sigma'_h$ kPa	$\frac{\Delta u}{s_u}$	$\frac{\Delta\sigma'_h}{s_u}$	$\frac{\Delta u}{\sigma'_{vo}}$	$\frac{\Delta\sigma_h}{\sigma'_{vo}}$	Prior to testing u_c kPa	σ_{hc} kPa	σ'_{hc} kPa	k_o
7.99	A	465	62	177	115	1.75	200	262	477	415	225 (246)	-252 (-231)	-452 (-431)	1.04	-0.97 (-0.93)	3.61	-3.93 (-3.75)	87	241 (264)	154 (177)	1.34 (1.54)
	B								510	448	456 (467)	-54 (-43)	-254 (-243)	0.96	-0.55 (-0.52)	3.90	-2.21 (-2.11)	82	-	-	-
14.26	A	420	112	307	195	1.45	280	392	429	317	352 (355)	-77 (-74)	-357 (-354)	0.75	0.86 (-0.84)	1.63	1.83 (-1.82)	178	653 (656)	475 (478)	2.44 (2.45)
	B								650	538	684 (688)	34 (38)	-246 (-242)	1.28	-0.59 (-0.58)	2.76	-1.26 (-1.24)	181	429 (432)	248 (251)	1.27 (1.29)
20.18	A	520	165	435	270	1.30	355	520	20*	-145	FAILED	-	-	-0.28	-	-0.54	-	327	-	-	-
	B								-49*	-214	929 (940)	978 (989)	623 (634)	-0.41	1.20 (1.22)	-0.79	2.31 (2.35)	361	860 (871)	499 (510)	1.85 (1.89)
26.18	A	650	214	564	350	1.2	410	624	1274	1060	1835 (1798)	561 (524)	151 (114)	1.63	0.23 (0.16)	3.03	0.43 (0.33)	460	1555 (1516)	1095 (1056)	3.13 (3.02)
	B								1591	1377	2028 (2028)	437 (437)	27 (27)	2.12	0.04 (0.04)	3.93	0.08 (0.08)	439	1636 (1636)	1197 (1197)	3.42 (3.42)

NOTES:

1. u_i is the maximum value recorded immediately after the drive
2. σ_{hi} is the maximum value recorded immediately after the drive
3. u_c and σ_{hc} are values on morning of the test
4. Total pressure values are corrected for cross-sensitivity. Uncorrected values are shown in brackets.
5. All values given in kPa.
6.* Pore pressures increased during set-up period to a maximum 15 days after driving.

PIEZOMETERS

Distance from pile wall (m)	Depth (m)	STATIC – Values Before Drive $\frac{r}{rp}$	u_o kPa	s_u kPa	σ_{vo}' kPa	MAXIMUM – Values During Drive u_{max} kPa	Δu_{max} kPa	$\frac{\Delta u_{max}}{s_u}$	$\frac{\Delta u_{max}}{\sigma_{vo}'}$	MAXIMUM – Values After Drive u kPa	Δu kPa	$\frac{\Delta u}{s_u}$	$\frac{\Delta u}{\sigma_{vo}'}$
1	8.1	3.62	66.4	465	115	364	298	0.64	2.59	90	24	0.05	0.21
2	8.1	6.24	61.4	465	115	281	218	0.47	1.90	119	56	0.12	0.49
1	20.4	3.62	132	520	270	955	823	1.58	3.05	661	529	1.02	1.96
2	20.4	6.24	156	520	270	945	789	1.52	2.92	634	478	0.92	1.77

Fig. 18. Pressure cell readings, OC pile

2.1.4.10. Load–deflection response

The pile head load–deflection stiffness reduced from about 225 MN/m at 50 per cent of the peak load to 167 MN/m at peak load, and down to 50 MN/m at the final residual load, which is approximately 10 per cent lower than the peak. While the API (1984) Para 2 skin friction method underestimated both peak and residual frictional capacity, the pile head stiffness using this method agreed very closely with the measured stiffness for loads up to 75 per cent of the peak load.

2.2. OC pile

2.2.1. Installation

The pile was installed to the target penetration of 30 m by driving with the BSP HA 40 hydraulic hammer. Plug level measurements inside the pile indicate that the pile was continuously coring but approached plugging at 13 m, though full plugging was not achieved. It is estimated that the displacement ratio below 13 m penetration was 80 per cent (compared to 100 per cent for a fully plugged pile).

The total earth pressure and pore pressure cells provided continuous measurements of these quantities as the pile was driven to the target penetration. Selected values of measured pore pressures and total pressures including static, end of driving and prior to testing are given on Fig. 18. The cells are roughly located at elevations corresponding to approximately 8, 15, 21 and 27 m below the ground level, when the pile tip is at the target penetration of 30 m. Thus the first cell that enters the soil is Cell 27.

2.2.1.1. Total earth pressure and pore pressure measurements

Pore pressures generated by driving were continuously increasing with depth and were within a large range, varying from negative values to values as high as 2000 kPa. Negative pore pressures of up to 80 kPa were measured at several depth intervals and these appear to be associated with readings from the cell nearest to the pile tip. These negative pore pressures were also recorded in the cone penetration tests. There was an 'abrupt increase in the pore pressures at about 20 m in the underlying Oxford Clay where values of up to 2000 kPa were measured. Excess pore pressures generated by driving are on average 0.5 to 0.7 s_u in the Lowestoft Clay (Stratum I) and 2 s_u in the Oxford Clay (stratum where values correspond to the design profile II), based on the UU tests shown in Fig. 33. At the end of drivingthe ratio of maximum excess pore pressure to effective overburden pressure was between 1.6 and 3.9 in the Lowestoft Clay and between 3.9 and 4 in the Oxford Clay. Away from the pile the same ratio was between 2.6 and 3 at 1m from the pile wall, and between 1.9 and 2.9 at 2 m from the pile wall. Overall, the measured pore pressure profile for the pile appears to be reasonably consistent with the

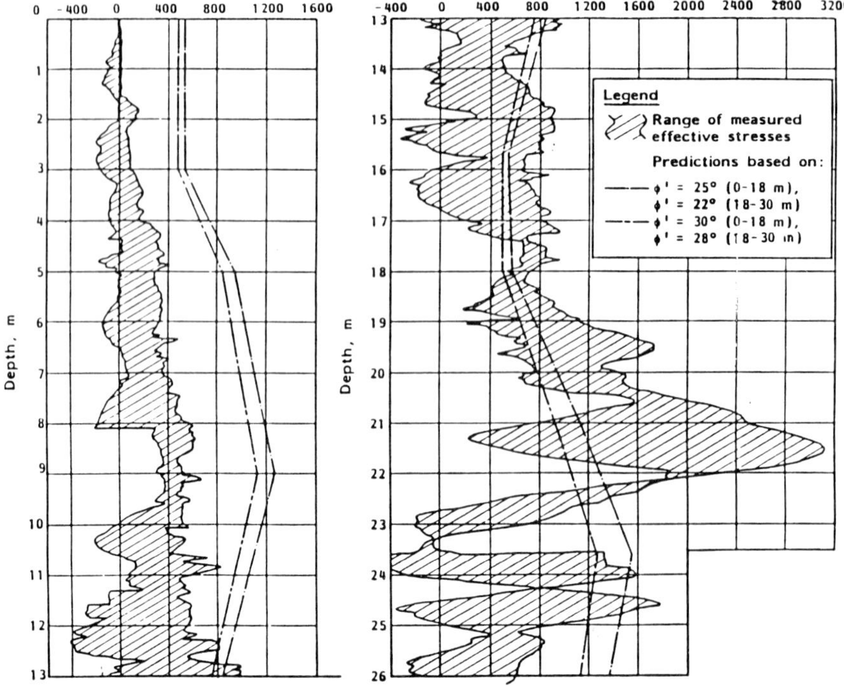

Fig. 19. Effective stresses during driving, OC pile

trends in the strength profile as higher pore pressures were measured in the stronger clays.

Total earth pressure was consistently increasing with depth. An abrupt increase in earth pressure was measured in the Oxford Clay as similarly displayed by the pore pressure measurements. Values of 2000 kPa and greater were measured in the Oxford Clay.

The total earth pressure (including ambient pore pressure) is estimated to increase from about 1.75 times the CIU strength (1.0 times UU) at 4 m penetration to 3 CIU (1.6 times UU) at 20 m penetration. Between 20 and 25 m, average total earth pressure is 2000 kPa or about 5 times the average CIU strength (3 times UU strength). For a full displacement pile, predictions based on cavity expansion and the Cam clay model give total radial stresses of the order of 6 s_u for an OCR range of between 8 and 32, Randolph *et al.* (1979) which is in very good agreement with the normalised ratios in the Oxford Clay using CIU strength data.

In general, the effective radial stress was more or less constant, at 200 kPa between ground level and 8 m, increasing to 400 kPa between 8 and 17 m, thereafter showing a large increase with depth to an average value of near

1700 kPa at 22 m. Below 22 m, the effective stress was reducing with penetration with an average value of 400 kPa, see Fig. 19.

Based on a cavity expansion approach, Randolph *et al.* (1979) the predicted ratios of the effective radial stress to undrained shear strength are on average 2.6 in the Lowestoft Clay and 2.9 in the Oxford Clay. These predictions are found to grossly overestimate the measured data between ground level and 13 m but are in reasonable agreement below this depth. This is expected, however, and attributed to the effect of the installation residual stresses in the shallower soils which is not accounted for by the theory.

2.2.2. *Consolidation stage*

The OC pile load test was performed 130 days after driving. At that time, the percentage dissipation of excess pore pressures, assessed from the pore pressure measurements on the pile wall, was between 80 and 95 per cent in the Lowestoft Clay and between 77 and 84 per cent in the Oxford Clay as shown in Fig. 20. The estimates in the Oxford Clay are based on the data from the pair of cells at 27 m. The other pair of cells at 21 m showed a continuous rise in pore pressure up to 15 days from driving and this was followed by dissipation. All other cells recorded maximum pore pressures on or shortly after installation. Estimates of dissipation percentage from the 21 m cells are somewhat uncertain. However, based on the response of piezometers preinstalled at the same depth as these cells, it is estimated that the percentage dissipation is unlikely to be lower than 65 to 70 per cent and is probably of the order of 80 per cent. Excluding these uncertain estimates, excess pore pressures of 10 per cent and 20 per cent are estimated to have remained in the Lowestoft Clay and the Oxford Clay respectively, at the time of load testing.

Away from the pile wall, data from the piezometers preinstalled 1 m and 2 m from the pile show 100 per cent dissipation in the Oxford Clay and between 65 and 70 per cent in the Lowestoft Clay.

Predictions of excess pore pressure dissipation periods were based mainly on consolidation coefficients estimated from laboratory tests. This is because the *T-Z* probe pore pressure data were inconsistent and the cone penetrometer dissipation tests gave a very wide range of consolidation coefficients varying between 2 and 100 m^2/year. In general the measured dissipation periods were within the predictions and corresponded to the higher end of the consolidation coefficients assessed from the laboratory test data. However the backfigured consolidation coefficients based on the measured installation excess pore pressures and dissipation periods are smaller than those estimated from the oedometer tests in the Lowestoft Clay but are comparable in the Oxford Clay.

During consolidation, the measured total earth pressure was nearly constant for all cells except those at 27 m. These showed reductions of about 15

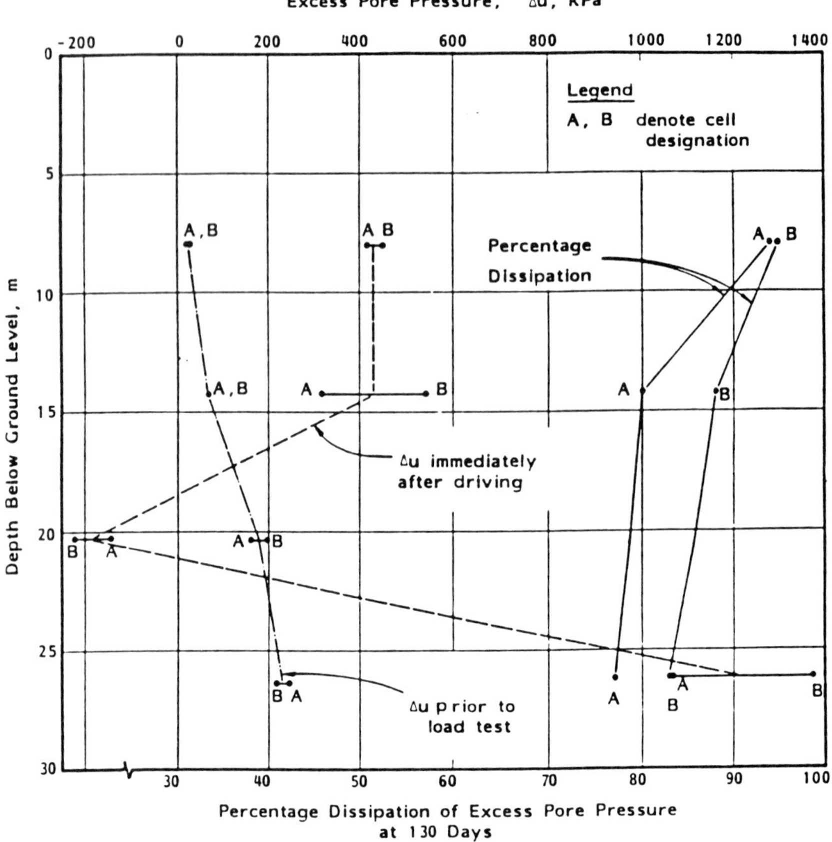

Fig. 20. Excess pore pressures after driving and before load test, OC pile

to 20 per cent after the set-up period of 130 days. An elastic type analysis would predict the total radial stress to remain constant during consolidation but incorporation of soil non-linearity effects results in a reduction of about 20 per cent, and this is in agreement with the measurements.

At the end of consolidation, taken as 130 days, the 'effective radial stress was found to be between 1.9 times the CIU strength at 14 m increasing to 3CIU at 26 m (if UU strengths are used, the corresponding ratios are 0.85 to 1.8). Lower ratios were obtained at 8 m indicating the effects of the installation process. The increase in the effective radial stress during consolidation was about 1.7 CIU(0.9 to 1.0 UU), which is in agreement with analytical predictions.

The ratio of the final (end of consolidation) effective radial stress to the effective overburden pressure designated k, is found to increase with depth from 1.3 at 8 m to 1.9 at 14 m and to 3.3 at 26 m. These values are higher than the k_0 values estimated from oedometer and k_0CU triaxial tests but somewhat

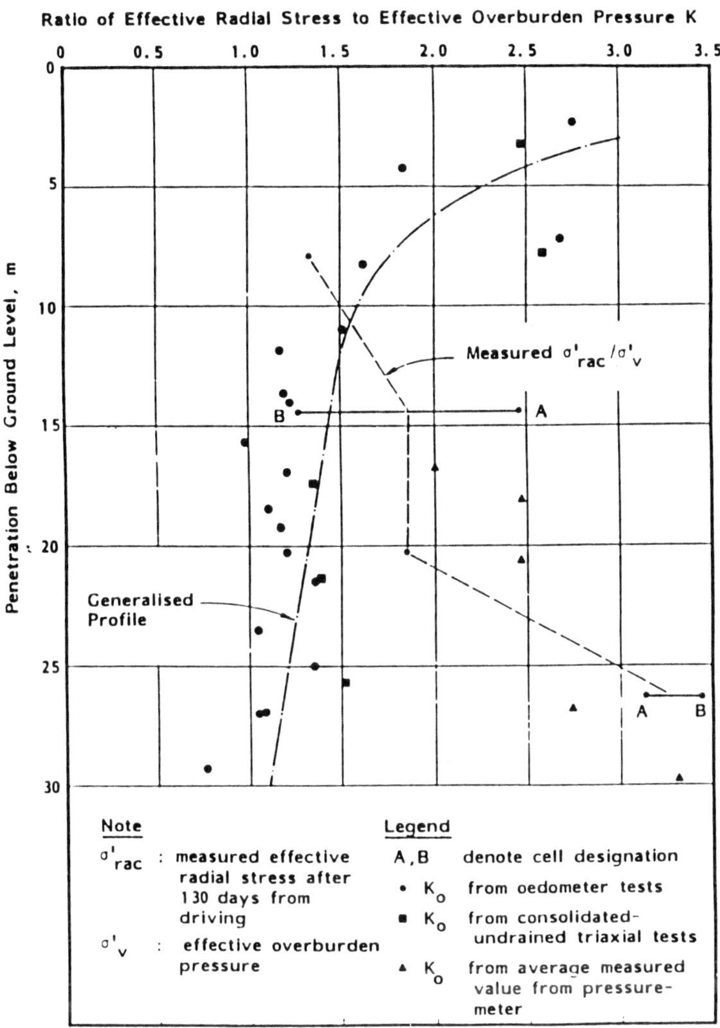

Fig. 21. Comparison of measured and predicted k values, OC pile

lower than the pressuremeter average values. The measured k value at 26 m exceeds the passive coefficient estimated to be between 2.8 and 3 (ignoring cohesion). Published experimental findings show that at the end of consolidation the ratio of the final effective radial stress to the original effective vertical stress lies between k_0 and $2k_0$. Below 10 m penetration, the measured k values are about 1.3 to 2.7 times the estimated k_0 design profile as shown in Fig. 21.

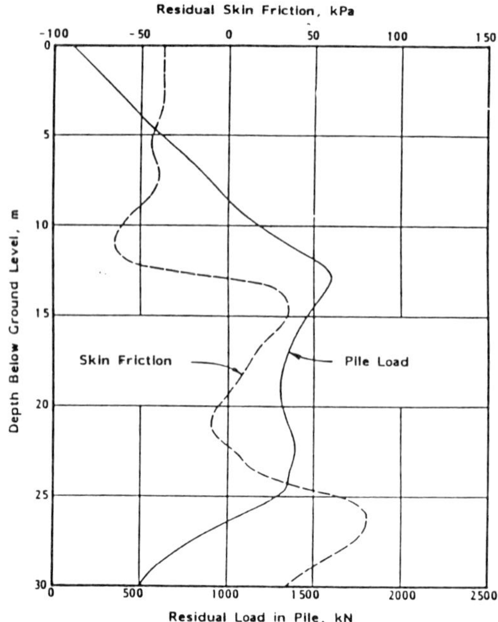

Fig. 22. Residual pile load prior to load test, OC pile

2.2.3. Installation residual stresses

The measured profile of residual stresses for the OC pile prior to pile load testing is presented in Fig. 22. Both the mobilised skin friction and distribution of load have been plotted versus penetration. As shown, large negative residual stresses occur over the upper 13 m of the pile. These negative friction values range between 40 and 65 kPa with an average value of 45 kPa corresponding to about 0.12 times the average undrained shear strength over this depth interval. As shown subsequently, the residual stresses in the upper part of the pile lead to reduced friction values throughout the test. The research residual load at the tip is about 0.5 MN, or approximately 25 per cent of the computed ultimate end bearing value assuming an undrained shear strength of 475 kPa at the pile tip.

2.2.4. Static pile load test data

The measured peak load was 16.13 MN and the pile top displacement was 28 mm. At the final residual (post-peak) stage the load reduced to 14.21 MN (88 per cent of peak) and the top displacement was 129 mm. The field load versus pile head deflection curve is given in a companion paper, Cox *et al.* (1993b).

At the peak load of 16.13 MN, the load carried by the tip is estimated to be 1.45 MN or about 9 per cent of the total load. The average measured unit skin

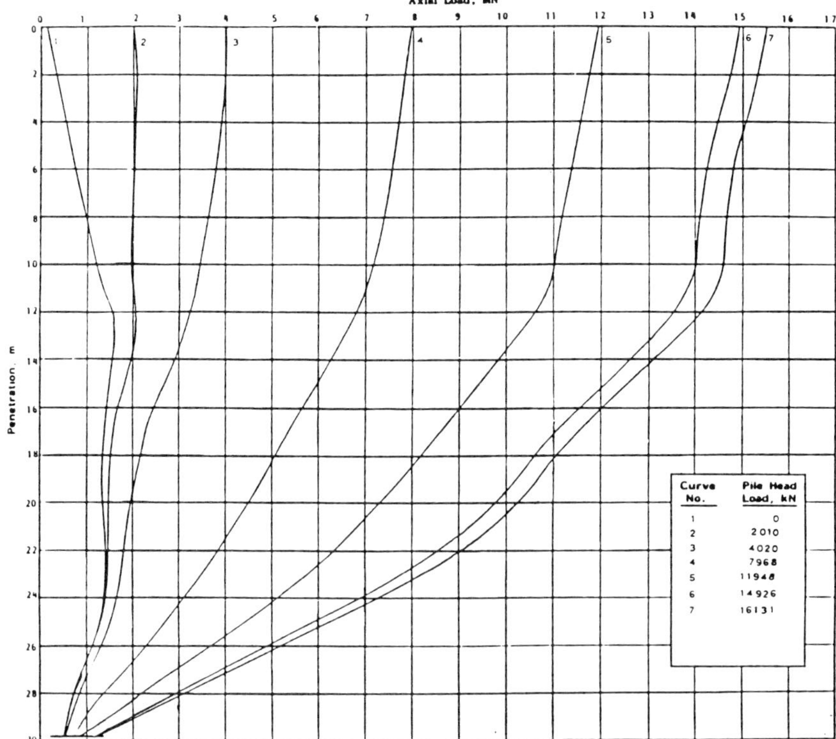

Fig. 23. Axial load distributions: static load test, pre-peak loads (OC) pile

friction is 209 kPa, giving a conventional friction/strength α value of 0.44 based on the UU data. At the final residual (post-peak) load, the average unit skin friction is 173 kPa giving an α value of 0.37 and the load carried by the tip is 2.07 MN.

As noted in section 2.2.2 it is estimated that average percentage dissipation at time of testing, 130 days after installation, was about 80 per cent. This value is believed to be within the current API pile test database which includes information from pile tests performed a few days to a few hundred days after pile installation.

2.2.4.1. *Results of static load test*

Results of the static load test are given in Figs 23 to 30 as axial load distributions, skin friction distributions along the pile, the shaft load transfer (*T-Z*) curves and the end bearing load transfer (*Q-Z*) curves.

The axial load distribution curves on Figs 23 and 24 were developed by applying very small adjustments to the curves resulting from a direct reduction of field data. These small adjustments give a smooth curve that eliminates abrupt and unrealistic changes in slope of the load curve.

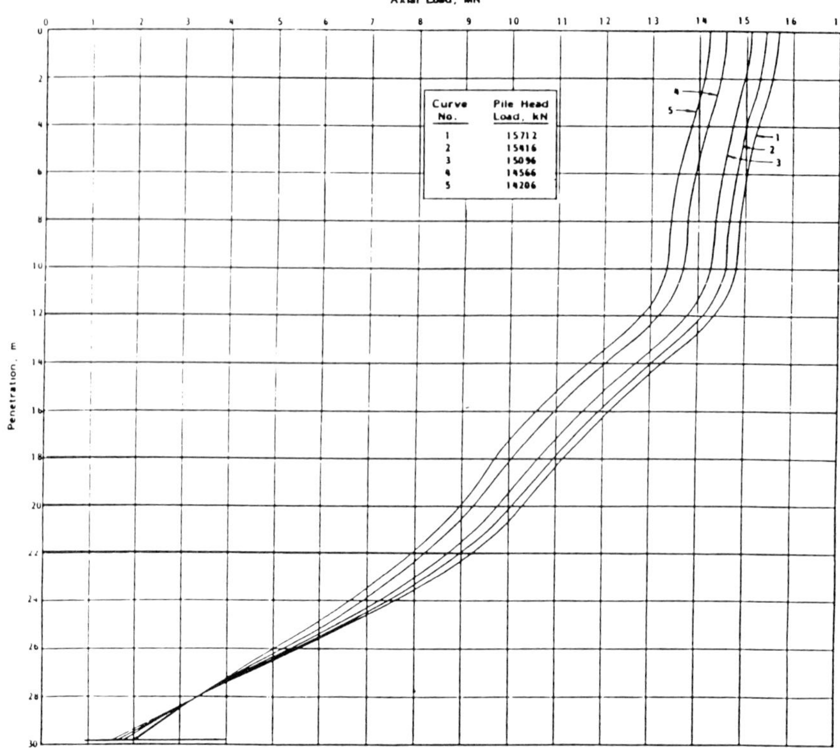

Fig. 24. Axial load distributions: static load test, post-peak loads (OC) pile

Skin friction values shown on Figs 25 and 26 are a closed form mathematical solution using load curves on Figs 23 and 24. Section 3.6 describes the method used.

Figure 25 presents the skin friction distributions for load levels up to the maximum pile head load of 16.13 MN. In addition, the skin friction is shown for the residual load system which existed at the start of the test. Top load on the pile for this residual system was 0.106 MN corresponding to the dead weight of the load cells, jacks and pile cap. Five curves of skin friction distribution at load levels beyond the peak pile load are presented in Fig. 26.

Load transfer curves were generated for 28 penetrations along the embedded length of the pile. Representative curves are presented in Figs 27 to 29. Load at the pile tip versus tip movements is plotted in Fig. 30.

2.2.4.2. Results of pressure cell measurements

Total and pore pressures measured at 15 m during the load test are presented versus time in Fig. 31. Total pressure readings are shown uncorrected and corrected for cross-sensitivity.

The pore pressure recorded by the piezometers situated adjacent to the

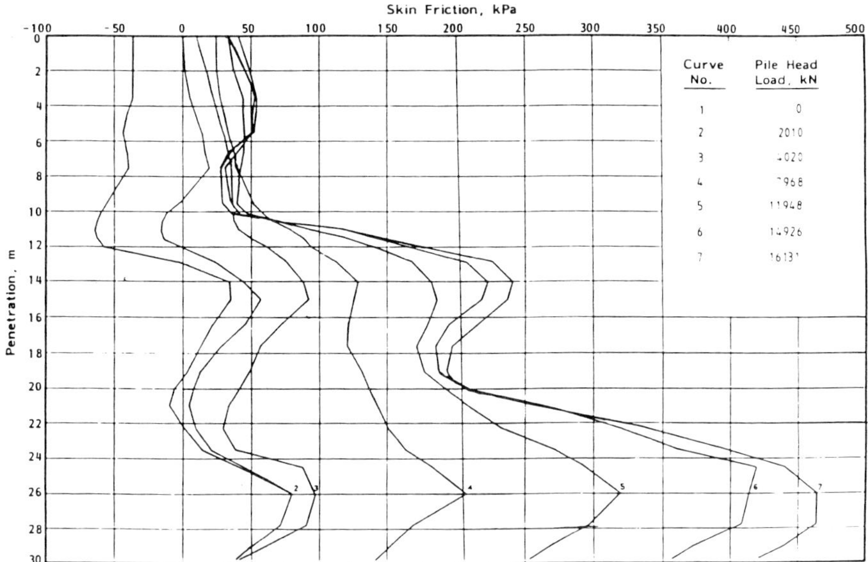

Fig. 25. Skin friction versus penetration: static load test, pre-peak loads (OC pile)

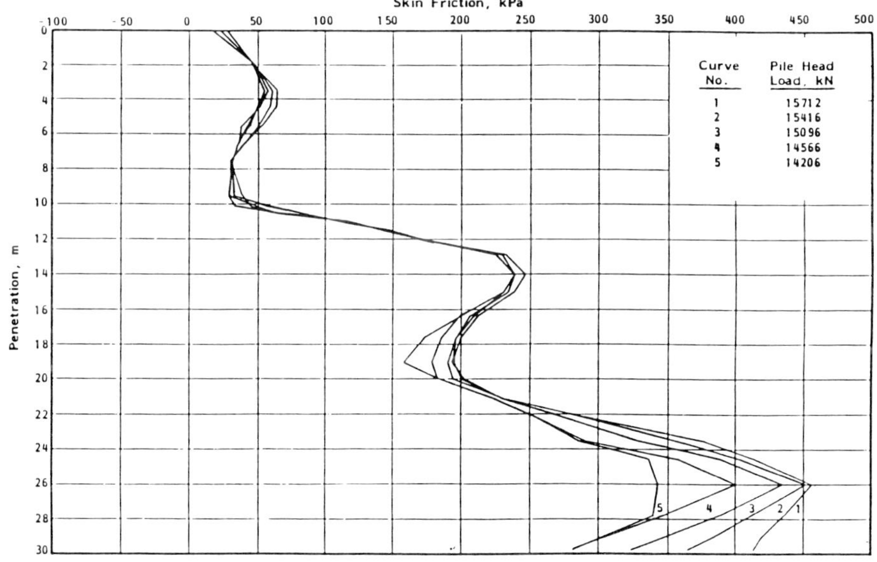

Fig. 26. Skin friction versus penetration: static load test, post-peak loads (OC pile)

pile showed changes in the order of 2 and 12 kPa, for the instruments at 8.1 m and 20.4 m, respectively. These are shown on Fig. 32.

2.2.4.3. *Skin friction curves*

Selected profiles of mobilised unit skin friction versus penetration are presented in Fig. 33. The undrained shear strength profile assessed from the UU data is also shown as well as the strength profile from the CIU data, so that comparison can be made between the trends in the mobilised friction profiles and the strength profiles. The skin friction profiles have been selected consecutively to correspond to residual, low pre-peak load, 50 per cent of peak, peak, and final post-peak before the pile was unloaded. The influence of the residual installation stresses down to 10 m in reducing skin friction is clearly apparent. Below 10 m depth, the comparison between the trends in the mobilised friction at load approaching the peak and the strength data is favourable.

Unless noted otherwise the strength profile definition is based only on results of unconsolidated-undrained, UU, triaxial tests conducted on high quality push samples. This selection is consistent with the recommendations of API RP2A (1984).

Prior to pile loading and as a consequence of the driving process, negative residual stresses were mobilised over the upper 13 m of the embedded pile length. These stresses are on average about 45 kPa, corresponding to 0.12 times the average UU strength over this depth interval. The negative stresses were eliminated at loads in excess of 12 per cent of peak. At the peak load, the average unit skin friction over the upper 10 m of the pile is 42 kPa corresponding to an average α of 0.11, assuming an average undrained shear strength of 375 kPa over the same depth interval. Over the lower third of pile (20 to 30 m), the average unit skin friction is 448 kPa, compared to an average undrained shear strength of 615 kPa giving a conventional α value of 0.73.

At the final post-peak load of 14.21 MN, there was virtually no change in the mobilised friction in the upper 15 m, from the values mobilised at peak load. However significant reduction in friction occurred in the Oxford Clay below 18 m, clearly confirming the strain softening characteristics of the clay as have been measured in the laboratory tests. Between 20 and 30 m, the average unit skin friction is 295 kPa representing some 35 per cent reduction from the average value at peak, and the corresponding α is 0.48.

2.2.4.4. *T-Z curves*

Measured *T-Z* curves are shown in Figs 27 to 29. In the Lowestoft Clay, *T-Z* curves in the upper 5 m show that the response was rather irregular with maximum friction being mobilised at displacements of about 20 mm. The curves for penetrations of 7.5 to 10 m showed well defined peaks followed by large reductions in friction to residual values of about 65 to 75 per cent of

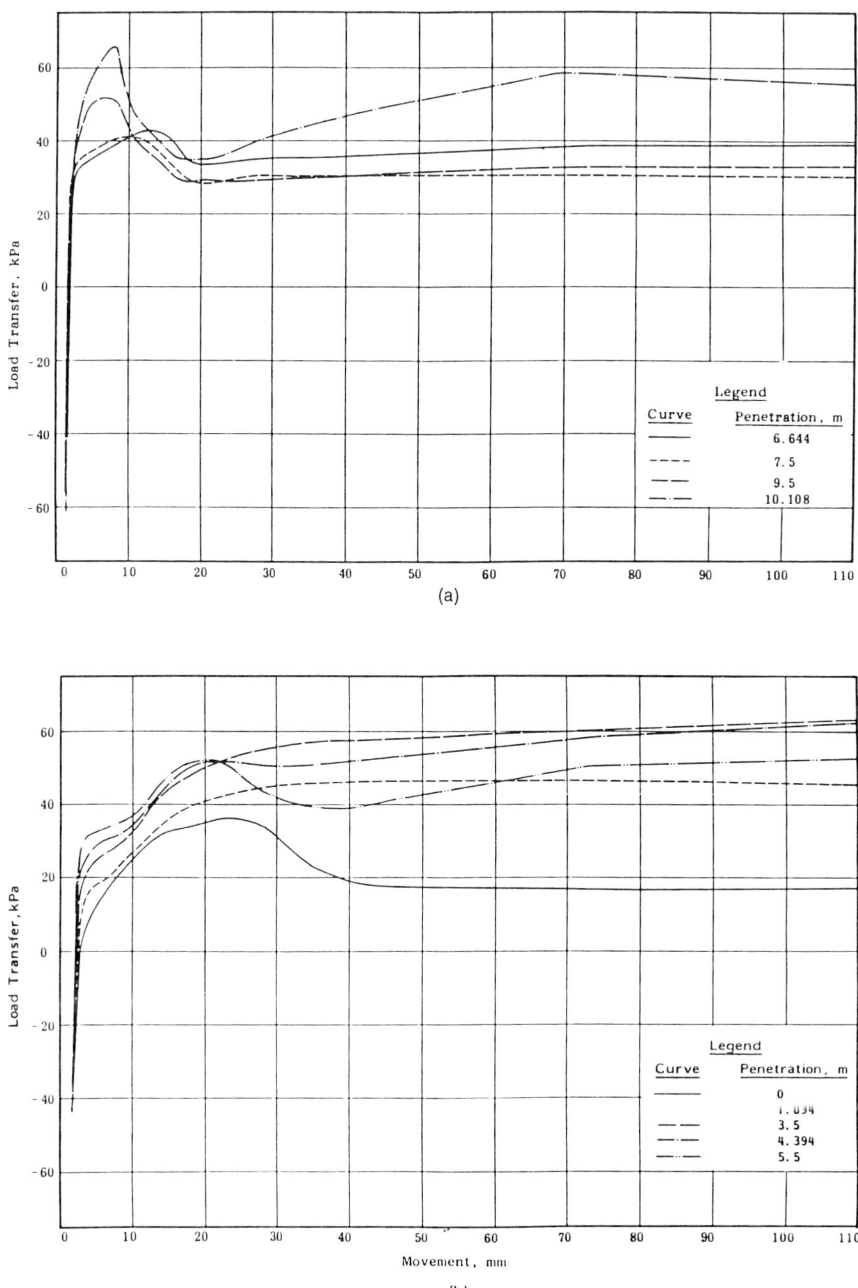

Fig. 27. Load transfer curves, static load test (OC pile): (a) top and (b) bottom

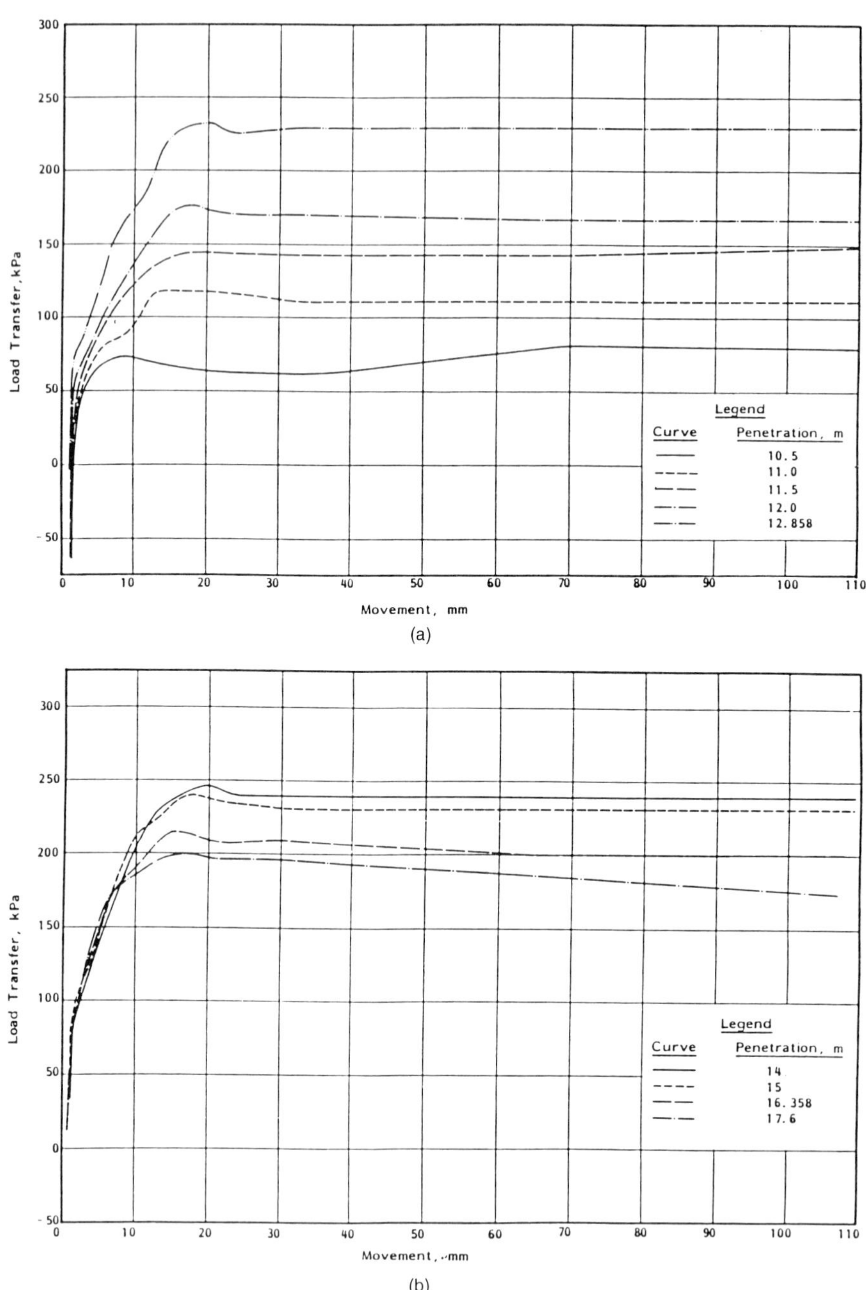

Fig. 28. Load transfer curves, static load test (OC pile): (a) top, (b) bottom and (c) facing page

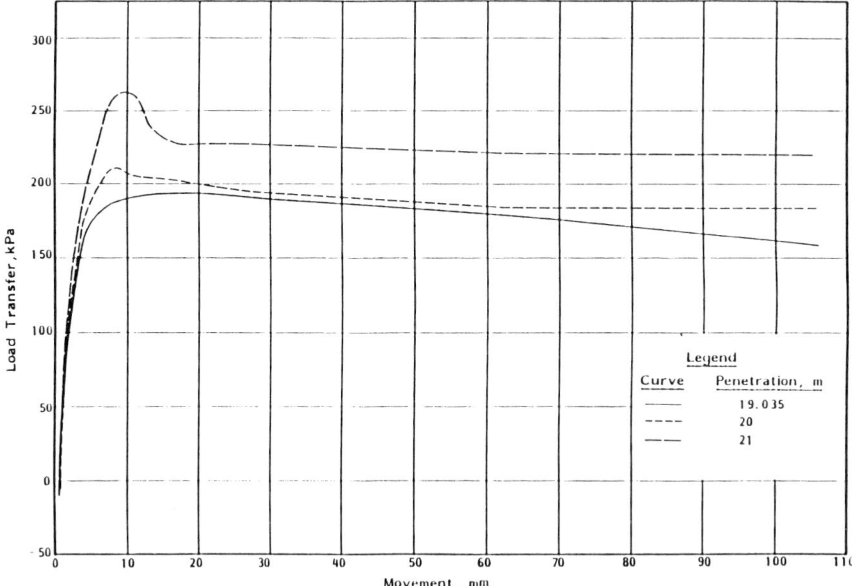

Fig. 28(c)

peak. Peak frictions appear mobilised at movements of 8 to 12 mm. Over the lower part of the Lowestoft Clay, the critical displacements are between 15 and 20 mm. Overall maximum frictions appear mobilised at displacements of 10 to 20 mm (1.3 to 2.5 per cent of pile diameter) much larger than expected in heavily overconsolidated clays for this pile size. In the Oxford Clay, the curves exhibit distinct peaks with residual to peak values typically in the range of 70 to 80 per cent. The maximum skin friction throughout this stratum appears to be mobilised at displacements of 8 to 10 mm (1.3 per cent of pile diameter). The observed degradation of skin friction at large displacements is consistent with the direct shear test results which indicated reductions of between 30 and 50 per cent.

In view of the scatter in the measured UU strength data and the discrepancies between the UU and CU strengths, an effective stress approach similar to Meyerhof's (1976) is ultimately preferable. Values of α were obtained by dividing the measured shaft friction over a segment length by the effective overburden pressure acting at the middle of that segment.

Values of β at peak reduced from over 2 near ground level to as low as 0.25 at 10 m. Below 10 m penetration, the value of β ranged between 0.8 and 1.4. Over the embedded pile length the average value of β is about 1.0.

It is worth noting however that the trend in the measured skin friction profile in the deeper soils, where the effects of residual installation stresses

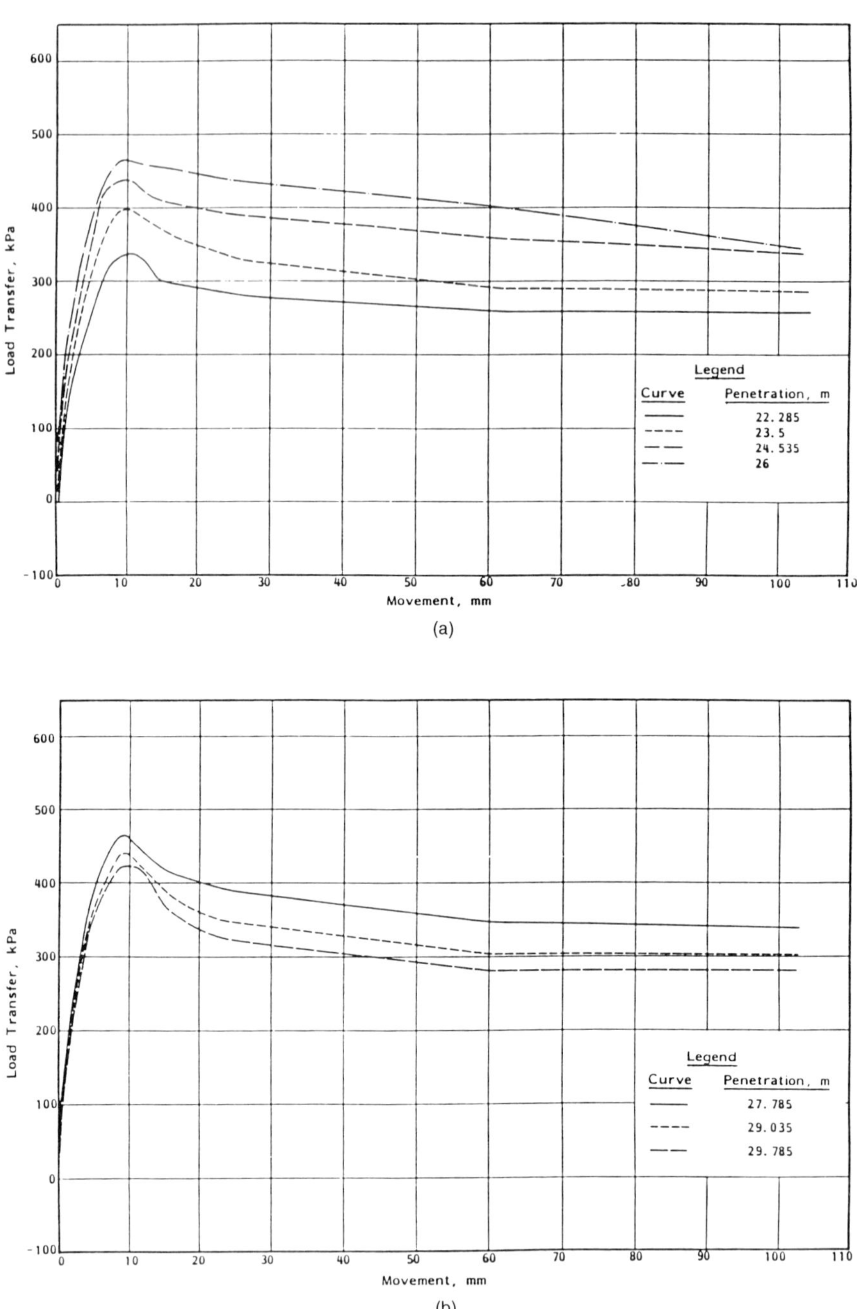

Fig. 29. Load transfer curves, static load test (OC pile): (a) top and (b) bottom

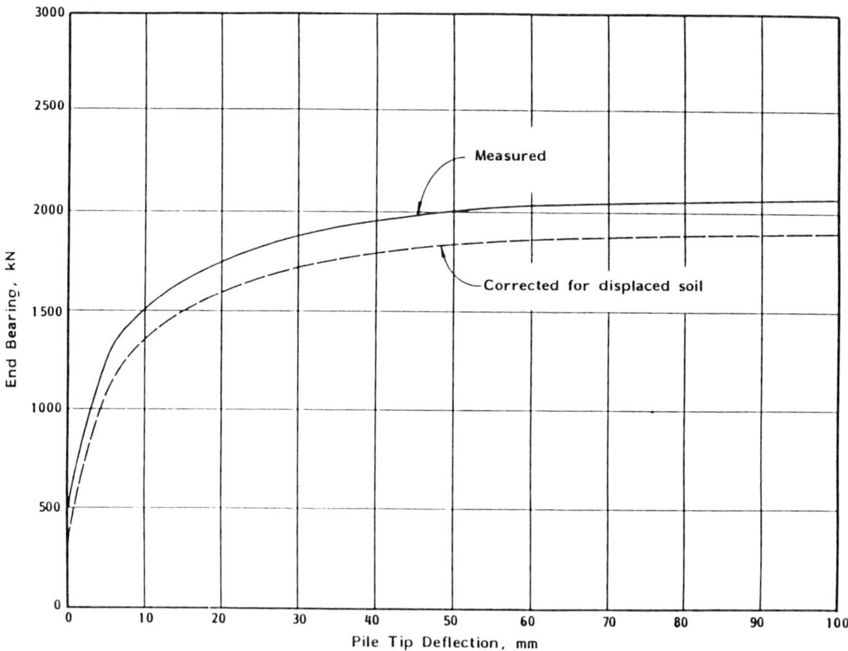

Fig. 30. End bearing versus pile tip deflection: static load test, OC pile

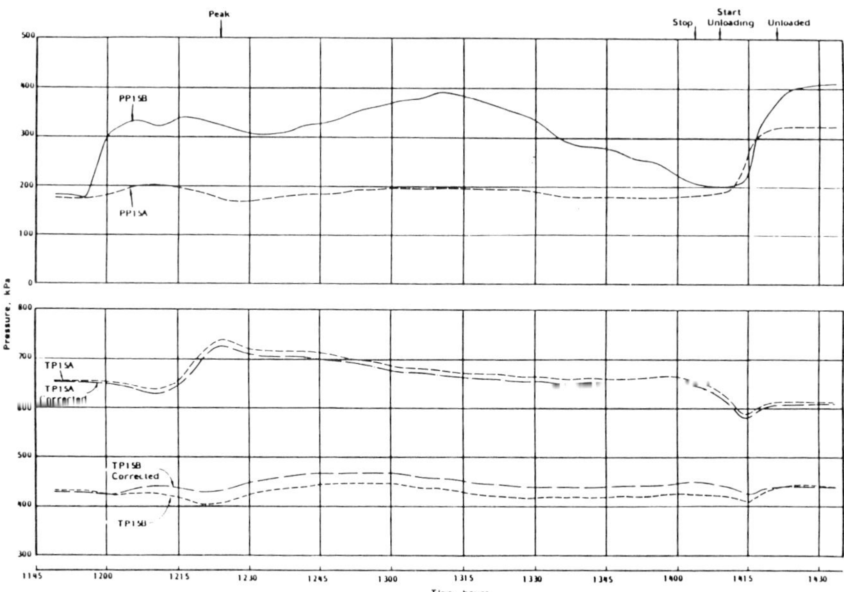

Fig. 31. Total pressure cells reading versus time: static load test, pressure cells at 15 m (OC pile)

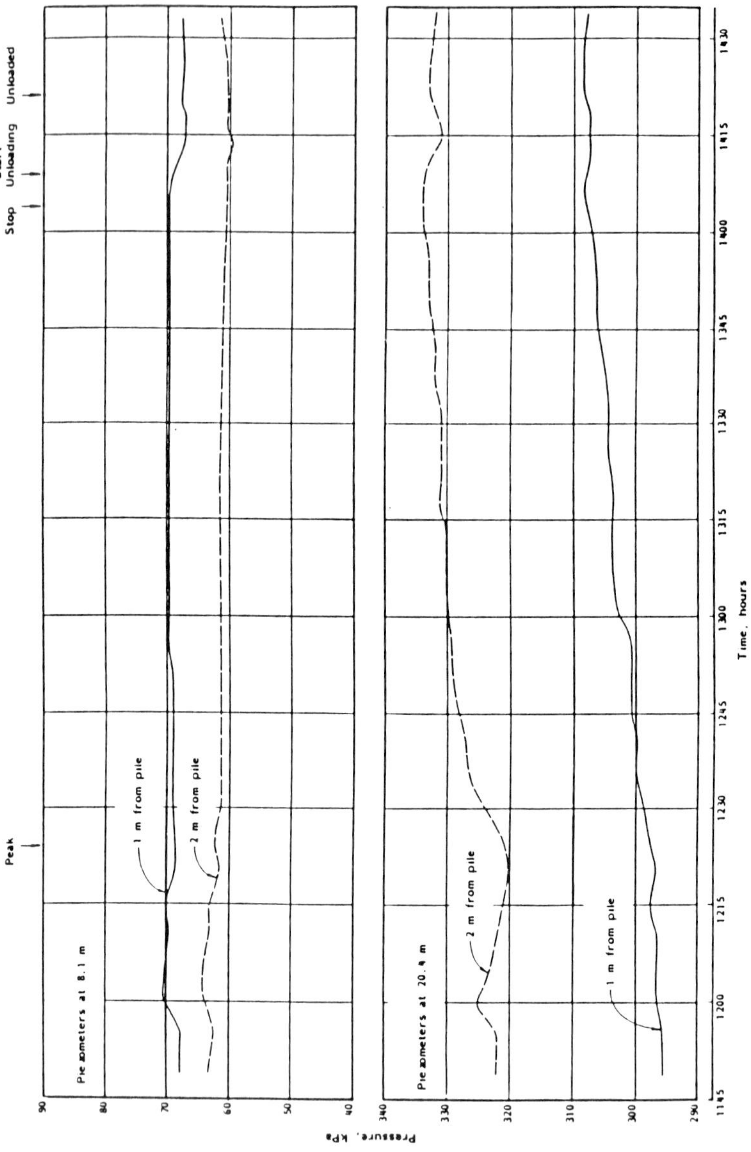

Fig. 32. Piezometer readings versus time: static load test, piezometer at 8.1 m and 20.4 m (OC pile)

are minimal tends to indicate a reasonable agreement with the trend in the undrained shear strength profile. This is likely to be masked by an approach such as β. Also, the measured friction data tend to suggest that the incremental approach of α is appropriate.

2.2.4.5. *End bearing*

The measured tip load versus tip deflection is given on Fig. 30. The residual component of end bearing was 0.5 MN. At peak load, the measured tip load was 1450 kN (9 per cent of total) and at the final post-peak the tip load increased to 2075 kN (15 per cent of total load). The tip movement at peak was 8.6 mm (1.1 per cent of pile diameter), and at the final residual was 102 mm. Allowing for the weight of the displaced soil, the mobilised tip bearing

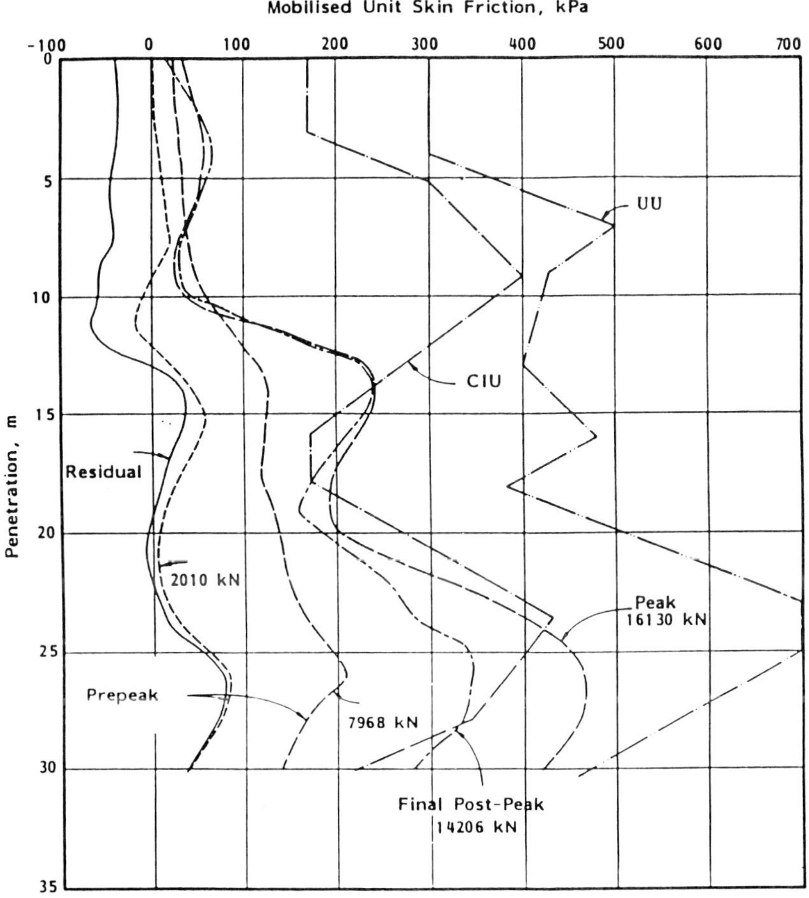

Fig. 33. Mobilised unit skin friction compared with undrained shear strength (OC pile)

Table 5. Predicted 'rigid' frictional capacity, OC pile

Method	Predicted Frictional Capacity, MN	Average Peak Skin Friction KPa	Predicted/ Measured
API(1984) Para 2	16.97	236	1.14
Meyerhof- ß(1) (1976)	12.84	179	.86
Meyerhof- ß(2) (1976)	18.78	261	1.26
Parry Swain (1977)	14.11	196	0.95
Esrig & Kirby (1979)			
1) ∅ from Direct shear tests	16.77	234	1.13
2) δ from Direct shear tests	18.38	256	1.24
Kraft et al (1981)			
λ	11.52	160	0.78
ß	16.66	232	1.12
Randolph & Wroth (1982)			
1) D_R = 1.0	14.93	208	1.00
2) D_R = 0.22	13.43	187	0.90
Randolph & Murphy (1985)	13.34	186	0.90
T-Z probe	13.10	182	0.88
Measured	14.84	209	

capacity at peak load is 1290 kN and at the final post-peak is 1915 kN. At this load the computed factor N_c is 8.85. Thus the value of N_c at the ultimate load conforms to the conventional value of 9, although the residual component of end bearing has been included in the measurements. The measured curve is stiffer than normally assumed in conventional offshore pile design. However the ultimate value is mobilised at about 10 per cent of pile diameter, which is consistent with current practice.

2.2.4.6. *Measured versus predicted shaft capacity*

Predictions of shaft friction were made using API RP2A (1984), Meyerhof (1976), Parry and Swain (1977), Esrig and Kirby (1979), Kraft *et al.* (1981a), Randolph and Wroth (1982) and Randolph and Murphy (1985), and using skin friction measurements made with the *T-Z* probe. Two sets of assumptions in the input parameters were used in Esrig and Kirby (1979) and in Randolph and Wroth (1982) procedures. These assumptions are indicated in Table 5. Application of Kraft *et al.* (1981a) procedures consisted of both the lambda (λ) and the alpha (α) methods. The static *T-Z* data derived from the load test were also compared with formulations currently used in offshore practice.

Peak skin friction values computed based on the 'rigid' pile assumption for all methods are given in Table 5. Peak skin friction profiles are shown on Figs 34a and 34b.

The comparison in Table 5 indicates that for a 'rigid' pile, Randolph and Wroth's method gave the closest prediction assuming a full displacement pile. The test pile actually approached plugging at 13 m penetration and from static capacity considerations the method predicts pile plugging at about 10m penetration, which is reasonably close.

The capacities noted above correspond to those associated with the peak load transfer integrated over the pile length. For a long compressible pile in a strain softening soil, peak capacity will be affected by the amount of softening in the *T-Z* curve, and by the relative soil–pile stiffness.

It is incorrect therefore to make a direct comparison between the computed rigid pile skin friction profiles and the measured profile for the test pile shown on Figs 34a and 34b. It is possible however to estimate the reduction from the rigid pile capacity due to strain softening and pile compressibility effects using simplified procedures such as those by Murff (1980) and Randolph (1983).

2.2.4.7 Comparison of measured and predicted load-deflection response

Predicted load–deflection curves for the pile head are presented on Figs 35a and 35b. As for the NC pile these curves are results of computations of a subgrade reaction numerical solution similar to that described by Coyle and Reese (1966). To develop the predictive curves, the *Q-Z* curve as measured was used while the *T-Z* curves used varied depending on the predictive method. Results of these computations should then serve as a check on the predictive skin friction only, as the end bearing has been taken as the same throughout. The results are compared in Table 6.

The comparison in Table 6 shows that the majority of the methods predicted the measured peak capacity to within less than 15 per cent. The scatter in the predicted residual capacity however is greater, varying between overprediction by 13 per cent and underprediction by 25 per cent.

Overall, the closest prediction of peak capacity is given by the Randolph and Wroth (1982) method assuming a full displacement pile and this is followed by Esrig and Korby (1979) method based on ϕ' values determined from direct shear tests on undisturbed specimens, and by Kraft *et al.* (1981a) α method. The last two methods also predicted the residual capacity accurately, but provided pre-peak load deflection stiffnesses greater than the measured.

2.2.4.8. Pore pressure

At the early stages of loading, the pore pressures showed a small increase and this was followed by a continuous decrease as the peak load was approached. At the peak load, the measured changes in pore pressure from the values before the load test ranged between -0.45 to +0.45 times CIU as both reductions in the deeper soils and increases in the shallower soils were measured. This variation corresponds to pore pressure changes of 0.65 to 1.4 times the effective overburden pressure, σ'_v, and is at variance with assump-

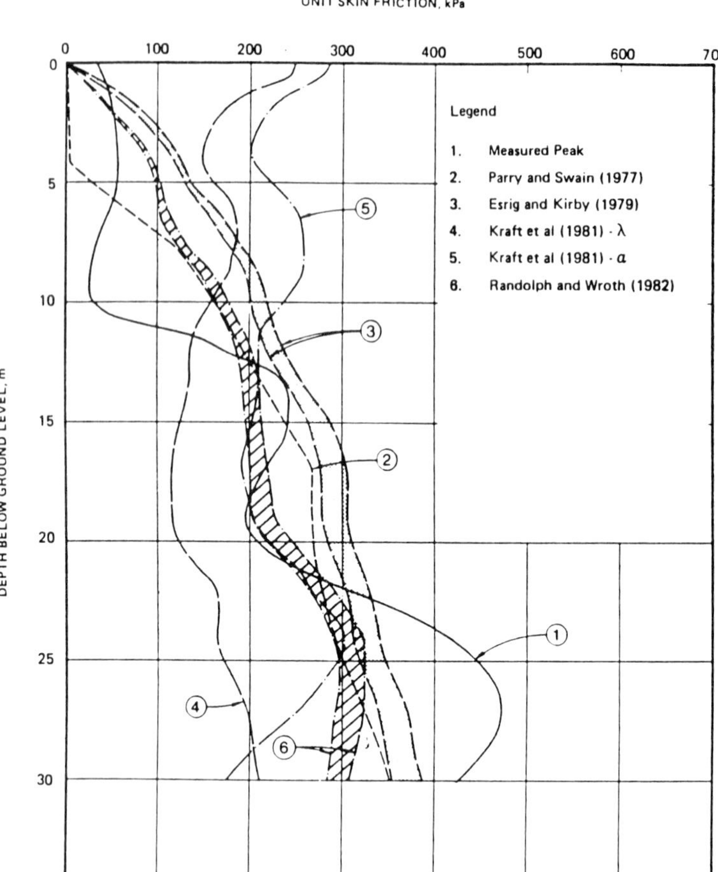

Fig. 34. Comparison of peak skin friction profiles for a 'rigid' pile: (a) above and (b) over page

2.2.4.9. Earth pressure

The measured total earth pressures showed that during load testing, little variation occurred in the shallower soils while in the deeper soils large reductions were observed. At peak load, the average reduction is about 0.27 σ'_v at depth while at mid pile length there was an increase of 0.2 σ'_v.

2.2.4.10. Effective stresses

Curves showing the effective stress variation versus time during the load test are shown in Fig. 36. During the load test there was a reduction in effective radial stress prior to achieving the peak load. However at peak there

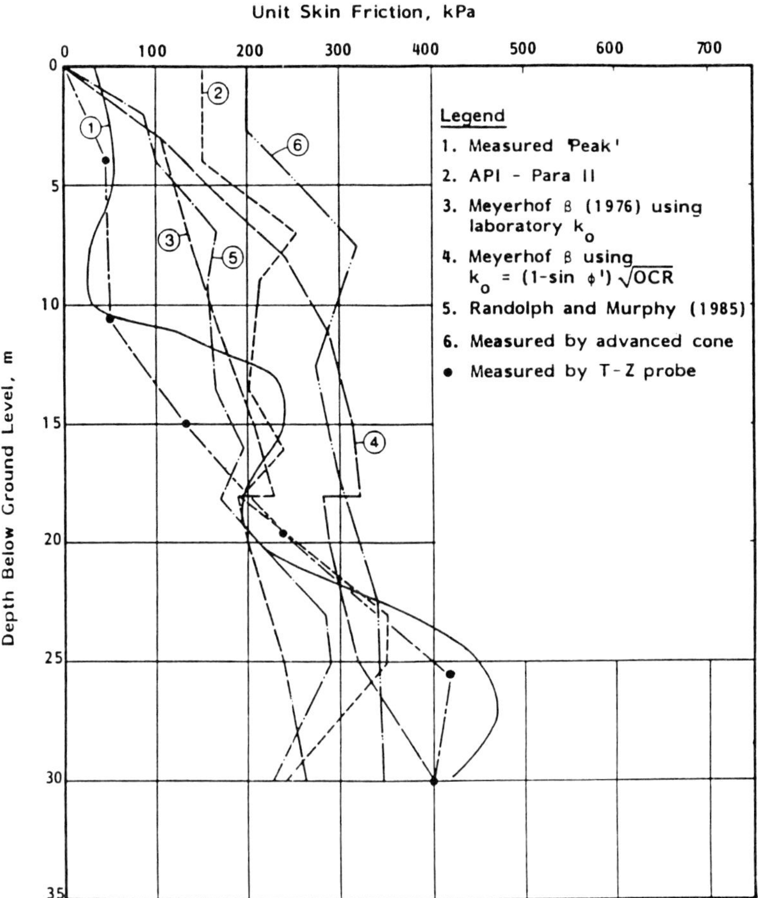

Fig. 34(b)

was only a small reduction from the value at the start of the test. At depth, large differences in effective stresses were measured with cells at the same elevation during the post-peak stage indicating pile bending.

The ratio of the measured effective radial stress to the effective overburden pressure, designated k_s, varied from between 1.7 to 3.13 at peak and between 1.8 to 2.5 at the final post-peak load. These ratios apply to the clays over the lower half of the pile. In the shallower clays the values of k_s at peak are found to be comparable to the measured k values at the 'end' of consolidation. The measured k value at depth is higher than the estimated passive coefficient and this is attributed to cohesion and development of negative excess pore pressures during loading.

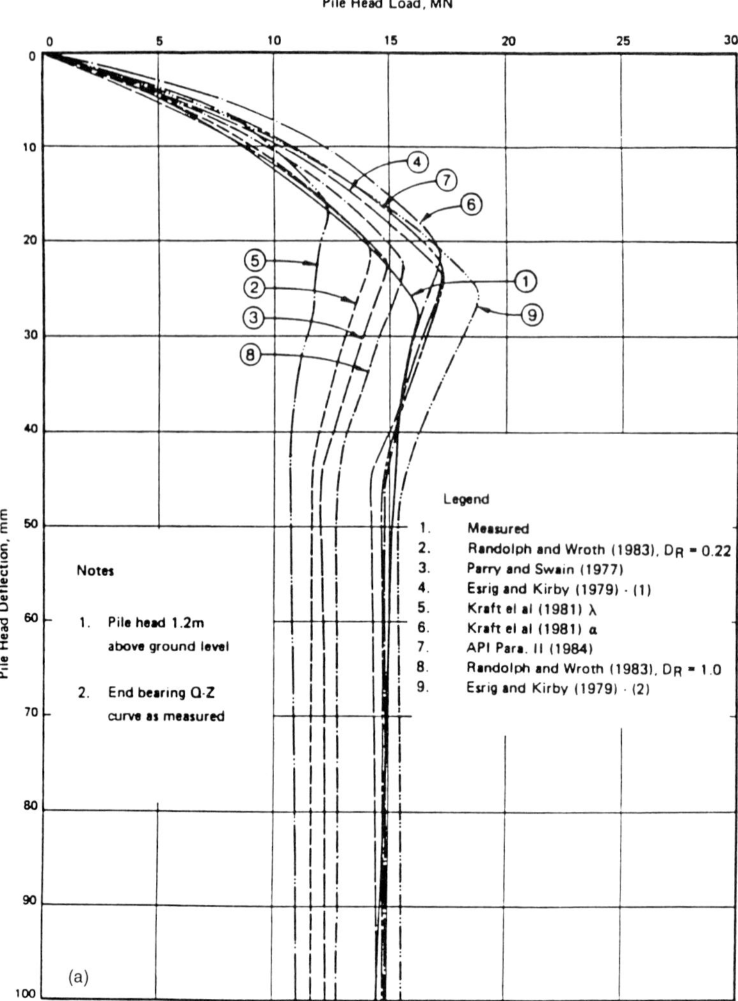

Fig. 35. Comparison of measured and predicted pile head load–deflection response (OC pile): (a) above and (b) over page

2.2.4.11. Friction angles

The mobilised interface angles at peak load are fairly uniform at about 23°. At final post peak load there were large variations in the backfigured angles which are found to be between 23° and greater than 45°, with average values of about 33°. This wide range obviously indicates pile bending due to the large displacement experienced.

The estimated friction angles at peak load compare favourably with the results of direct shear tests which gave interface angles of 20.5° to 26° at peak. However at final residual load and for the reason noted earlier the interpreted

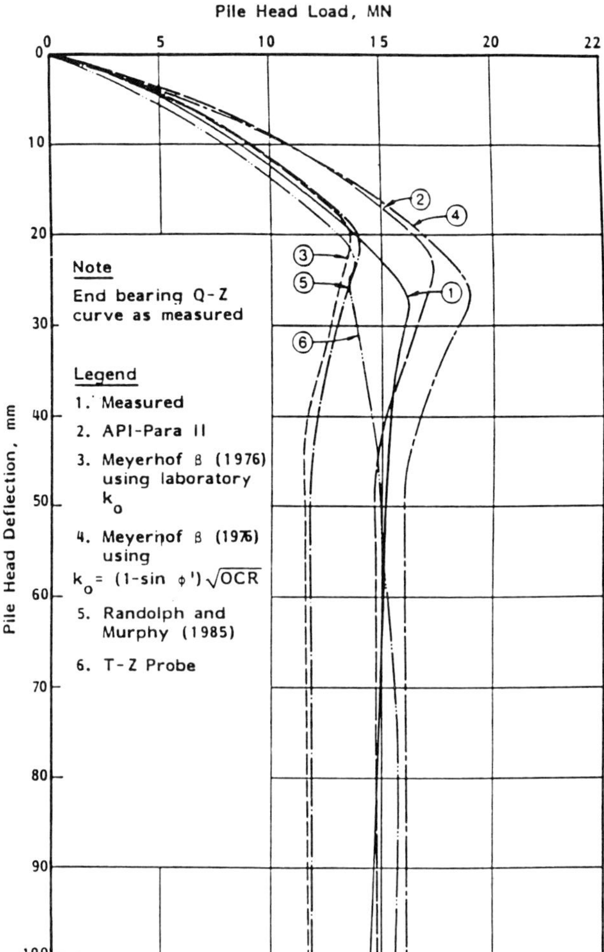

Fig. 35(b)

angles are much higher than the direct shear test results of 11.5° to 17.5° at residual (the ring shear tests gave 16° at the residual for both strata).

2.3. *Additional tests*

Additional testing carried out at both sites included a creep test, a second static test, a one-way cyclic test, a rapid load test, and finally a post-cyclic static test. Results of the cyclic tests are presented in a companion paper, Cox *et al.* (1993b). Limited interpretation has been undertaken.

Table 6. Comparison of measured and predicted load–deflection response (OC pile)

Predictive Method	Peak Capacity MN	Pile Head Deflection mm	Residual Capacity MN	Predicted/Measured Peak	Residual
API (1984) Para 2	17.4	24.0	14.6	1.08	1.03
Meyerhof - β (1)	13.6	20.7	11.5	0.84	0.81
Meyerhof - β (2)	19.1	26.7	16.0	1.18	1.13
Parry & Swain	14.8	22.2	12.2	0.92	0.86
Esrig & Kirby					
1) ø'from Direct Shear Tests	17.3	23.7	14.2	1.07	1.00
2) δ from Direct Shear Tests	18.8	25.3	15.4	1.17	1.08
Kraft et al					
λ	12.3	17.3	10.7	0.76	0.75
β	17.2	20.7	14.8	1.07	1.04
Randolph & Wroth					
1) D_R = 1.0	15.5	22.2	12.7	0.96	0.89
2) D_R = 0.22	14.1	20.6	11.6	0.88	0.82
Randolph & Murphy	14.0	21.3	11.7	0.87	0.82
T-Z probe	13.8	23.2	15.7	0.86	1.11
Measured	16.1	28.1	14.2	-	-

Note : The measured end bearing curve has been used in the predictions.

2.4. *API data base*

The API data base for piles in clays, after Clarke *et al.* (1985) is reproduced on Fig. 37. The original data had been presented by Vijayvergiya and Focht (1982).

For completeness, we have revised the West Sole pile load test data points originally presented by Vijayvergiya and Focht. This revision has been made in view of the results of the site investigation carried out at West Sole in 1978, which revealed higher undrained shear strengths than were given by the initial 1968 investigation. The 10 West Sole points shown on Fig. 37 correspond to the tensile capacities and revised shear strength profile reported by Clarke *et al.* The compressive frictional capacities have therefore been assumed equal to the tensile capacities. Results of the NC and OC pile load tests have been superimposed on the data base on Fig. 37.

Appraisal

The load tests at the NC and OC sites were carried out on fully instrumented piles, that were carefully selected to reproduce the compressibility characteristics of most offshore piles. Control was exercised throughout all test stages by virtue of the extent of continuity of the measurements. Installation stresses were measured prior to load testing, hence the measured pile load distributions are considered to be reasonably accurate. The data from the two tests are therefore considered to be of high quality and the quality control is believed to be better than on any other test in the API data base.

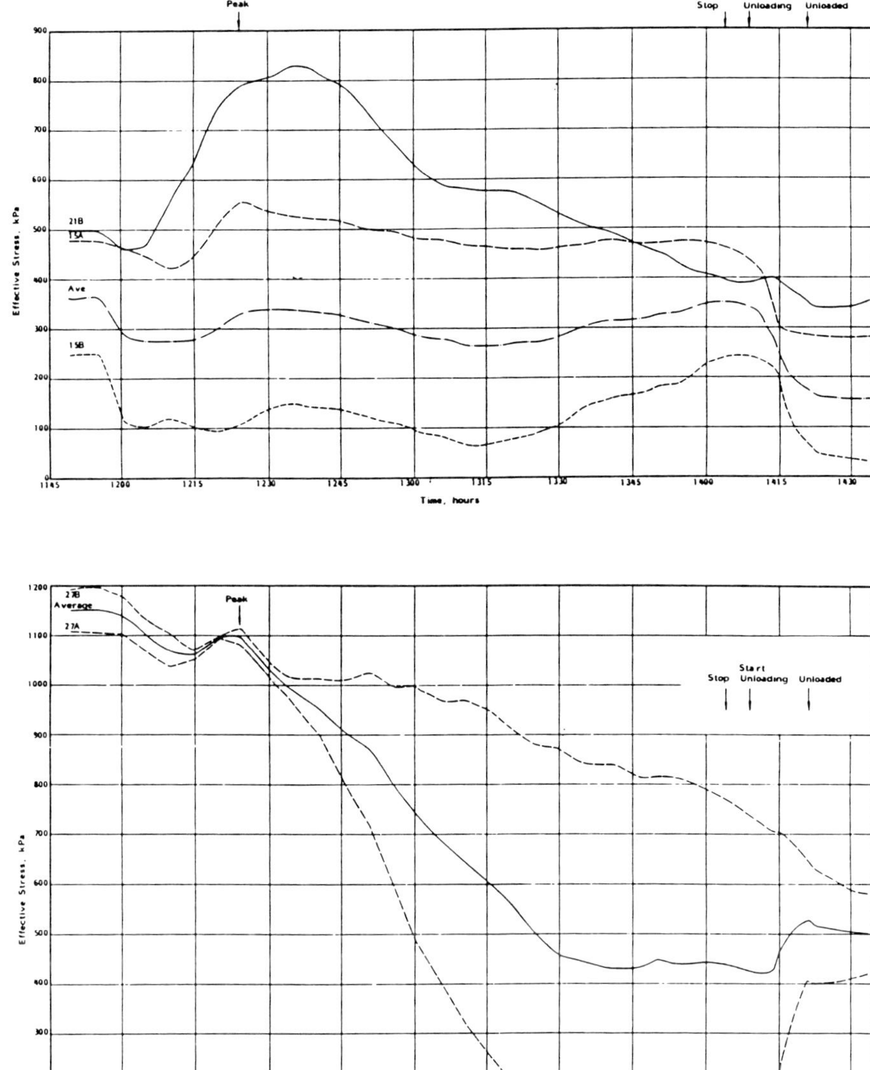

Fig. 36. Effective stress versus time: static load test, pressure cells at 15 m and 21 m (OC pile)

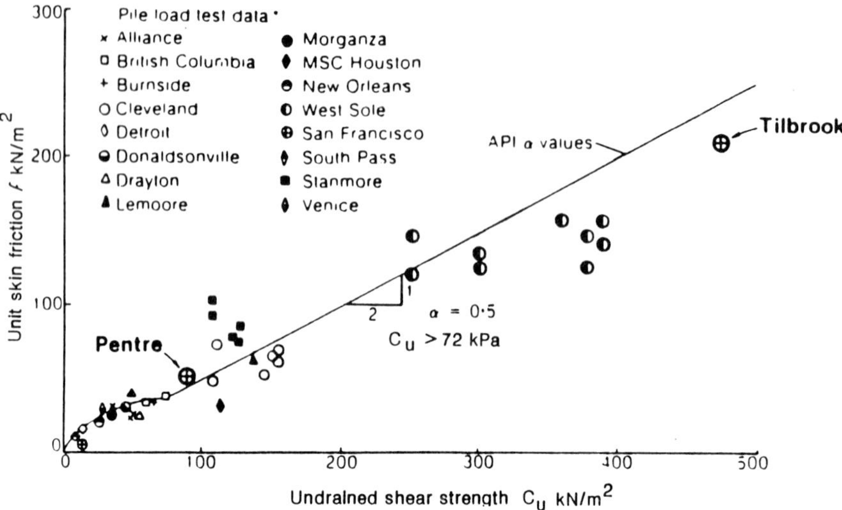

Fig. 37. API data base for piles in clays (after Clarke et al.)

3.0. Methods for data reduction

Herein we present the equations and procedures used in reducing field data from the load tests.

3.1. Areas of pile cross-section

The cross-sectional areas of the piles are required for confirmation and calculation for pile stiffness, AE, for computing pile elongations and compressions under load. During calibrations of the pile sections load deflection data for each section enabled the area for stiffness, A_e, to be calculated as follows.

$$A_e = \frac{PL}{\Delta LE} \tag{1}$$

Where P is axial load, L is unstrained pile length, ΔL is change in length of pile under load P, and E is Young's modulus of elasticity.

A value for A_e can also be determined from the nominal dimensions of the pile, channels, cable brackets, strain modules, pressure cells and cable exit reinforcements. There are a large number of changes in cross-sectional areas along the length of each section, hence for a pile of n section changes

$$\Delta L = \frac{PL}{A_e E} = \frac{PL_1}{A_1 E} + \frac{PL_2}{A_2 E} + \; + \frac{PL_n}{A_n E}$$

from which $A_e = L/(\sum L_i/A_i)$ (2)

The pile sections were calibrated prior to completion of the welding of the channel covers, therefore a correction to the calibrated area was made in the following manner

$$A_e = A_c\,(A_2/A_1) \tag{3}$$

where A_e= effective area for stiffness of the section of the completed pile; A_c= area for stiffness determined from the calibration data Eq. 1; A_1 = area for stiffness determined from nominal dimensions, Eq. 2, for the incomplete sections; A_2 = area for stiffness determined from nominal dimensions, Eq. 2 for the complete sections. In Eq. 1 the accuracy of A_e is dependent on the accuracy of measuring the load P, length L and change in length ΔL. For consideration of the stiffness of piles a value of E does not have to be assumed in Eq. 1 since the value of A_e can be determined from Eq. 1.

$$A_e E = \frac{PL}{\Delta L} \tag{4}$$

3.2. Equilibrium equations for pile

In Fig. 38 a pile is shown under a loading system where all vectors are defined positive downwards. The vectors have the following definitions:

PT = top load sensed by the load cells
LW = weight of load cells
JW = weight of load jacks
CW = weight of the pile cap or load head
WW = weight of water inside pile down to top of soil plug
PW = weight of pile including channels, instruments and cables
SW = total (saturated) weight of soil plug
SF = total skin friction on pile
PB = soil reaction at the bottom of the pile

From an equilibrium equation it can be shown that

$$\text{SF} = -(\text{PT} + \text{LW} + \text{JW} + \text{CW} + \text{WW} + \text{PW} + \text{SW} + \text{PB}) \tag{5}$$

For a compression test the skin friction, SF, and the bottom soil reaction, PB, both act upward and are negative. If during a compression test we say that the soil plug and the water above the plug are not supported by skin friction on the internal pile wall, but the weight of the water and plug go

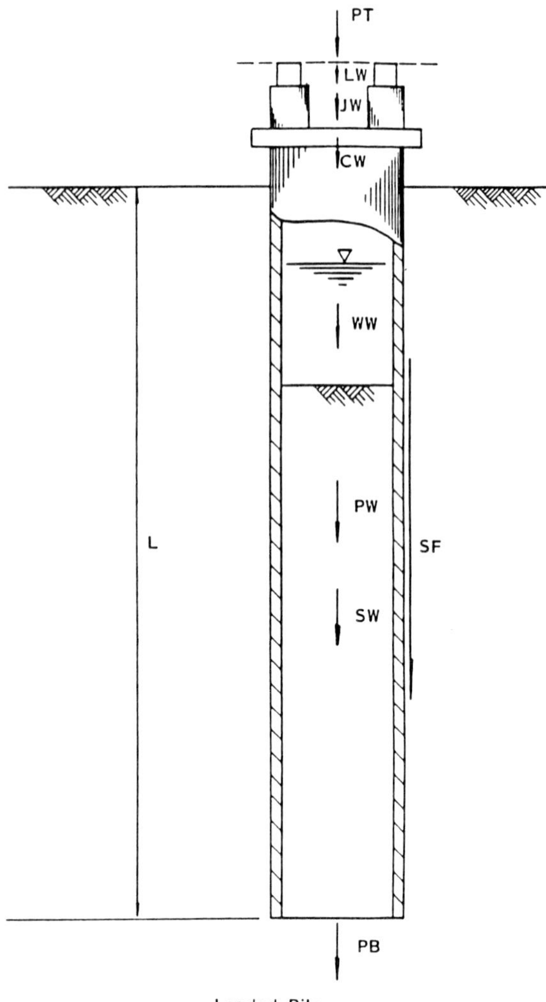

Fig. 38. Parameters for pile equilibrium

direct to the bottom of the pile and are carried by the soil, then PB would be equal to:

$$PB = P + WW + SW \tag{6}$$

where P= load in pile wall at the tip of the pile as determined (extrapolated) from strain gauge data.

If we substitute Eq. 6 into Eq. 5 and recognise for a compressive test PB is negative, we get the expression used for reducing data for the test piles.

$$SF = -(PT + LW + JW + CW + PW - P_L) \tag{7}$$

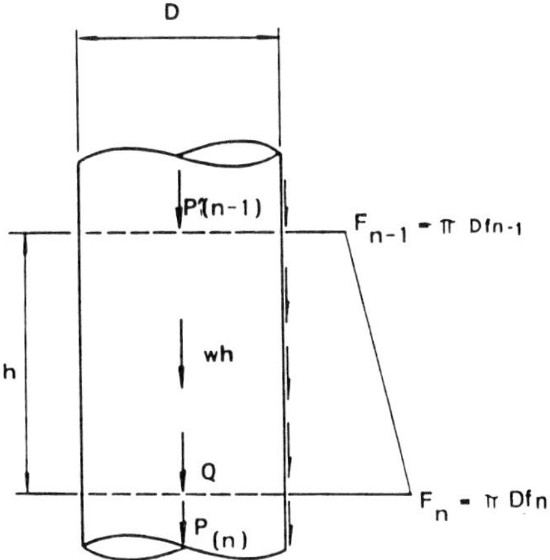

Fig. 39. Equilibrium of pile segment

3.3. Equilibrium equations for pile segments

Figure 39 shows a pile segment length h. All force vectors on the segment are shown in the positive direction (downwards). The skin friction at top of the segment is $f_{(n-1)}$ and at the bottom f_n. The total force of the skin friction on the segment is:

$$SF_h = \frac{h}{2}(F_{n-1} + F_n) \tag{8}$$

Other force parameters are: P_{n-1} = force in pile wall at top of segment; P_n= force in pile wall at bottom of segment; w'' = uniformly distributed load representing weight of steel, instruments and cable, and, where appropriate, the weight of the soil plug per unit length of pile; Q = a concentrated load at the bottom of the segment (if appropriate). The equilibrium equation for the full segment is:

$$P_{n-1} + P_n + wh + Q + SF_h = 0 \tag{9}$$

For a compression test P is negative and Eq. 9 can then be written as:

$$SF = -(P_{n-1} = P_n + wh + Q) \tag{10}$$

The terms P_{n-1} and P_n are determined from strain gauge data. The terms wh and Q are known from the physical description of the pile. We now have the equations to evaluate the amount of load that is shed to the soil in friction for each pile segment.

3.4. Axial movement

The axial movement of a pile reflects the movement at the pile top as the pile is loaded and the elastic change in length along the pile in response to

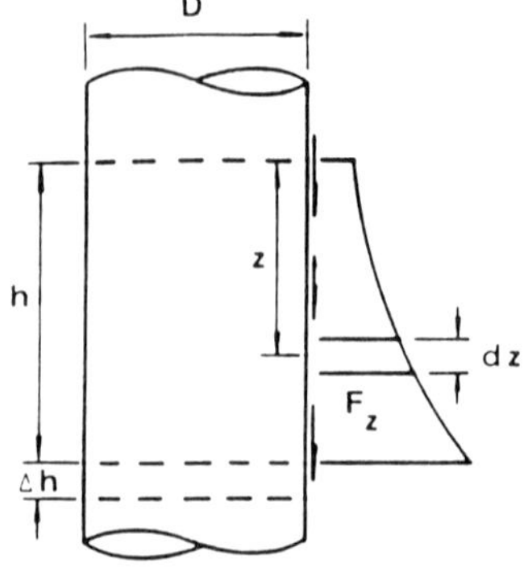

a) General Case

$$F_z = \pi D f_z$$

$$\Delta h = \int_0^h \frac{z F_z dz}{AE}$$

$$= \frac{1}{AE} \int_0^h z F_z dz \qquad (11)$$

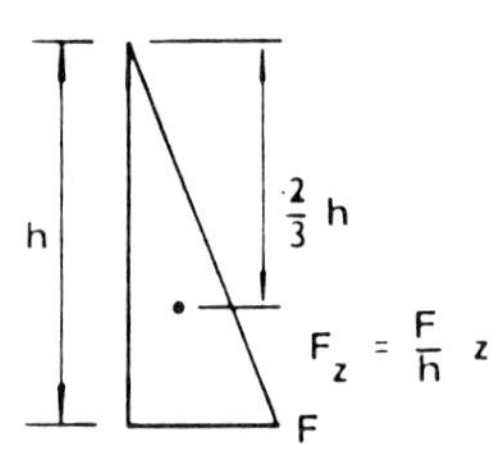

b) Triangular, Case 1

$$\Delta h = \frac{F}{hAE} \int_0^h z^2 dz$$

$$= \frac{Fh^2}{3AE}$$

$$= \frac{F (h/2)}{AE} \left(\frac{2}{3} h\right) \qquad (12)$$

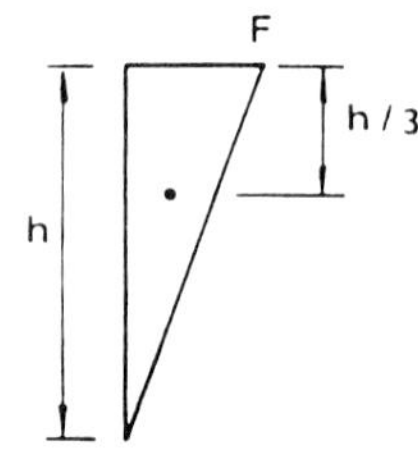

c) Triangular, Case 2

$$\Delta h = \frac{F (h/2)}{AE} \left(\frac{h}{3}\right) \qquad (13)$$

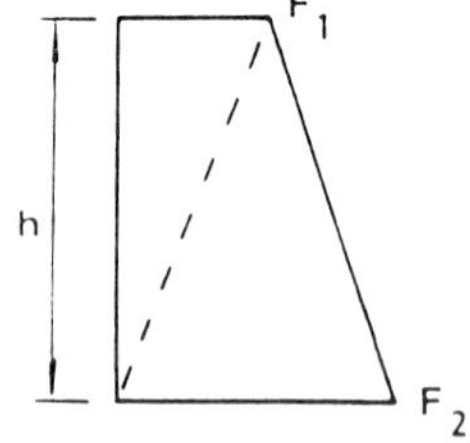

d) Trapezoidal

$$\Delta h = \frac{h^2}{6AE} (F_1 + 2 F_2) \qquad (14)$$

Fig. 40. Changes in length of pile segment

the top load gravity loads and soil reactions. Fig. 40 shows the development of the formula to be used in evaluating the elastic pile movements caused by friction. (Eqs 11 to 14 are given in Fig. 40).

$$\Delta h = \frac{h^2}{6AE}(F_1 + 2F_2) \tag{14}$$

3.5. *Movement of pile segment*

The elastic movement Δh, of a pile segment is indicated in Fig. 41. The load in the pile wall at the bottom of the segment, P_n, and any concentrated load on the pile at the bottom of the segment, Q, each produce a contribution to movement proportional to the length of the segment, h. A uniformly distributed load, w, produces a movement determined from concentrating a load wh at the middle of the segment. The contribution of the skin frictions were determined previously, Eq 14. The contribution of the pile segment to total movement of the pile is:

$$\Delta h_n = \frac{h}{\mathrm{AE}}(P_n + Q + \frac{wh}{2} + \frac{h}{6}(F_{n-1} + 2F_n)) \tag{15}$$

If DT is the movement at the top of the pile as shown in Fig. 41, then the movement at the bottom of the ith segment in the pile will be:

$$h_i = \mathrm{DT} + \sum \Delta h_n \tag{16}$$

3.6. *Methods for determining skin friction*

In Fig. 43 axial load distributions are shown for a pile hanging in air and for a pile under a compression test and under a tensile test. It is commonly assumed that the slope of a load distribution curve is equal to the skin friction of the soil on the pile. This cannot be correct since a pile hanging in the air has no skin friction yet the axial load distribution curve has a slope corresponding to the distributed gravity loads along the pile.

We can now examine the effect of gravity loads in Fig. 43. From the top of a segment of length h lines are drawn with a slope equal to the slope of the pile hanging in air. This slope over the length h defines the weight of the pile over this length. On the tension side it can be seen that the pile weight subtracts from skin friction and on the compression side it adds to skin friction. It should be understood if there are Q loads on the pile, a discontinuity in the axial load curve will occur at the elevation of the Q load. For tension loading a Q load causes the axial load curve to move to lower tension below the point of application, hence decreasing skin friction. The opposite is true for compression loading. Three of the several methods than can be used for determining skin friction are described below.

Method 1. If we assume linear variation of skin friction for each segment, then there are two skin frictions that describe the skin friction: one at the top of the segment and one at the bottom of the segment. Unfortunately, we only have one equilibrium equation for each segment, Eq. 9, and can only solve

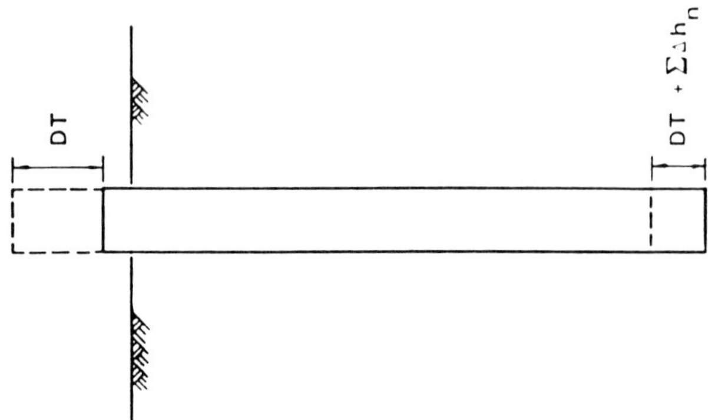

Fig. 42. Total movement of pile

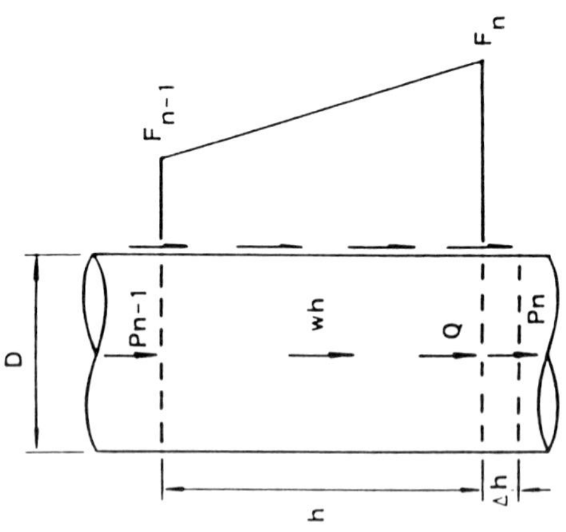

Fig. 41. Elastic movement of pile segment

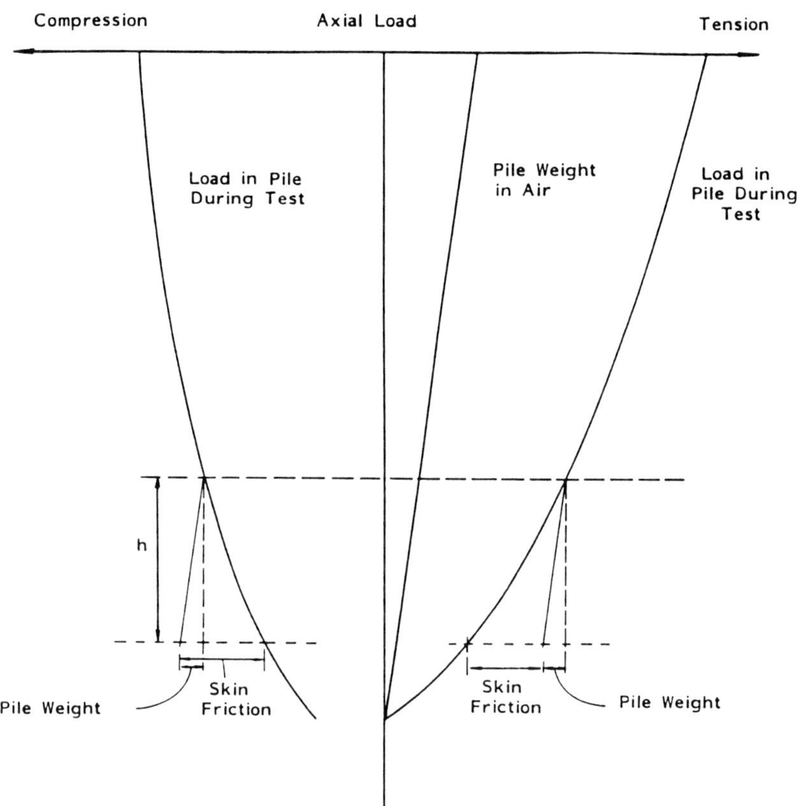

Fig. 43. Axial load distributions in pile

for one unknown. If we can accurately estimate a value for skin friction at the ground line, then we can solve for the skin friction at the bottom of the first segment.

Unless there is a discontinuity between skin friction at the bottom of the first segment and the top of the second segment, we then can solve for skin friction at the bottom of the second section. We continue this process to the bottom of the pile. Method 1 is illustrated in Fig. 44a.

Method 1 can produce trends in skin friction distributions but Method 2 described below is a better procedure.

Method 2. In this method a plot is made of the average skin friction at the centre of each segment, as shown in Fig. 44b. The average value is:

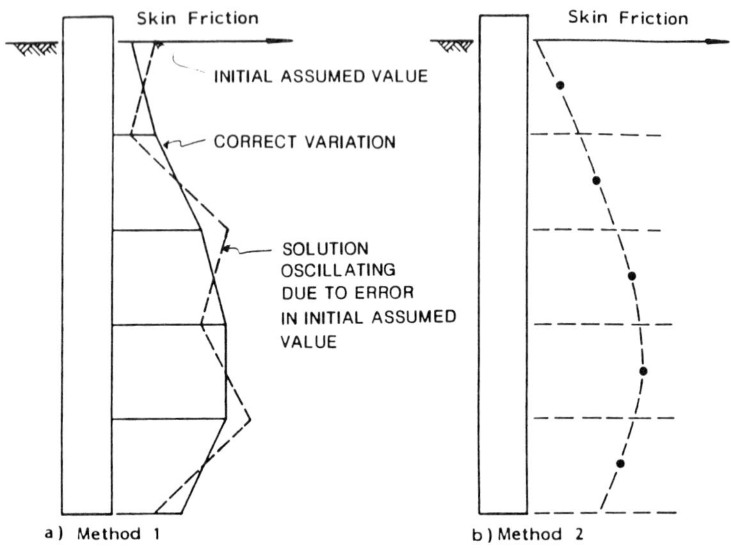

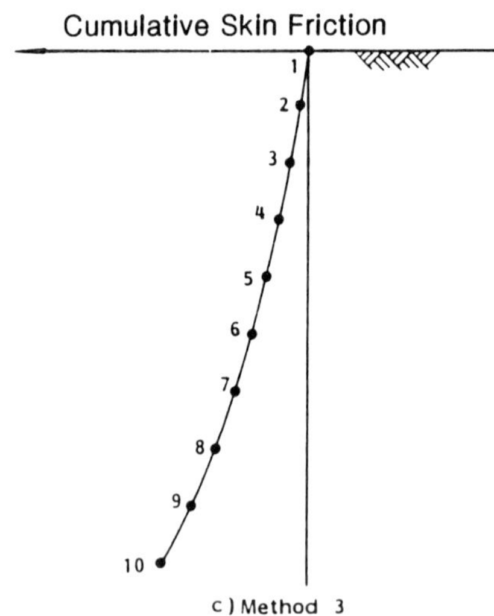

Fig. 44. Methods of solving for skin friction

$$f_{avg} = \frac{SF}{\pi D_h^h} \tag{17}$$

The value of SF is determined from Eq. 10. The plotted points are then connected by a curve which represents skin friction with depth.

In Method 2 any error in a value for average skin friction in a segment is not carried on to other segments.

Method 3. In this method we first generate a curve of cumulative values of total skin friction on each segment. This curve is a load curve with the influence of gravity loads and any concentrated loads removed. The slope of this curve is skin friction.

There are numerous ways to determine the slope of the load curve. One is Method 3 that makes use of a third-degree polynomial that passes through four consecutive points on the curve. The polynomial is then differentiated to give skin friction.

For example, in Fig. 44c a load curve representing cumulative skin friction is given for ten values of cumulative friction. A third degree polynomial is derived for the four points 1 through 4. The polynomial is then differentiated to give values of slope at points 1 and 2.

A second polynomial is derived for points 2 through 5. It is differentiated and a slope computed for point 3. The next polynomial is for points 3 through 6 and a slope calculated for point 4.

The procedure continues to the bottom of the pile where the polynomial through points 6 through 9 is used to derive slopes at 7 and 8. The last polynomial through points 7 through 10 is used for deriving slopes at points 9 and 10.

Method 3 is a systematic procedure for determining skin frictions by analytical means. It adapts reasonably well to smooth load curves representing cumulative skin friction.

The proof of the correctness of any methodfor determining skin friction is to enter the values of skin friction into a programme that also considers pile gravity loads and *Q* loads and generate a load curve. This generated load curve is then compared with the load curve determined from strain sensing instruments. When there is reasonable matching of the two load curves, the skin friction distribution is acceptable.

3.7. Pile movements

Pile movements were generated by applying the skin friction distributions and the gravity loads to the unstressed pile. The resulting movement at each segment along the pile was then added to the movement at the top of the pile as measured by LVDTs to compute the overall movement of each segment.

References

API RP 2A (1984), *Recommended practice for planning, designing and constructing fixed offshore platforms*. American Petroleum Institute, Dallas, Texas, Fifteenth Edition.

API RP 2A (1987), *Recommended practice for planning, designing and* Constructing *fixed offshore platforms*. American Petroleum Institute, Dallas, Texas, Eighteenth Edition.

AZZOUZ, A. S. and MORRISON, M. J. (1988), Field measurements on model pile in two clay deposits. *J. Geotech. Engng Div., Am. Soc. Civ. Engrs*, Vol. 114, No. l, pp. 104-121.

BRIAUD, J. L. and TUCKER, L. (1984), Piles in sand. A method including residual stresses. *J. Geotech. Engng Div., Am. Soc. Civ. Engrs*, Vol. 110, No. 11, pp. 1666 - 1680.

CLARKE, J., RIGDEN, W. J. and SENNER, D. W. F. (1985), Reinterpretation of the West Sole Platform WC pile load tests. *Géotechnique*, Vol. 35, No.4, pp. 393-412.

COYLE, H. M. and REESE, L. C. (1966), Load transfer for axially loaded piles in clay. *J. Soil Mech. & Fndn Div., Am. Soc. Civ. Engrs*, Vol. 92, No. SM2, pp. 1-26.

COX, W. R, SOLOMON, I. J., and CAMERON, K. (1993a), Instrumentation and calibration of two 762 mm diameter pipe piles for axial load tests in clay. This volume.

COX, W. R, CAMERON, K., and CLARKE, J. (1993b) Static and cyclic axial load tests on two 762 mm diameter pipe piles in clays. This volume.

ESRIG, M. E. and KIRBY, R. C. (1979), Advances in general effective stress method for the prediction of axial capacity for driven piles in clays. *Proc. 11th Offshore Technology Conf.*, Houston, Vol. 1, pp. 437-448.

HOLLOWAY, D. M., CLOUGH G. W. and VESIC, A. S. (1978), The effects of residual driving stresses on pile performance under axial loads. *Proc. 10th Offshore Technology Conf.*, Houston, Paper No. 3306.

KARLSRUD, K. and HAUGEN, T. (1985a), Axial static capacity of steel model piles in overconsolidated clay. *Proc. 11th Int. Conf. on Soil Mech. and Fndn Engng*, San Francisco, vol. 3, pp. 1401-1420.

KARLSRUD, K. and HAUGEN, T. (1985b), Behaviour of piles in clay under cyclic axial loading—results of field model tests. *Proc. Conf. on behaviour of offshore structures*. Elsevier, Amsterdam.

KRAFT, L. M., FOCHT J. A., and AMERASINGHE, S. F.(1981a), Friction capacity of piles driven into clays. *J. Geotech. Engng Div., Am. Soc. Civ. Engrs*, Vol. 107, No. GT11, pp. 1521-1541.

KRAFT, L. M., RAY, R. P. and KAGAWA, T. (1981b), Theoretical *T-Z* curves. *J. Geotech. Engng Div., Am. Soc. Civ. Engrs*, Vol. 107, GT 11, pp. 1543-1562.

KRAFT, L. M. (1982), Effective stress capacity model for piles in clays. *J.*

Geotech. Engng Div., Am. Soc. Testing and Mater., Vol. 108, No. GT11, pp. 1387-1404.

LAMBSON, M. D., CLARE, D. G., SEMPLE, R. M. and SENNER, D. W. F (1993), Investigation and interpretation of Pentre and Tilbrook Grange soil conditions. This volume.

MANSUR, C. I. and HUNTER, R. H. (1970), Pile tests—Arkansus River Project. *J. Soil Mech. & Fndn Div., Am. Soc. Civ. Engrs*, Vol. 96, No. SM5, pp. 1545-1582.

MASSARCH, K. R. (1978), New aspects of soil fracturing in clay. *J. Geotech. Engng Div., Am. Soc. Civ. Engrs*, Vol. 104, No. GT8, pp. 1109-1123.

MEYERHOF, G. G. (1976), Bearing capacity and settlement of pile foundations, *J. Geotech. Engng Div., Am. Soc. Civ. Engrs*, Vol. 102, No. GT3, pp. 1-29.

MURFF, J. J. (1980), Pile capacity in a softening soil. *Int. J. Numerical and Analytical Methods in Geomechanics*. Vol. 4, pp. 185-189.

PARRY, R. H. G. and SWAIN, C. W. (1977), A study of skin friction on piles in stiff clay. *Ground Engineering*, Vol. 10, No. 8, pp. 33-37.

POULOS, H. G. (1987), Analysis of residual stress effects in piles. *J. Geotech. Engng Div. , Am.Soc. Civ. Engrs*, Vol. 113, No. 1, pp. 143-157.

RANDOLPH, M. F. (1983), Design considerations for offshore piles. *Proc. Conf. on Geotechnical Practice in Offshore Engineering*. ASCE, Austin, Texas, pp. 422-439.

RANDOLPH, M. F. (1985), *RATZ computer program — load transfer analysis of axially loaded piles.* Cambridge University, Engineering Department.

RANDOLPH, M. F., CARTER, J. P. and WROTH, C. P. (1979), Driven piles in clay — the effects of installation and subsequent consolidation. *Géotechnique*, Vol. 29, No. 4, pp. 361-393.

RANDOLPH, M. F. and MURPHY, B. S. (1985), Shaft capacity of driven piles in clay. *Proc. 17th Offshore Technology Conf.*, Houston, Vol. 1, pp. 371-378.

RANDOLPH, M. F. and WROTH C. P. (1982), Recent developments in understanding the axial capacity of piles in clay. *Ground Engineering*, Vol. 15, No. 7, pp. 17-32.

SOLOMON, I. J., COX, W. R., CLARKE, J., and POSKITT, T. J. (1993), The development of instrumentation for drivability testing and load testing of large diameter piles. This volume.

VIJAYVERGIYA, V. N. and FOCHT, J. A. (1972), A new way to predict the capacity of piles in clay. *Proc. 4th Offshore Technology Conf.*, Houston, Vol. 2, pp. 865-874.

Discussion

Part 1

B. HOSPERS, Shell Internationale Petroleum Maatschappij BV, Netherlands

On the soil investigation at the Pentre site, the location of the test pile was on the edge of the soil investigation area, and having seen the variation in shear strengths, I wonder whether one should have done more soil investigation around the location of this pile test.

D. G. CLARE, Ove Arup & Partners

The test pile location at the Pentre site was moved because there was a failure during the construction of one diaphragm wall panel anchor. The variation in shear strengths that I showed at Pentre were primarily a function of the method of measurement and not location. If you compare the UU tests for the three boreholes that were put down at the site, you would find that the shear strength profiles were similar, showing similar ranges and variations, and so I think, on balance, that it was not considered necessary to do any further testing at Pentre due to the rather small relocation of the test pile.

Dr D. J. MURFF, Exxon Production Research Company, USA

The paper states that there were some driven samples taken, but I could not determine whether there was any difference in strengths or whether these samples were tested.

D. G. CLARE, Ove Arup & Partners

There was only a very small number taken at Pentre in the very weak soils in the top 10–15 m in one borehole and those were, if my memory serves me correctly, 64 mm outside diameter liner samples. They were not subject to laboratory testing as I recall, being recovered from above the pile test stratum. They were just a few samples that John Johnson (Director, McClelland Limited) suggested we take with the liner sampler to see the effect on sample recovery and subsequent handling. We were experiencing some difficulties in handling the extruded piston samples of the upper, very weak soils. In the event, sample handling was not greatly improved and we just continued with the piston sampler, changing to push sampling when the soils became too strong to use the piston sampler efficiently.

K. KARLSRUD, Norwegian Geotechnical Institute

I had the opportunity to review quite a bit of the soil test data, from both Pentre and Tilbrook, and very extensive investigations have been carried out at both sites. I have two comments and observations. First, regarding Pentre, I think we all saw from the undrained shear strength profiles that there was enormous variation in undrained shear strengths between UU tests, CAU tests, CIU tests and so forth. Not only is there variation in the different types of tests, but there is also very large scatter in the UU data, which of course makes it rather difficult to come up with a unique interpretation in terms of alpha methods. We have to know when we start the discussion what are the properties of the material in situ before we drive the pile into the ground. What are the true undrained shear strength characteristics? For Pentre, you presented triaxial compression tests which we believe give us a representative shear strength for that mode of failure, but it certainly does not give us a representative shear strength for the simple shear type of failure which we will have along a pile wall. For this type of silty, or clayey silt material the anisotropy of the undrained shear strength is quite considerable. Typically we would see that triaxial extension would be maybe half or even less than the shear strength from triaxial compression, and simple shear somewhere in between.

I also think we should keep in mind the very large coefficient of consolidation of the Pentre soils. I was very intrigued by the vane shear strength results when I first saw them. I had never seen anything like that in such a material. I think you are right that there must be some pore pressure dissipation going on during the vane shear tests, which again emphasises the large coefficient of consolidation or the large permeability of the material.

Regarding Tilbrook, you did not show the triaxial test results. A number of isotropically as well as anisotropically consolidated tests were run and they consistently gave lower undrained shear strengths than the UU tests. We are talking about maybe 30, 40, 50% lower values than the UU tests. To us it is quite unusual that triaxial consolidated tests give lower shear strength than UU tests. Is this possibly due to you having allowed water to suck into the sample before/during the consolidation phase; not putting on a back pressure first? I would like some of our British colleagues that have had more experience with these very stiff clays to comment on this issue.

Dr R. JARDINE, Imperial College of Science, Technology & Medicine

It is interesting to compare the stress-strain and strength characteristics of the Pentre soil to the Magnus clay that was discussed in session 1. Figs 3 and 6 in Paper 4 show that, at low OCRs, the Magnus clay is contractant in undrained shear. In contrast, Figs 23 to 25 show that the 'normally consolidated' Pentre soil dilates in UU, CAU and CIU tests, except when

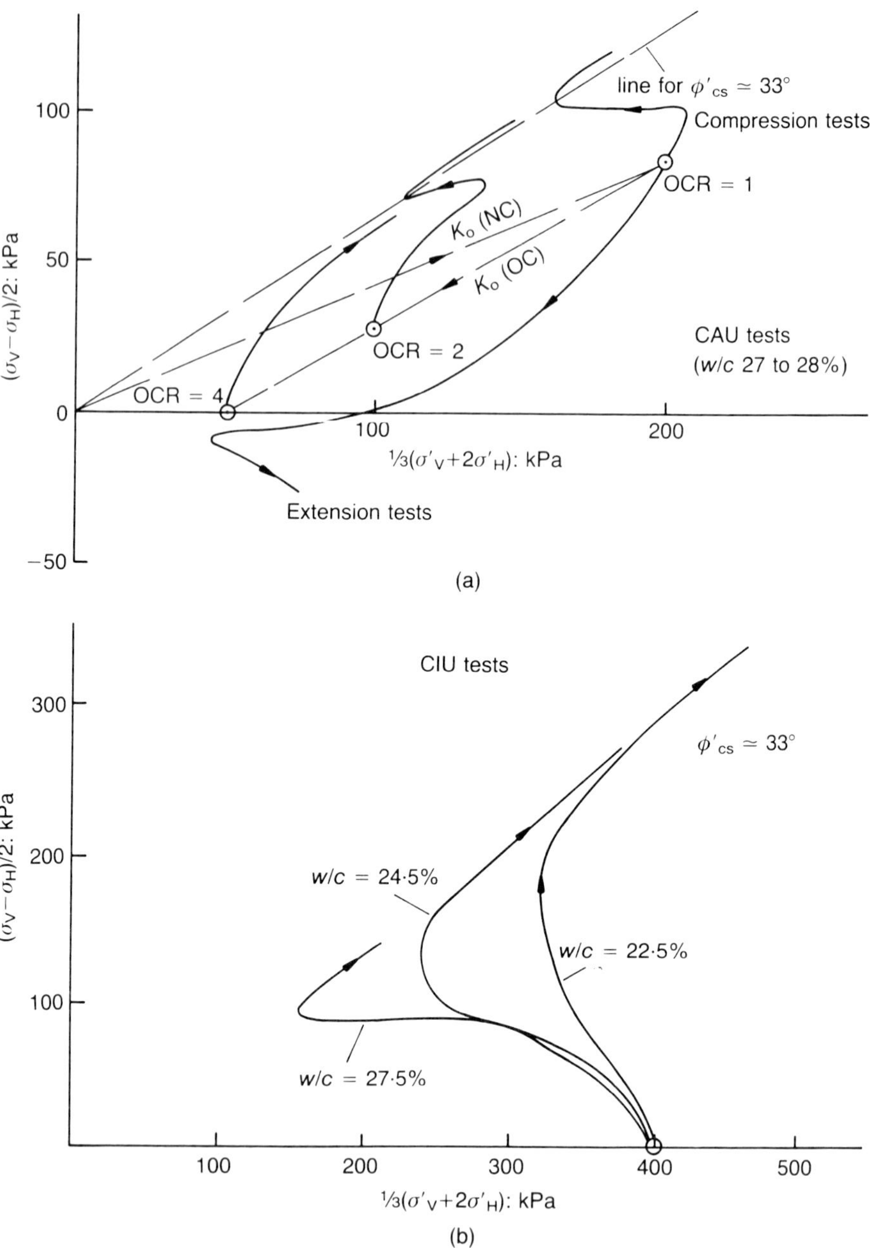

Fig. 1. Effective stress paths for CAU and CIU triaxial tests on HPF4 silt (after Ovando-Shelley, 1986)

consolidated to elevated stresses. Such a response suggests that the soil's mechanical behaviour may be more akin to that of a silt than a clay.

Figure 1 of this discussion shows data from CAU and CIU triaxial tests performed by Evando-Shelley (1986) as part of his PhD study at Imperial College on a rock-flour silt (termed HPF4). His results match the Pentre data, showing close agreement at similar water contents, OCRs and stress levels. One particular point of interest is that anisotropically consolidated HPF4 shows a higher ultimate C_u/p' ratio in compression tests at OCR = 1 than does Magnus clay. This is because it dilates. However, it is very weak in extension: CAU samples can hardly cross the isotropic axis without developing very large strains. This anisotropy and sensitivity could affect markedly the stresses developed during pile installation.

A second factor to consider is permeability. It was reported that the pore pressures set up during pile driving at Pentre dissipated very quickly: installation was probably partially drained. The approximate permeabilities of Magnus clay, HPF4 and the Pentre soil as measured at σ'_v = 400 kPa are listed in Table 1: Pentre's permeability falls far closer to that of the silt than that shown by the North sea low plasticity clay.

Table 1.

Soil	k: ms^{-1}
Pentre	1.10^{-8}
HPF4 silt	1.10^{-7}
Magnus clay	3.10^{-11}

Figure 2 compares the grading curves of the three materials. Is it reasonable to conclude that the Pentre soil appears to classify as being closer to the silt? These observations do not detract from the value of the Pentre pile test. We often meet clay-silts offshore, and it appears from Pentre - and Kjell Karlsrud's data - that they may need to be treated differently to 'standard' clays. If so, we need to be able to classify our materials in an agreed way and to set diagnostic tests which identify soils that are likely to give us particular problems.

Dr R. HOBBS, Lloyds Register

Could the Authors confirm that the Pentre site is considered representative of North Sea normally consolidated silty clays?

D. G. CLARE, Ove Arup & Partners

In the selection of the test site by my colleagues, you can see that a number of sites were examined. At some the clays were not deep enough, at some the clays were not strong enough, some clays were too plastic and some did not have the right stress history. So if one wants to find a site on land, the choices are somewhat limited with the selection criteria that were set. The source rocks from which the Pentre soils were derived are different to those which produced the soils in the Cyrus/Andrew area of the Witch Ground bed and in the Norwegian Trench, and therefore the mineralogy is different between the Pentre site and those two areas of the North Sea. So the question as to whether it is a suitable site for the test pile must be whether it does satisfy the basic characteristics considered to govern the engineering behaviour of soil. The engineering behaviour of a clay soil is deemed to be governed by its plasticity, its stress history and its strength and the Pentre soil does satisfy the requirements of the site selection. However, the soil does have some peculiarities in terms of its high consolidation rate and the stress paths which indicate that its behaviour is rather similar to a clayey silt. I doubt whether, if my colleagues had searched very much more, they would have found any better site on which to conduct this pile test in the UK.

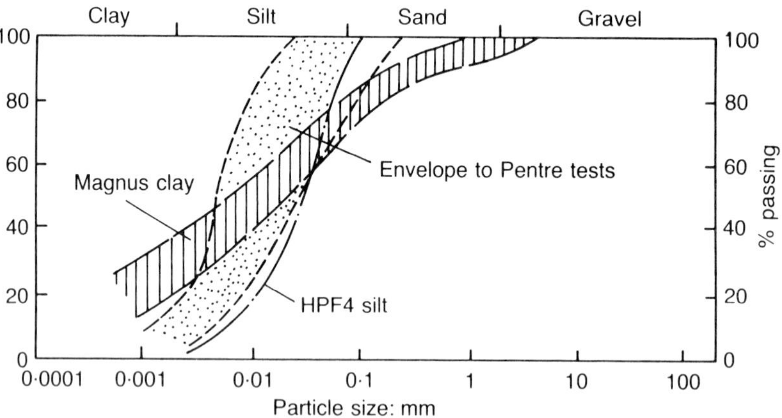

Fig. 2. Particle size distribution

Dialogue between Dr R. M. Semple (Dames & Moore, USA) and D. G. Clare (Ove Arup & Partners, UK)

RMS: The silt size fraction of the Pentre pile is comprised predominantly of clay minerals?

DGC: That is correct, yes.

RMS: Is it possible that the operational PI of this material is actually lower than the measured value when you compare it with the body of clays and characterising those by PI?

DGC: Well there are certainly clay minerals throughout the silt fraction. In determining the plasticity, those minerals in the other fractions other than clay size would be taken into account.

RMS: In terms of controlling or affecting the behaviour of the material in situ, would it really be affected? Those clay minerals would behave as silts not as clays and would not behave as a colloidal material.

DGC: Well it would depend upon the stress level as to whether the particles, I suppose, would contract as perhaps an NC clay would do or dilate as a silt might be expected to do.

RMS: I think it is at least possible that the PI is less than what was measured. If we are thinking in terms of classifying soils it could actually be less.

DGC: So you are suggesting that the measured PI is greater than the operational, because, if you tested the plasticity of the clay fraction you would get a very much lower value as some plasticity of the soil is coming from the silt fraction.

RMS: No, it is a concept - if I classify or try and capture the engineering properties of the material by its plasticity index, then in fact for a material like this we would put it in the wrong category.

DGC: Yes, I think that is a fair comment.

RMS: Because the clay minerals are actually silt size materials and are not colloidal.

DGC: That is right, a large proportion are very large plates of illite-mica and chlorite.

Dr U. MIRZA, Kvaerner, Earl & Wright

I am not really worried by the scatter of the results. If you look at the LI of the Pentre site they are all over the place. It does not mean I cannot design a pile. My point is that you can appreciate that there will be fast consolidation of soil like at the Pentre site which is behaving overall like a clayey silt material, considering the values of the coefficient of consolidation of between 100–150m^2/year. I do not think that the whole profile would have behaved overall like a clayey silt. There were indications that there were differences in the pore pressure response along the different elevations of the pile. There were also differences in the K values at the end of the consolidation. We had large differences. So while perhaps there were zones which were richer in silt at some depths, there were also other zones that were richer in clay and did behave like a clay, like any essentially normally consolidated North Sea clay.

Professor M. RANDOLPH, University of Western Australia

I would like to address the problem of residual stresses and forces which play an important role in the interpretation of strain gauge data from instrumented pile tests. From previous experience, residual strains appear in piles during driving, even where the residual forces are zero, due to locked in fabrication stresses. To illustrate this problem I would like to look at the dynamic stress-wave data from the Tilbrook pile, taking the initial strain gauge readings from three blows as the start, mid-point and end of driving. I shall restrict attention to the average of the four strain gauges around the pile, although each strain gauge behaved rather similarly. Referring to Fig. 3, the first profile is at the start of driving when the pile is embedded only over the last 3 m. Since I am not sure of the zero offsets on these gauges, this profile should really be taken as the effective 'zero' force. If we then look at the changes of measured force in the pile at 15 m penetration, when the upper 15 m of pile is still protruding from the ground, it appears that close on 1 MN of residual force has magically accumulated over the upper 18 m of the pile.

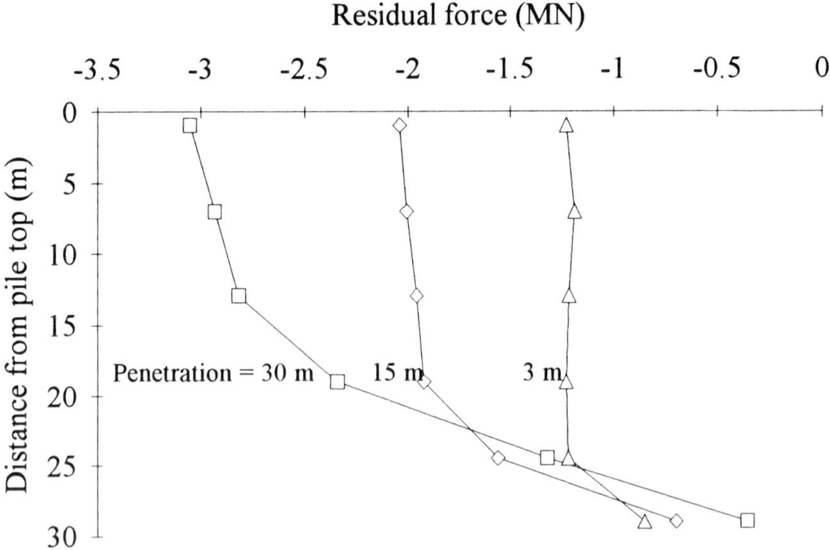

Fig. 3. Residual forces recorded by strain gauges during driving of the Tilbrook pile

Similarly, when the pile is fully installed, we have a profile of large tensile residual forces, with a very large residual force at the pile top — a little unexpected! Whether these residual forces recorded by the strain gauges are due to the release of residual strains locked in during fabrication of the pile or some other reason, I do not know. I would welcome some comment as to whether other strain gauge data showed the same phenomenon and how the readings from the static load tests were subsequently adjusted in order to interpret the data.

Professor J. BURLAND, Imperial College of Science, Technology & Medicine

I would like to ask Dr Cox a question on instrumentation. Something that you did not deal with in your paper was the work on the calibration of results and the reference to cross sensitivity, which looks very significant. I would like you to say a bit more about how it was dealt with in the processing of the results and whether this cross sensitivity refers only to the effects of axial load when you are calibrating the pile; and whether you checked for the effects of friction on the total pressure cells as well?

Dr W. R. COX, Fugro McClelland Marine Geoscience Inc., USA

It would be helpful to restate that no radial pressures were applied during calibration of the pile sections under axial compression loads. Signals from the pressure cells were monitored during calibration to establish magnitudes of cross sensitivity. As stated in Paper 10 (Cox, Solomon and Cameron), cross sensitivity values for the total pressure cells in the NC pile with 15 mm wall thickness were unexpectedly high, but were low in the OC pile with 30 mm wall thickness. Cross sensitivities of the pore pressure cells were negligible. We believe that after the piles were driven, the radial pressure produced by the soil inside and outside the pile would reduce the magnitude of cross sensitivities that were observed during calibration, but we have no data to guide us as to how much reduction this would give. As stated in Paper 14 (Gibbs, McAuley, Mirza and Cox), all data from total pressure cells were corrected for the cross sensitivity values that were measured during calibration. The effects of the corrections are shown in Fig. 11 for the NC pile and in Fig. 31 for the OC pile.

We did not make a correction for friction on the face of the total pressure cells. During calibration of the total pressure cells we did not subject them to friction loads so we had no basis for making a correction for friction.

Part 2

Dr M. A. STROUD, Ove Arup & Partners

This is the climax in a way of a lot of hard work and effort put in by a great number of people and it is always a very exciting time to see the data. As often happens when one produces a lot of data like this, there is a lot of satisfaction in seeing things that have turned out much as you expect, and there are also some surprises inevitably. The surprises are worth thinking about and when we look at the data, there are two sorts of question that we can ask.

- Is the data reasonable and is the instrumentation that we have put in working properly?
- If it is working properly, and we have convinced ourselves it is, then what does this new information tell us?

The first question really takes us back to the beginning of the testing period, immediately after installation, and the second will take us to the very end.

On completion of driving of the NC pile in particular, the pore pressures in the ground appeared to be very similar to the total stresses — in other words the effective stresses were very small. If we look at the top curves there, the *P* represents pore pressure, the *T* represents total pressure, so that in the top plot in the early stages immediately after installation of the pile, the pore pressures appear to be greater than the total pressure. That is quite an interesting phenomenon. The same is true in the other suite of data which accompanied that figure, recovering five of the total pressure cells in the pile we get a rather similar picture. The pore pressure for one gauge is something like 200 kilopascals higher than the total stress would indicate from the total stress measurement. That suggests that the effective stress is negative, which is a new phenomenon and I think that the pore pressures tend to drift down to the ambient pore pressure which is well established. It does tend to indicate that there is something of a problem with the total pressure cell measurement. It may well be that in fact the effective stresses are a good bit higher in the early stages immediately after driving. The residual skin friction in the pile, which is quite significant at the top, tends to suggest that there is some lateral effective stress there locked in. So that, the first question is: what confidence do the investigators have in the total stress measurements? Because if they are not operating properly at the beginning of the tests in that situation, then what confidence do we have that they are operating sensibly later on?

The second question I want to ask is one that relates to the bottom line, the end of the test, and it is summarised in Fig. 15 of the Gibbs *et al.* paper. It summarises unit skin friction against depth below ground level. The results from the pile show that they are very much to the left of all the predictions made by eminent people, and another curve is the result of the *t-z* probe which seems to match, as well as anything, the measured performance. So that is the situation for the NC pile. If we look at the situation for the OC pile, the situation is in many ways very similar and again the *t-z* probe gives a very good match to the measured performance of the pile itself. One wonders in this situation how significant the whip phenomenon that has been mentioned today is, because that does not really apply to the *t-z* probe as I understand it. So are we looking at something else that is causing a reduction in the stresses on the shaft of the pile?

Dr U. MIRZA, Kvaerner, Earl & Wright

As far as the low effective stresses that were measured, I believe the results that we presented are compatible with several other measurements, and I do not know whether it has been mentioned in this report, but on the Haga site NGI had also measured similar low effective stresses during installation. You can measure pore pressure without many problems. With total pressure you nearly always have problems. You need something along the line of the PLS cell of MIT which can actually measure total pressure to accuracies of something like 0.5%. We feel the pore pressure measurements were quite reasonable. In fact using predictive effective stress methods such as the ones developed by Kraft, the ESM1, ESM2, ESM3, you can predict reliably the pile capacity from the excess pore pressure that one measures. I feel that there was nothing wrong with the pore pressures measured. I think on the total pressure front, in reply to the point that the pore pressure was actually higher than the total pressure, it might suggest that the cross sensitivities of the total pressure cells need confirmation. There were certainly anomalous or near anomalous data with some of these cells, bearing in mind that there were some losses in the cells during driving — I think it is only five of the eight total pressure cells that survived driving, so that was unlucky. We had seven going for the pore pressure. But to sum up, we do feel that the total pressure cells have given us a good idea about the effective stresses, certainly at one of the cell locations, given that we only have five left, and that at 43 m. The rest were not as reliable, and as we have said in the paper, there were some, cell number 52b, for example, which drifted, so I think it might be argued that the total pressure measurements are difficult to make sense of. It is a common problem. As far as the comparison between the *t-z* probe and the measured capacity, there is a difference between a rigid capacity and a capacity of a compressible pile like the one dealt with here. I cannot really say why the measurements should actually fit well. You have to bear in mind that what one is doing is putting lines through just like we did, just connecting the points. The measurements were actually done at isolated locations. What happens in between, we do not know. But overall, to compare a peak friction which you get from a compressible pile with a friction you would get from a rigid pile is incorrect.

K. KARLSRUD, Norwegian Geotechnical Institute

Before the large pile Pentre test was carried out (actually the test had been carried out but we had no idea about the results), I was asked by Statoil to make a prediction of the pile capacity using the principles that we have developed at NGI. Predicted capacity of pile: 5 MN was a best estimate and 6.3 MN was a high estimate. I think that fits pretty well with the results, but it does not really prove the method. Back to the issue of the total pressure measurements that was raised here. Measurement of total earth pressures I can agree is more difficult than measurement of pore pressure. In NGI's pile tests which include four soft clay sites, we have measured in some cases negative effective stresses immediately after pile installation, up to 20 kPa, never anything like 200 kPa. Effective stresses after pile installation in soft, normally consolidated clays of medium sensitivity are practically close to zero. We have measured from -20 to +20 kPa in effective stresses immediately after pile installation. From then on they increase. One further aspect which I would like to raise is that when we talk about strain softening and getting to residual capacities, I think we should be very precise in talking about such effects. The tests were deformation controlled at a constant rate of displacement. This type of strain softening behaviour, which I think maybe is getting a little bit too much attention, is related to that specific type of loading condition and that is not the type of loading condition that we actually have on real piles, they are load controlled more than displacement controlled.

Dr U. MIRZA, Kvaerner, Earl & Wright

The Heerema Kontich tests, for example, were not deformation controlled, they were load controlled, yet there were certainly residual capacities. I think also some of the tests by Tomlinson, i.e. the CIRIA ones, showed degradation and were not deformation controlled either.

K. KARLSRUD, Norwegian Geotechnical Institute

There is some strength softening effect I agree, but I think the point is overstated in the sense that maybe we are talking about a 5 or 10% influence of pile capacity. There are other issues that influence pile capacity by +/- 50% and that is the issue we should be concentrating upon.

Dr A. BOND, Geotechnical Consulting Group

Dr Cox asked for some comments on the pore pressure dissipation curves that he observed on the OC pile (Paper 13, Figs 15–16). At Imperial College we performed a series of instrumented pile tests in normally and overconsolidated clays and we measured similar effects in the overconsolidated clays as those observed on the OC pile. Fig. 1 shows pore pressure dissipation curves recorded in London Clay. There are three pairs of curves for leading, following and trailing instruments. The leading instruments are located eight radii above the pile tip, the following instruments twenty-five radii, and the trailing instruments fifty radii. As the figure shows, the instruments at the leading positions record pore pressures that rise up very rapidly from negative values at the end of installation, to reach a peak after a few seconds, and then dissipate with time. These results are for a 100 mm diameter closed-ended pile.

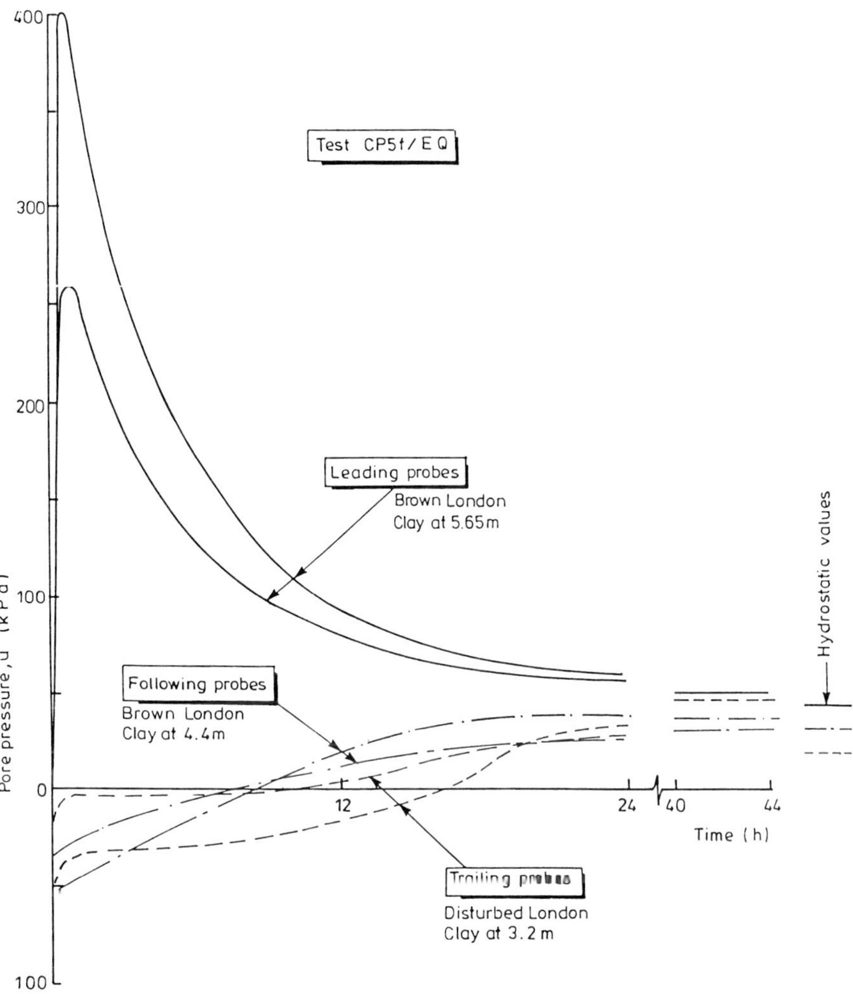

Fig. 1. Pore pressure dissipation curves for a pile installed in London Clay (Bond and Jardine, 1991)

One of the consequences of these results concerns the radial effective stress acting on the pile during equalisation. Fig. 2 summarises the results of tests at Canons Park, Cowden and Bothkennar. The response of the two overconsolidated soils, the London Clay and Cowden till is similar in that the radial effective stresses drop after installation. In the London Clay they drop and remain virtually steady. In the Cowden till they drop dramatically and rise with time. If this behaviour occurred over the full length of the pile, then clearly there is a potential for shaft capacity to reach a minimum shortly after installation. Full dissipation in the London Clay took around four days and at Cowden about two weeks, for 100 mm diameter piles. The equivalent

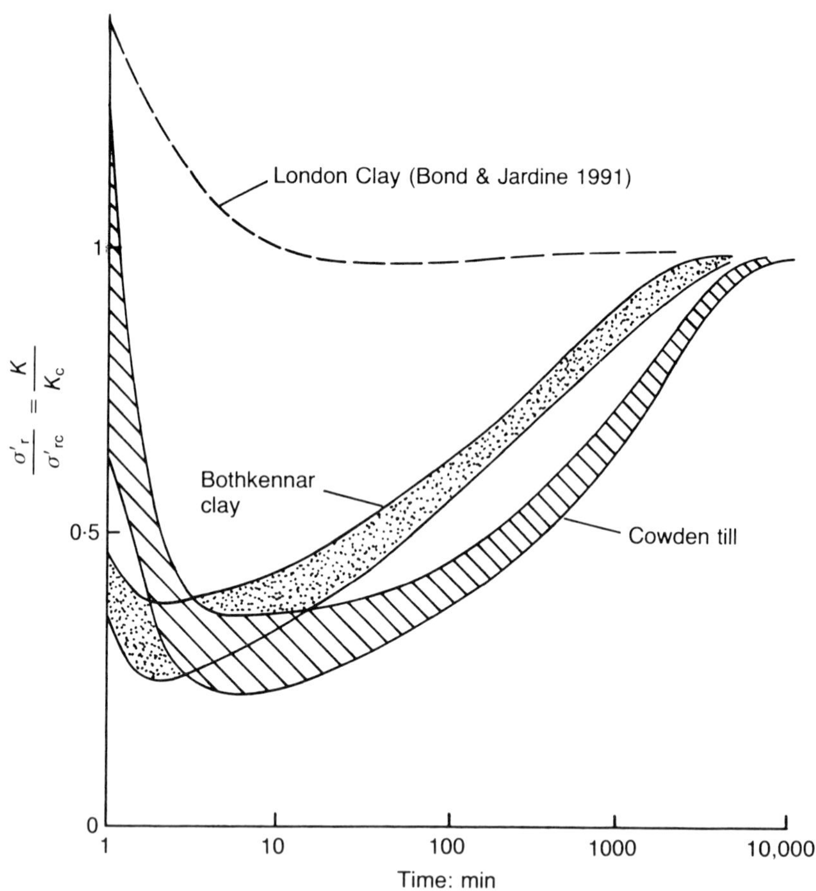

Fig. 2. Variation in radial effective stress with time following pile installation (Lehane, 1992)

period for the large diameter piles would be about 6 months to 2 years, so I am not surprised that Dr Cox did not observe full dissipation over the timescale of the OC tests. I would suggest therefore that the K_c values for the OC piles are underestimated, in other words, the fully equalised K_c values would have been higher than was measured at the start of the pile test.

Reference: Bond, A. J. and Jardine, R. J. (1991), Effects of installing displacement piles in a high OCR clay, *Géotechnique* **41**, No. 3, 341-363.

Dr J. D. MURFF, Exxon Production Research Company, USA

How certain are you that the pile at Pentre was plugged during the load test. Was the analysis decision that you had full end bearing?

Dr W. R. COX, Fugro McClelland Marine Geoscience Inc, USA

The pile plug was 1.7 m above the casing. I cannot remember if we went and measured the plug after the pile test. I do not think so. We had no really convenient way to get in.

Dr J. D. MURFF, Exxon Production Research Company, USA

If you redistribute some of the end bearing through the skin friction, you might end up having a closer match.

Dr W. R. COX, Fugro McClelland Marine Geoscience Inc, USA

That is very true. The division between how much of the skin friction is on the outside and inside is still not understood. We do not have any readily available methods for measuring it. There has been some work on it, for example at MIT, but normally we put the gauges on the outside face or inside face, depending on whether you are outside or inside the pile. It would be nice to have on both so that we really know how much is inside.

J. PELLETIER, Shell Development, USA

We had a similar problem during the Beta Pile Test, but were able to determine the amount of end bearing from our two-way cyclic tests. Did you do any two-way cyclic tests?

Dr W. R. COX, Fugro McClelland Marine Geoscience Inc. USA

No. The cyclic tests were an afterthought, added on to the test programme at a late date. We would have liked to have done two-way tests but this was not possible.

Dr M. A. STROUD, Ove Arup & Partners

On the subject of whip and the flexibility of the pile, I get the impression from some of the later papers in this volume, that there were one or two of you who believed that it was not quite as significant as professor Poskitt would have.

Dr W. R. COX, Fugro McClelland Marine Geoscience Inc. USA

I want to say something on the discussion of the performance of the pore pressure cells at 21 m. Professor Poskitt's information indicates that the pile did whip, and if the pile whips you can imagine there is a greater hole and whipping at depth too, not just at the surface. So there is a hole larger than the diameter of the pile and, if so, at the last blow of the hammer if the piles bent over, then certainly on one side we can have negative pressure which is what that one cell said and on the other side because it is in an oversized hole would not show perhaps as large a pore pressure as you might think. That is one reason I might suggest that they behave that way, perhaps there are others. In fact the main text does not discuss these. I am inclined to think that our whole thinking process, and I confess I am part of this group, when we think about residual and so on, we think about a pile that is completely in contact with the soil. In fact Professor Poskitt's information says, that it is not the case. You are getting separation from the soil and the pile at considerable depth.

Dr M. A. STROUD, Ove Arup & Partners

Professor Poskitt mentioned the lateral acceleration, and said that represented obviously some lateral movement. Has Professor Poskitt been able to make an estimate of the lateral movement there?

Professor T. J. POSKITT, Queen Mary and Westfield College, University of London

Measurements were made of lateral accelerations and estimates made of lateral displacements. These were analysed after completion of the main project and have been reported in *Ground Engineering*, October 1990.

D. E. SHARP, BP Exploration

Professor Poskitt showed back-figured soil resistance to pile driving. How did that compare to actual predictions before driving?

Professor T. J. POSKITT, Queen Mary and Westfield College, University of London

Mr Sharp has asked how our predictions for driving compared with what actually occurred. Our calculations for the OC pile were much more comprehensive than for the NC pile. The reason for this was that we felt installation problems were more likely at the OC site than at the NC site. In fact the resistance at the NC site was found to be so low that it posed serious problems concerning the sensitivity of the measuring system.

On the whole it was found that total resistance could be found with a degree of confidence, however the distribution of this resistance on the pile was more difficult to predict. This was not entirely unexpected since it is known from formula such as Engineering News and Hiley that this is the case.

Generally both piles drove more easily than expected.

Dr U. MIRZA, Kvaerner, Earl & Wright

Why is the back-figured quake for both piles for both sites the same?

Professor T. J. POSKITT, Queen Mary & Westfield College, University of London

Dr Mirza has raised the interesting point of why the quakes are so insensitive to soil type. It is interesting to note that this can also be deduced from load test data. In his paper, "A new method for single pile settlement prediction and analysis", *Géotechnique* **42** No. 3, September 1992, Dr W. G. K. Fleming presents a wide range of cases back-analysed using the hyperbolic law. In a discussion of this paper, I have shown that the parameters of the hyperbola can be related to quake. These also show a remarkable insensitivity to soil type.

15. The axial tension test of an instrumented pile in overconsolidated clay at Tilbrook Grange

J. CLARKE, BP Engineering, BP International Limited, and
M. M. LONG and J. HAMILTON, Ove Arup & Partners

The paper describes the results of a constant rate of extraction and a creep test on a 762 mm diameter steel pipe pile embedded in over-consolidated clay. The load take-out profile down the pile and local soil resistance/displacement response (T-Z) plots are presented. The results of the axial tension test are compared with those of an axial compression test carried out on a similar pile on the same site.

Introduction and background to test programme

The results of the compression pile test at Tilbrook Grange are reported by Cox *et al.* (1993) and Gibbs *et al.* (1993). While the result of the compression pile test had shown that the existing predictive methods were satisfactory, the project and its sponsors had hoped to prove that measured capacities were higher. The lower than 'anticipated' ultimate static capacity achieved therefore resulted in much discussion between the sponsors of the compression test programme. The discussions centred around reasons for the apparent shortfall on 'anticipated' pile capacity and the possibility of performing an additional axial pile test at Tilbrook. The general consensus of opinion was that the static pile capacity achieved was probably strongly influenced by two separate phenomena, namely:

(*a*) Effects of pile whip during driving resulting in an oversized hole. The upper soils materials are heavily over consolidated and it was the considered view of the majority of the participants that soil 'creep' back to the pile wall had not fully occurred. This affect was suggested by the low 'load take out' and unit skin friction values for the upper 10 m of soil.
(*b*) Lack of excess pore water pressure dissipation in the Oxford Clay stratum (approximately 73 per cent at time of test).

The possibility of performing an additional axial test at Tilbrook was discussed in some detail. The calculated driving resistance by Queen Mary College (QMC) during the driving trial (12 MN) and the 'retap' resistance some 8.5 months later (22 MN) was suggested as evidence of a significant set up (80 per cent) occurring over that period, Poskitt *et al.* (1993). This should be compared with the driving resistance on the compression test pile (12 MN) and its static compressive capacity some 4.5 months later of 16.1 MN. While no direct comparison can be made between dynamic and static resistance, there was sufficient evidence from the dynamic measurements to indicate that a new axial test on the driving trial pile could result in an increase in measured pile capacity.

Three possible options for performing an additional axial test were considered.

1. *Retest the compression test pile at some later date.* Pile instrumentation would remain hooked up and a retest on this pile would permit 'load take out' to be determined. This option was briefly discussed but rejected because of the lengthy delay necessary to allow the pile to fully set up prior to retesting. In addition, it was felt by some parties that the effects of pre-shearing associated with a now completed programme of static, creep and cyclic tests could invalidate the retest capacity.

2. *Perform a static compression test on the adjacent driven trial pile.* In order to perform a compression test on this pile, an additional 4 reaction piles would need to be installed. It was considered that the process of post installing these piles could disturb the surrounding soil and create large excess pore pressures in the vicinity of the driving trial pile. In addition, the trial pile was fabricated in grade 50 steel (nominal yield stress of 340 N/mm^2) compared with grade RQT 701 (nominal yield of 770 N/mm^2) used for the compression test pile. Yield of the trial pile therefore would occur at around 27 MN and hence, allowing for bending etc, it would not be possible to load the pile in excess of 20 MN. This option was therefore rejected.

3. *Perform a static tensile test on the adjacent driven pile.* This option was discussed in some detail and was generally considered to be the best option given the time constraint on site usage and the technical problems associated with options 1 and 2 above.

It was considered that a test of this nature would allow a direct comparison to be made of the skin friction measured in the compression test pile (set up 4.5 months) and the tensile skin friction in the driving trial pile (set up one and a half years). Whilst it was acknowledged that the tensile skin friction may be less than its compressive counterpart, there is evidence to suggest that the difference (if any) may be small. The reinterpretation of the West Sole Platform 'WC' pile load tests, Clarke *et al.* (1985), indicated that compressive and tensile shaft frictions in very stiff OC clays are not significantly different.

The conclusion from the discussions was that the participants strongly supported the need for an additional low cost axial test and that the preferred option was to perform a tensile test (option 3 above).

The tension test was carried out on 22 September 1988. A creep test was carried out the following day.

Site conditions

The site is located at Tilbrook Grange in Cambridgeshire, England, at National Grid reference TL 009 710. The site ground level was assumed to have a nominal value of +100.0 m. In the vicinity of the pile ground level was lowered to 98.4 m (see Fig. 1).

Ground conditions at the site are described in detail by Lambson, Clare, Senner and Semple (1993). The following is a brief description of the soil conditions. The succession of strata beneath the site can be summarised as follows:

Stratum	**Thickness** (m)
Lowestoft Till	17.1 - 18.6
Oxford Clay	23.9 (proved)
Cornbrash	-

Groundwater level is close to ground surface. Information from the previous test project indicated hydrostatic water pressures from about 2 m to 25 m depth below ground level. Below this pore pressures are lower than hydrostatic due to presumed underdraining of the Oxford Clay by the Cornbrash.

The Lowestoft Till is a very stiff to hard grey silty clay with scattered sand to boulder sized flints, and occasional chalk and mudstone fragments. It is a clay of intermediate plasticity with a liquid limit in the range 35 per cent to 45 per cent and plastic limit of 15 per cent to 20 per cent. Natural moisture content is about 15%. The averaged undrained shear strength over the stratum thickness is about 400 kPa, see Fig. 1.

The Oxford Clay is a hard dark grey fissured clay with occasional silt and sand pockets and some shell fragments. It has high to intermediate plasticity with liquid and plastic limits of about 55 per cent and 20 per cent. Natural moisture content is about 18 per cent. It has an undrained shear strength varying between 300 and 1000kPa, with an average of around 600kPa. The site displays an apparent overconsolidation ratio, OCR, of some 20 to 50 in the upper 7 m. OCR decreases with increasing depth to a value of about 10 at 25 m.

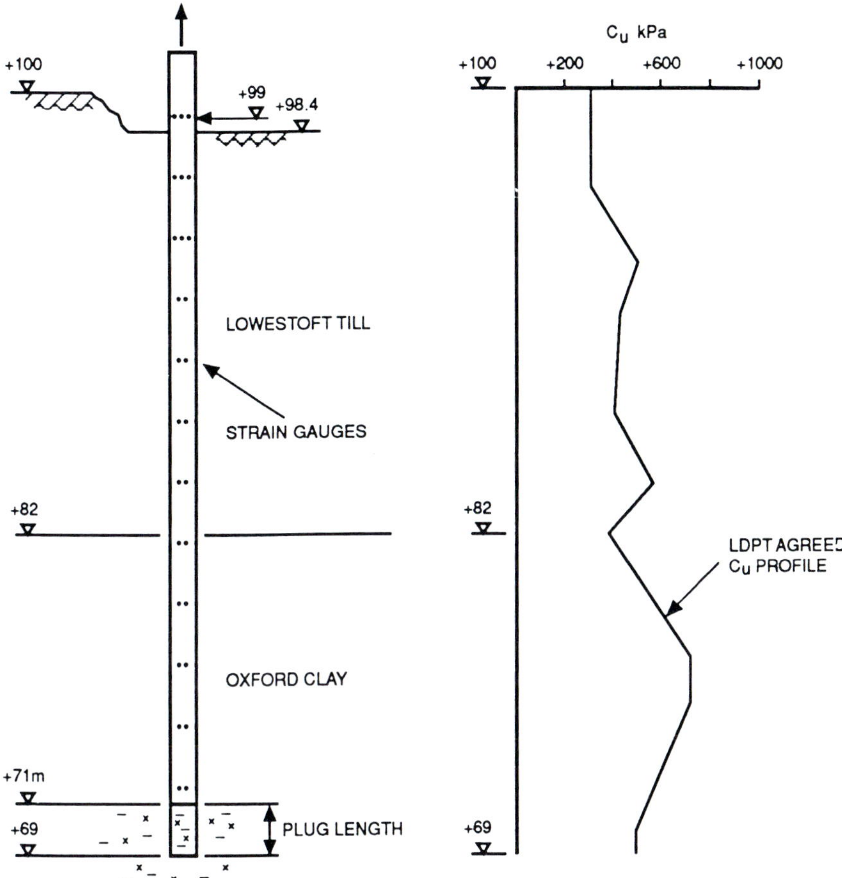

Fig. 1. Undrained shear strength profile and pile sensor locations

Trial pile

The trial pile was a 762 mm OD, 35 mm wall thickness steel pipe pile. It had been driven open ended to a depth of 30.5 m, i.e. + 69.5 m between January 14 and 19 January 1987 as part of a trial driving contract. It was driven a further 0.5 m, i.e. to +69.0 m, on 5 October 1987. The plug of soil from within the pile was augered out on 21 August 1991 to a depth of 29 m, i.e. an elevation of +71m, see Fig. 1.

Test set up

The test set up is shown on Fig. 2. The design load was 30 MN. An economical design was achieved by re-using, as far as possible, the test frame built for the earlier compression test. The existing main beam span of 9 m set the layout for the reaction pads. Analyses were carried out to confirm the interactions between the pads and the test pile were negligible and so shallow footings were acceptable. These footings were 12 m long by 2 m wide and built of in situ reinforced concrete.

The tension load was applied using four 7.5 MN capacity jacks with two placed at either end of the main beam. A uniform displacement hydraulic control system was used to lift the main beam which was bolted to the pile head using twenty-four 50 mm diameter Macalloy bars.

Instrumentation and data acquisition

The instrumentation was installed, following augering out of the pile, during the 24 - 30 August 1988. Whilst it would have been advantageous to have installed the instruments before pile driving so that residual stresses due to pile driving could be monitored, this was not possible, as the pile was driven as part of an earlier pile driving exercise.

Four load cells were used to measure the pile head loads. The load cells were individually calibrated at the National Engineering Laboratory and the deviations from the intended loads were negligible. Four displacement transducers (LVDTs) were used to record vertical displacement at the pile head.

Ailtech weldable strain gauges were installed down the inner wall of the trial pile to measure vertical strain at the positions shown on Fig. 1.

Zero readings were taken on site for a number of hours before the start of the test. Zero reading drift was negligible. During the test, instruments were read at 0.5 Hz with over sampling at 2 Hz. Signals from certain instruments were displayed on chart recorders during the test to provide control and to monitor progress. There were no instrument failures during the test. Only two instruments produced questionable data. The high quality of the data recorded is illustrated by the consistency of the results.

Precise levels were taken on the reference beams which confirmed they were stable during the test. Levels were also taken on the reaction frame. At peak load, during the test, there was an 11.8 mm differential movement of the main beam and a 19.5 mm central settlement of the reaction pads. The pile head movement was also checked using a precise level.

Tension test procedure

The constant rate of extraction (CRE) tension test on the trial pile was undertaken on the 22 September 1988. The weight of the reaction system was initially taken on the jacks before the nuts on the Macalloy bars were hand tightened. The reaction beam was then lifted uniformly to extract the pile. The rate of pile extraction achieved was 0.67 mm/min up to a test load of 14.8 MN and thereafter 1.06 mm/min.

Fig. 2. Tension test set-up

The following day a creep test was carried out. The tension load was held constant while the creep movements were observed.

Tension test results

Axial capacity

Plots of the test load versus time are shown on Fig. 3. The test load versus displacement response at the pile head is shown on Fig. 4.

The pile head loads were determined by subtracting the weight of the reaction system (0.38 MN) from the test load. The peak loads were as follows:

- Peak Test Load = 16.58 MN
- Peak Pile Head Load = 16.20 MN

The pile head displacement was taken to be the average of the four transducers equally spaced around the pile head. The pile head displacement at peak load was 41.5 mm. The pile was then further extracted to a total displacement of 96.0 mm, by which time the pile head load had dropped to 15.2 MN. On removal of the load a residual displacement of 77.3 mm remained giving a pile elastic recovery of 18.7 mm.

Load distribution and skin friction along pile

The load distribution and skin friction along the pile were calculated from the strain gauge data. The calculations were carried out assuming the same residual stress profile as measured in the compression pile. Whilst no allowance was made for the redriving sequence carried out on the test pile; the pile properties were similar and it was assumed that the residual stress profile would also be similar. For comparison purposes a zero residual stress profile was also assumed.

The load distribution down the pile at the peak load is given in Fig. 5. The plots of load distribution for prepeak loads is given in Fig. 6 and for post-peak loads in Fig. 7. Corresponding skin friction plots are given in Figs 8 to 10. It can be seen that a substantial portion of the load is taken on the bottom 20m of the pile. The readings from one set of strain gauges close to ground level appear to be slightly high although there is no apparent error or inconsistency in the data.

The accuracy of the strain gauges is quoted to be +3 per cent by the manufacturers. This would imply a maximum potential error of 0.5 MN at peak load. The actual accuracy can be checked by comparing the result from the top level of strain gauges with the applied load measured by the load cells. At about peak load the load cell measured 16.03 MN whereas the top strain gauges indicated 16.52 MN. The difference of 0.49 MN is within the ±3 per cent quoted accuracy. Calculations were carried out on the sensitivity of the

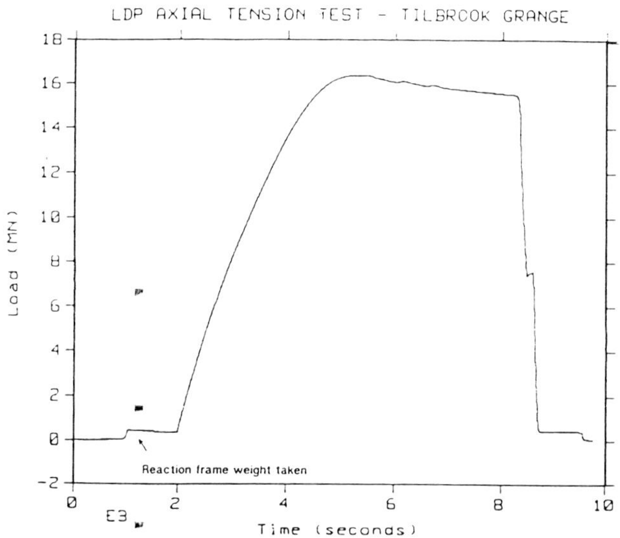

Fig. 3. Pile test load versus time

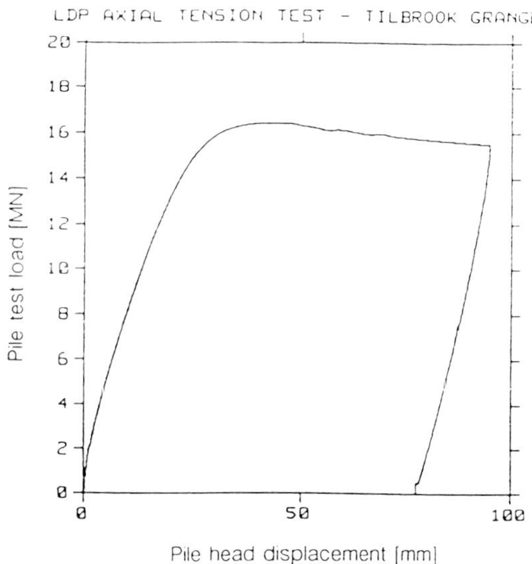

Fig. 4. Load versus displacement for tension test

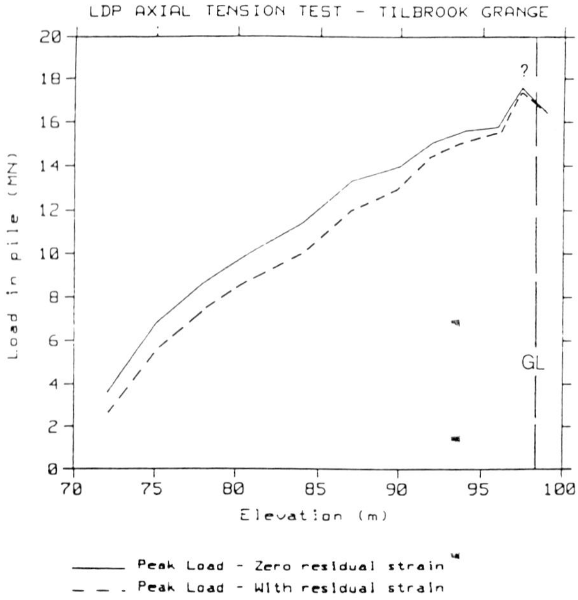

Fig. 5. Load distribution in pile at peak load with and without residual strain

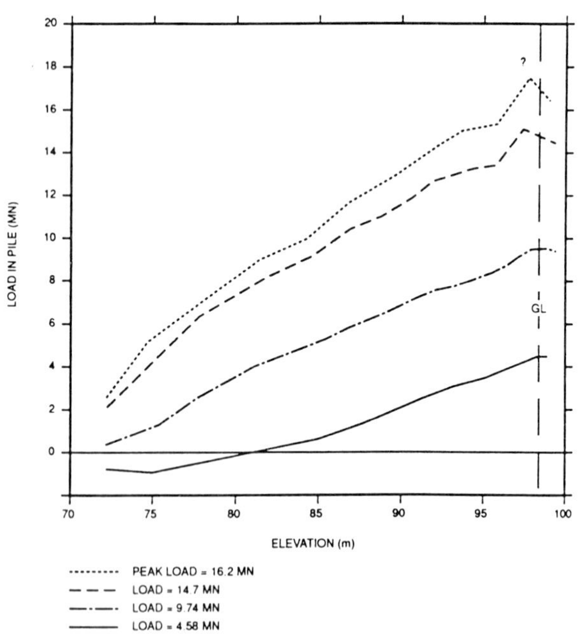

Fig. 6. Load distribution pre-peak with residual strain

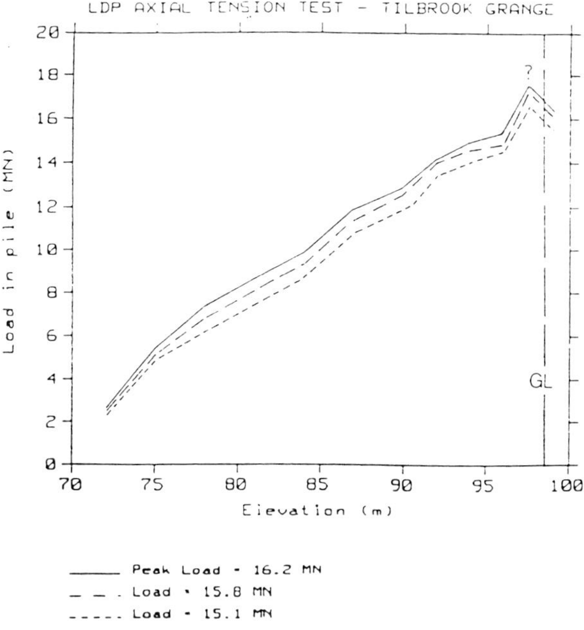

Fig. 7. Load distribution post-peak with residual strain

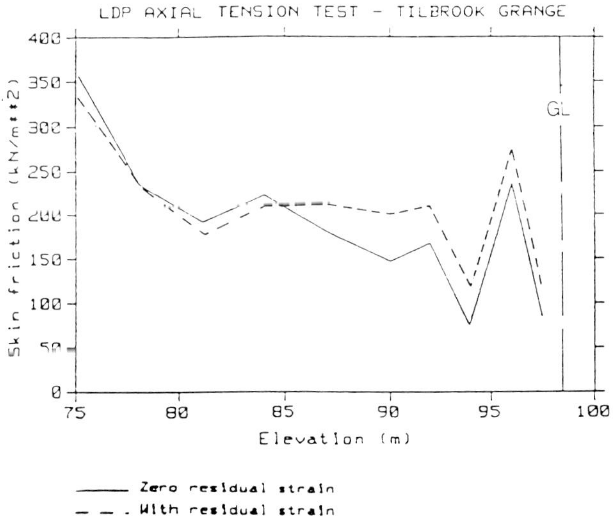

Fig. 8. Skin friction in pile at peak load with and without residual strain

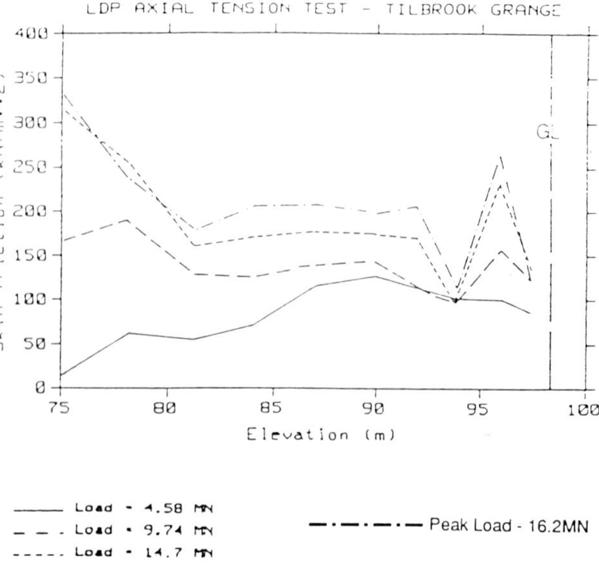

Fig. 9. Skin friction pre-peak with residual strain

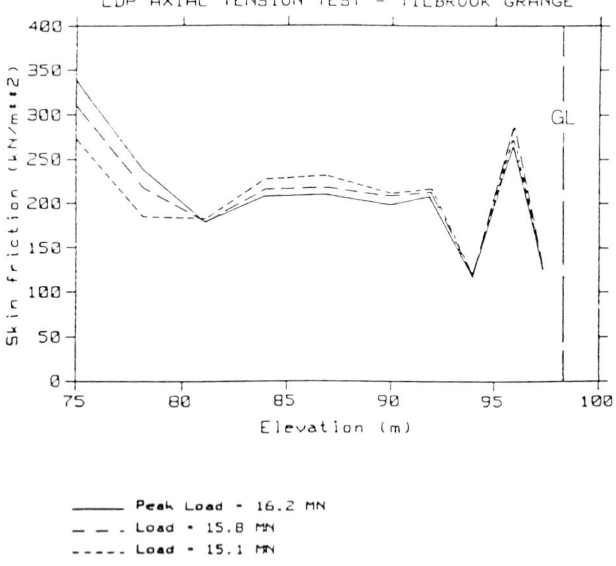

Fig. 10. Skin friction post-peak with residual strain

load distribution plots for variations in the strain readings. These calculations show that a maximum difference of about 8 per cent occurs between the load calculated by using output from one strain gauge and that calculated by averaging the results from both strain gauges at any level. Some pile bending or local variations in wall thicknesses may account for this variation.

Soil resistance (T-Z)

The build up of soil resistance (*T*) with pile movement (*Z*) has been calculated from the strain gauge data at six typical levels. The profiles for pile elevations 75.1 m, 81.1 m, 87.0 m, 92.0 m, 94.0 m and 96.0 m are shown on Fig. 11. The measured profiles for the Lowestoft Clay and the one at the top of the Oxford Clay stratum (81.1 m elevation) show no degradation with increasing displacement. This is at variance with the measured response on the compression pile as reported by Gibbs *et al.* (1993). Peak friction in the Lowestoft Clay appears mobilised at movements of 20 to 35 mm (2.5 to 4.5 per cent of pile diameter). This is much larger than expected for a pile of this size in heavily overconsolidated clays. In the Oxford Clay, the curve for elevation 75.1 m exhibits a distinct peak with a residual to peak value of approximately 80 per cent. The maximum skin friction at this elevation is mobilised at a

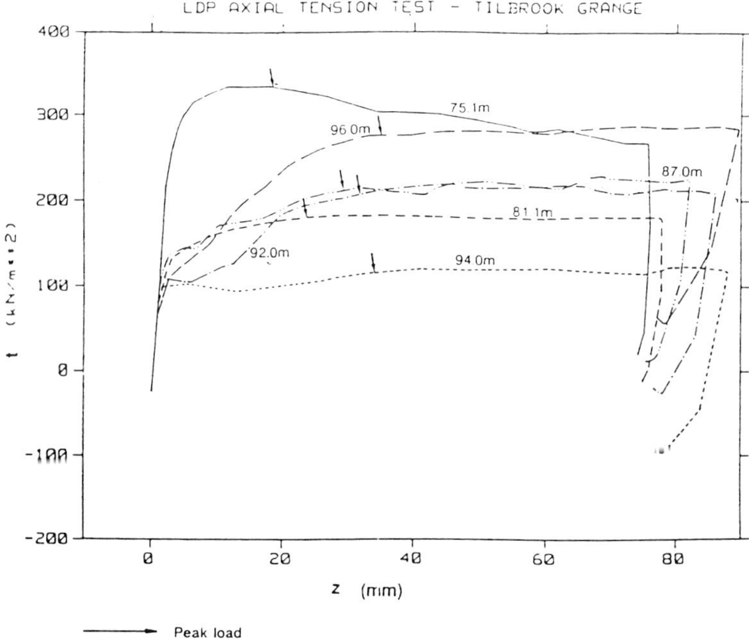

Fig. 11. T-Z profiles with residual strain

displacement of around 13 mm and is similar to that observed in the compression pile test.

Calculations show that the difference between calculating soil resistance using a single strain gauge or the average of the pair at any level are negligible up to loads of 10 MN and increased to about 25 per cent at peak load. However, as the average of two strain gauge readings give good agreement of measured load, it can be assumed that the average will also give a reliable estimate of soil resistance.

Creep test

In the creep test the pile head was loaded at a rate of 0.47 mm/min to 10.8 MN (2/3 of the peak load achieved in the static test), which caused an immediate pile head movement of 11.9 mm. This load was then maintained for a period of 4 hours and 36 minutes, causing a further pile head movement of 1.1 mm. Instead of unloading at the end of the creep test, additional load was again applied to the pile. This second peak pile head load value was 14.0 MN. Since the beginning of the creep test, the pile had been extracted 41.1 mm. On removal of the load the residual displacement was 24.3 mm. The pile elastic recovery was equal to 16.8 mm. During both the tension and creep tests the pile had been extracted a total of 101.6 mm.

Plots of pile test load versus time, and pile test load versus displacement for the creep test are shown on Figs 12 and 13.

Comparison with compression pile test

The results of the compression pile testing are detailed in Cox *et al.* (1993) and Gibbs *et al.* (1993). The load versus displacement plots for both piles are shown on Fig. 14. The peak test load was similar for both piles (16.2 MN and 16.13 MN). However some 48 per cent more movement (41.5mm as opposed to 28.1 mm) was required to mobilise maximum load in the tension pile. It would appear that an insignificant increase in shaft capacity resulted from either a further reduction of excess pore water pressure and/or soil flow back in these heavily OC clays reducing gapping effects.

The ultimate base resistance in the compression test was 1.45 MN. The higher shaft capacity in the tension test is probably due to the effect of this pile penetrating an extra 1 m into the Oxford Clay. It may be also partly due to further dissipation of excess pore pressure and closing of the soil gap.

Figure 15 shows that load take out was substantially greater in the Lowestoft Till layer for the tension test, i.e., an extra 2.5 MN over the upper 10-11 m. This difference is shown more clearly in terms of skin friction in Fig. 16. In contrast load take out and hence magnitude of skin friction for the bottom two-thirds of the pile are less for the tensile pile test than the compression test. The reason for this difference in behaviour between the tension and compression pile tests may be due to residual stress effects.

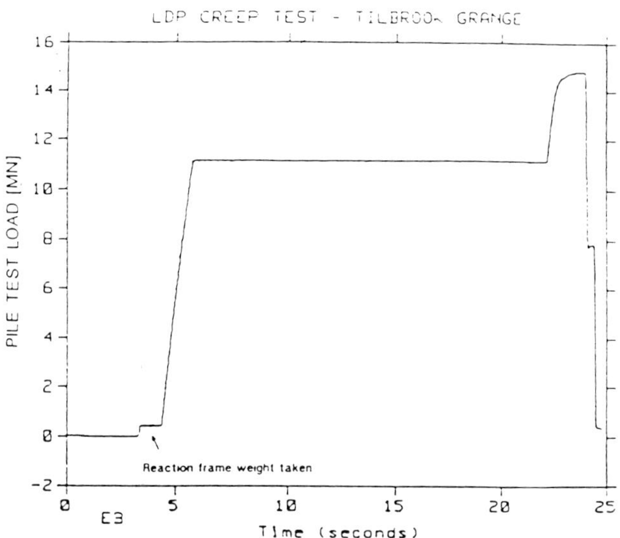

Fig. 12. Load versus time for creep test

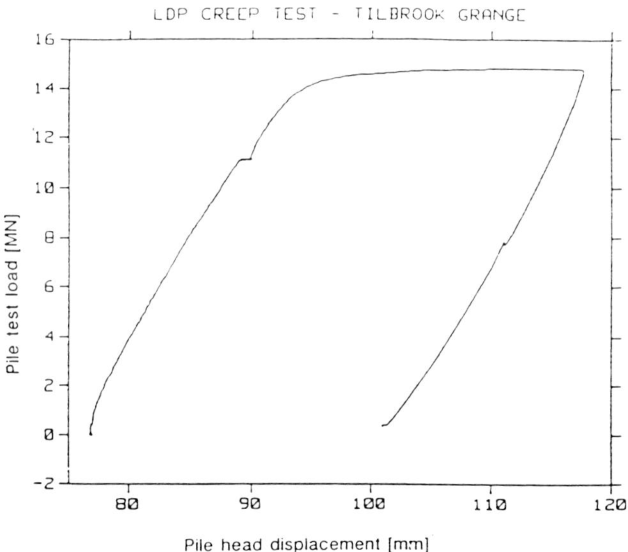

Fig. 13. Load versus displacement for creep test

Measurements, detailed by Gibbs *et al.* (1993) showed that, in the compression test pile, over the upper 12.5 m the residual friction is negative. This results from the ground resisting rebound of the pile. During the static compression test, the skin friction over this upper portion of pile is reversed. This strain reversal may account for the very low skin friction observed in this section of pile at the conclusion of the compressive pile test. In comparison, the bottom section of the pile starts in compression and remains so during the test. As a result, the ultimate skin friction over this bottom portion is very high.

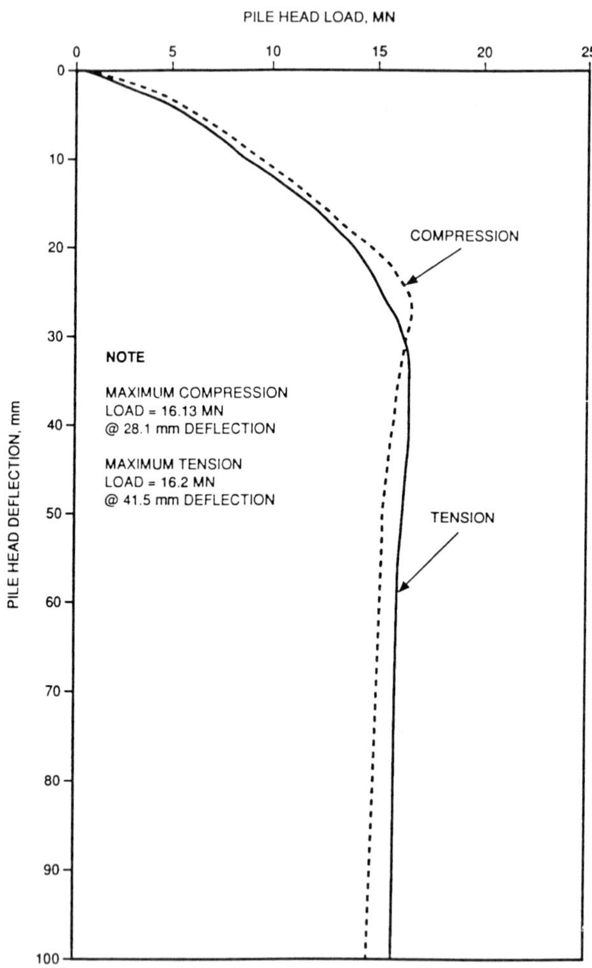

Fig. 14. Load versus displacement

Figure 16 also shows the results of the tension test. The same residual profile was used to interpret the tension test as that measured in the compression pile (the tension pile was only instrumented just prior to testing). It is evident from this figure, that over the upper portion of the pile, the soil skin friction which starts in tension and remains so for the completion of the test, achieves a greater ultimate skin friction than that measured for the compression pile test. In contrast, the bottom portion is affected by strain reversal and consequently the magnitude of the ultimate skin friction is less than that measured during the compression test.

Figure 17 shows the position of the Tilbrook pile load tests relative to the API (1987) method, a derivative of the original Randolph and Murphy (1985) method. This method is shown to give a good prediction for hard overconsolidated clays. On an overall pile basis, the tension and compression data points overlie each other. However, it would appear that the similarity in ultimate shaft capacity may have been fortuitous, due partly to strain reversal effects and the different pile penetrations.

The Lloyd's Register method as presented by Hobbs (1993), limits skin friction to that which can be sustained by passive pressure acting on the pile wall. This method gives a good correlation with the compression pile skin

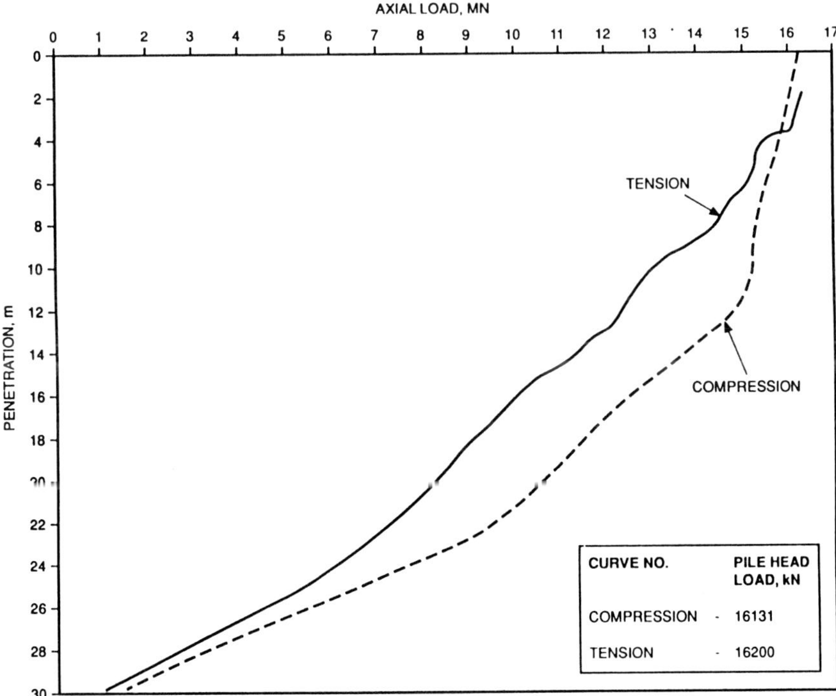

Fig. 15. Load distribution on piles at peak load

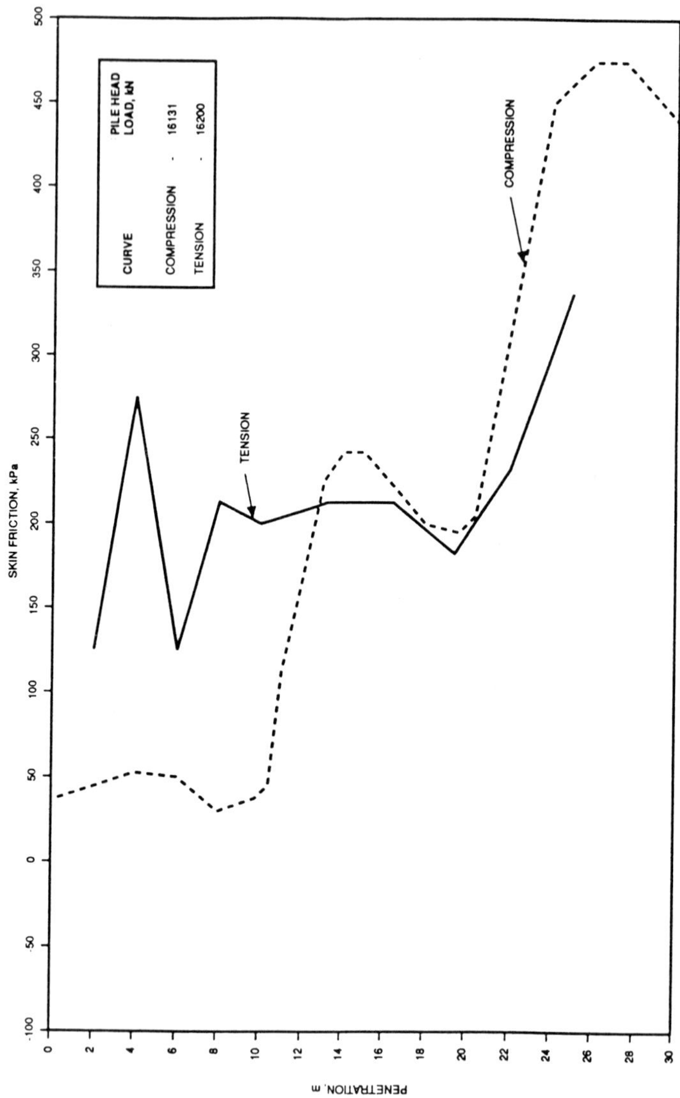

Fig. 16. Skin friction in piles at peak load

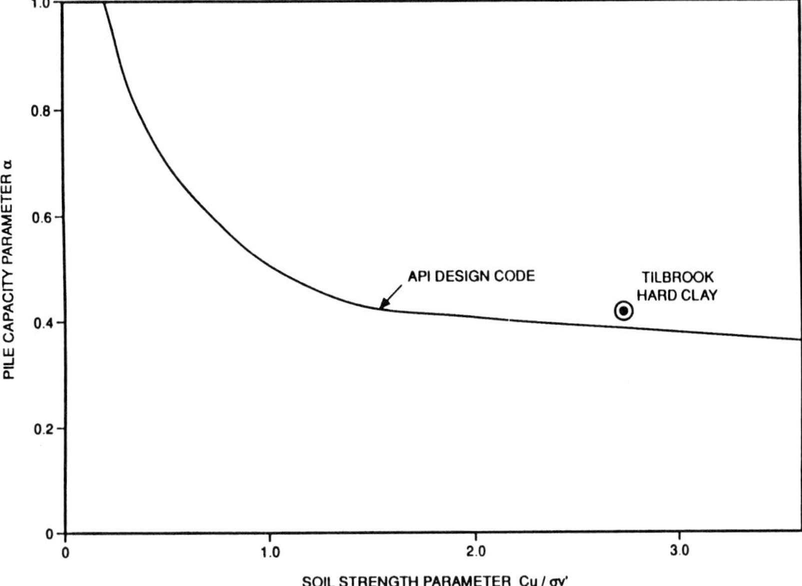

Fig. 17. Comparison with API design code

friction profile for the upper 10 m of the pile, but significantly underestimates the skin friction profile for the upper third of the tension pile. It is difficult to extrapolate from one data point, however, for piles dictated by tensile requirements the skin friction limitations for the near surface soils proposed by Hobbs may not apply.

Acknowledgements

The Authors wish to thank the sponsors of the Tilbrook Tension Pile Test Project for permission to publish this paper. The Tilbrook Pile Test was a joint industry funded project and the contributions both financially and technically, of the following are gratefully acknowledged.

- BP Development Limited
- Britoil Plc
- Exxon Production Research Co
- Statoil
- Fugro-McClelland
- UK Department of Energy

The authors also wish to acknowledge the contributions to the project by the following:

- Ove Arup & Partners
- Fairhurst Structural Monitoring
- Arbil Equipment Ltd
- Roadworks Ltd

References

API RP2A (1987). *Recommended Practice for Planning Designing and Constructing Fixed Offshore Platforms.* American Petroleum Institute, Dallas, Texas, 18th edition.

CLARKE, J., RIGDEN, W. J. and SENNER, D. W. F. (1985). Reinterpretation of the West Sole Platform 'WC' Pile Load Tests. *Géotechnique* **35,** No. 4, 393-412.

COX, W. R., CAMERON, K. and CLARKE, J. (1993). Static and cyclic axial tests at Pentre and Tilbrook Grange. This volume.

GIBBS, C., MCAULEY, J., MIRZA, U. and COX, W. R. (1993). Reduction of field data and interpretation of results. This volume.

HOBBS, R (1993). The impact of axial pile load tests at Pentre and Tilbrook on the design and certification of offshore piles in clay. This volume.

LAMBSON, M. D., CLARE, D. G., SEMPLE, R. M., and SENNER, D. W. F. (1993). Investigation and interpretation of Pentre and Tilbrook Grange soil conditions. This volume.

POSKITT, T. J., YIP-WONG, K. L. and COX, W. R. (1993). Measurement of axial strain and pile wall displacement and the determination of skin friction during the driving of instrumented piles. This volume.

RANDOLPH, M. F. and MURPHY, B. S. (1985). Shaft capacity of driven piles in clay. *Proc. Seventeenth Offshore Technology Conference,* Houston, Vol. 1, pp. 371-378.

16. Cyclic lateral loading of an instrumented pile in overconsolidated clay at Tilbrook Grange

M. M. LONG, Ove Arup & Partners, M. D. LAMBSON, and
J. CLARKE, BP Engineering, BP International Ltd, and
J. HAMILTON, Ove Arup & Partners

Introduction

The opportunity arose during discussions regarding undertaking an additional axial load test on a pile at Tilbrook Grange (Clarke *et al.* 1993), to further extend the test programme and include a lateral load test. It was recognised that the existing axial compression and tension test piles could be used to react against each other allowing static, one way cyclic or two way cyclic tests to be performed, dependent on the sophistication of the reaction system used.

The need for this test arose because the database for laterally loaded piles in very stiff overconsolidated clays was limited and the design methods in use were generally thought to be conservative. In addition, there was a trend within the offshore industry towards more frequent use of vertical piles to support offshore structures. With the consequential increases in pile wall thickness (100 mm is not uncommon), the development of a more soundly based lateral load–deflection (p-y) design method for stiff clays becomes more important. The database (with some exceptions) then applicable is presented as Table 1 which is reproduced from Gazioglu and O'Neill (1984).

A number of the original participants to the large diameter pile test project (compression test) agreed to fund the lateral pile test at Tilbrook Grange. Findings at the University of Houston (Dunnavant, 1986) had indicated that there was no significant difference in soil–pile interaction under one-way and two way cyclic lateral loading in stiff clay materials. Based on these findings, and after consultations with the participants, it was decided to test the two Tilbrook piles under one-way cyclic loading using one pile as a reaction for the other.

The pile lateral load test was carried out in October 1988, following the tension axial load test. Lateral load test data was recorded for both the former compression and tension test piles. Greater emphasis was placed on the lateral performance data from the former tension pile and these are the results presented in this paper.

Table 1. Data base synopsis (after Gazioglu and and O'Neill, 1984)

Site	No of Tests	Pile Head Conditions	Soil	Pile Properties
Sabine	2 static 2 cyclic	Free Restrained	Soft Clay	12.75 inch OD Steel Pipe Piles
Austin	1 static 2 cyclic	Free	Medium Stiff Clay	12.75 inch OD Steel Pipe Pile
Houston	2 static 2 cyclic	Free	Stiff Clay	10.75 inch OD Steel Pipe Pile, 48 inch OD Steel Pipe Pile
Manor	3 static 1 cyclic	Free Restrained	Very Stiff Clay	25.25 inch OD Steel Pipe Pile 6.625 inch OD Steel Pipe Piles
Harvey	1 static 1 cyclic	Restrained	Soft clay	6.625 inch OD Steel Pipe Piles
Hamilton (Flooded Site) Hamilton (Dry Site)	4 static	Free	Medium Stiff Clay Stiff Clay	4.50 inch, 8.62 inch, 12.75 inch, 16.00 inch OD Steel Pipe Piles
Manfredonia	1 static 1 cyclic	Free	Very Soft Clay	48 inch OD Steel Pipe Piles
Chalmette	1 static 1 cyclic	Free	Very Soft Clay	Octagonal Pile with Diameter varying from 33 inches to 22 inches
Biwako-Ohashi Bridge	1 static	Free	Soft Clay	59 inch OD Steel Pipe Pile
St Gabriel	1 static	Free	Soft Clay	10 inch OD Steel Pipe Pile filled with concerete

The object of this paper is to provide information on the test equipment, procedure and results of the pile lateral load test at Tilbrook Grange. After an outline of the soil conditions, a description is given of the test set-up and instrumentation. The test procedure and results are then presented. Finally comparisons between the measured and predicted behaviour are made.

Site conditions

The site is located at Tilbrook Grange in Cambridgeshire, England, at National Grid Reference TL009 710. For the test programme, the ground surface datum was set arbitrarily at +100 m. Soil conditions at the site are described in detail by Lambson *et al.* (1993). The succession of strata can be summarised as follows:

Stratum	Datum level, m From	To	Description
Lowestoft Till	100.0	82.9-81.4	Very stiff to hard dark grey silty clay
Oxford Clay	82.9-81.4	59.8-	Hard dark grey fissured clay
Cornbrash	–	–	–

The Lowestoft Till is a very stiff to hard dark grey silty clay with numerous sand to medium gravel sized chalk fragments and with occasional gravel to cobble sized flints. The clay is of intermediate plasticity with a liquid limit in the range 35 to 45 per cent and a plastic limit of about 15 per cent. Natural moisture content is close to the plastic limit. The Oxford Clay stratum comprises a hard dark grey fissured clay with occasional silt partings and shale fragments. It is of high to intermediate plasticity with liquid and plastic limits of about 55 per cent and 20 per cent, respectively. Natural moisture content is about 18 per cent. The undrained shear strength profile developed for pile axial load testing at the Tilbrook Grange site is given in Fig. 1(a). The upper portion of this was modified for the lateral load test following additional site investigation.

Two further boreholes, designated 207 and 208, were drilled and sampled to 8m penetration at locations close to the test pile position. They were advanced using a truck-mounted rotary drilling system. Borehole 207 was drilled using a hollow stem flight auger with sample tubes pushed into the soil. In borehole 208 samples were obtained by using a double core barrel. Unconsolidated undrained (UU) triaxial tests were performed on recovered samples. Fig. 1(b) details the revised undrained shear strength profile for the upper 10 m of soil at Tilbrook Grange. It was derived using UU results from the five soil borings together with cone penetrometer test results for the main site investigation.

Groundwater level is close to the ground surface, with piezometers indicating hydrostatic water pressures present between elevations +98 m and +75 m. Below this pore pressures are lower than hydrostatic due to underdraining of the Oxford Clay by the Cornbrash. To facilitate the lateral load test, a pit

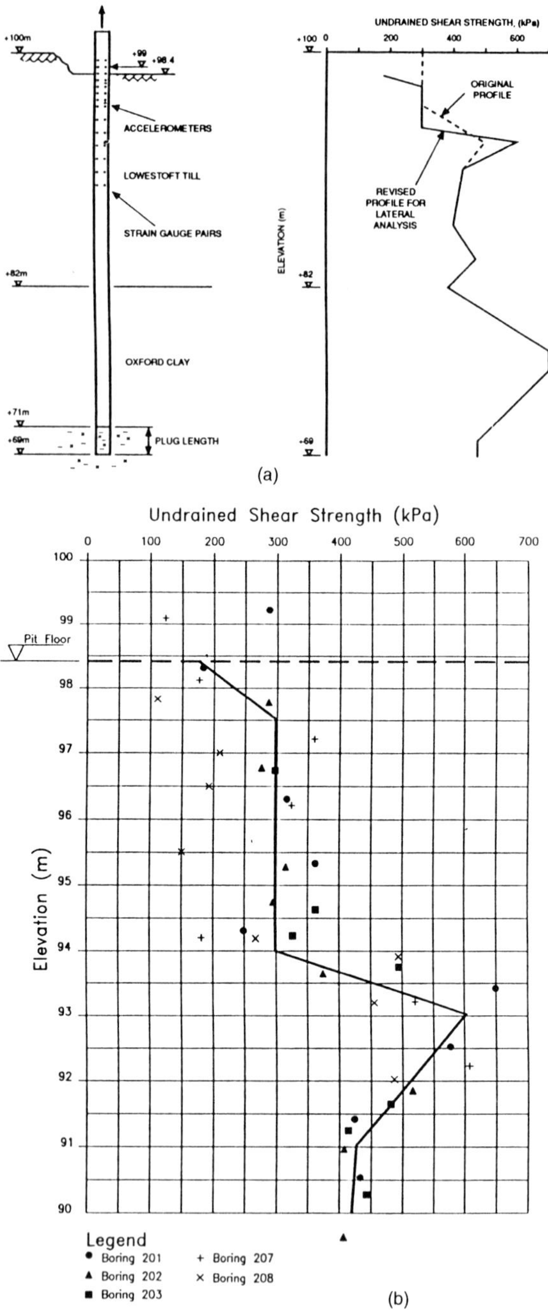

Fig. 1. (a) (top) Undrained shear strength profile and sensor locations and (b) (bottom) revised profile for Tilbrook Grange

was excavated to a depth of 1.6 m around both pile and flooded for a period of one month prior to the pile lateral load test.

Test pile

The lateral load test pile was a 762 mm OD 35 mm wall thickness steel pipe pile. It had been driven open ended to a depth of 30.5 m, i.e. toe at elevation +69.5 m, between 14 January 1987 and 19 January 1987 as part of a trial driving contract. It was driven a further 0.5 m, i.e. to +69.0 m elevation, on 5 October 1987. The plug of soil from within the pile was augered out on 21 August 1991 to a depth of 29 m, i.e. an elevation of +71 m, as shown in Fig. 1(a). Previously the lateral load test pile had been used as a trial pile for driveability studies and as a tension test pile (Clarke *et al.* 1993).

Test set-up

The test set-up is shown in Fig. 2. The former tension/trial test pile and the compression test pile were jacked apart using a central steel strut. The test design load was 3 MN. One jack was used at the trial pile end. The test was displacement controlled.

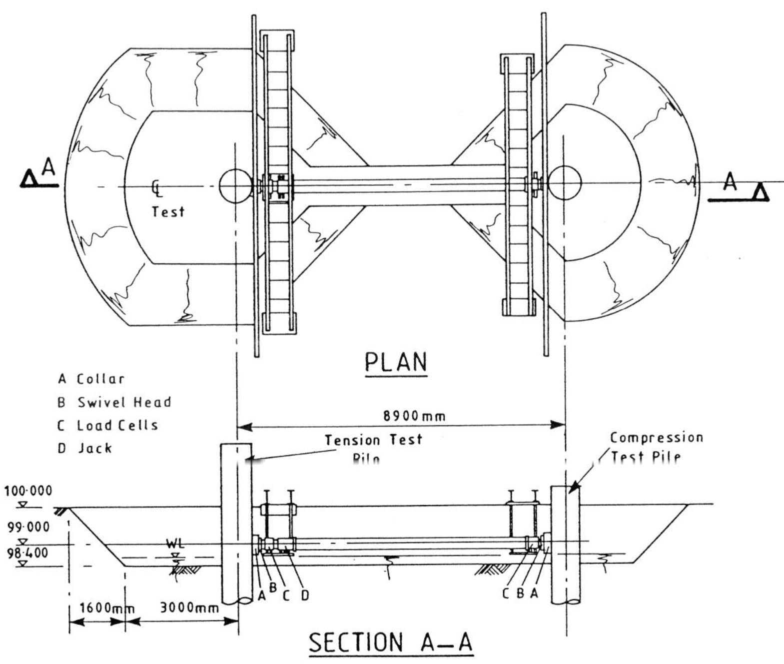

Fig. 2. Test set-up

The electro-hydraulic circuit used to control the test is shown in Fig. 3. A function generator was used to produce the saw tooth cyclic wave form required. This wave form was fed to the jack via a MOOG servo-valve. The movement of the jack was checked using a displacement transducer (LVDT).

Instrumentation and data acquisition

The instrumentation was installed between 24 and 30 August 1988, following augering out of the pile. It would have been advantageous to have installed the instruments before pile driving so that residual stresses due to pile driving could be monitored. However as the pile was installed as part of a pile driving trial this was not carried out.

Load cells adjacent to both piles were used to measure the pile head loads. The loads cells were individually calibrated and deviations from the intended load were negligible.

Ailtech weldable strain gauges were installed down the inner wall of the trial pile to measure vertical strain at the positions shown on Fig. 1(a). Sundstrand QA-700 accelerometers were also installed at four levels to measure slope angles directly.

Several hours of zero readings were taken on site before the start of the test. Zero reading drift was negligible. During the test, instruments were read at 0.5Hz with oversampling at 2Hz. Signals from certain instruments were displayed on chart recorders throughout the test to provide control and to monitor progress. There were no instrument failures during the test. Only two instruments produced questionable data. The high quality of the data recorded is illustrated by the consistency of the results.

Precise levels were taken on the reference beams which confirmed they were stable during the test. The pile head movement was also checked using a theodolite.

Lateral cyclic test procedure

The lateral cyclic test was displacement controlled. Prior to the test the relative advantages of load control and displacement control had been considered. Although it was agreed that load control was a better representation of the random wave loading encountered offshore, displacement control was chosen because:

- The test is easier to control. Bending moment need only be checked on the first cycle to ensure the pile is not overstressed.
- It can be ensured that sufficient displacement is allowed between loading increments to overcome too soft a response in the first cycle.

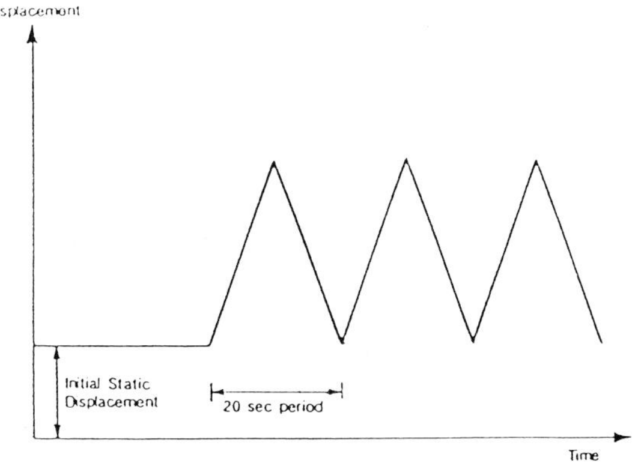

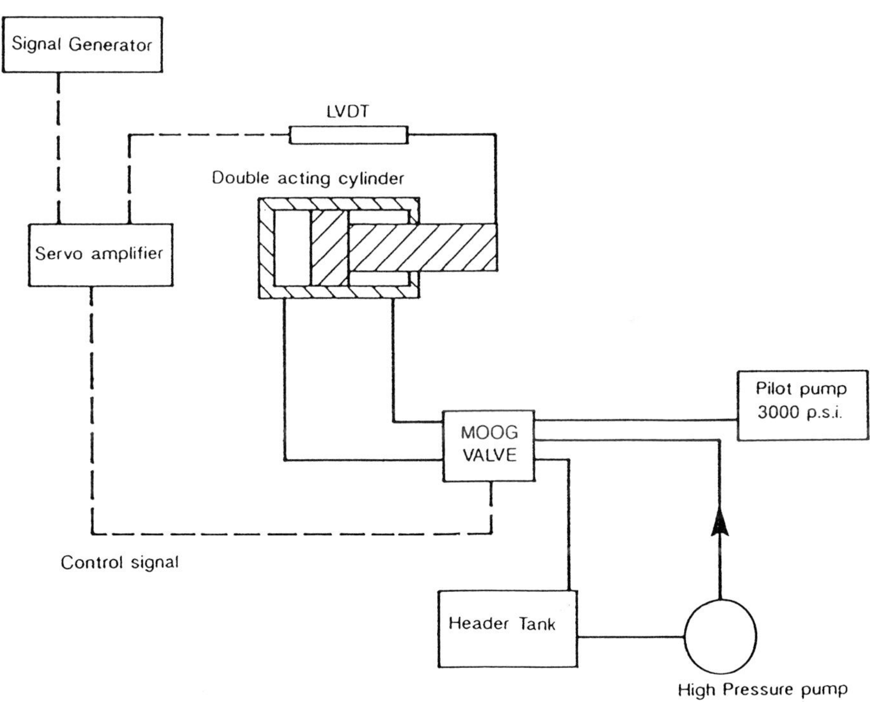

Fig. 3. Electro-hydraulic circuit for lateral test

- Limited test results, at the time, suggested that the cyclic load deflection curve after a number of cycles depends very little upon whether the test was load controlled or displacement controlled.

During the test the minimum and maximum displacements were held as close as possible to the values specified. The minimum values increased slightly over the testing period since the pump response was marginally slower at higher loads. The period of application and removal of load for increments 1 to 8 was 20 s. The period of application and removal of load for increment 9 was 26 s. The test comprised one static and eight cyclic increments in the following sequence.

Increment	Number of cycles	Displacement Min.	Displacement Max.	Maximum force
1	Static	0	4.3	0.29
2	100	2.4	4.1	0.28
3	100	2.9	6.2	0.37
4	100	3.9	9.3	0.47
5	100	4.0	15.0	0.61
6	100	4.0	24.0	0.74
7	120	5.0	37.2	1.05
8	120	7.0	54.0	1.23
9	136	30.8	104.5	1.83

The number of cycles was increased for increments 7, 8 and 9 to allow further stabilisation of the load applied.

Test results

The plot of applied load versus time for increment 5 is shown in Fig. 4. The peak load of 0.61 MN was generated in cycle 1 and the load stabilised at about 0.54 MN after about 90 cycles. The stabilisation of the load after a maximum of 120 to 130 cycles is contrary to the conclusion made by Dunnavant (1986) who found that cyclic degradation in *p-y* characteristics did not stabilise after 200 cycles of loading.

The load versus displacement for three individual cycles in increment 3 and increment 8 are shown in Figs 5 and 6 respectively. Fig. 7 , which shows terminal cycle load versus displacement response, suggests that there is a transition from petal shaped hysteresis loops in the early increments to banana shaped loops in the later increments. This indicates the change from full soil contact to postholing where soil contact is lost at shallower depths and small displacement.

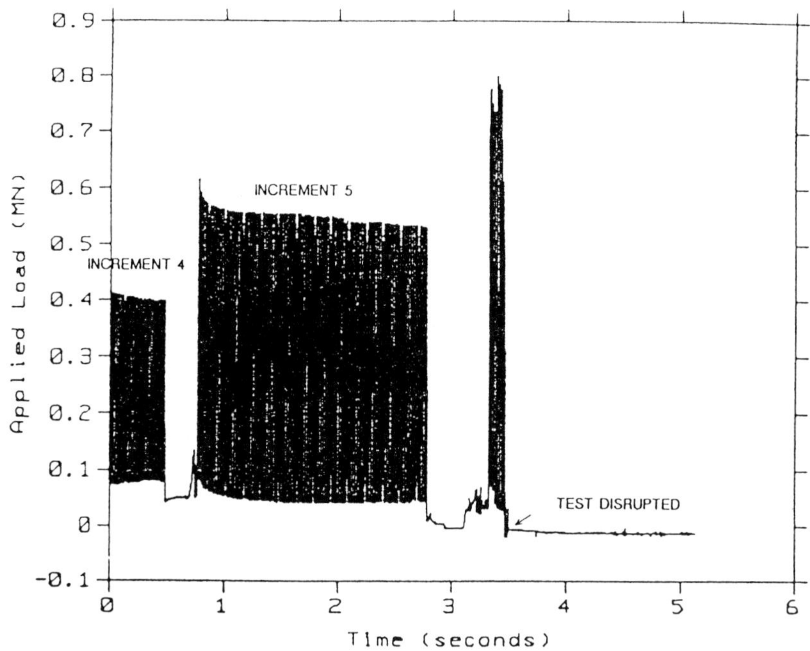

Fig. 4. Applied load versus time

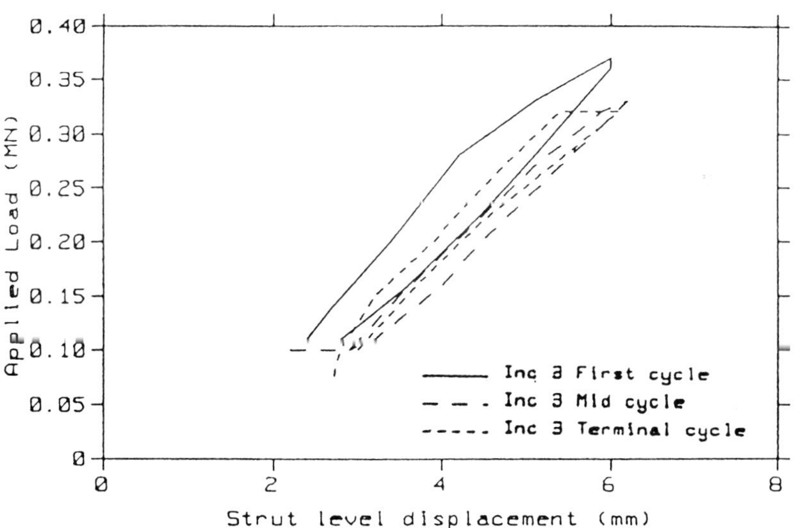

Fig. 5. Individual load–displacement cycles—increment 3

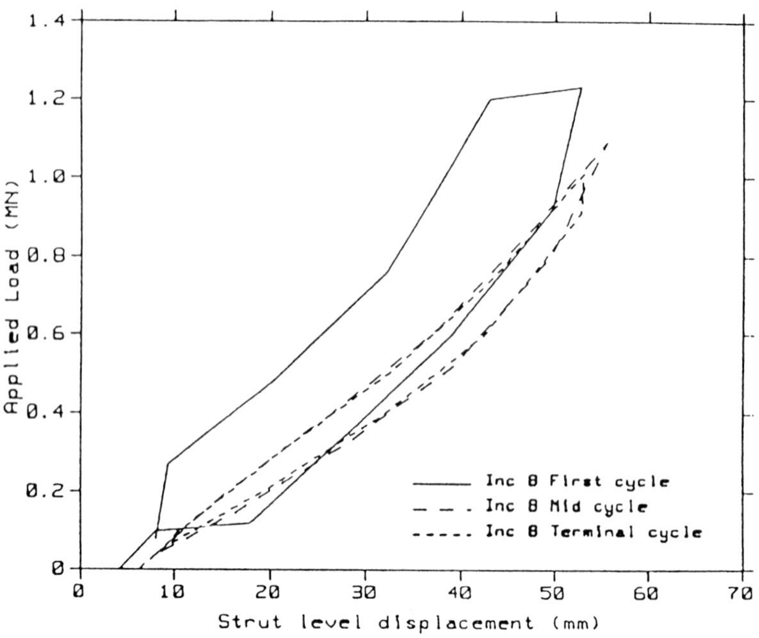

Fig. 6. Individual load–displacement cycles—increment 8

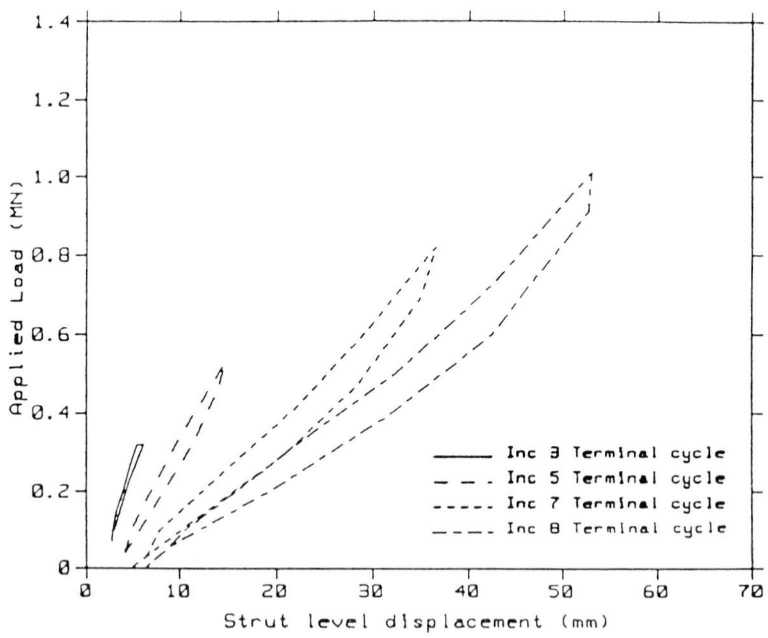

Fig. 7. Terminal cycle load–displacement response—increments 3, 5, 7, and 8

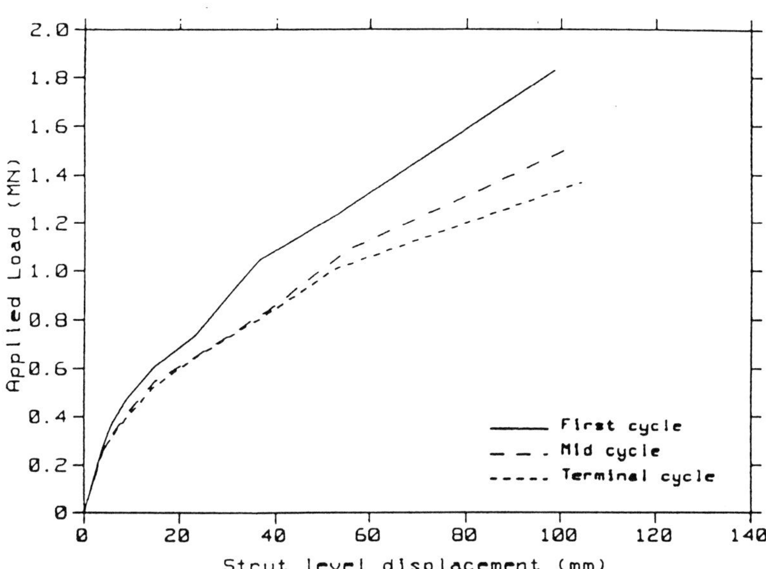

Fig. 8. Applied load versus strut level displacement

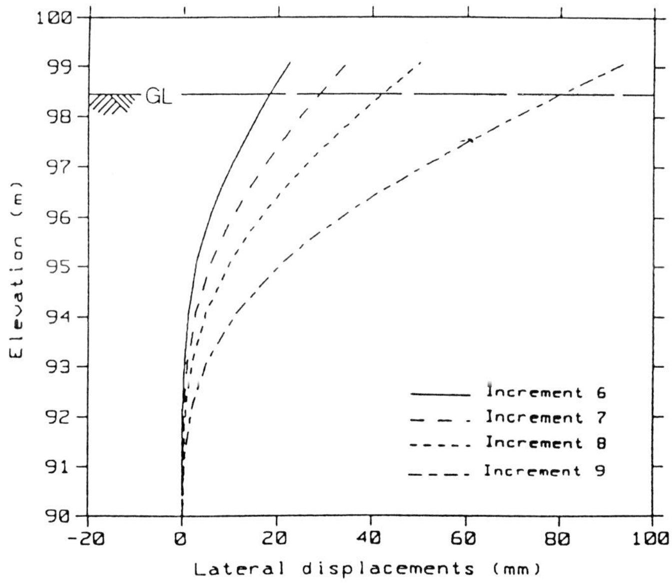

Fig. 9. Pile deflection profile for increments 6–9 (terminal cycles)

The envelope of peak load–displacement points for the first mid and terminal cycle of each load increment is given in Fig. 8.

The pile deflection profiles for the terminal cycle of increments 6 to 9, are given in Fig. 9. The comparison of the movements calculated from strain gauges and measured movements from the LVDTs at the top of the pile show small errors, increasing to about 10 per cent for the final increment.

Bending moment profiles at the first and terminal cycle of load increments 6 to 9 are shown in Figs 10 and 11. These show that the maximum moment reduces and its location moves down the pile with an increasing number of cycles. This phenomenon is due to soil erosion and post-holing effects.

Observations during the test

During the test the area around both piles was submerged to simulate offshore conditions. Water movement occurred into and out of the gaps formed by the moving piles. The velocity of the water was sufficiently high to cause softening and erosion of the soil adjacent to the pile. The formation of the gap was first clearly apparent during increment 3. The clouding due to water movement was more excessive on the tension face of both piles.

During the sixth increment the water was removed from the area around the former compression test pile in order to observe the soil degradation effects. During this increment cracks were developing in the soil on the compression face of the pile and a clear gap was evident on the tension face. The clouding effects on the test piles were similar on both faces. During increments 7, 8 and 9 the water level around the test pile could not be maintained due to the flow of water into the pile through a small hole in the pile at 5 m depth.

The structural breakdown of the clay is consistent with the findings of Reese *et al.* (1989) who reported that North Sea type silty clays are susceptible to gapping.

At the end of increments 8 and 9, a 10 mm diameter rod was lowered into the gap between the soil and the tension face of the test pile. It indicated gap depths of at least 1.6 m and 2.6 m respectively. Following the test, the gap around the trail pile was grouted. While the soil around the pile was being excavated for the bend test the thickness of the grout was measured at 9 vertical locations around the pile. Thicknesses of the grout up to 90 mm were measured. A plan of the grout thickness is shown in Fig. 12.

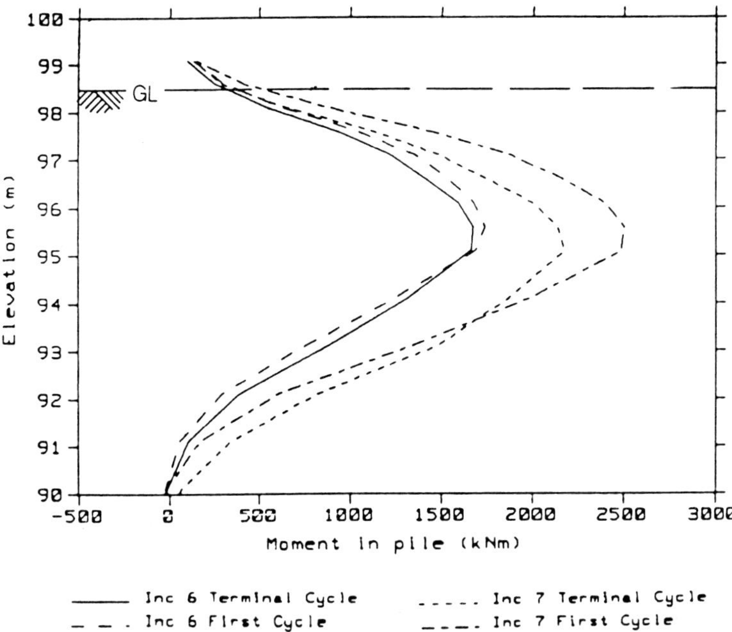

Fig. 10. Bending moment profile for increments 6–7

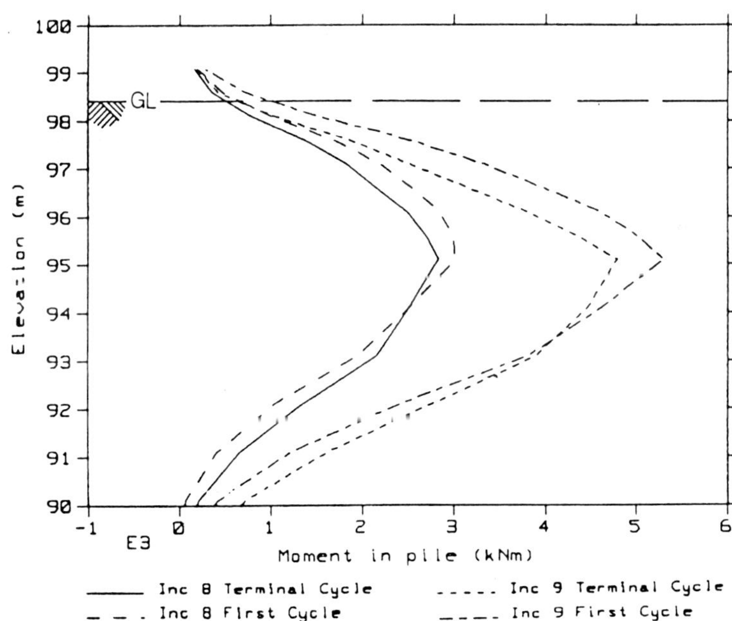

Fig. 11. Bending moment profile for increments 8–9

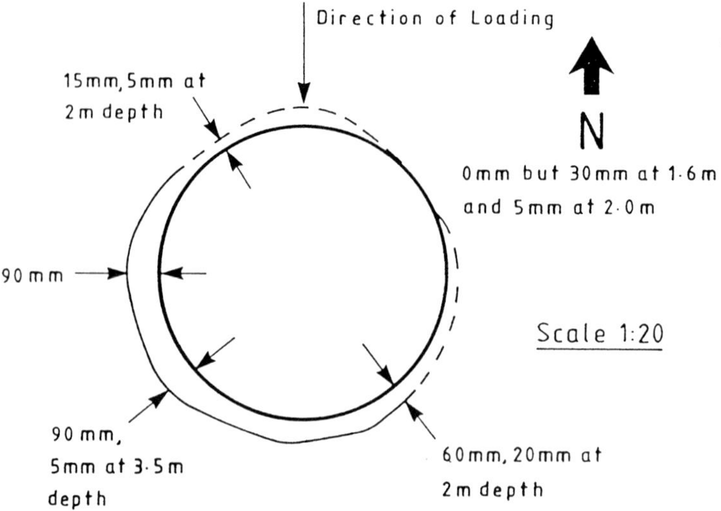

Fig. 12. Grout thickness measurements

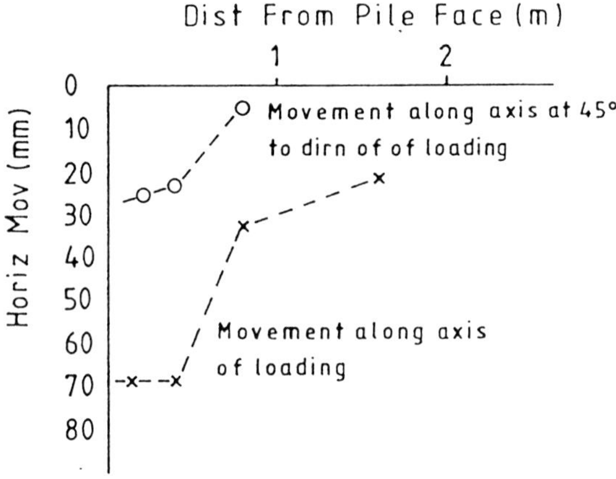

Fig. 13. Ground movements

Ground surface movement was also monitored by surveying 10 steel rods, before and after the lateral test. Maximum horizontal ground movements up to 69 mm were measured close to the pile using these rods, see Fig. 13.

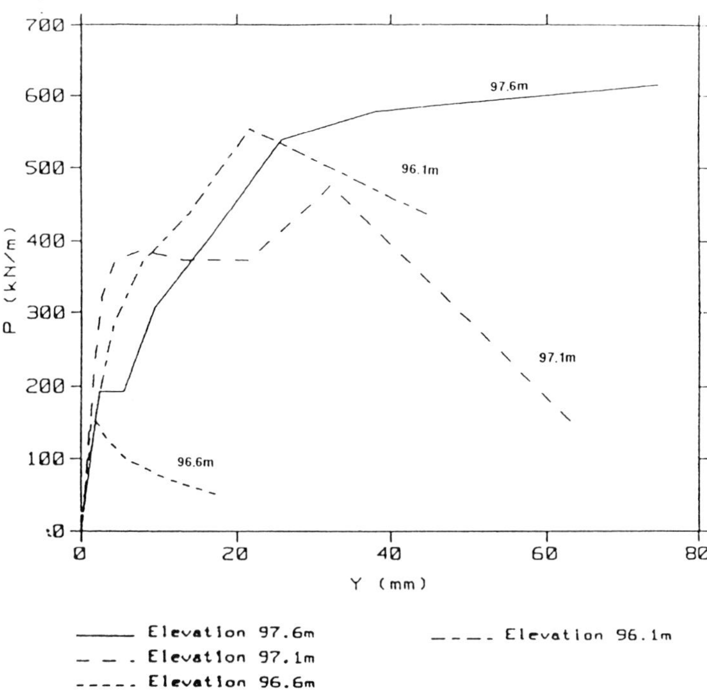

Fig. 14. Terminal cycle p-y curves at +97.6 m, +97.1 m, +96.6 m and +96.1 m

Bend test

The purpose of the bend test was to check the flexural stiffness (EI) of the trial pile alone in order to obtain values for use in the lateral test analyses. The bend test was carried out after the lateral cyclic test.

The soil around the front of the pile was excavated to a level of 94 m (5 m below the axis of the loading strut). The LVDTs on the compression pile and on the trial pile were relocated to measure the lateral movements of the pile at 100.3 m, 99.3 m, 98.3 m, 97.8 m, 97.3 m, 96.8 m, 96.3 m and 95.8 m. The trial pile was then laterally loaded in 9 approximately equal static increments to a load of 0.96 MN. At this point the stroke of the jack was exceeded.

The measurements from the strain gauges and the known applied moments were then used to calculate EI. The values calculated were generally between 1274 MNm^2 and 1120 MNm^2. In a general EI reduced slightly with increasing stress level and also decreased from the values of the first few load increments as the soil adhesion at the back of the pile was eliminate. Excavation on this side of the pile was not achieved on site.

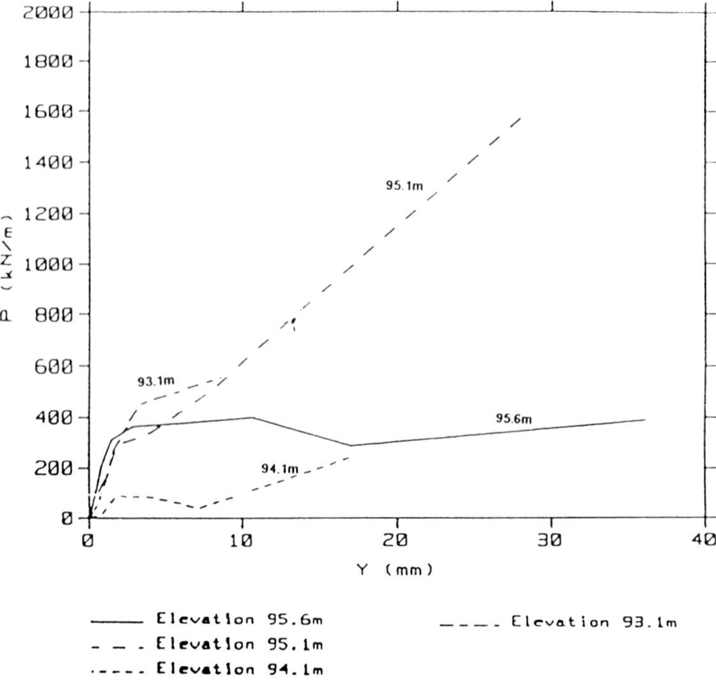

Fig. 15. Terminal cycle p-y curves at +95.6 m, +95.1 m, +94.1 m and +93.1 m

The average of the stiffness values for increments 7, 8 and 9 was 1136 MNm^2. This value was used in the analyses of the lateral test. The theoretical EI value for the pile is 1085 MNm^2 and is within 5 per cent of that measured.

The curvature of the pile was also calculated from the measured displacements and was used to estimate EI. It was found that the movement of the pile at the base of the excavation was larger than expected so that the rotation of the pile above this level was very small and the EI values calculated were very erratic.

Development of *p-y* curves from test data

Values of soil resistance per unit pile length (p) and corresponding local displacements (y) shown in Figs 14 and 15 have been developed from the strain gauge data by a double integration to produce displacement, and by double differentiation to produce the soil pressure. Calculations were carried out using numerical approximation methods. The curves are an envelope of 8 points, one from each load increment.

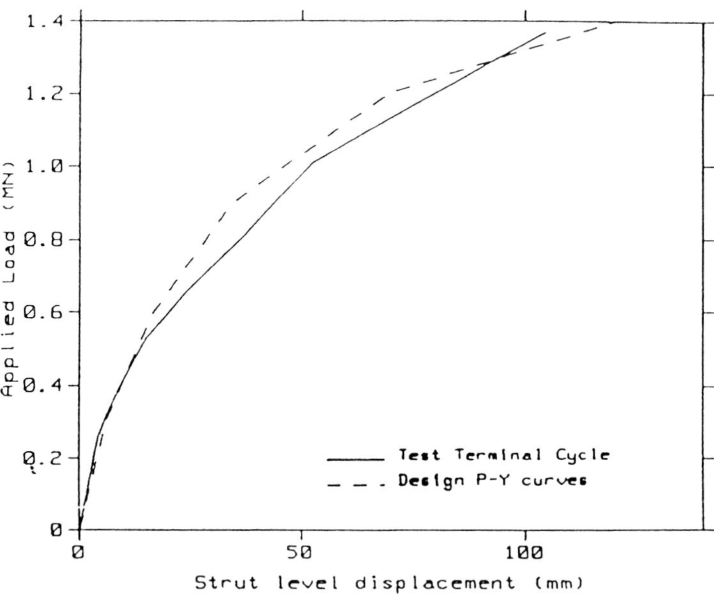

Fig. 16. Load versus deflection from design p-y curves

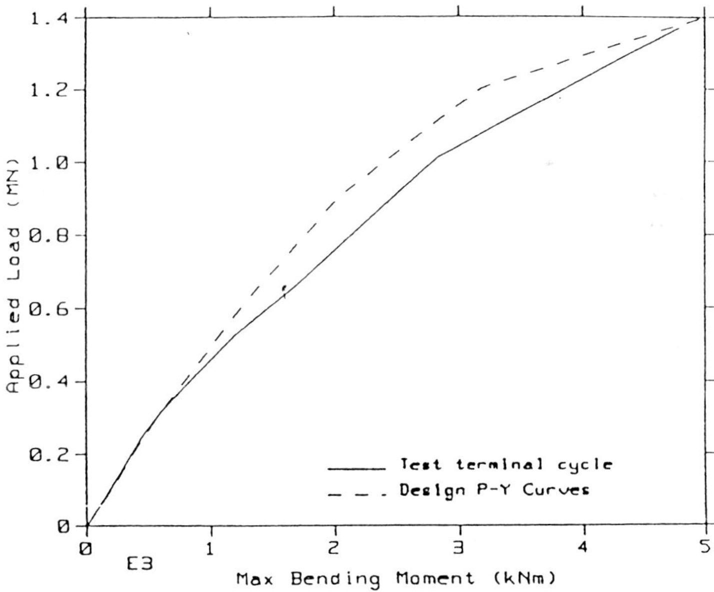

Fig. 17. Load versus bending moment from design p-y curves

The *p-y* curves obtained from the test results have been used to predict the pile load deflection behaviour. A numerical solution using the finite difference technique for analysing beams/columns (Matlock and Reese, 1960) was employed. Figs 16 and 17 show that load versus deflection behaviour and the load versus bending moment behaviour predicted using these derived *p-y* curves agrees well with the measured values. This confirms the adequacy of using the beam/column approach to predict laterally loaded pile behaviour once the *p-y* curves have been determined.

Published methods for determining of *p-y* curves in cohesive soil

Present day practice for analysing laterally loaded piles usually requires the development of a family of uncoupled, non-linear unit load transfer functions, known as *p-y* curves. A full discussion on the published methods to determine *p-y* curves in clay soils is beyond the scope of this paper. However, an outline is given here to set the context. The first *p-y* curves for analysing piles subject to cyclic loading in an offshore environment were developed by Matlock (1970) and Reese, Cox and Koop (1975).

These methods, known as the soft clay method and the stiff clay method, respectively were based on the back analysis of pile behaviour. They are the methods recommended by API RP2A (1989). Additionally the DNV (1977) guidelines for generating *p-y* curves are based on the stiff clay criteria. The soft clay method was subsequently reviewed by its author who introduced a revised procedure for stiff clays above the transition depth known as the modified Matlock method (Matlock 1979).

Sullivan, Reese & Fenske (1980), proposed the unified method, applicable to all clays. In this method a structured parameter selection determines a soft clay like or a stiff clay like method.

Studies by Stevens and Audibert (1979) among others, have suggested that the reference deflection, used to construct *p-y* curves by the soft or stiff clay methods is not linearly dependant on the pile diameter, but is a function of the square root of the pile diameter.

Gazioglu and O'Neill (1984) proposed the integrated clay method which was based on analysis of load tests performed at five test sites. The integrated method took account of the concern introduced by Stevens and Audibert, allowed for the effects of relative pile/soil stiffness and also attempted to incorporate continuum effects using concepts from Randolph (1981). The characteristic shape of the *p-y* curves produced by the integrated method is similar to that produced by the modified Matlock method.

More recently Dunnavant (1986) and Dunnavant & O'Neill (1989) have introduced a revised *p-y* model based on the back analysis of independent full scalepile lateral cyclic load tests. The model which they term the site criterion produces *p-y* curves similar in shape to those developed using the modified Matlock method. The method appears to be particularly appropriate to large diameter non-displacement piles. A comparison between the *p-y* curves derived from the Tilbrook Grange pile lateral load test with those developed from published criteria is given by Hamilton and Dunnavant (1993).

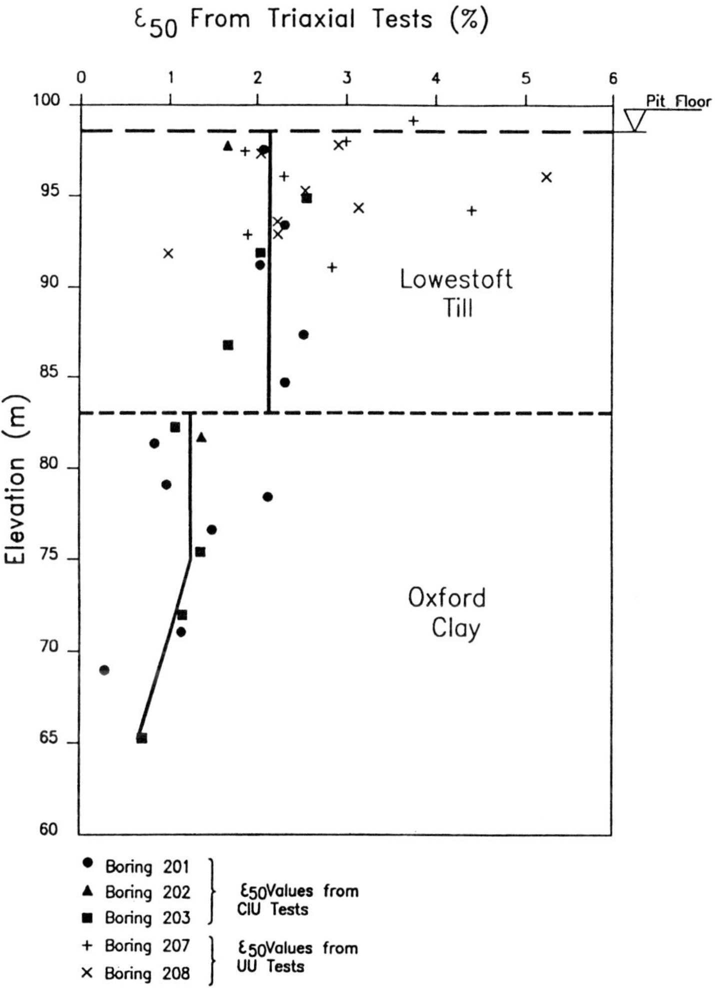

Fig. 18. Tilbrook Grange ε_{50} *values*

Comparison of measured and predicted displacement and bending moments

For the single pile lateral load–deflection analyses the soil reaction was represented as a series of non-linear Winkler type springs, or *p-y* curves, acting at discrete points down the pile length. These were used together with a numerical solution for analysing beam/columns (Matlock and Reese, 1960) to predict pile displacements and bending moments for a given lateral load. Before discussing the results of the analyses, this section first describes the input parameters required to generate *p-y* curves and justifies the form of the curve used.

Selection of ε_{50}

An important input parameter needed to generate *p-y* curves for clays is ε_{50}. This is defined as the value of strain required to mobilise one half of the maximum undrained shear strength of the soil. ε_{50} values are used to define the critical displacement, y_c, and hence the effective stiffness of the *p-y* curve. Fig. 18 shows ε_{50} values measured from laboratory triaxial test specimens for the Tilbrook Grange clays. It can be seen that values of ε_{50} measured during isotropically consolidated undrained (CIU) triaxial compression tests show less scatter than those measured during unconsolidated undrained (UU) triaxial compression tests. Also shown in Fig. 18 is the selected trend line for ε_{50} values, as used in analyses. An ε_{50} value of 2.2 per cent (0.022) was chosen for the Lowestoft Till material. This is the mean value from the CIU test measurements. It should be recognised that this is over five times the nominal value recommended by Sullivan *et al.* (1980) for sites with an equivalent undrained shear strength. For the Oxford Clay values of ε_{50} were judged to reduce from 1.3 per cent to 0.9 per cent at the pile toe elevation.

Selection of *p-y* curve

In accordance with current industry practice, the forms of cyclic *p-y* curve proposed by Matlock (1970) for soft clay and subsequently modified by him for stiff clay (Matlock, 1979), have been used in analyses. The stiff clay method proposed by Reese *et al.* (1975) was not used since it is derived from pile lateral load tests in a very stiff, heavily overconsolidated fissured and jointed clay of high plasticity; conditions which are considered unrepresentative of North Sea sediments. For comparison purposes, ε_{50} values derived from CIU triaxial test results and those recommended by the unified method (Sullivan *et al.* 1980) were used. The latter case may be representative of the situation where no site specific data are available. Additionally the effect of having better definition to peak load transfer was considered. Current industry practice here is for three points to peak ($0.21Y_c$, Y_c and $3Y_c$). The effect of

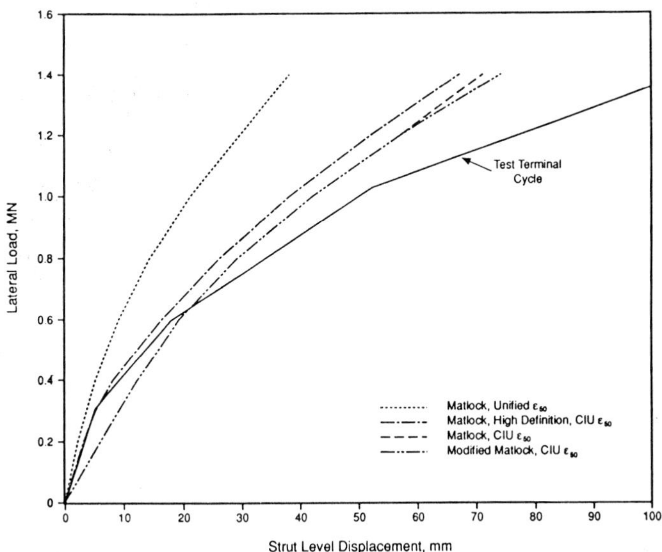

Fig. 19. Comparison of measured and derived load versus displacement characteristics

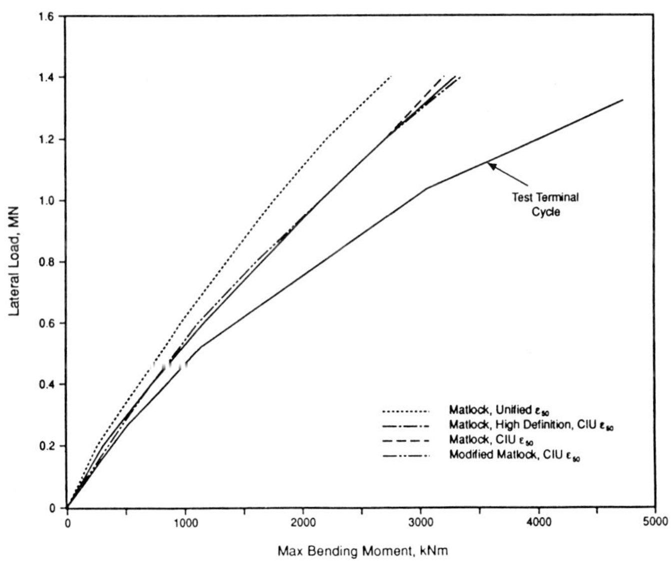

Fig. 20. Comparison of measured and derived load versus bending moment characteristics

including six points to peak and hence, giving higher definition to load transfer at small displacements was examined.

Results of analyses

Results from the analyses are given in Figs 19 and 20, which compare the derived lateral load displacement and lateral load bending moment characteristics, respectively with measured values. Fig. 19 shows that the calculated displacement at the maximum applied lateral load (load increment 9) is grossly underestimated (65 per cent error) if ε_{50} values from the unified method are used. The margin of error is reduced to 35 per cent if ε_{50} values from CIU tests are used. However for the design of offshore pile foundations, lateral displacements are normally limited to about 0.025 pile diameters in order to satisfy serviceability criteria. Within this limit (about 20 mm displacement for the test pile) the Matlock derived *p-y* curves using CIU ε_{50} values model the lateral load displacement response very well. Better definition of the *p-y* curve slightly stiffens the general response, and improves the predicted displacement at small deflections, but makes little overall difference within the serviceability limit. The modified Matlock method for stiff clays gives the softest prediction overall, but its effect is limited to relatively large displacements due to the magnitude of ε_{50} . Within the serviceability limit, there is no difference between the Matlock and modified Matlock formulation displacement prediction.

Figure 20 shows that the predicted maximum bending moments are all less than those measured. There is very little difference between predicted bending moments from analyses using CIU based ε_{50} values. They underpredict the maximum bending moment by about 33 per cent at the maxi mum applied lateral load level, and by about 20 per cent at a serviceability limit load level. Predicted bending moments from analyses using the unified method based ε_{50} values are some 45 per cent below the measured value at the maximum applied lateral load level. The position of the maximum bending moment was fairly well modelled in all instances. Its location moved down the pile with increasing lateral load level. The maximum bending moment occurred at about 2 m and 4 m below the point of load application for load increments 2 and 9, respectively.

Conclusions

(*a*) The pile lateral load test in the overconsolidated very stiff to hard clays at Tilbrook Grange was undertaken to supplement limited database information. The high quality test equipment, procedure and results obtained achieve this aim.

(*b*) The numerical solution for analysing beam/columns may be used to predict laterally load pile behaviour if appropriate, *p-y* curves are used.

(*c*) For the upper stiff clays at Tilbrook Grange, measured values of ε_{50} are typically 0.022. This is significantly greater than previously recommended design values which may be used in the absence of test data.

(*d*) Use of the *p-y* criteria proposed by Matlock (1970), together with measured values, results in overall pile lateral load displacement predictions which are in very good agreement with measurements up to typical serviceability displacement limits (about 0.025 pile diameters). At these limits maximum bending moment is underpredicted by up to 20 per cent. At lateral loads close to the ultimate, the Matlok criteria underpredict both displacement and maximum bending moment by about one-third. However, in all cases, the vertical position of the maximum bending moment is predicted well.

Acknowledgements

The authors wish to thank the sponsors of the Tilbrook Grange Lateral Pile Test for permission to publish this paper. The test was a joint industry funded project. The financial and technical contributions of the following organisations are gratefully acknowledged:

- BP Petroleum Development Ltd.
- Britoil plc
- Exxon Production Research Company
- Statoil
- Fugro McClelland Ltd.
- UK Department of Energy

The assistance of Todd Dunnavant, Jed Hamilton and Trevor Evans in providing significant input to the early stages of the project is gratefully acknowledged.

The authors also wish to acknowledge the contributions to the project by the following organisations:

- Arbil Equipment Ltd.
- Ove Arup & Partners
- Fairhurst Structural Monitoring
- Roadworks Ltd.

References

API RP2A(1989) *Recommended practice for planning, designing and constructing fixed offshore platforms.* American Petroleum Institute, Dallas, Texas, 18th Edition

CLARKE J., LONG, M. M., and HAMILTON J. (1993) The tension test at Tilbrook Grange. This volume.

DET NORSKE VERITAS (1977), *Rules for the design, construction and inspection of offshore structures.* Norway.

DUNNAVANT, T. W. (1986) *Experimental and analytical investigation of the behaviour of single piles in overconsolidated clay subject to cyclic lateral loads.* PhD thesis, University of Houston.

DUNNAVANT, T. W. and O'NEILL, M. W. (1989) Experimental *P-Y* model for submerged stiff clay. *J. Geotech. Engng Div. Am. Soc. Civ. Engrs*, Vol. 115, No. 1, Jan, pp. 95-114

GAZIOGLU, S. M. and O'NEILL, M. W. (1984) Evaluation of *P-Y* relationships in cohesive soils. *Proc. Symp. on Analyses and Design of Pile Foundations* ASCE, pp. 192-213

HAMILTON, J. M. and DUNNAVANT, T. W. (1993) Analysis of behaviour of the Tilbrook Grange lateral test pile. This volume.

LAMBSON, M. D., CLARE, D. G., SENNER, D. W. F. and SEMPLE, R. M. (1993) Investigation and interpretation of Pentre and Tilbrook Grange soil conditions. This volume.

MATLOCK, H. (1970) Correlations for design of laterally loaded piles in soft clays. *Proc. 2nd Annual Offshore Technology Conf.* Paper No. 1204, Vol. 1 pp. 577-588

MATLOCK, H. (1979) Fugro McClelland Limited. Internal memorandum

MATLOCK, H. and REESE, L. C. (1960) Generalised solutions for laterally loaded piles. *J. Soil Mech. & Fndn Engng Div., Am. Soc. Civ. Engrs*, Vol. 86, No. SM5, pp. 63-91.

RANDOLPH, M. F., (1981) The response of flexible piles to lateral loading, *Géotechnique*, Vol. 31, No. 2, pp. 247-259

REESE, L. C., WANG, L. C. and LONG, J. H. (1989) Scour from cyclic lateral loading of piles. *Proc. 21st OTC*, Houston, Vol. 2, pp. 395-402.

REESE, L. C., COX, W. R. and KOOP, F. D. (1975) Field testing and analysis of laterally loaded piles in stiff clay. *Proc 7th Annual OTC.* Vol. 2, pp. 671-690.

STEVENS, J. B. and AUDIBERT, J. M. E. (1979) Re-examination of *P-y* curve formulations. *Proc. 11th Annual OTC.* Paper 3402, Vol. 1, pp. 397-403.

SULLIVAN, W. R., REESE L. C. and FENSKE, C. W. (1980) Unified method for analysis of laterally loaded piles in clay. *Numerical Methods in Offshore Piling*, ICE, London, pp. 135-146.

17. NGI's pile tests at Tilbrook and Pentre — review of testing procedures and results

K. KARLSRUD, S. B. HANSEN, R. DYVIK and B. KALSNES,
Norwegian Geotechnical Institute

The Norwegian Geotechnical Institute (NGI) has carried out a series of axial pile load tests on driven closed-ended steel pipe piles at the Tilbrook Grange and Pentre test sites in the UK. The primary objective was to make comparisons to the larger scale pile tests carried out at the same sites, the results of which are presented in detail in this volume. The tests at Pentre were also part of a larger test programme on piles in soft clay carried out by NGI, and which involved similar but more extensive and also larger scale tests at two test sites in Norway (Ref. 1). The tests at Pentre were carried out in the spring of 1987 and those at Tilbrook Grange two years later. A very significant amount of data were collected. The scope of this paper is limited to giving a factual presentation of test procedures and the most essential test results. Complete data are given in Refs 2 and 3 or reports referred to therein. These reports also present interpretation and evaluation of the test results.

Soil conditions

Very comprehensive soil investigations were carried out at both sites as part of the LDP project. The results are summarized in a paper in this volume (Ref. 4). NGI carried out some supplementary oedometer and direct single sheer test on the Pentre soils (Ref. 3), but none on Tilbrook materials. Due to the thorough treatment in Ref. 4, only a brief description of the soil conditions will be given herein, but some soil parameters are also referred to in subsequent sections.

Tilbrook Grange

At Tilbrook Grange there are two distinct soil strata:

(1) 0–18 m Lowestoft Till, a very stiff silty clay containing frequent fragments of chalk and flint. The water content is typically 16 per cent, 0–3 per cent below the plastic limit, and the plasticity index 20–25 per cent.

(2) 18–30 m Middle Oxford Clay, a very stiff fissured clay with scattered silt partings and shell fragments. The water content is about 18 per cent, 3–6 per cent below the plastic limit, and the plasticity index 30–40 per cent.

Undrained shear strengths from UU-triaxial tests lie typically in the range 400–500 kPa in the Lowestoft Till and 450–700 kPa in the Middle Oxford Clay. Values up to 1000 kPa have, however, been measured around a depth of 24 m. The UU strengths have been found to be most representative of the in-situ shear strength of these deposits, in contrast to CIU and CAU triaxial tests which typically gave 30 to 50 per cent lower values (Ref. 2). The estimated overconsolidation ratio decrease with depth from 35 to 10 through the Lowestoft Till and 10 to 5 through the Middle Oxford Clay.

Pentre

The Pentre deposit is basically a clayey silt, but with a varying clay content (0 to 25 per cent). Within the depth range 15 to 35 m of interest to NGI's pile tests, the natural water content generally tends to decrease with depth from around 30 per cent to 25 per cent, but with local values up to 33 per cent in clayey zones. The plasticity index varies within the range of I_p - 10 to 23 per cent. In relation to the liquid and plastic limits (w_L, w_p) water contents seem to go from a few per cent below w_L at 15 m to a few per cent above w_p at 35 m, suggesting softer and more sensitive material in the top.

The various types of undrained tests on Pentre soils show considerable variation, corresponding to normalized undrained shear strength ratios ranging from $s_u/\sigma'_{vo} = 0.28$ for UU tests to $s_u/\sigma'_{vo} \approx 0.70$ for in-situ vane tests. Interpretation of undrained strengths from CIU and CAU triaxial tests are difficult due to strong dilation at large strains. This may be partly a result of sample disturbance effects, which is a significant problem for such soft, silty materials. The CIU/CAU triaxial compression tests typically gave s_u/σ'_{vo} of 0.31 at moderate strains when the Mohr–Coulomb failure line is reached, climbing to $s_u/\sigma'_{vo} \approx 0.50$ at large strains (20 per cent) as a result of dilation. Two direct simple shear tests by NGI gave on average $s_u/\sigma'_{vo} = 0.35$ at large strain and 0.29 at moderate strain.

Based on these data it is suggested that a representative in-situ undrained shear strength may correspond to $s_u = 0.33\ \sigma'_{vo}$, for a simple shear mode of failure, which is most relevant for pile analyses. Typical triaxial compression and extension values may be respectively 30 per cent higher and lower. The UU strengths will definitely be too low to be representative due to sample disturbance effects.

It is suggested (Ref. 4), that the deposit is overconsolidated with OCR of about 1.6 at 15 m decreasing to 1.4 at 38 m.

The geological origin, mineralogy and understanding of the physical properties of the Pentre silts may well warrant further studies, also in light of the rather unusual pile test results obtained at this site.

Test program

Test piles

Two piles (A5, A6) were installed and load tested at Pentre and four at Tilbrook (A, B, C, D), as illustrated by the schematic layout in Fig. 1. Except pile D at Tilbrook, all were closed-ended steel pipe piles of 219 mm diameter. These test piles had 220 mm long instrument units acting as joining pieces between the uninstrumented pile segments (Fig. 3). The instrument units and pile segments were assembled by, screw joints. The instrument units had clockwise threads at the top and anticlockwise at the bottom thus avoiding twisting the instrument cables during assembly. Above the instrumented portion of the piles, segments of various lengths were added to obtain the desired lengths. The steel pipes making up the uninstrumented parts of the piles at Pentre had a wall thickness of 8 mm. The steel pipes making up the uninstrumented parts of the piles at Tilbrook had a wall thickness of 16 mm except for the following parts: The lower 2.5 m of pile A (8 mm); the upper 5 m of pile B (25 mm); the upper 10 m of pile C (25 mm) and the upper 5 m of pile D (32 mm).

All the test piles were driven through augered casings to avoid interference with the upper soil layers and to make sure that only specific soil layers were in contact with the piles. The lower portions of piles B and C were telescopic to avoid the pile end effects during testing. This approach simplified the data interpretation.

At Tilbrook the 17.5 m × 273 mm diameter pile D acted as casing for pile B, but was also load tested as an uninstrumented open ended pile. Pile D had a driving shoe at the tip (Fig. 1) with inside clearance which prevented any tendency for plugging.

Installation method

At Pentre an ordinary drop hammer pile driving plant was used to install the piles. The instrumented part of each pile was preassembled on the ground before being lowered into the augered casings. The remaining extension pieces were added during driving.

At Tilbrook a hydraulic pile driving hammer was used. This allowed all piles to be preassembled on the ground before installation. A steel tower was erected as a lateral support during the first stages of driving to prevent buckling of the piles.

Loading arrangement

Figure 2 shows the loading arrangement. It consisted of a load frame equipped with a hydraulic power plant and a hydraulic cylinder. The cylinder used at Pentre had a capacity of 100 tons. At Tilbrook the plant was

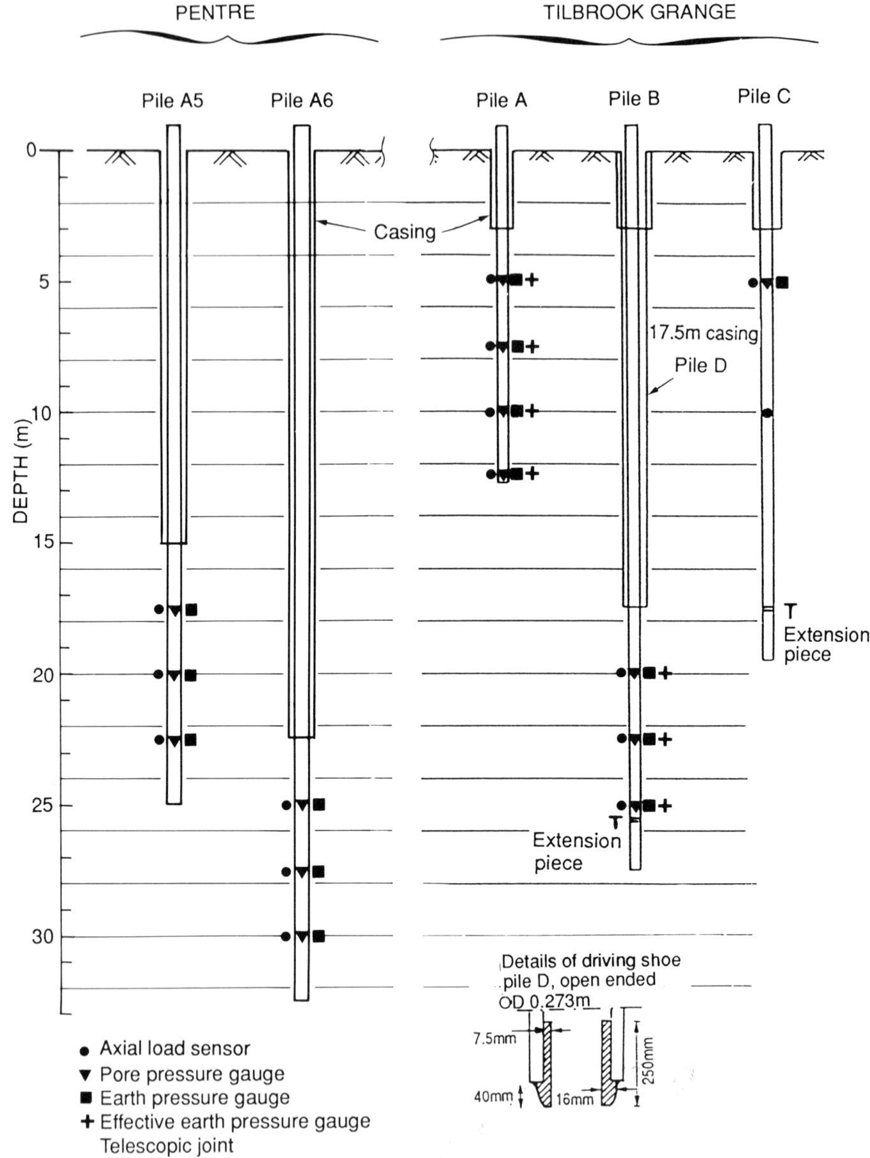

Fig. 1. Schematic layout of test piles

updated to 200 tons capacity by using a larger cylinder and arrangement increasing the hydraulic plant pressure. The load frame ran on rails fixed to concrete foundations. This arrangement allowed the frame to be moved quickly from one test pile to the next. The loading arrangement could apply static or cyclic loads to the pile top. Parameters such as cyclic period, maximum and minimum loads could be set as desired.

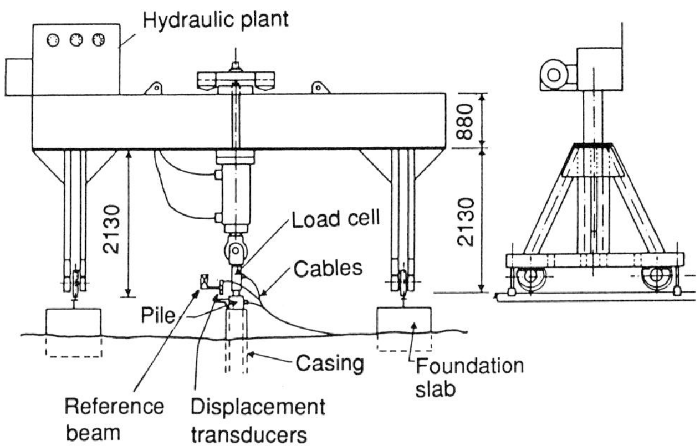

Fig. 2. Loading arrangement

Pile instrumentation

Type of instruments

The parameters that were monitored during the tests were:

- Axial force applied to the pile top
- Displacement (vertical) of the pile top
- Axial forces along the pile
- Pore pressures
- Total earth pressures
- Effective earth pressures (Tilbrook only)

Figure 3 shows sections through the instrument units (the instrument units used at Pentre were modified for use at Tilbrook). At Pentre each instrument unit was equipped with 2 vibrating wire (VW) load sensors, 2 total earth pressure and 2 pore pressure gauges – all of the VW membrane type based on the Geonor M 600 unit. The load cell measuring load at the pile top was also of the VW type. Displacement transducers at the pile top were of the rectilinear potentiometer type. Due to their short range (50 mm) these transducers had to be reset occasionally during cyclic tests. At Tilbrook each instrument unit had 4 resistance type strain gauges in addition to the 2 VW

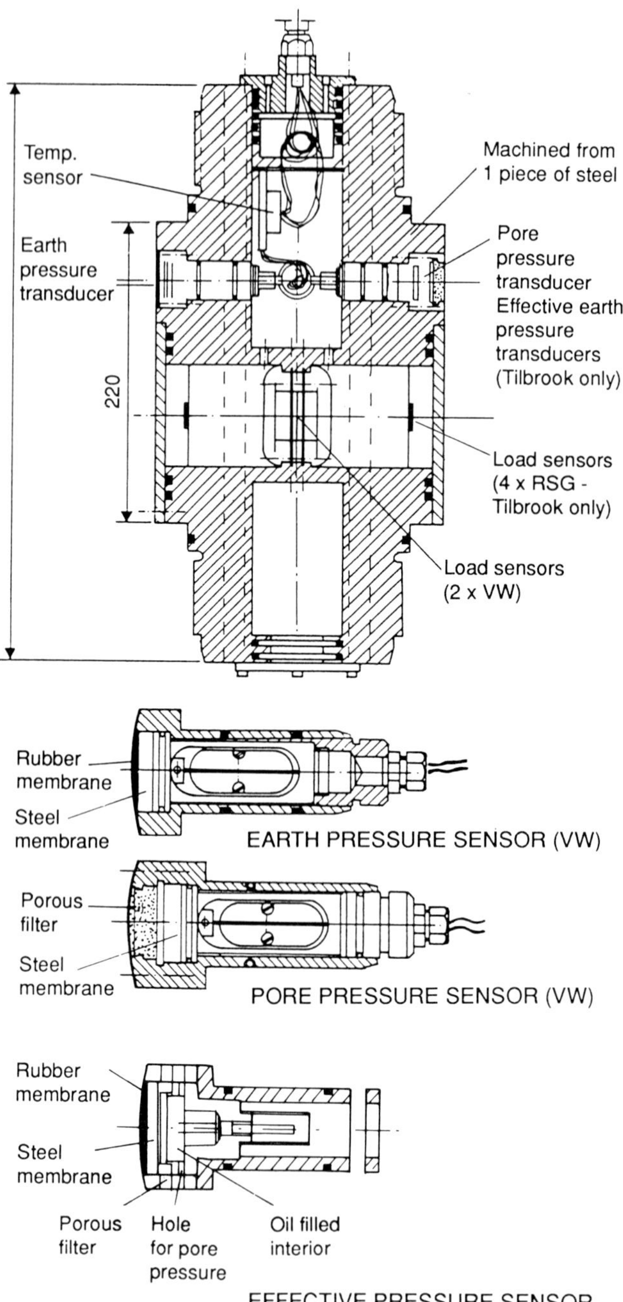

Fig. 3. Instrument units

type to measure axial loads. Furthermore, 1 total earth pressure and pore pressure gauge was replaced by two gauges specially designed to directly measure effective earth pressures (Fig. 3). These were also membrane type but allowed pore pressures to act on the backside of the membrane through filters and holes in the housing. In addition to the one instrument unit, pile C also had a section instrumented with resistance strain gauges to monitor axial load only.

The displacement transducers at the pile top were of the magneto restrictive type (Temposonic) with a range of 600 mm. The load cell measuring load at the pile top was of the VW type.

Calibration of instruments

All instruments and load cells were calibrated in the laboratory. The pore pressure, total and effective earth pressure sensors were calibrated at NGI using standard calibration techniques. The load sensors in the instrument units and the load cells at the pile top had to be calibrated at DnV's (Det norske Veritas) laboratories due to the large loads involved (up to 200 tons).

Prior to the tests at Pentre and Tilbrook extensive repeated acceleration tests were performed on single sensors as well as assembled instrument units, to ensure that they would withstand heavy pile driving.

Before the tests at Pentre the piezometer filters were saturated in the laboratory by applying vacuum to remove air from the filters and then adding evacuated water. The filters were kept saturated by special clamps until they were submerged in the water in the casings during installation. At Tilbrook silicon oil was used to saturate the piezometer and effective pressure sensor filters.Vacuum was used as for Pentre. To keep the filters saturated until submerged a rubber membrane was applied around each instrument and sealed off while submerged in the silicone oil.

Zero point readings for all the sensors were taken before installation and after pile retrieval (when possible). The zero points taken after pile retrieval were the ones preferred for use during data reduction.

Performance of instrument systems

As anticipated some instruments stopped functioning after pile driving was completed. The fallout at Pentre was minimal. Out of a total of 39 sensors only 2 failed (one load sensors and one pore pressure sensor).

At Tilbrook the failure rate was higher due to the higher driving resistance (number of blows was up to 1600 per m and maximum driving energy 35 kNm). Out of a total number of 87 sensors, 13 failed during pile driving. Six of the failed sensors were of the vibrating wire type, seven were resistance strain gauges. Some of the remaining sensors were either disregarded or treated with caution during data reduction. One factor which may have contributed to some of the interpretation problems after the tests may have been the high content of pebbles and small rocks (chalk and flint) in the clay. Point loads caused by such fragments on the relatively small diameter membranes on the total and effective pressure sensors would certainly distort the measured values.

Data acquisition and data reduction

The main part of the data acquisition system was the NGI Fastscanner data logger. This is a microprocessor based logging system mainly intended for vibrating-wire sensors, but analogue signal are catered for as well. Data was stored on a Tandberg TDC3000 digital cassette recorder. All sensors were scanned three times every second.

All data were later run through a computer program, especially written for this application, on NGI's PRIME mainframe computer and time charts produced for all the measured parameters.

Test results, Tilbrook

Observations after installation and reconsolidation

Figure 4 shows pore pressures measured along the pile shafts shortly after the end of pile driving. The installation excess pore pressure, Δu_i, generally increase with depth. The data along pile A seem most consistent, whereas the two data points for pile B show considerable scatter. Increasing pore pressure during the first phase of reconsolidation suggests incomplete filter saturation and too low pressures initially for the lowest gauge on pile B. Disregarding that value, normalized access pore pressures are typically:

$$\Delta u_i / s_u = 0.2 \text{ to } 1.3 \text{ (0.6 average)}$$

$$\Delta u_i \ / \ \sigma'_{vo} = 0.7 \text{ to } 2.9 \text{ (1.5 average)}$$

These values must generally be characterized as low compared to both theoretical predictions (cavity expansion methods typically give = 3.4 to 3.7 for such stiff clays, Refs 2 and 5) and previous measurements in softer clays (measured $\Delta u_i / s_u$ typically lies in the range 8 to 4, decreasing with OCR, Ref. 1).

At onset of pile loading 2 months later, there was still some excess pore pressure. Fig. 5 shows actual degree of pore pressure dissipation versus time for selected gauges. Dissipation was apparently much faster for gauge PP2 along pile A than for PP2 along pile B, especially during the last phase. The observed dissipation rates are in fair agreement with predictions also shown in Fig. 5. These predictions were based on unloading/reloading type coefficient of consolidation, C_r and a log-linear initial excess pore pressure distribution (Refs 1 and 2). Fig. 6 shows effective earth pressures against the pile shafts measured both immediately after installation and at onset of pile loading. Both directly measured values from the two effective earth pressure

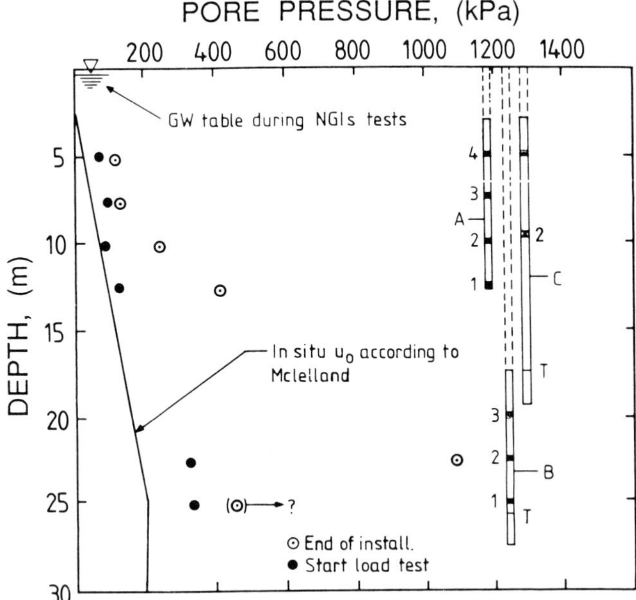

Fig. 4 Pore pressures, Tilbrook

gauges on each instrument unit, and values taken as the difference between the diametrically opposite total earth pressure and pore pressure gauges, are shown. Although there is a fair amount of scatter, the data suggest a general increase in effective earth pressure with depth, and the following is noted:

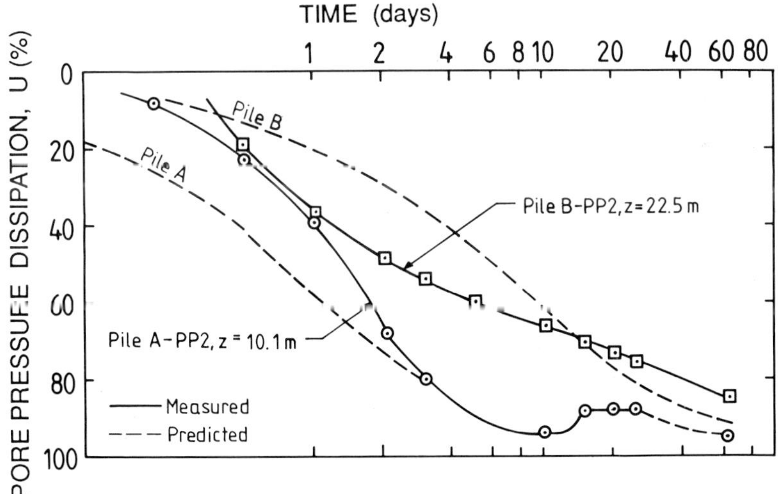

Fig. 5. Pore pressure dissipation at selected levels, Tilbrook

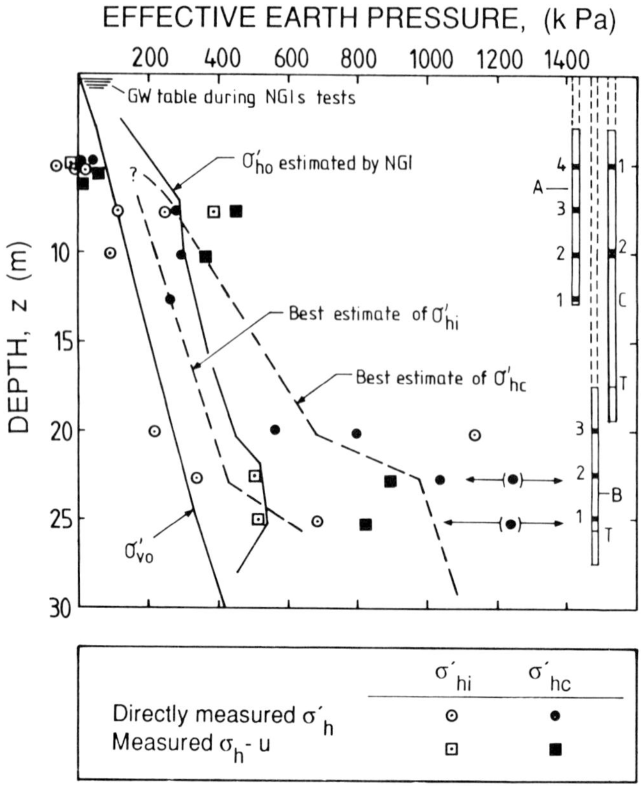

Fig. 6. Effective earth pressures, Tilbrook

- Immediately after installation, the effective earth pressure ratio, $\sigma'_{hi}/\sigma'_{vo}$, was typically in the range 0.8 to 2.0 (average about 1.5).
- At onset of pile loading, the effective earth pressure ratio had increased to typically $\sigma'_{hc}/\sigma'_{vo} = 2.0$ to 3.5 (average 2.7).
- The above neglects the most extreme values, and also disregards all data from the uppermost gauges on piles A and C. These showed very low effective stresses, which may be due to pile whipping effects.

The build up of effective stresses during the reconsolidation phase was generally a direct result of reduction in pore pressures. Total earth pressure changes were small. This observation is in strong contrast to what has been measured by NGI and others in softer clays (Ref. 1), where a significant reduction in total earth pressure has been observed during reconsolidation.

As a final comment regarding accuracy of the data, it may be noted that both earth and pore pressure gauges were designed to measure values up to 1800kPa. Thus, for the pressure measurements in particular, a relatively small error in full scale value is significant when compared to the in-situ pore pressures ranging from 30 to 200 kPa.

Load testing programme

Following pile driving, the piles were left in the ground for 2 months before onset of load testing. Fig. 7 illustrates the typical loading sequence for all piles. There were two almost identical test series, consisting of an initial static, a cyclic and a post-cyclic static test. There was a waiting period of 3 days in between each series.

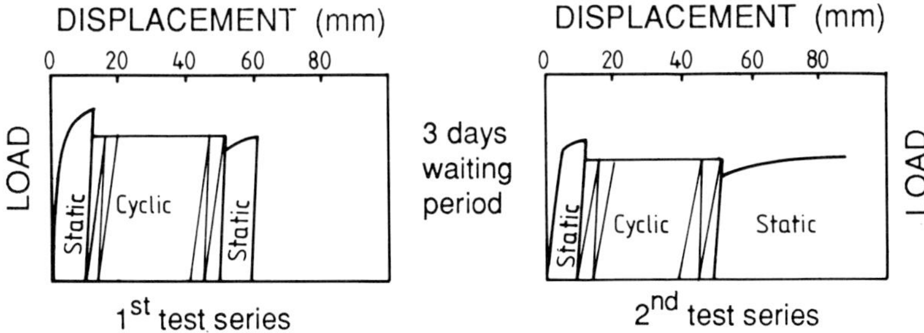

Fig. 7. Loading sequence, Tilbrook

All tests were load-controlled tensile type tests. The static tests were incremental with about 2 min duration on each step and 15–20 steps to peak load. The initial static tests were stopped once a peak load was approached, whereas the post-cyclic static tests were carried to relatively large displacements. In the cyclic tests the load amplitude was generally kept constant, with a peak (max) load being 85 to 90 per cent of the capacity obtained from the initial static tests, and a minimum load equal to zero. The cyclic period was 10 s. From 60 to 200 cycles were applied.

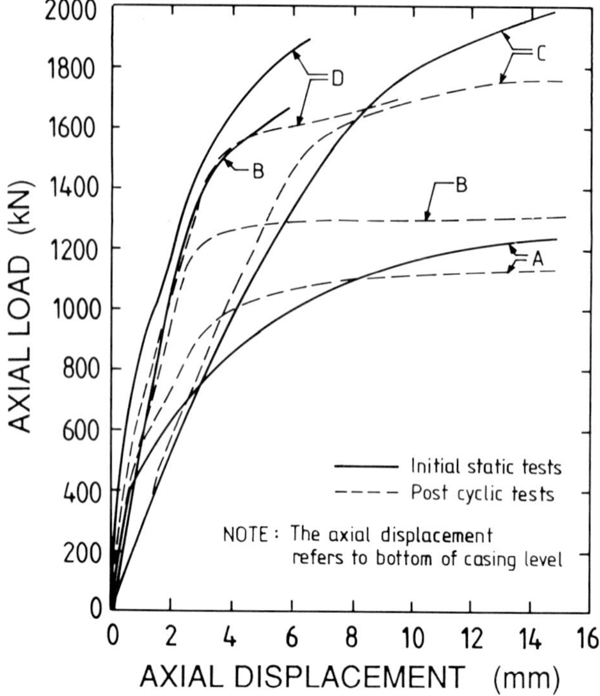

Fig. 8. Load–displacement curves for static tests, first series, Tilbrook

along the casing has been subtracted from displacement measured at pile top). Furthermore, the displacements correspond to the end of each load increment.

The post-cyclic tests, as well as all the static test in second test series showed a stiffer and more elasto-plastic type response than the first initial static tests. For all the tests, the ultimate static pile capacity, Q_{us}, was consistently defined when the rate of displacement exceeded 1 mm/min. Table 1 summarises all the capacities.

Table 1.Summary of static capacities, Tilbrook

	Ultimate capacity, Q_{us} (kN)			
	Pile A	Pile B	Pile C	Pile D
1st Initial Static	1236	1670	1993	1888
1st Post Cyclic	1116	1289	1768	1648
2nd Initial Static	1143	1382	1950	1656
2nd Post Cyclic	-	1209	1808	1352

Figure 9 shows measured distribution of loads down along piles A and B during the lst initial static tests, whereas Fig. 10 summarises deduced ultimate shaft friction values, τ_{us}. The shaft friction follows closely the trend of predictions made with the old and the new API design guidelines (Refs 6 and 7). The locally high shaft friction at 22.5 to 25 m along pile B coincides nicely with locally very high UUstrengths.

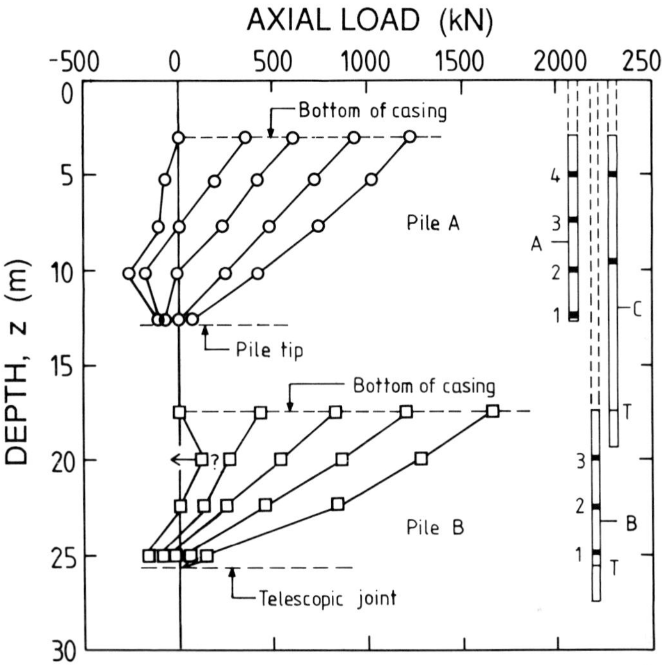

Fig. 9. Load distributions, initial static tests, Tilbrook

Since some excess pore pressures still remained when the pile tests were carried out (Fig. 4) it is estimated that the shaft friction values could have increased by another 5 per cent or so for piles A, C and D in the Lowestoft Till, and by 7 per cent for pile B in the Oxford Clay.

Local shear stress versus displacement (τ-z) curves calculated from the initial static tests are shown in Fig. 11. Note that several curves start with non-zero τ due to remaining residual stresses after driving (Fig. 9).If all curves had started from $\tau = 0$, they would have fallen closer. For a $\tau = 0$ starting point, the curves for pile A typically show a yield point at $\tau/\tau_{us} = 0.65$ and local displacement of $z = 2$ mm (corresponding to 0.5 to 1.0% of the pile diameter).

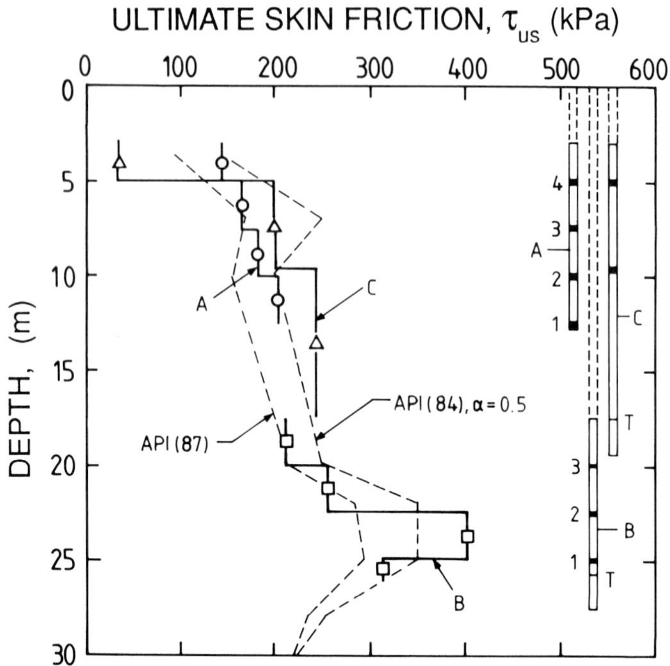

Fig. 10. Ultimate skin friction distributions, initial static tests, Tilbrook.

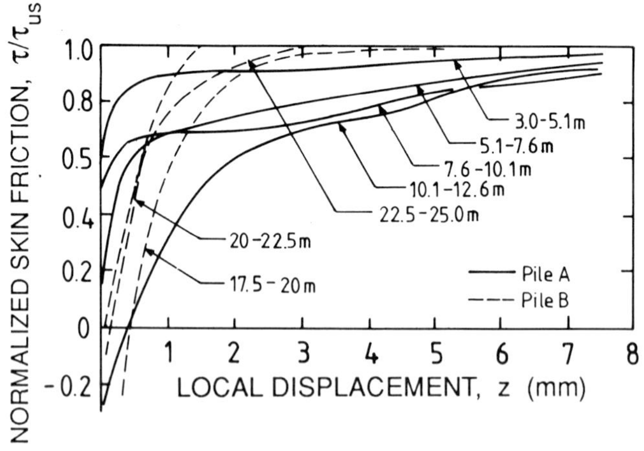

Fig. 11. τ-z curves, initial static tests, Tilbrook

The curves then climb gradually to ultimate at 8–10 mm displacement. For pile B in Oxford Clay, the yield point seems higher, at $\tau/\tau_{us} \approx 0.85$ and $z = 1$ – 2 mm, and ultimate is reached at z = 2–4 mm. The more dilatant strain hardening type τ-z curves for pile A is probably related to the higher overconsolidation ratio and lower plasticity of Lowestoft Till compared to Oxford Clay.

One purpose of pile C was to compare the shaft friction to pile A, over the 3–10 m depth range and thus, reveal if there was any length of pile driven effects. (For pile C 9.7 m of pile was driven beyond 10 m, compared to 2.7 m for pile A). The total load carried by pile C from 3 to 10 m was 7 per cent smaller than for pile A. That was mainly due to very low shaft friction along the upper 2 m of pile C, which could be due to pile whipping effects during driving. In the 5–10 m depth range, the shaft friction along pile C was 15 per cent larger than for pile A.

Another purpose of pile C was to have a direct comparison to the perfectly open-ended non-plugging pile D, which had precisely the same embedded test length. The ultimate capacity of pile D was 5 per cent smaller than for pile C (Table 1). Since the pile diameter was 25 per cent larger, the average shaft friction for the open pile D was 30 per cent lower than for the closed pile C. It is possible that whipping of pile B when it was driven inside the casing pile D, reduced lateral stresses and shaft friction along pile D. Thus, the suggested 30 per cent difference in shaft friction between closed and open pile penetration is probably too high.

Cyclic pile capacities and response

During the cyclic tests, the peak cyclic load amplitude ranged from 85 to 88 per cent of the ultimate static capacities defined by the initial static tests, Table 2.

Table 2.Summary of cyclic capacities, Tilbrook

Pile	Cyclic failure load, $Q_{max\ cy}$	$Q_{max\ cy}/Q_{usi}$	No. of cycles to failure, N_f
A, 1st	1070	0.87	60
B, 1st	1470	0.88	35
B, 2nd	1190	0.86	240
C, 1st	1700	0.85	35
C, 2nd	1665	0.85	45
D, 1st	1670	0.88	20
C, 2nd	1415	0.85	45

Cyclic failure developed by an increasing rate of accumulated displacement, whereas the cyclic displacement amplitude remained essentially constant throughout. This type of failure mode in one-way (unidirectional) cyclic loading agrees with previous cyclic pile tests carried out by NGI (Refs. 1 and 8). To be consistent with the failure criteria used for the static tests, cyclic failure was defined when the rate of accumulated displacement exceeded 1mm/min. The accumulated displacement then ranged from 2 to 6mm, and the number of load cycles to failure was with one exception in the range N_f = 25 to 60. For the cyclic load level in question, this number of load cycles to failure also agrees closely with NGI's previous cyclic pile tests (Refs. 1 and 8).

Effect of previous load history on capacities

By comparing the different capacities in Table 1 and 2 the following observations are made:

- The post-cyclic capacities were in both test series typically 14 per cent smaller than the initial static, and about the same as the cyclic failure load applied. (This is more clearly seen in Table 6.2). Note that the cyclic tests went on to 40 mm of accumulated displacement. The reduction was largest for pile B (about 20 per cent), suggesting that the Oxford Clay is more prone to strain softening and/or cyclic degradation than the Lowestoft Till.
- The capacities in the second test series were on average 10 per cent lower than in first test series. The reductions was again most pronounced for pile B embedded in Oxford Clay. This observation is in strong contrast to results of pile tests in soft clays, where a fairly substantial gain in capacity has been measured from one test series to the next (Refs 1 and 9).

Shaft friction in relation to effective stresses

During the first initial static test there were relatively small changes in measured lateral effective stress against the pile shafts, but some increase was measured at most levels, on average by 20 per cent along pile A and 6 per cent along pile B.

Although there is considerable scatter and variation with depth, the normalized effective skin friction values are typically as follows for piles A and B:

	τ_{us}/σ'_{hc}	$\tan\delta = \tau_{us}/\sigma'_{hf}$	δ: degrees
Pile A	0.62	0.49	31.7
Pile B	0.35	0.33	19.2

The δ-values above are *apparent* friction angles mobilized along the pile shaft, and given by the expression $\tan\delta = \tau_{us.}/\sigma'_{hf}$. For comparison, the peak friction angle from triaxial tests on these clays is typically ϕ'–27°. It is a common misunderstanding to assume that tan δ is a direct reflection of the true interface friction angle. That depends, however, on the complete state of effective stress in soil elements against the shaft (Ref. 2).

Test results, Pentre

Observations during installation and reconsolidation

Earth and pore pressure were recorded during the last 6 m of the pile driving on both piles, and are presented in Fig. 12. It is seen that the effective installation stresses σ'_{hi}, were generally small. At the end of pile installation $\sigma'_{hi} \approx 0$ for pile A5, and for pile A6 it corresponds to σ'_{hi} = (0.11 to 0.30) σ'_{vo}, with the highest value at the lowest level. This agrees with NGI's earlier experience in soft clays (Refs 1 and 9). The pressures were generally larger for the leading instrument units than the trailing ones. This may partly be due to pore pressure dissipation taking place in the soil around the pile shaft during installation, which was very quick at this site. Before the last approx 1m of pile was driven, there was an approx. 5 min break in pile driving. This caused a drop in pressures at most instrument levels, which is also a consolidation phenomenon. Furthermore, driving of pile A5 had to be stopped

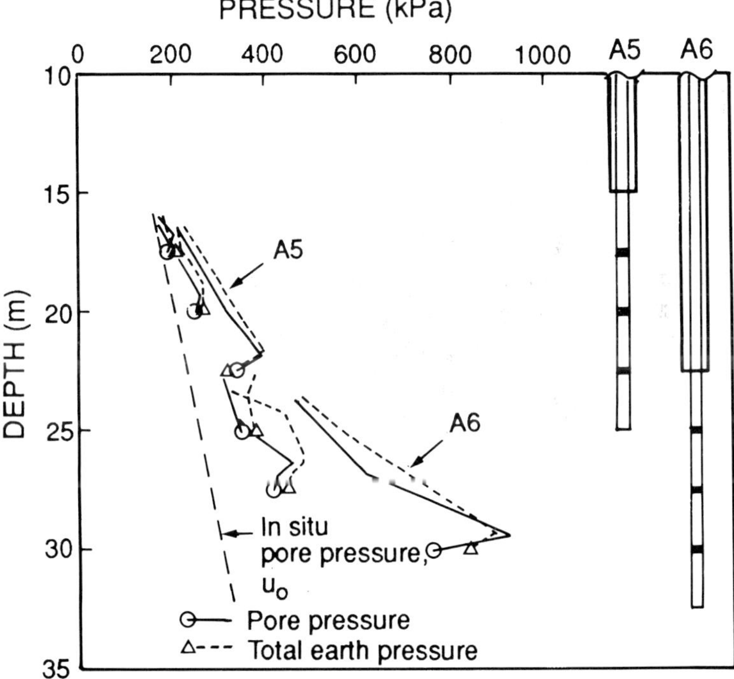

Fig 12. Earth and pore pressures during last 6 m of pile driving, Pentre

when the tip penetration had reached 19.5 m. Thus, pressures above this level are definitely reduced as a result of consolidation.

When considering the excess pore pressures during installation, the data for the leading instrument during driving probably best reflects an undrained response. Table 3 summarizes these installation excess pore pressures. The $\Delta u_i/\sigma'_{vo}$ or $\Delta u_i/s_u$ values must be characterized as low down to 25.0 m depth, but are more typical of soft clays at 27.5 and 30.0 m.

Table 3.Summary of maximum excess pore pressures during driving,Pentre

Pile	Depth (m)	σ'_{vo} (kPa)	S_u* (kPa)	Δu_i (kPa)	$\Delta u_i/\sigma'_{vo}$	$\Delta u_i/s_u$
A5	20.0	175	58	110	0.63	1.91
A5	22.5	198	65	108	0.55	1.67
A6	25.0	220	73	265	1.20	3.63
A6	27.5	243	80	417	1.72	5.21
A6	30.0	265	87	665	2.51	7.60

*Herein assumed $s_u = 0.33\ \sigma'_{vo}$

The excess pore pressures remaining at end of pile installation dissipated quickly. Ninety per cent dissipation was reached after about $^1/_2$ to 6 hours. This rapid dissipation is in agreement with the high permeability and coefficient of consolidation of the Pentre silts compared to more clay type materials. (The permeability is more than 100 times larger than what is typical of clay deposits).

During reconsolidation, the total horizontal stress against the shaft also decreased considerably. Fig. 13 shows the effective stress levels at onset of pile loading when all excess pore pressures had dissipated. The horizontal effective stress level, σ'_{hc}, is much smaller than the estimated in-situ effective stress, σ'_{ho}. For pile A5 the effect stress ratio $k'_c = \sigma'_{hc}/\sigma'_{vo}$was in the range 0.13 to 0.24, and around 0.29 for the upper two instrument levels along pile A6. It is only the lowest instrument level along pile A6 which shows close k'_c to the assumed k'_0 of 0.6. Similar results were obtained by some of NGI's identical pile tests at the Lierstranda test site in Norway (Ref. 1). These very low k'_c values must be a result of relatively large volume changes in the remoulded

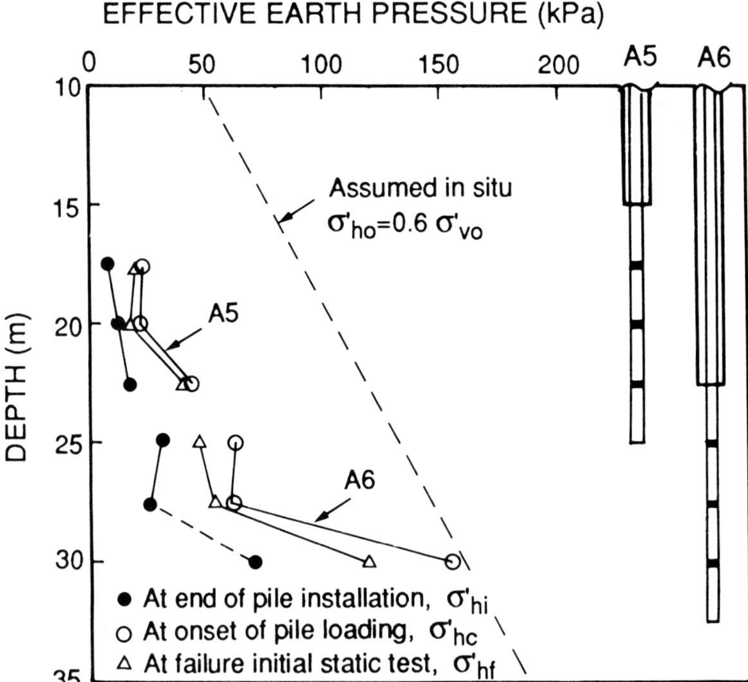

Fig. 13. Effective earth pressure

zone around the pile, and a cylindrical arching type effect. Such observations have also been made or inferred from pile tests in loose sands and calcareous sands.

Load testing programme

The piles were load tested 31 days after pile driving. The loading program consited of two series on each pile with one day waiting period in between. The two test series consisted of an initial static test immediately followed by a pure one-way cyclic test, apart from the second series on pile A5 which was limited to an initial static test. Note that unlike at Tilbrook, there were no post-cyclic static tests. The loading procedure was otherwise much the same as at Tilbrook.

Static pile capacities and response

Fig. 14 shows load–displacement curves for the initial static tests. Note that loads and displacements refer to the uppermost instrument level, and not what was measured at ground surface. The reason for this is that one did not succeed in augering out all the soil inside the casings. Respectively 3.5 m and 8.5 m of liquid material was left in the casing for piles A5 and A6 when the piles were driven. This liquid material must have filled a substantial amount of the annulus between pile and casing. Thus, an unknown amount of load was taken as friction through the casings. The load distribution curves shown in Fig. 15 tend to confirm this assumption.

The ultimate load was reached at about 15 mm displacement in the first tests and 10 mm in the second tests. The selected normalized τ-z curves in Fig. 16 suggest more variation, the ultimate shaft friction being reached at a local displacement ranging from 6 to 15 mm, corresponding to 2.7 to 6.8 per cent of the pile diameter, which must be considered large. Fig. 17 presents ultimate shaft friction values, τ'_{us} deduced from the loads in the instrument units. The shaft friction between the lowest instrument unit and the pile tip was computed assuming that a resistance corresponding to

$$Q_{tip} = 9\ s_u\ \text{Area}$$

was taken as suction at the pile tip. (Assuming no tip resistance would increase the shaft friction at this level by respectively 14 and 18 kPa for A5 and A6).

The deduced shaft friction values show considerable variation with depth, but are on average very low. Compared to an assumed typical in-situ undrained shear strength of $s_u = 0.33\ \sigma'_{vo}$, the average apparent α-values are:

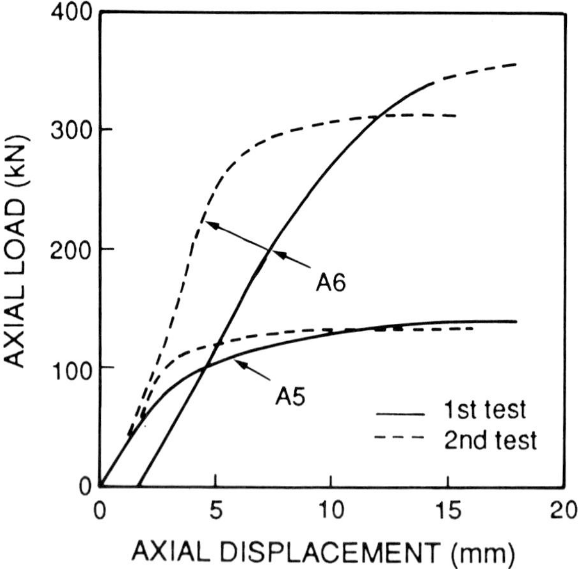

Fig. 14. Load–displacement response at level of uppermost instrument unit, Pentre

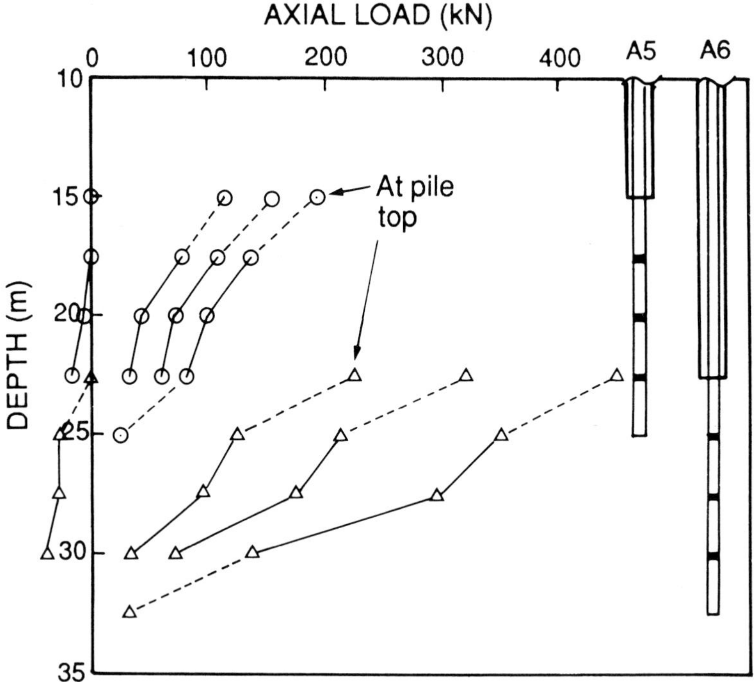

Fig. 15. Load distributions, initial static tests, Pentre

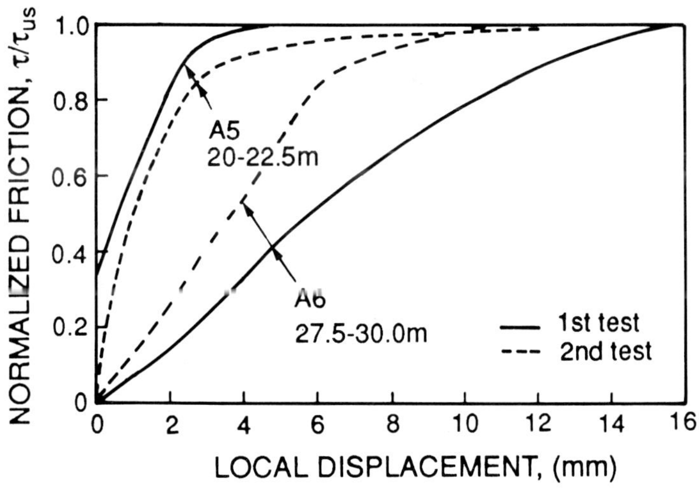

Fig. 16. Selected normalized τ - z curves, Pentre

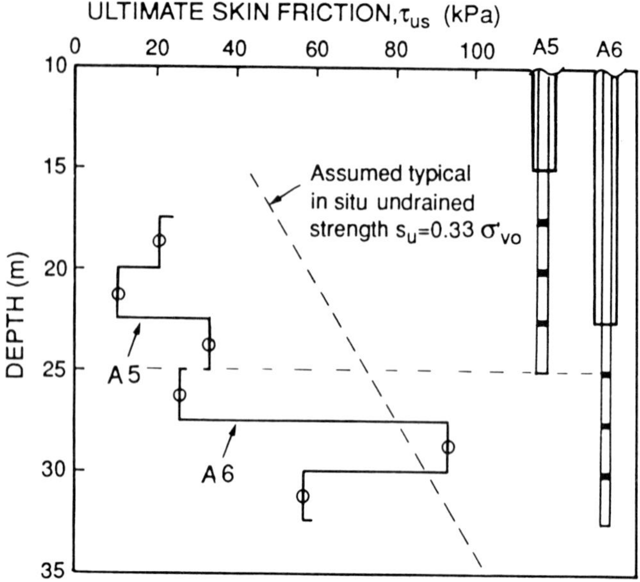

Fig. 17. Ultimate shaft friction, first tests, Pentre

Pile A5: $\alpha = \tau_{us}/s_u = 0.35$ (0.18 to 0.49)

Pile A6: $\alpha = \tau_{us}/s_u = 0.69$ (0.35 to 1.11)

For comparison, the current API design guideline (Ref. 8) would give α= 0.88 if one treated the Pentre soils as clay.

NGI has previously measured similar and even lower shaft friction values at the Lierstranda test site (Ref. 1), where as mentioned earlier one also measured very low effective earth pressures, σ'_{hc}, after reconsolidation. The Lierstranda clay deposit is also silty, but not quite like Pentre.

Cyclic capacity and response

In these one-way cyclic tests failure also developed by a rapidly progressing accumulated displacement, whereas the cyclic displacement amplitude remained constant.

Due to the problem of load sheading through the casings, and because it is more uncertain if one can count upon suction at the pile tip, the cyclic behaviour is best assessed by looking at average shear stresses between the top and bottom instrument unit. Table 4 presents the average cyclic shear stresses imposed at failure along the middle 5 m of the piles in relation to the ultimate shaft friction from the static tests. The applied ratio of $\tau_{max,cy}/\tau_{us}$ lies in the range 0.88 to 0.9 and gave failure after 10 to 60 load cycles. (Failure defined as when rate of displacement exceeded 10 mm/min) .

Effects of previous load history

As shown by Fig. 14 and Table 4, pile A5 showed practically no difference in capacities between the first and second test series. For pile A6 the capacity in the second series was about 12% lower than the first.

Table 4. Average shaft friction at failure, Pentre

Pile (m)	Test	Static	Cyclic			
		τ_{us} (kPa)	$\tau_{min\ cy}$ (kPa)	$\tau_{mas\ cy}$ (kPa)	$\tau_{max\ cy}/\tau_{us}$	N_f
A5 17.5-22.5	1st Static	16				
	Cyclic		5	15	0.94	30
	2nd Static	15				
A6 25.0-30.0	1st Static	60				
	1st Cyclic		7	54	0.90	10
	2nd Static	53				
	2nd Cyclic		5	47	0.88	60

Shaft friction in relation to effective stresses

When failure was reached in the lst static tests it was generally observed a reduction in effective stress against the shaft, as shown in Fig. 13. This reduction was primarily a result of pore pressure increase, whereas the total earth pressures remained essentially constant.

The pore pressure data suggested that some dissipation took place during the time when the load was maintained at each load level. Thus, the response may be regarded as partially drained. It is difficult to assess precisely how much smaller the skin friction could have been if there had been no pore pressure dissipation or perfect undrained respones, but it is not believed to be a very significant factor.

The ultimate shaft friction values were otherwise typically as follows in relation to effective stresses:

	τ_{us}/σ'_{hc}	$\tan\delta = \tau_{us}/\sigma'_{hf}$	δ, degrees
Pile A5	0.64	0.77	37.6
Pile B6	0.59	0.74	36.5

Reference is made to the discussion of the Tilbrook tests on the significance of the interface friction angle, δ, but for comparison the Pentre silts have a peak friction angle of about 33°.

Concluding comments

The scope of this paper was mainly limited to presenting factual data, but further evaluation and interpretation can be found in the references. Still, a few main observations are emphasized:

- The pile tests in the extremely stiff clays at Tilbrook gave results which seem to tie nicely in with general experiences in softer clays.
- The test results in the soft clayey Pentre silts were in many respects unusual. In particular the very low shaft friction compared to the in-situ undrained strength of the deposit, which seems to tie in with very low effective stresses against the pile shafts. It may be argued that the Pentre soils are not typical of offshore marine clay deposits. On the other hand, similar low capacities have been obtained by NGI in a soft silty marine clay deposit at Lierstranda in Norway. Thus, offshore pile design in soft silts on silty clays may be an aspect of concern and warrants further studies.

Acknowledgements

Finally, NGI and the authors would like to thank the many companies that gave financial and technical support to these pile testing programs. For the Tilbrook tests this included: Amoco, BP and the Department of Energy, UK. The two latter in addition to the following supported the Pentre tests: Agip, Conoco, Elf, Esso, Mobil, Norsk Hydro, Saga, Shell and Statoil.

References

1. NORWEGIAN GEOTECHNICAL INSTITUTE. *Design of offshore piles in clay-field tests and computational modelling*. Final report, summary and recommendations. Report 52523-28, April 1989.
2. NORWEGIAN GEOTECHNICAL INSTITUTE. *Pile load tests in stiff clay – Summary and evaluation of NGI's pile tests at Tilbrook Grange*. Report 885032-2, Oct. 1989.
3. NORWEGIAN GEOTECHNICAL INSTITUTE. *Design of offshore piles in clay – field tests and computational modelling. Summary, interpretation and analyses of the pile load tests at the Pentre test site*. Report 52523-27, April 1988.
4. LAMBSON, M. D., CLARE, D. G., and SEMPLE R. M. *Investigation and interpretation of Pentre and Tilbrook Grange soil conditions*. This volume.
5. RANDOLPH, M. F., and WROTH, C. P. An analytical solution for the consolidation around a driven pile. *Int. J. Numerical and Analytical Methods in Geomechanics*, 1979, Vol. 3, No. 3, pp. 217–229.
6. AMERICAN PETROLEUM INSTITUTE. *Planning, designing and constructing fixed offshore platforms*. API recommended practice, RP2A, 13th edition, 1982.

7. AMERICAN PETROLEUM INSTITUTE. Planning, designing and constructing fixed offshore platforms. API recommended practice, RP2A, 17th edition, 1987.
8. KARLSRUD, K., NADIM, F., and HAUGEN, T. Piles in clay under cyclic loading: Field tests and computational modelling. *Proc. 3rd Int. Conf. on Num. Meth. in Offshore Piling*, pp. 165–190. Nantes, France, May 21–22, 1986.
9. KARLSRUD, K., and HAUGEN, T. Axial static capacity of steel model piles in overconsolidated clay. *Proc. Int. Conf. on Soil Mech. and Fndn Engng*, 11, San Francisco, 1985.

Discussion

J. CLARKE, BP Engineering

In answer to the question "Were the reference beams for the vertical displacement independently referenced?" asked by a delegate, the answer is yes.

Dr U. MIRZA, Kvaerner, Earl & Wright

A most welcome contribution to our understanding of laterally loaded piles. My first question relates to the fact that the pile head was free. How does that compare with the fixed head pile analysis that we normally assume? What I am saying is: would you expect to have had any differences had you tested a fixed head pile? Are the *p-y* of free head piles applicable to fixed head piles? Although I bear in mind what we learned about Magnus: that the piles were not completely fixed. Secondly, I wonder why should ε50s obtained from consolidated undrained tests give the best correlation. As a pile designer, should I go for CIUs or CAUs or K_0CUs or UUs, or what should I do? Thirdly, while we are usually interested in the storm condition, do you have any idea what would have happened at large deflections? I think you have taken the test, correctly so, something like 105/110 mm, while sometimes we are interested in large pile movements especially, for example, if you have a structure to dock against piles, in which case you would really be looking to estimate deflections at very large loads. Is it possible to give an indication on which method you would be recommending to do this sort of thing?

M. M. LONG, Ove Arup & Partners

The pile was free headed and all of our work was based on a free headed pile. I am sure the results may be different for a fixed head, but I do not know by how much.

Before I answer the outstanding two questions, can I just go back to your first point again on the free headed and fixed headed situation. I would refer you to the database which is in the paper. We wanted to extend that database and most of the pile tests were of that type.
[U. Mirza agreed.]

There are a few tests on restrained headed piles, but they are very small piles, approximately 6" diameter. It is very difficult and expensive to try and create a fixed head pile test and that is why we did not conduct it. On looking at the ε50 value data closely, there did not appear to be any relationships between

$\varepsilon 50$ and say shear strength. Perhaps one of the reasons for a better correlation between the CU tests and $\varepsilon 50$ was that the action of the CU test takes out the bedding errors in the triaxial cell and thus you get a better fit.

U. MIRZA, Kvaerner, Earl & Wright

I think the message you are getting through is that $\varepsilon 50$s from Sullivan, Reese and Fenske in the 1979 ICE, i.e. the unified method, did not fit your work.

M. M. LONG, Ove Arup & Partners

Yes, it should be treated with caution.

Dr J. D. MURFF, Exxon Production Research Company, USA

My recollection is that in Reese's tests in stiff clay in 1966, or whenever they were, he actually did use or attempt to use a fixed headed or, at least, a partially restrained head pile. I know he did that on some of his tests and he did not observe any significantly different *p-y* behaviour. With regard to large displacements, we have carried out some tests in the centrifuge on piles undergoing large lateral displacements, although they were all on softer clays. We are quite interested in the problem of total platform collapse resistance and in may cases consider that in our designs or re-analyses. Matlock's cyclic *p-y* criteria severely degrade with large displacements, with even a few inches of displacement. We were of the opinion that this criterion might not be appropriate for collapse resistance assessments. Our tests were on remoulded, consolidated material, so some of the sensitivity of natural soils might not have been there. We saw fairly small degradations in *p-y* curves' ultimate resistances, of the order of 80% or so of the static values, at continued lateral displacements of over a diameter, with a fairly rigid pile so that the whole pile was moving back and forth.

Dr W. R. COX, Fugro-McClelland Geoscience's Inc., USA

With reference to Mr Long's paper, I wanted to draw your attention again to what they did after the lateral load test. They excavated the soil around the pile, set up a reference column and put LVDTs, a displacement measuring system, and determined an actual EI. We spent a lot of money instrumenting these piles, installing them, doing the site investigations, and we too often walk away from a wonderful laboratory that is waiting for additional examination. I know that it is not that easy in the UK. I became quite educated as to the problems, of obtaining planning councils' permission to install piles, working with landowners, etc. In a place like Texas, you can get a site in which you can do a pile test relatively cheaply and it is much easier to leave the pile there. The point is still strongly to be made that, if possible, we should allow money in these programmes, such that we can go back and examine why certain things happen. For large diameter piles, for example, it is possible to

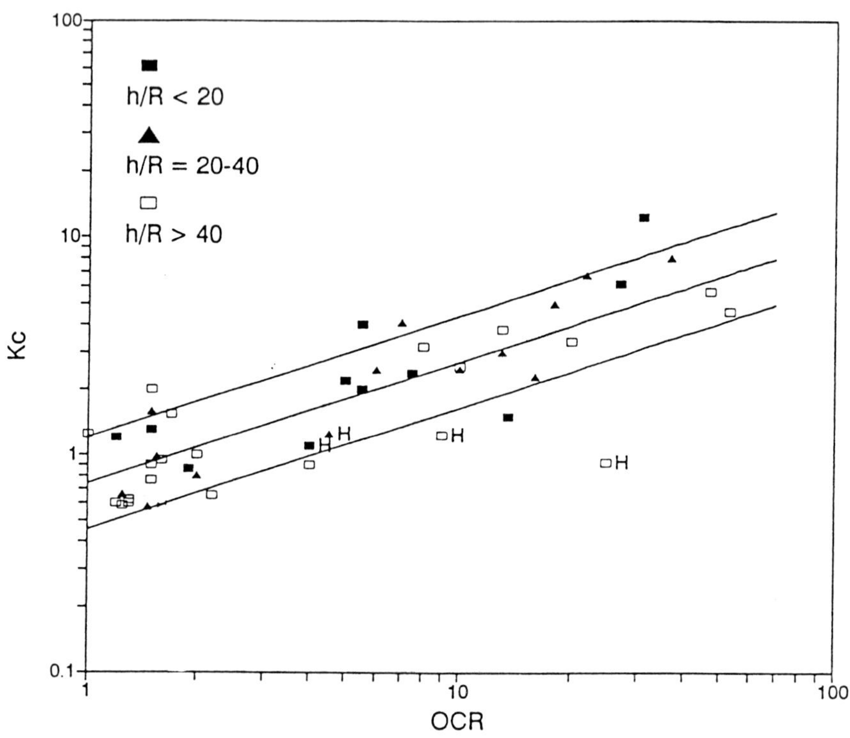

Fig. 1. Imperial College K_c–OCR database (after Bond et al., 1992)

go down inside the pile and remove instruments and have a look at them. Another important point is the importance of the hydraulic erosion in stiff clays. In the test that we did in Manor, Texas in 1967 it was a highly erodible material — a stiff clay with a great deal of secondary structure with fissures

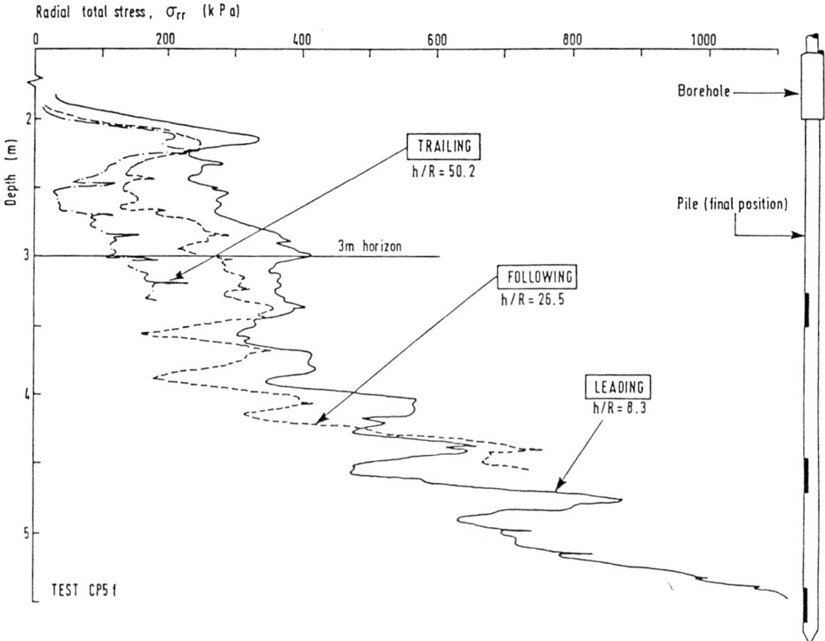

Fig. 2. Variations in radial total stress with depth at three instrument positions for pile installation in London Clay (after Bond and Jardine, 1991)

and gypsum enclosures — and as you saw at Tilbrook, and I could not determine completely, our water at the Manor site was much clearer and you could see particles of soil the size of the end of your thumb that were being washed out as the pile moved back and forth. I was particularly pleased to see that you went to the trouble to show us that, because that does project to me what really is going on. Yet erodibility is not a geotechnical property that we determine during a conventional site investigation. it is not quantified in any way, we do not say this is a very easily erodible stiff clay. Therefore, when considering lateral loading, one must take that into account. We look at strength and all the other properties but not the erodibility.

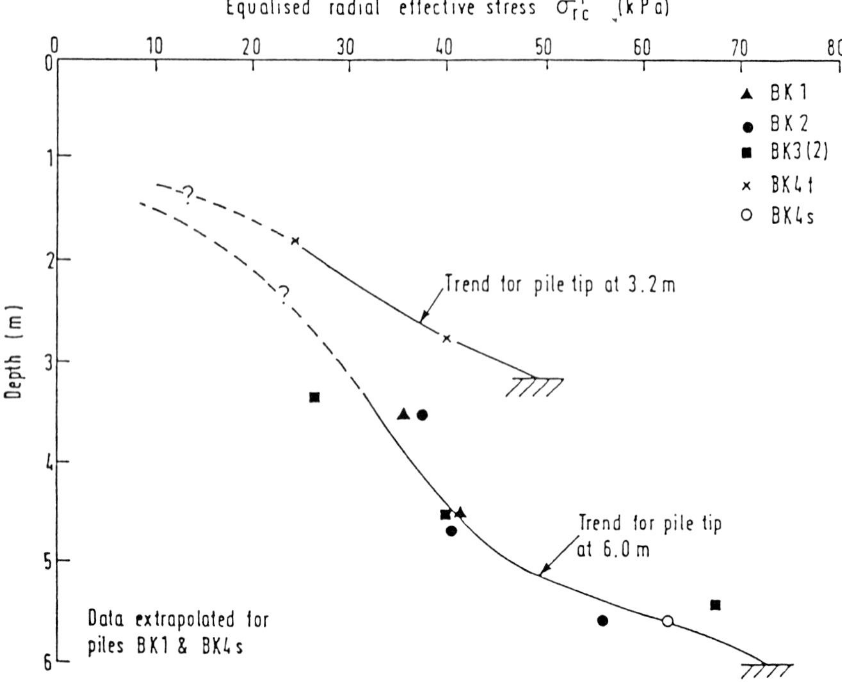

Fig. 3. Variations in σ'_{rc} with depth for piles installed to two different depths at Bothkennar (after Lehane and Jardine, 1992 , Wroth Memorial Symposium paper, in Predictive Soil Mechanics, Thomas Telford, London)

Dr R. HOBBS, Lloyds Register

Mr Karlsrud mentions in his paper that for the sleeved piles at Pentre, some soil was left in the annulus, so that some friction may have been developed. Could he tell us what effect this might have had on the interpretation, and why perhaps there were no gauges on the pile just below the sleeve?

K. KARLSRUD, Norwegian Geotechnical Institute

We have, of course, measured the load at the pile top. From the bottom of the sleeve to the top load sensors we have not made any interpretation of the shaft friction because of the uncertainty of how much load transfer there is, if any, through the sleeve. So all the results I showed are relevant for where we have the load sensors, further down along the pile, and even to the pile tip, you can make an estimate of the tip suction resistance. The reason for not including it was because we are trying to run a very tight budget on these pile tests. We would have liked to have had as many as possible, but cost is always something you have to consider.

Dr R. JARDINE, Imperial College of Science, Technology & Medicine

Mr Karlsrud's database for K_c and OCR makes a very interesting contribution to the discussion. Fig. 1 presents an exactly comparable but independent diagram which reflects the perspective gathered through the Imperial College tests with instrumented piles, combined with the limited set of experimental data published by others; one example is Mr Karlsrud's data from Haga, which are indicated 'H'.

However, the two diagrams tell different stories. In general, we anticipate rather higher stresses than Mr Karlsrud. This divergence reflects a number of factors. One is a bias in the soil types considered. The NGI plot contains far more sensitive low plasticity clay-silts than does ours. We have discussed the possible effects on capacity of low clay contents in session 2. Sensitivity is a factor that the MIT group (Baligh, Whittle *et al.*) identified some time ago. The tendency for sensitive clays to develop lower K_c values than insensitive ones is also clear in our database.

Another important missing parameter is the effect of pile length — or more probably relative pile tip position (h/R) on K_c. Our tests in London Clay, Bothkennar clay-silt, Cowden till and even sand, all show that the shear and radial stresses developed in any given horizon during pile installation decay as the pile tip advances to greater depth. This is shown in Fig. 2, which reproduces the traces followed by three levels of on-pile instrumentation during installation by fast jacking in London Clay. Pile whip was not a factor during these tests. The MIT strain path approach helps to explain some of the features observed close to the pile tip, but we believe that other factors to do with the cyclic nature of loading during practical pile installation are also playing an important role. The h/R ratio influences the long term radial effective stresses, as illustrated on Fig. 3 by data gathered at Bothkennar. Multiple tests with fully reliable instrumentation confirm that the stresses remaining after full equalisation in any given layer reduce systematically as h/R increases.

Bond *et al.* (in the 1992 SUT Offshore Site Investigations and Foundations Behaviour Conference) show that accounting for sensitivity and h/R reduces considerably the scatter in the K_c database, so making effective stress analyses of pile shaft capacity more reliable.

18. Analysis of stress-wave data from pile tests at Pentre and Tilbrook

M. F. RANDOLPH, Department of Civil and Environmental Engineering, The University of Western Australia

Introduction

The main analysis of the driving performance of the 762 mm diameter piles tested at Pentre and Tilbrook has been presented at this conference by Poskitt *et al.* (1993). This present paper describes additional analysis of some of the stress-wave data, with a view to assessing the integrity and consistency of the data and estimating profiles of shaft friction at different stages of penetration of the pile. The author has had access to stress wave data from three blows for each of the main test piles, taken from near the start of driving, at mid-penetration and at the end of driving. The three blows have been analysed using a stress-wave analysis program, IMPACT (Randolph, 1990), in order to obtain estimates of the soil resistance profile at each stage.

Accurate estimation of the drivability of piles with a particular hammer is an important aspect of pile design, particularly for large diameter offshore piles. The drivability depends critically on the shaft and end-bearing resistance offered by the soil, and on the extent to which a plug will form within the pile. Improved understanding of the factors that affect the shaft friction along the pile during installation, and quantification of any degradation in friction due to the cyclic action of many hundreds of blows, will lead to more accurate estimates of drivability, and may also influence design methods for the long term pile capacity. Before describing the analysis of the stress-wave data at Pentre and Tilbrook, it is appropriate to consider current approaches for estimating the soil resistance during installation.

It is now widely accepted that the shaft friction available for a pile driven into clay soils is a function of the horizontal effective stress and the interface friction angle between pile and clay. In other words, the shaft friction is determined by the effective stresses acting along the pile–soil interface, and may differ significantly from the local (undrained) shear strength of the clay. In spite of this, it is often convenient to express the shaft friction, τ_s, as a ratio of the in situ undrained shear strength, s_u (specifically, the shear strength measured in unconsolidated undrained triaxial compression tests), although such correlations should take account of the strength ratio (s_u/σ'_v). The principal difficulty with a direct application of the effective stress approach to estimate the long term shaft friction lies in the complexity of the total and effective stress changes that occur around the pile during installation, re-consolidation of the surrounding soil, and then loading of the pile.

In drivability studies, it is still customary to estimate the shaft friction during driving in terms of the undrained shear strength of the clay. Two approaches are common: Toolan and Fox (1977) recommend taking the shaft friction during driving equal to the remoulded shear strength of the clay; Semple and Gemeinhardt (1981) use an approach which modifies the long term static shaft friction (as estimated using procedures such as the API guidelines) by a factor which is a function of the overconsolidation ratio (OCR), varying from 0.5 for OCR = 1, to unit for OCR of 10 or greater.

An exception to the above approaches is that proposed by Heerema (1980) which takes account of the importance of the radial effective stress around the pile. Heerema's approach the radial effective stress is assumed to vary exponentially along the pile, from a maximum value near the tip to a low value near the ground surface. The maximum value is an empirical function of the shear strength of the clay at the level, and the penetration of the pile. An important feature of his approach is the observation (from back-analysis of driving data) that the shaft friction at a given level appears to decrease as the pile is driven further. This observation is termed friction fatigue being attributed to the two way plastic shearing cycles undergone by the clay adjacent to the pile shaft.

Assuming only limited drainage occurs during pile installation, the effective stress state around the pile will be that associated with the in situ clay after severe remoulding. Depending on the orientation of the major and minor principal stresses, the radial effective stress may be greater or less than the mean effective stress. Near the tip of the pile it may be expected that the radial stress will exceed the vertical stress, while further back along the shaft, indications from the strain path method (Baligh 1986) are that the radial stress reduces to less than the vertical stress.

Extreme values for the radial stress, and hence the shaft friction, may be estimated from Mohr's circle for the stress state in an element of soil close to the pile (see Fig. 1). For the case of normally consolidated or lightly overconsolidated clays where any cohesion may be ignored, it may be shown that the shaft friction will lie between limits of about 0.5 and 0.9 times the remoulded shear strength, for typical values of friction angle and interface friction angle, ϕ'. For more heavily overconsolidated clays, the shaft friction may be expressed in terms of the mean effective stress of the clay adjacent to the pile, with typical ratios being 0.2 to 0.4.

Under the action of the cyclic shearing that occurs in the soil close to the pile shaft, the interface friction angle may reduce as more of the pile is driven past a particular location, although the high strain rates will limit the extent to which any residual shear planes may form. Also, the mean effective stress

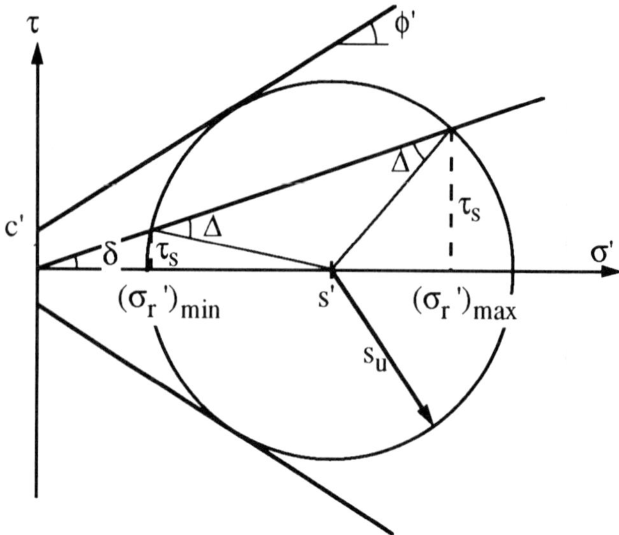

Fig. 1. Mohr's circle of stress for element of soil adjacent to pile

may vary due to any drainage that takes place during pile installation. Overall this would be expected to lead to increases in mean effective stress, since positive pore pressures are generated around the pile. However, it is possible that local flow of water into the cyclically sheared zone of soil may occur, which would lead to reduced mean effective stress immediately around the pile.

Within the framework of the above ideas, the author has analysed stress wave data from the Pentre and Tilbrook piles, and compared the deduced soil resistance at time of driving with estimates made prior to pile installation. The geometries of the two piles, showing the approximate positions of the instrumentation as shown in Fig. 2.

Pentre pile

The Pentre site is a deep deposit of essentially normally consolidated silty clay, with shear strengths varying from 35 kPa at 15 m depth to 150 kPa at 60 m depth. The test pile was cased over the upper section, so that the effective embedment was over the depth range 15 to 55 m. The main section of the pile consisted of a 762 mm diameter, by 15 mm wall thickness, pipe (Fig. 2). However, internal channels to protect the instrumentation cables increased the net steel area to 43200 mm^2, which has been modelled as an equivalent 18.5 mm wall thickness Young's modulus and density of the steel have been

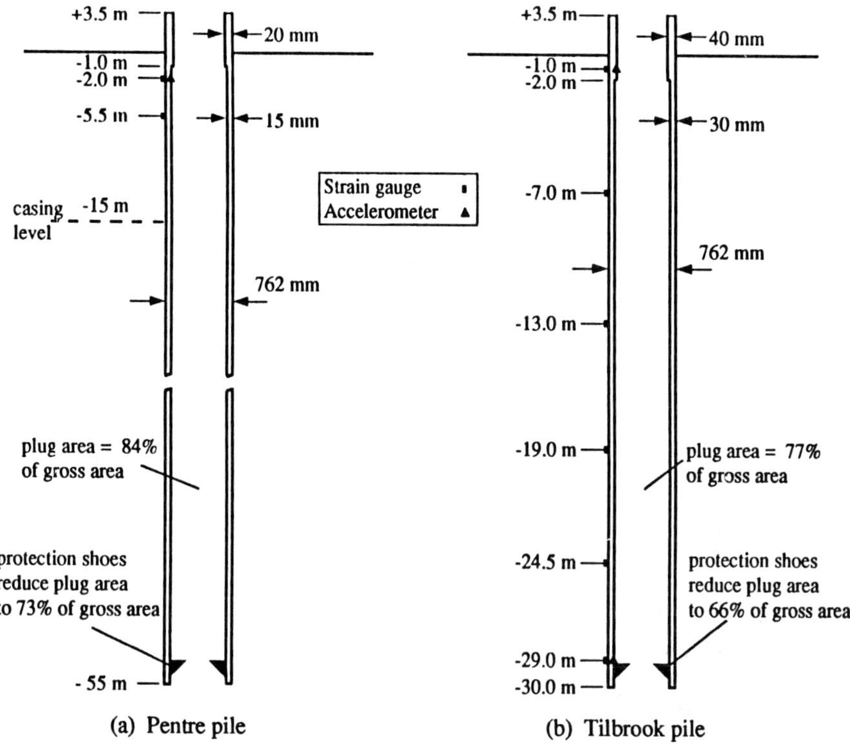

Fig. 2. Geometries of the Pentre and Tilbrook test piles (not to scale)

taken as E = 210 GPa, ρ = 7.8 Mg/m^3 respectively, giving a wave speed of c = 5190 m/s. Protection shoes at the bottom of the instrumentation channels protruded by a total of 87000 mm^2 into the interior of the pile. These shoes will have provided additional bearing resistance at the pile base, and also will have led to increased internal shaft friction in the leading 0.35 m of the pile.

Stress-wave data were only available at two locations, at depths of 2 m and 5.5 m below ground level (distances of 53.0 and 49.5 m from the tip of the pile). At both of these locations, dynamic strain measurements were made at four positions around the pile, and these have been averaged in order to calculate values of force. However, accelerometers were only positioned at the upper location, and only one of the axial accelerometers gave reasonable data. As such, it was not possible to eliminate effects of bending in the pile in estimating values of velocity. This may account for the unusual appearance of the stress wave data shown in Fig. 3, taken from Blow 10 (penetration of 21m, or 6m below the casing level). The force and velocity (multiplied by the

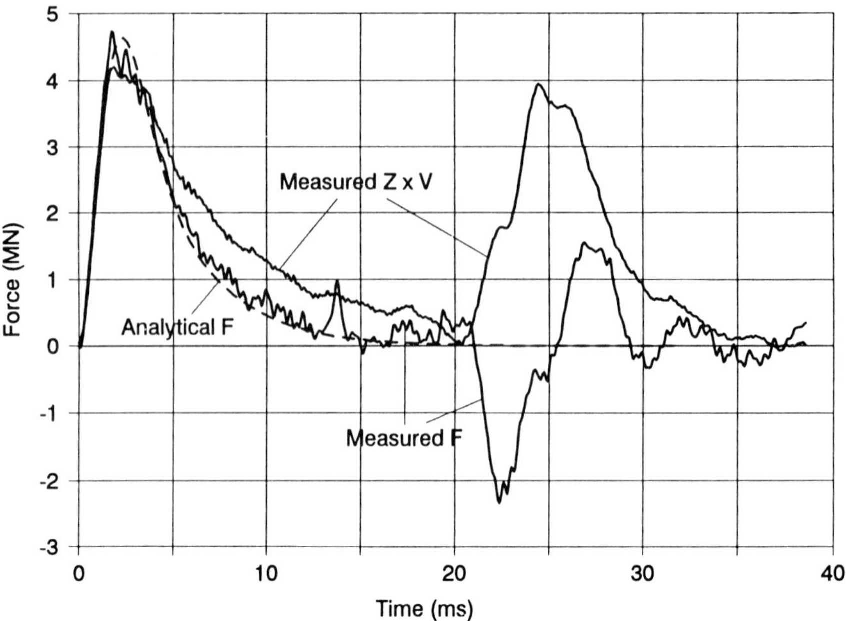

Fig. 3. Analytical modelling of impact force, Blow 10, penetration 21 m, Pentre pile

pile impedance, Z) responses should coincide up to the point of first reflection from the soil, which should not occur until a time of at least 18 ms after the start of the blow. Similar non-coincidence of the velocity response was found in all three blows analysed from the Pentre pile test. As such, it has not been possible to adopt the normal strategy, whereby one set of readings (for example, the measured force response) is used as input in a dynamic analysis, with soil parameters adjusted iteratively in order to provide a match to the other set of readings (the measured velocity response).

Instead, the force response has been simulated by modelling the hammer blow using the analytical solution developed by Deeks and Randolph (1992). That solution uses lumped masses for the ram (7 Mg) and anvil (0.87 Mg) and a cushion (stiffness 1.6×10^6 kN/m) to provide the theoretical response shown in Fig. 3. The impact velocity was set at 3.3 m/s, implying a delivered energy of 38 kJ. (It may be noted that direct integration of the measured force and velocity data implies a delivered energy of 41 kJ.) The fit between theoretical and measured force response is very good.

Analysis of the stress-wave data has been carried out with a view to fitting the measured force-time response up to the first main reflection from the pile tip. Inaccuracies in modelling the exact interaction of the hammer at the top of the pile limit the ability to match the signal after that point. The result of the fit for Blow 10 is shown in Fig. 4. This fit has been achieved with a static shaft friction varying from 30–45kPa over the 6m of pile penetration (shaft resistance of 0.54 MN) internal shaft friction of 0–20 kPa above the protection shoes, but 200 kPa below the shoes, and an end bearing resistance of 1.5 MPa (consistent with the cone resistance at a depth of 21 m). The total static resistance is therefore 0.87 MN. The dynamic resistance is affected by viscous and inertial damping, with the dynamic shaft friction taken as (Randolph, 1990)

$$\tau_d = \tau_s \left[1+\alpha\left(\Delta v/v_0\right)^{\beta}\right]$$

where Δv is the relative velocity across the shear zone adjacent to the pile, and v_0 is a reference velocity of 1m/s. In the present analysis α was taken as unity and β was taken as 0.2 (Litkouhi and Poskitt, 1980). The total dynamic resistance including inertial damping at the pile base, was about 2.44 MN.

Matching of the stress-wave data for the blow at the end of driving (Blow 2240, penetration of 55 m) is shown in Fig. 5. Again the main purpose has been to achieve a match to the overall shape of the force response up to the main reflection from the tip, without excessive refinement of individual soil parameters. It may be seen that the measured velocity response is not consistent with the 13 m of free-standing section of pile below the instrumen-

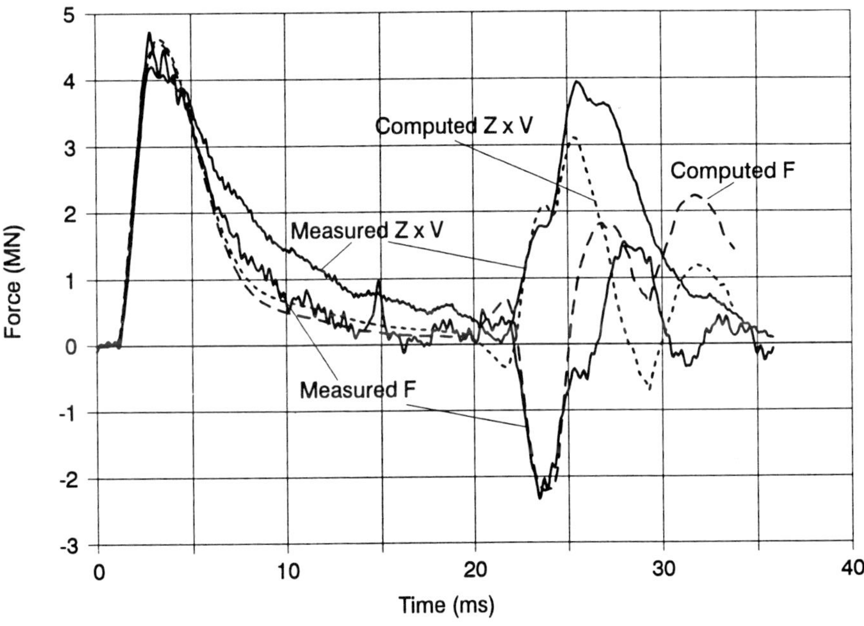

Fig. 4. Stress-wave match for Blow 10, penetration 21 m, Pentre pile

tation point (a wave return time of 5 ms). However, the computed velocity response stays reasonably parallel to the measured response, and the magnitude of the change in velocity on arrival of the tip relection is reasonably well predicted.

The match in Fig. 5 has been obtained with an impact velocity of 3.9 m/s (energy of 52 kJ) and shaft capacities of 1.3 MN (external) and 0.85 MN (internal). Approximately 50 per cent of the internal shaft resistance comes from below the protection shoes. An end-bearing resistance of 2.5 MPa gives a total static resistance of 2.27 MN, but a dynamic resistance of 4.89 MN.

A summary of the results of analysing the stress-wave data from the Pentre pile is shown in Table 1. The computed permanent sets agree very well with the measured results of approximately 17 mm and 8.5 mm for penetrations of 35 mm and 55 mm. However, there is something of an anomaly with the first blow analysed, at a penetration of 21 m (6 m below the bottom of the casing). The soil resistance at that depth is similar to that at a penetration of 35 m (20 m below casing level), hence the computed permanent sets are similar. When the pile was first driven, from a position with the tip at the bottom of the cased section, it suddenly plunged a few metres to a penetration of just over 20 m below ground level, after four blows. Over the next ten blows, the blow count fluctuated about 2–5 blows per 250 mm (sets of 50–100 mm per blow). It is possible that occasional blows (perhaps the one actually

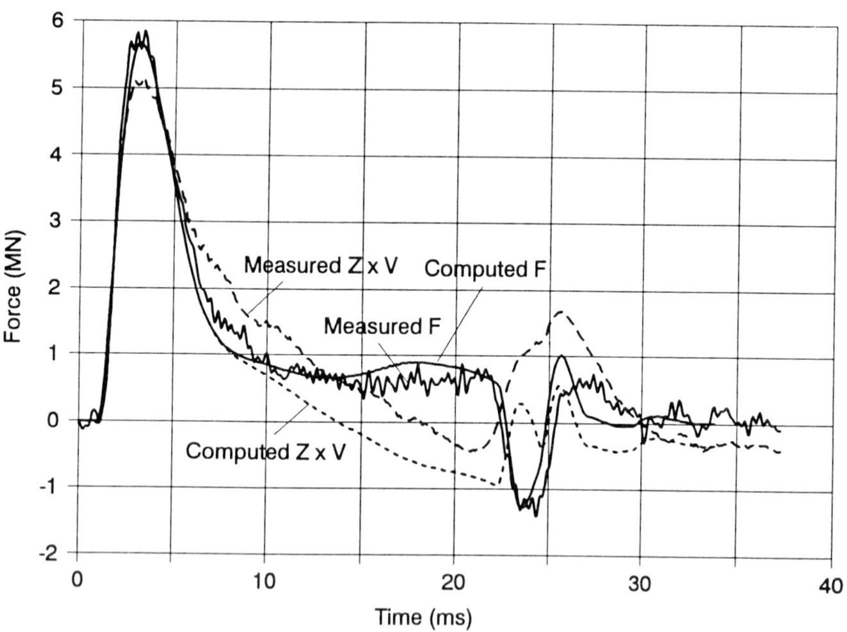

Fig. 5 Stress-wave match for Blow 2240, penetration 55 m, Pentre pile

analysed, Blow 10) encountered sporadic higher resistance, resulting in significantly lower sets. It is worth noting that double integration of the accelerometer data gave similar final displacements of about 27 mm for the two blows at penetrations of 21 m and 35 m. However, calculation of permanent sets from the accelerometer data is notoriously unreliable.

Table 1. Summary of analysis of stress-wave data from Pentre pile

	Static Soil Resistance				Dynamic Resistance	Energy Delivered	Permanent Set
	Shaft Resistance		Base	Total			
Penet'n (m)	External (MN)	Internal (MN)	(MN)	(MN)	Peak (MN)	(kJ)	(mm)
21	0.54	0.53	0.06	1.13	2.44	38.1	16.1
35	0.60	0.54	0.09	1.23	2.80	47.9	16.9
55	1.31	0.85	0.11	2.27	4.89	53.2	8.9

Fig. 6 shows the distribution of shaft friction assumed in the back-analysis of the three blows. Also shown, for comparison, is the profile of remoulded shear strength (taken from the original site investigation report for Pentre), factored by 50 per cent. The two features from this figure are (a) that the shaft friction at any given level appears to reduce considerably as the pile is driven further (consistent with the findings of Heerema, 1980), and (b) that the shaft friction in the lower region of the pile for the deeper penetrations is similar to 50 per cent of the remoulded shear strength. Of course the precise value and distribution of shaft friction deduced from the stress-wave data are sensitive to the amount of viscous damping assumed. Comparable matching of the force-time response to that shown in Fig. 4 is possible for a range of the parameters α and β in equation (1). As such, the shaft friction profiles shown in Fig. 6 should be viewed as indicative only.

Tilbrook pile

The Tilbrook site consists of very stiff overconsolidated clay, with shear strengths ranging from 300 to 500 kPa in the upper 20 m, increasing to peak strengths of 700 kPa at 25 m depth and then reducing to 500 kPa at 30 m. The test pile was driven from the ground surface to a penetration of 30 m. The main section of the pile consisted of a 762 mm diameter, by 30 mm wall thickness, pipe (Fig. 2). The wall thickness of the upper 6.5 m of the pile was increased to 40 mm and the instrumentation channels increased the net steel area to 99000 mm^2 and 69000 mm^2 respectively in the two sections. This has been allowed for by taking equivalent wall thicknesses of 44 and 34 mm. The steel properties have been taken as identical to the Pentre pile. Protection shoes at the base of the instrumentation channels protruded by a total of

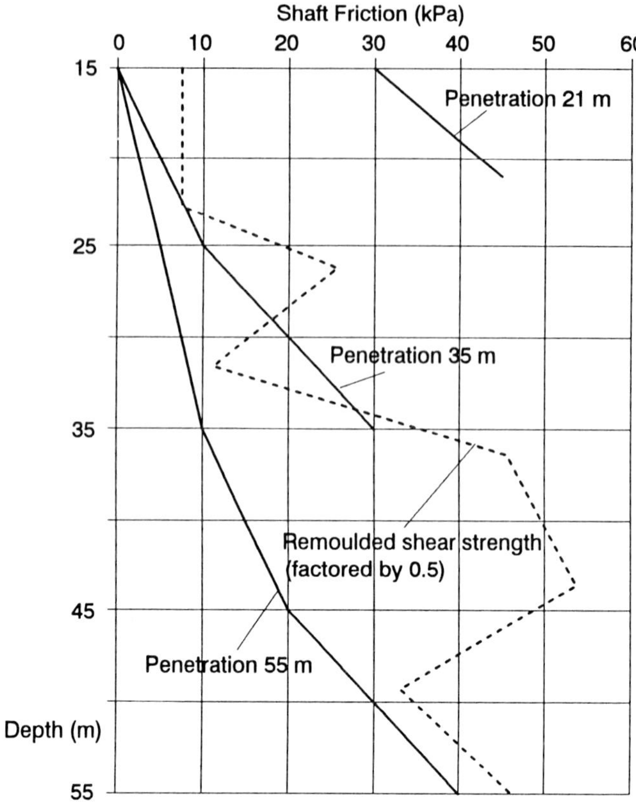

Fig. 6. Profiles of shaft friction deduced from stress-wave data for the Pentre test pile

87000 mm^2, which will have provided additional bearing area and have led to high internal shaft friction in the leading 0.35 of the pile.

Dynamic strain data were gathered at depths of 1 m, 7 m, 13 m,19 m 24.5 m and 29 m below ground level. At each location, strains were measured at four positions around the pile and have been averaged in order to calculate the force. Axial accelerometers were also positioned near the pile head (depth of 1 m) and at 0.5 m from the pile tip (depth of 29.5 m). The accelerometers showed significant effects of drift, resulting in non-zero terminal velocities at the end of each blow. Fig. 7 shows accelerometer data for a blow near the end of driving (Blow 2850, penetration 30 m). While the signals appear to return satisfactorily to zero at the end of the blow, the deduced velocity data show final velocities of -4 m/s and -1.4 m/s (see Fig. 8). Even if the velocity responses are rotated to give terminal velocities of zero integration of the corrected responses does not yield sensible displacement data (see Fig 9). Similar problems were encountered in the other two blows analysed.

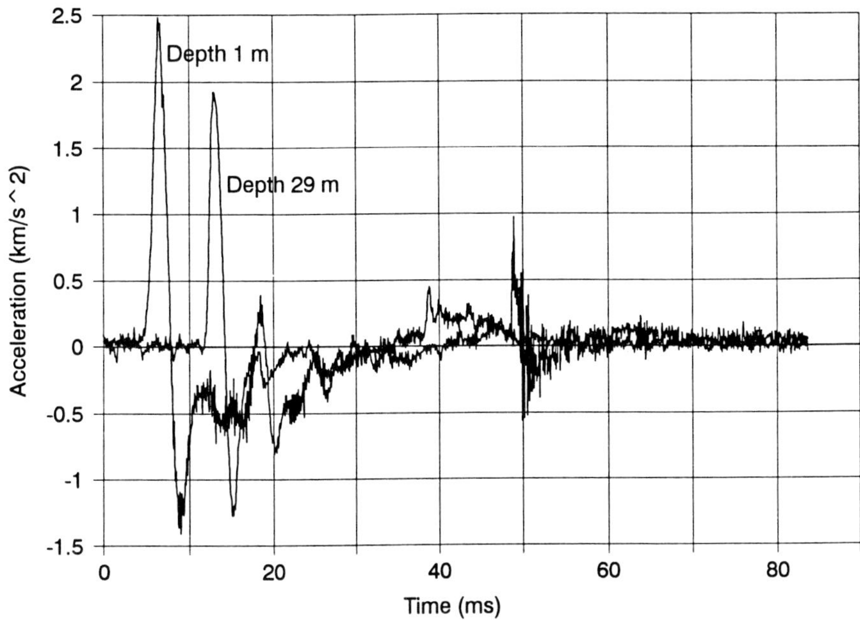

Fig. 7. Accelerometer data from Blow 2850, penetration 30 m, Tilbrook pile

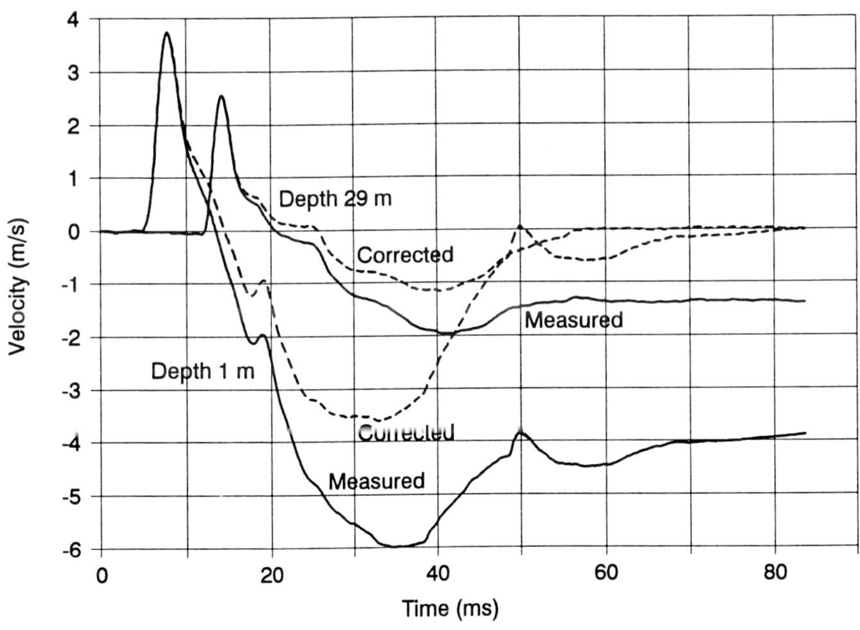

Fig. 8. Velocity responses from Blow 2850, penetration 30 m, Tilbrook pile

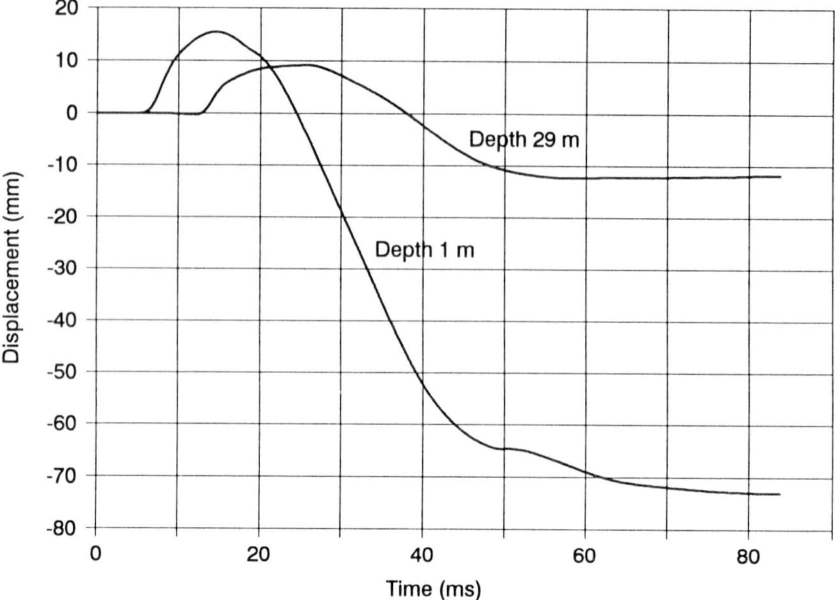

Fig. 9. Displacements from corrected velocities, Blow 2850, penetration 30 m, Tilbrook pile

A further problem with the accelerometer data may be seen from Fig. 10, which shows the measured force and (uncorrected) velocity response at depth of 1m for a blow near the start of driving (Blow 8, penetration 3 m). The first reflection due to soil resistance should not come from the soil until a time of 10 ms after the start of the blow. However, the data show separation of the force and velocity signals very early on, with the force response lying above the velocity response, implying resistance to driving. In reality, the velocity response should lie above the force response during the first part of the blow. This is because the uppermost strain gauges and accelerometers lie above the 10 mm change in wall thickness (see Fig. 2). This results in partial reflection of the stress-wave and, since the pile decreases in area, the velocity will increase relative to the force.

Figure 10 shows the theoretical force and velocity responses obtained using the analytical hammer model of Deeks and Randolph (1992). The hammer impact has been simulated tkaing ram and anvil masses of 40 Mg and 6.1 Mg, a cushion stiffness of 7.5×10^6 kN/m and a cushion damping of 2400 kNs/m. (Note that a Bongassi cushion was used at Tilbrook, compared with a steel cushion at Pentre, hence the need to allow for damping in the cushion). The effect of the change in section is evident from the computed force and velocity responses.

Computed and measured force responses at depths of 7 m and 19 m above the pile tip are shown in Fig. 11. Agreement between the two pairs of curves is reasonably good up to the main reflection from the pile tip, but deteriorates after that. This appears to be due to limitations in the dynamic model of the soil response at the pile base, and in the bottom section of the internal soil plug. The shaft friction over the 3 m penetration has been taken as uniform at 100 kPa, with a limiting end-bearing stress of 4 MPa. No attempt has been made to refine the soil parameters on an element-by-element basis in order to improve the fit to the measured force responses at different locations down the pile.

Matching of the stress-wave data for the blow at the end of driving (Blow 2850, penetration of 30 m) is shown in Fig. 12. Again the quality of fit to the data deteriorates after the main reflection of the tip. However, the overall force response is reasonably well matched, particularly in view of the need for analytical modelling of the hammer impact. As pointed out earlier, the measured velocity data do not appear trustworthy, and the lack of fit between computed and measured velocities is not surprising.

The computed and measured force responses at depths of 7 m and 25 m are shown in Fig. 13. The two sets of curves agree reasonably well, confirming that the assumed distribution of soil resistance is approximately correct. It

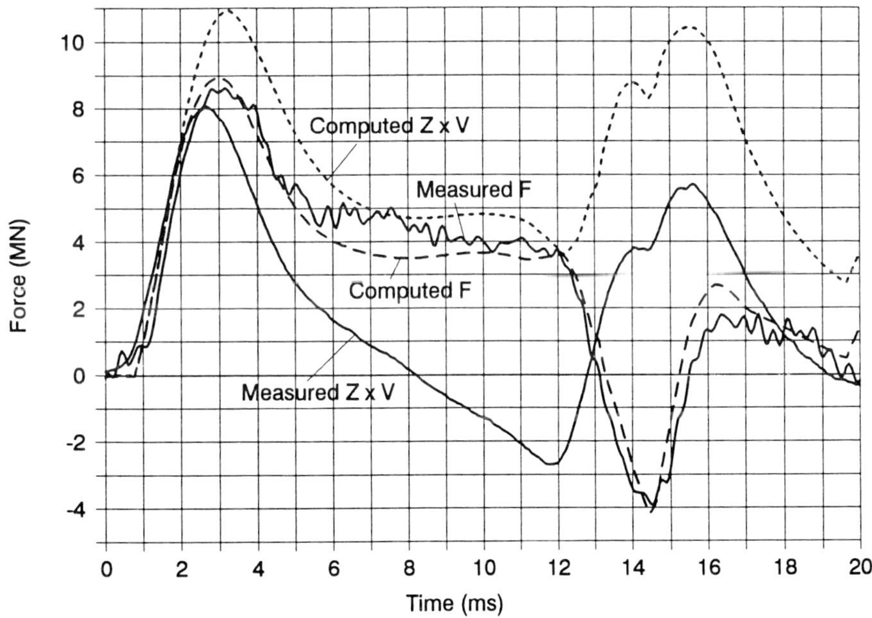

Fig. 10. Stress-wave match for Blow 8, penetration 3 m, Tilbrook pile

should be pointed out, however, that it has been necessary to adjust the zero point of the measured strain gauge data. All the strain gauge data showed residual strains at the start of each blow that varied typically between –100 and –200 microstrain, equivalent to tensile forces of 1.5 to 3 MN. The sign and magnitude of these residual forces is difficult to comprehend, since it would be expected that compressional residual forces would be locked into the pile at the end of each blow. The analyses shown in Figs 12 and 13 have been conducted by applying the (analytical) hammer blow to the pile top three times in succession, allowing a stable residual force system to develop in the pile and soil.

Thus, the force at a depth of 25 m starts off at 4 MN compression (Fig. 13) and then increases during the blow to a maximum of 14 MN. At the end of the full blow (lasting 80 ms), the force returns to the initial value of 4 MN. The measured force data for the depth of 25 m have been shifted from an average initial force of about –1.3 MN (uncorrected) to a value of 2.0 MN. This residual force differs from the computed value of 4 MN, but has been chosen in order to give the same peak force during the blow.

A summary of the key features of the stress-wave analysis for the Tilbrook pile is shown in Table 2. The tabulated values of delivered energy have been obtained by integration of the analytical force and velocity response, which

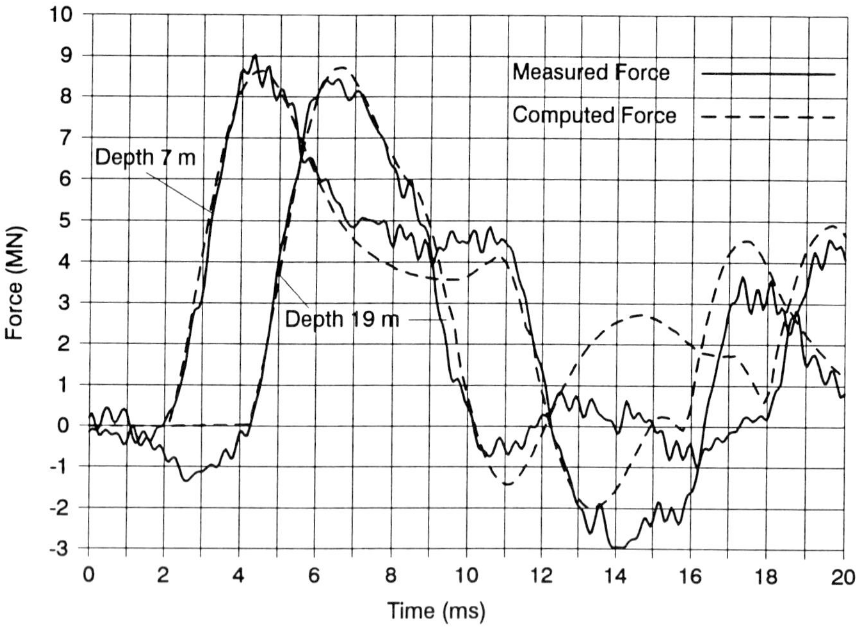

Fig. 11. Comparison of force responses for Blow 8, penetration 3 m, Tilbrook pile

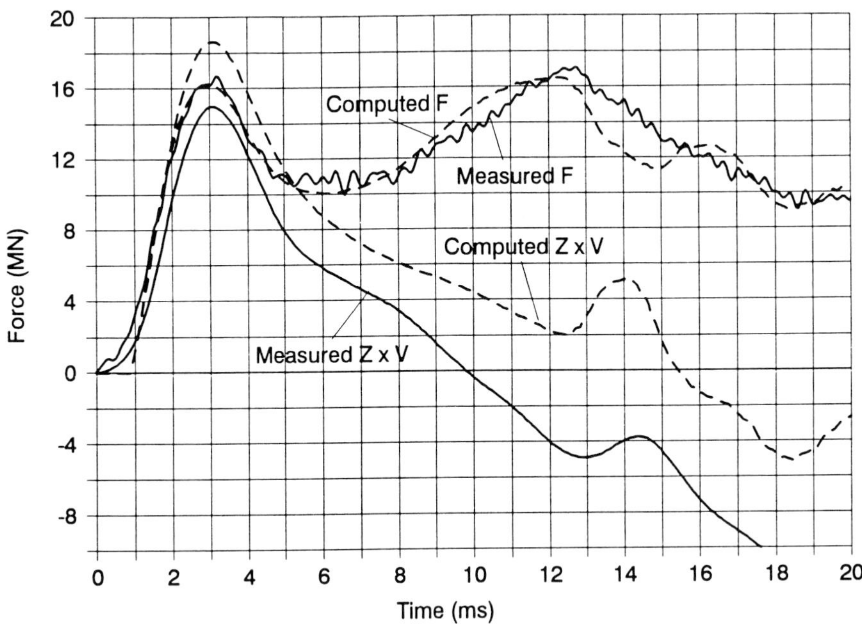

Fig. 12. Stress-wave match for Blow 2850, penetration 30 m, Tilbrook pile

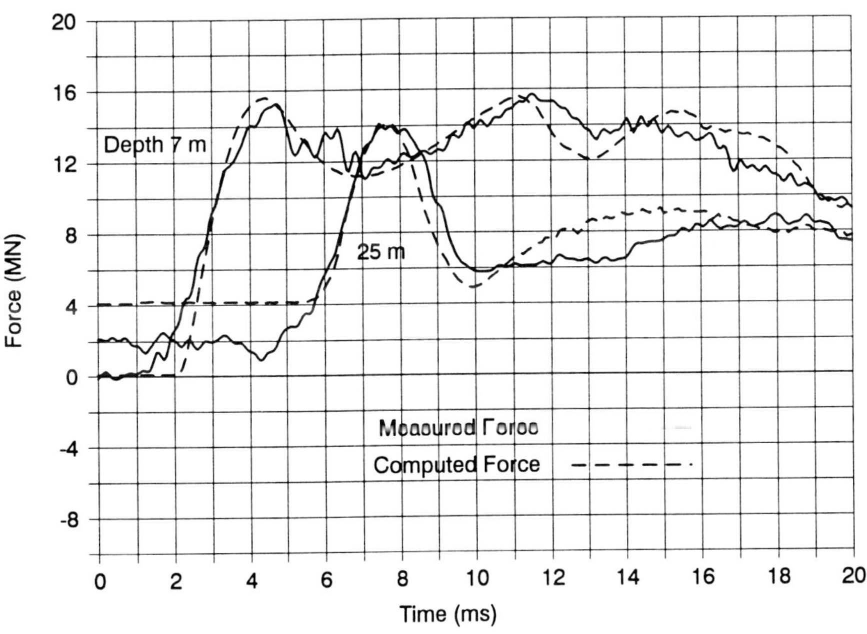

Fig. 13. Comparison of force responses for Blow 2850, penetration 30 m, Tilbrook pile

leads to rather higher values than integration of the measured data. However, in view of the marked zero drift on the accelerometers (see Fig. 8), the measured data will under predict the energy transferred. The computed energies (particularly for the first blow) are too high by comparison with the quoted drop heights of 0.2 m, 0.4 m and 1.0 m, which imply maximum available energies of 78.6 kJ, 157 kJ and 393 kJ for the three blows. However, given that the velocity signals are unreliable, the energy transferred to the pile for the first blow may be estimated by integrating F^2/Z(where F is the measured force at a depth of 1m, and Z is the pile impedance at that level) up to the point where the first reflections start coming back up the pile. This gives a value of just under 80 kJ for the first blow, which will be a lower bound on the total energy transferred over the full length of the blow. The indication is that the drop height may have been rather higher than the nominal recorded drop of 0.2 m.

As for the Pentre pile, the permanent set calculated from the shallowest blow is rather low by comparison with the measured blow count of 2–3 blows/250 mm. The permanent sets for the other two blows are in reasonable agreement with the measured blow counts of 28 blows/250 mm (penetration of 15 m) and 34 blows/250 mm (penetration of 30 m).

A summary of the shaft friction profiles deduced from the stress-wave analysis is shown in Fig. 14. As for the Pentre pile, it is clear that the shaft friction available at any given location gradually reduces as the pile is driven further into the ground. The final distribution of shaft friction may be compared firstly with the long term shaft friction profile deduced from the static load test, and secondly the profile of in situ mean effective stress (factored by 50 per cent in order to allow direct comparison). The remoulded shear strengths fluctuate between 400 and 500 kPa in the upper half of the soil profile and generally exceed 600 kPa for depths below 15 m. These values bear no relation to the deduced profile of shaft friction during driving. The limiting shaft friction will be determined by the radial effective stress acting around the pile, and this may be related to the mean effective stress. From the approximate comparison shown in Fig. 14 it would appear that the shaft friction during driving is broadly about half the in situ mean effective stress,

Table 2. Summary of analysis of stress wave data for Tilbrook pile

	Static Soil Resistance				Dynamic	Energy	Permanent
	Shaft Resistance		Base	Total	Resistance	Delivered	Set
Penet'n (m)	External (MN)	Internal (MN)	(MN)	(MN)	Peak (MN)	(kJ)	(mm)
3	0.72	0.87	0.31	1.90	3.77	95.0	41.7
15	2.99	4.11	0.39	7.49	10.80	163.0	12.4
30	8.77	3.00	0.62	12.39	18.73	347.0	7.4

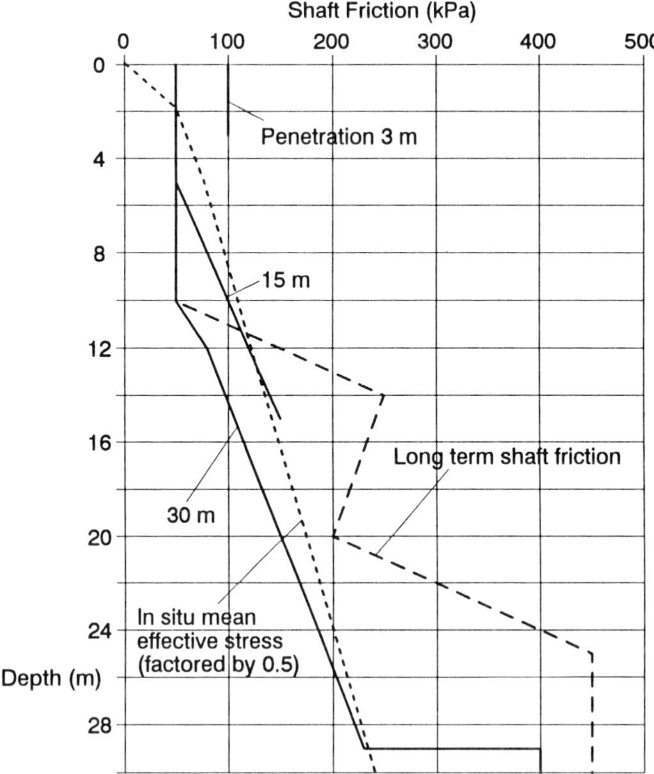

Fig. 14. Profiles of shaft friction deduced from stress-wave data for the Tilbrook test pile

although the ratio is lower at shallow depths and higher near the tip of the pile. Assuming that the mean effective stress in the soil adjacent tot he pile increases during shearing and remoulding of the soil (due to dilation effects), the ratio of shaft friction to the current mean effective stress will thus be rather less than 0.5, which is consistent with the Mohr's circle of stress for an element of soil just outside the shearing zone (see Fig. 1).

Conclusions

It is important in any field instrumentation exercise to assess the quality of the measured data and to allow for any shortcomings when using the data to deduce performance parameters. This is particularly true of dynamic stress-wave measurements obtained during pile driving, which is an area where special attention has to be paid to the design of the instrumentation mountings, cabling and any filtering of the signals.

The main principles used in assessing the quality of stress-wave measurements are:

(1) Demonstrating proportionality of force and velocity responses prior to reflections from the soil resistance arriving back at the instrumentation level.
(2) Checking that integration of the accelerometer data gives zero terminal velocity.
(3) Checking that double integration of the accelerometer data gives reasonable agreement with direct field measurements of temporary compression and permanent set.

The geometries of the test piles at Pentre and Tilbrook were relatively complex, with changes in wall thickness, channels to shield the instrumentation cables and protection shoes near the pile base. For the Tilbrook pile, the top instrumentation level was above the change in wall thickness, destroying proportionality of the force and velocity data, and rendering the task of assessing the integrity of the data that much more difficult. If possible, it would be better to locate key instrumentation below any change in cross-section of the pile.

In both the Pentre and Tilbrook pile tests, the velocity data have proved insufficiently reliable to serve as a basis for matching measured and computed stress-waves. Instead, analysis of the force data has been undertaken by means of an analytical model of hammer impact, with the soil parameters adjusted in order to match the measured force response at as many locations along the pile as possible. For the Tilbrook pile, significant residual forces were calculated at the end of each blow, and hence it was necessary to apply the analytical blow at least three times, in order to achieve a steady state residual force system. Matching of the force response with the measured data was then effected after the third imposed blow.

Ultimately, reasonable matches to the measured force responses were achieved for all three blows (for each pile) that the author has analysed. The deduced profiles of shaft friction at time of driving showed significant friction fatigue, with the shaft friction at any particular depth reducing with further penetration of the pile. For the Pentre pile, where the soil was close to normally consolidated, the shaft friction was of the order of half the remoulded shear strength of the soil in the lower part of the pile, but reduced below that value at shallower depths, due to the fatigue action of two-way plastic shearing. For the Tilbrook pile, in heavily overconsolidated soil, the remoulded shear strength was not a useful parameter for estimating the shaft friction during driving. A correlation with the in situ mean effective stress appeared more fruitful, in keeping with the principle that the shaft friction is determined by the radial effective stress acting around the pile, and that stress is closely related to the mean effective stress state in the soil immediately outside the shear zone.

In summary, as suggested by Heerema (1980), it is important to allow for friction fatigue in order to achieve accurate prediction of pile drivability. Such degradation of shaft friction during installation is consistent with results from the static loading tests on the piles at Pentre and Tilbrook, which indicate residual load transfer conditions in the upper part of the pile (Aldridge and Schnaid, 1992).

Acknowledgements

The author is grateful to BP International Ltd and the partners in the LDP test programme for permitting access to some of the stress-wave data. Particular thanks are due to Mr J. Clarke for his help in providing the data in digital form. The author acknowledges the value of comments by Dr M. Fahey during the preparation of this paper.

References

ALDRIDGE, T. R. and SCHNAID, F. (1993) Degradation of skin friction for driven piles in clay. This volume.

BALIGH, M. M. (1986) Undrained deep penetration, I & II. *Géotechnique*, **36**, No. 4, 471-501.

DEEKS, A., and RANDOLPH, M. F. (1992) Analytical modelling of hammer impact during pile driving. Paper in preparation.

HEEREMA, E. P. (1980) Predicting pile driveability: Heather as an illustration of the "friction fatigue" theory. *Ground Engineering*, 13, April, 15-20.

LITKOUHI, S., and POSKITT, T. R. (1980) Damping constant for pile driveability calculations. *Géotechnique*, **30**, No. 1, 77-86.

POSKITT, T. J., YIP-WONG, K. L., and COX, W. R. (1993) Measurement of axial strain and pile wall displacement and the determination of skin friction during the driving of instrumented piles. This volume.

RANDOLPH, M. F. (1990) *IMPACT: Dynamic analysis of pile driving*. User manual (2 vols). Geomechanics Group, The University of Western Australia.

SEMPLE, R. M., and GEMEINHARDT, J. P. (1981) Stress history approach to analysis of soil resistance to pile driving. *Proc. 13th Offshore Technology Conf.*, Houston, 1, 165-172.

TOOLAN, F. E., and FOX, D. A. (1977) Geotechnical planning for piled foundations for offshore platforms. *Proc. Instn Civ. Engrs*, Part 1, **62**, 221-244.

19. Analysis of behaviour of the Tilbrook Grange lateral test pile

J. M. HAMILTON and T. W. DUNNAVANT, Exxon Production Research Company, Houston, Texas

The results of the Tilbrook Grange lateral pile test have been used to judge the suitability of published P-Y criteria for use in analysis of laterally loaded piles in stiff, overconsolidated clays. The investigation was performed at two levels: (1) comparison of P-Y curves derived from the test results to P-Y curves synthesised using published criteria, and (2) comparison of observed gross pile behaviour to behaviour predicted using the various criteria. Consideration was given to the effects of using ε_{50} *values of 0.025 (measured) versus 0.005 (typically assumed for design in the absence of measurements). The derived P-Y curves show a moderate degree of post-peak softening (after cycling) down to 2.3 m depth (three diameters). No softening was observed below this depth. The P-Y curves synthesised using the Dunnavant and O'Neill method best match the derived curves. Additionally, overall pile head load–displacement behaviour was predicted best using the Dunnavant and O'Neill P-Y criteria.*

Introduction

Analysis of laterally loaded pile behaviour is usually accomplished using a beam-column idealisation in which the soil resistance is modelled as a series of uncoupled, nonlinear springs. The force-displacement behaviour of the soil springs (*P-Y* curves) is defined by criteria that are based mainly on analyses of instrumented pile load tests. Criteria have been proposed for soft clay (e.g. Ref. 1), sand (Ref. 2), stiff clay (Refs 3, 4, 5) and other materials, and these criteria have been used successfully in offshore engineering practice for twenty years.

Because of the expense of full scale load tests, the data available for developing and calibrating soil *P-Y* criteria are sparse. As a consequence the criteria are often based on results of a single load test, and therefore may not be universally applicable to all sites having the same broadly defined soil type. This is particularly true of criteria for stiff, overconsolidated clay soils, the characteristics of which tend to be less uniform from site to site. This was pointed out by Reese *et al.* (1975) (Ref. 4) about the soil at the Manor, Texas, site where the original field tests in submerged stiff clay were performed. In their concluding remarks they state that 'the development of methods for

Large-scale pile tests in clay. Thomas Telford, London, 1993

predicting *P-Y* curves for all manner of stiff clays is a complex problem and will probably require the performance of a number of full-scale tests.

There is a lack of consensus in offshore engineering practice as to the appropriate criteria for use in analysis of laterally loaded piles in stiff clay. Many believe the criteria proposed by Reese *et al.* (1975) to be overly conservative for the case of cyclically loaded piles because the tests on which they are based were conducted in a soil that was deposited in a salt-water environment but which was inundated during the tests using fresh water. The soil slaked substantially and suffered extreme degradation during the pile tests due to hydraulic erosion. The tests conducted by Dunnavant and O'Neill (1989) (Ref. 5) at the University of Houston were performed in a stiff fluvial clay and did not slake upon resaturation with fresh water. Consequently, the test results were not severely impacted by hydraulic erosion. The *P-Y* criteria derived from the University of Houston tests therefore exhibit much less severe degradation than do the Reese *et al.* criteria.

The lateral pile test at Tilbrook Grange offers an opportunity to check the applicability of existing stiff clay *P-Y* criteria for an additional (and considerably different) soil profile. The undrained shear strengths at the Tilbrook site are on average a factor of 2 to 4 times higher than those at the University of Houston or Manor sites. This paper describes the *P-Y* curves extracted from the measurements of load–displacement and bending moments acquired during the cyclic lateral test. The derived *P–Y* curves are compared to those predicted by the stiff clay criteria of Reese *et al.* and Dunnavant and O'Neill as well as the soft clay criteria of Matlock. Comparisons are also made of the measured and predicted pile head load–displacement curves and bending moment values.

Site and test description

Soil conditions, test set-up and instrumentation are thoroughly described elsewhere (Refs 6, 7) and will not be described in detail here. The basic data needed for lateral load analysis are pile dimensions and profiles of soil undrained shear strength, unit weight, and ε_{50} (strain at one-half maximum deviator stress in a UU test). The test pile was an open-ended steel pipe pile with the following properties:

Outside diameter	762 mm
Length	32 m
Wall thickness	35 mm
EI (from bend test)	1136 MN-m^2
Minimum yield stress	770 MPa

The ground surface elevation was arbitrarily set at 100 m for the test. A pit (to elevation 98.4 m) was excavated around the pile to contain ponded water during the lateral test. Lateral load was applied to the pile at elevation 99 m.

Profiles of undrained shear strength, soil unit weight and ε_{50} values are shown in Fig. 1. The ground surface inside the pit is indicated on the figure. A schematic of the test pile is also provided which shows the positions of the point of lateral load application and strain gauges for measuring bending moment. The symbols on the undrained strength profile represent unconsolidated undrained (UU) test results on samples from five separate borings in the immediate vicinity.

Two interpreted undrained strength profiles are shown. The dashed profile is the one originally selected by the joint-industry project for axial pile analysis. The solid line is the one used in this paper for analysis of the lateral pile test. The data plotted as open symbols are from the original site soil investigation, while the data plotted as solid symbols are from two supplementary soil borings drilled after the pile testing was completed at the site. The supplementary borings were taken specifically to obtain additional data that would be useful for interpretation of the lateral test results (Ref. 8). These data influenced our choice of the slightly different design strength profile.

Shown also in Fig. 1 is a plot of ε_{50} values from the UU tests performed during the supplementary soil boring programme. Values of ε_{50} were not reported for tests on samples from the original bearings. Note that the measured values average about 2.5 per cent strain which is roughly five times the nominal value recommended by Reese *et al.* for an average undrained shear strength equivalent to that at the Tilbrook site.

The test was a displacement controlled static-cyclic type in which the pile head displacement was increased to a new, significantly higher level following about 100 cycles of loading at the previous displacement value. The objective was to simultaneously measure a quasi-static resistance against which the degraded cyclic value (after 100 cycles) could be compared. The pile was subjected to one-way cyclic loading. A total of nine displacement increments was used as outlined in the following table:

Increment	Number of cycles	Displacement: mm Min.	Displacement: mm Max.	Maximum force: MN
1	1(static)	0	4.3	0.29
2	100	2.4	4.1	0.28
3	100	2.9	6.2	0.37
4	100	3.9	9.3	0.47
5	100	4.0	15.0	0.61
6	100	4.0	24.0	0.74
7	120	5.0	37.2	1.05
8	120	7.0	54.0	1.23
9	136	30.8	104.5	1.83

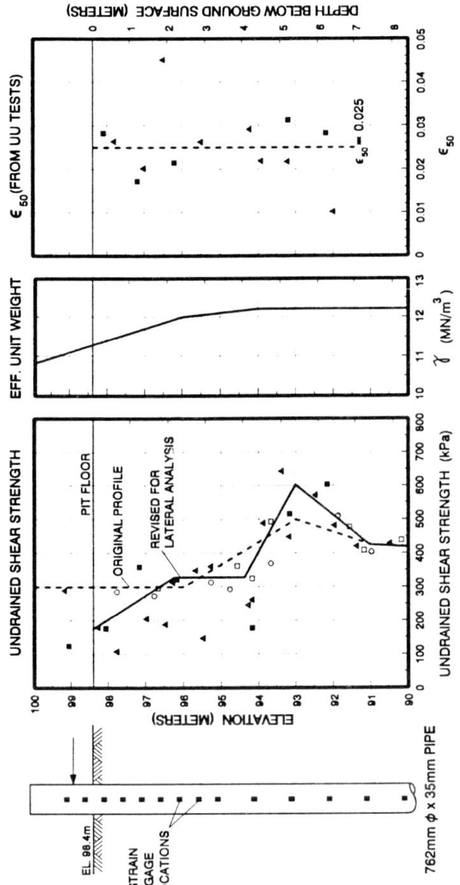

Fig. 1. Tilbrook lateral test pile layout and soil properties used in the analysis

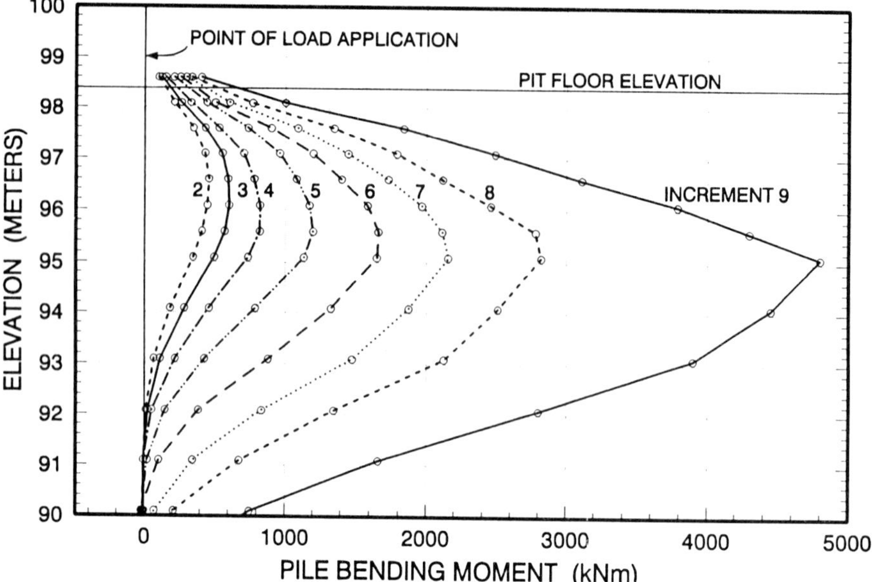

Fig. 2. Measured bending moment profiles for terminal cycles, increments 2–9

The period of application and removal of load for increments 1 through to 8 was 20 seconds. The period for increment 9 was 26 seconds. The pile was observed to post hole in the later increments. Gaps formed on both the tension and compression sides and water was expelled with significant velocity from the gaps as the pile was cycled.

Measurements were made of pile head load and displacement and bending moment during each cycle. Data used for the analysis in this paper was extracted from the project report of (Ref. 7). Shown in Fig. 2 is an example plot of bending moments recorded from the terminal cycles of several increments.

Derived *P-Y* curves

P-Y curves were extracted from the measured bending moment profiles using a proprietary computer program, PYGEN. Values of soil resistance per unit pile length (*P*) at a given displacement are determined from double differentiation of the moment (*M*) curve, while the corresponding local displacement (*Y*) is determined from double integration of the M/EI curve with the pile head displacement and rotation imposed. The computer programme calculates coefficients of multiple cubic spline polynomials fitted by least squares to the moment data. It allows variable spacing of tie points between the cubic polynomials to minimize the overall error of the fit. The slope of the moment vs. depth curve at the soil surface is constrained to equal the measured lateral force at the pile head.

P-Y curves were derived for the first and terminal cycles of each displacement increment. The derived *P-Y* curves are shown in Figs 3 and 4. The curves are reasonably smooth considering the numerical difficulty inherent in this process. There is a clear degradation of the lateral soil resistance from the first to the terminal cycle. Additionally, the degradation is more severe at the higher displacement levels. Meaningful *P-Y* curves could only be obtained to a depth of 4.3 m below the pit level (elevation 94.1 m). Below this level there is an inflection in the deflected pile shape (which shifts downward with each increment), displacements are very small and soil resistance goes through a sign change. The inability to accurately measure data to large depths is a problem that is inherent in lateral pile tests in stiff ground. The pile will generally yield before significant displacements can be achieved at depth. This is especially true of the subject test where the pile was relatively thin walled for a lateral test.

Comparisons with published *P-Y* criteria

The derived *P-Y* curves for the terminal cycles are compared to predicted curves developed from published criteria in Figs 5 to 9. Comparisons are made to *P-Y* criteria developed by Reese *et al.* (1975), Dunnavant and O'Neill (1989), and Matlock (1970). The Dunnavant and O'Neill curves were generated assuming 100 loading cycles. The predicted curves are shown for two values of ε_{50}; a value of 0.025 as measured in actual UU tests, and a nominal value of 0.005 which would typically be used for a stiff clay in the absence of triaxial test stress-strain data.

The criteria of Reese *et al.* overpredict the maximum mobilised soil resistance. The initial stiffness of the *P-Y* curves is better matched by the Reese curves in which ε_{50} is assumed to be 0.005. However, the method predicts much too severe degradation when ε_{50} = 0.005. The method models the measured gradual rate of softening somewhat better when the larger value of 0.025 is used for ε_{50}, although initial stiffnesses do not match well and the total degradation is again much too severe.

The criteria of Dunnavant and O'Neill on the other hand more closely predict the derived maximum soil resistance. This method matches the derived *P-Y* curves better (in both initial stiffness and rate of degradation) when an ε_{50} value of 0.005 is used. The predicted initial stiffnesses of the *P-Y* curves at the shallower depths are underpredicted when an ε_{50} value of 0.025 is used. The predicted rate of degradation matches fairly well at larger displacements but is too gradual immediately after the peak resistance has been achieved.

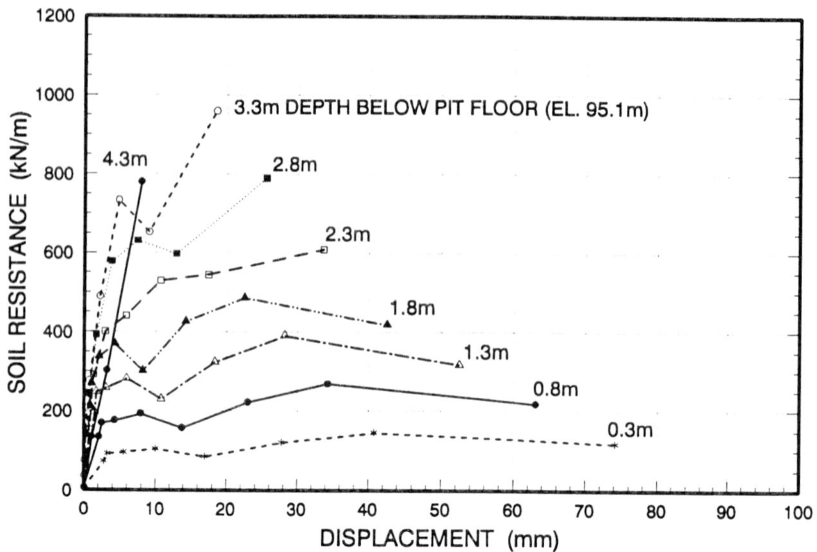

Fig. 3. Derived p-y curves from first loading cycles

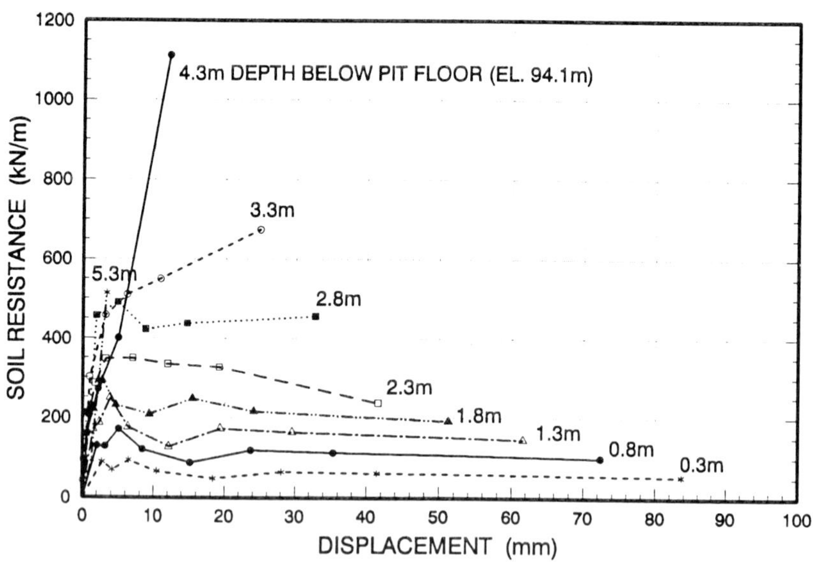

Fig. 4. Derived p-y curves from terminal loading cycles

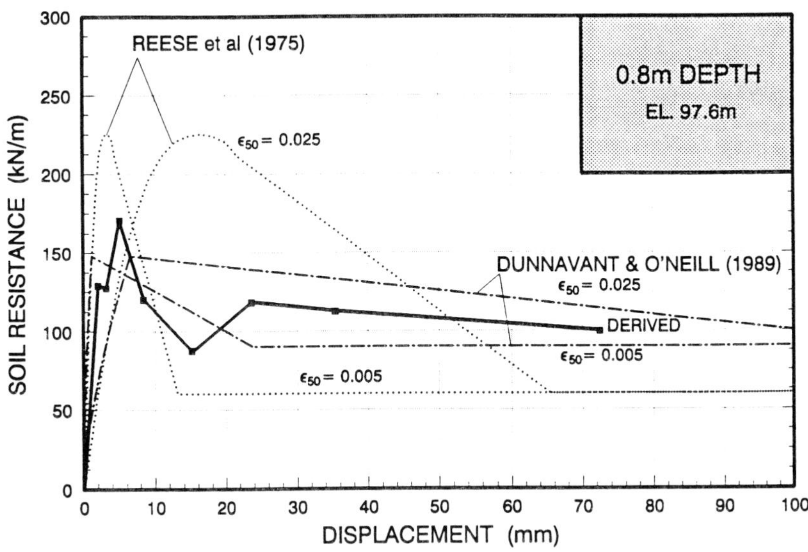

Fig. 5. Comparison of derived p-y curves with published stiff clay cyclic criteria—0.8 m depth below ground surface (pit floor)

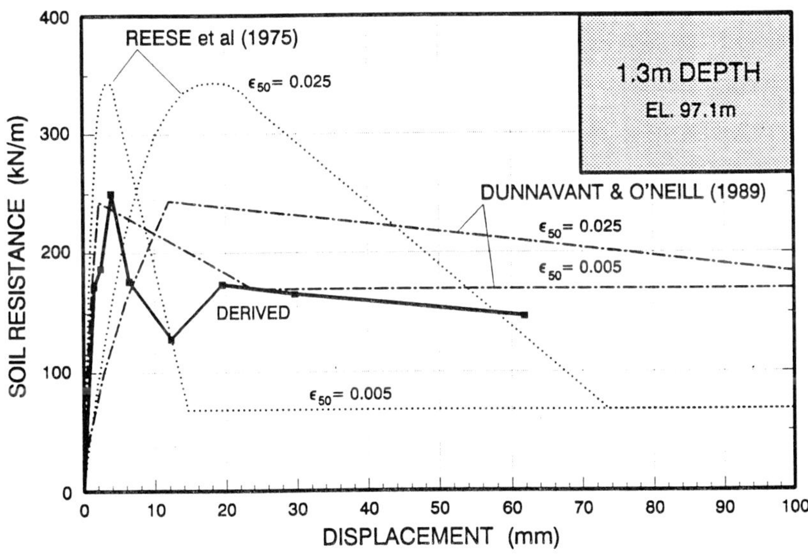

Fig. 6. Comparison of derived p-y curves with published stiff clay cyclic criteria—1.3 m depth below ground surface (pit floor)

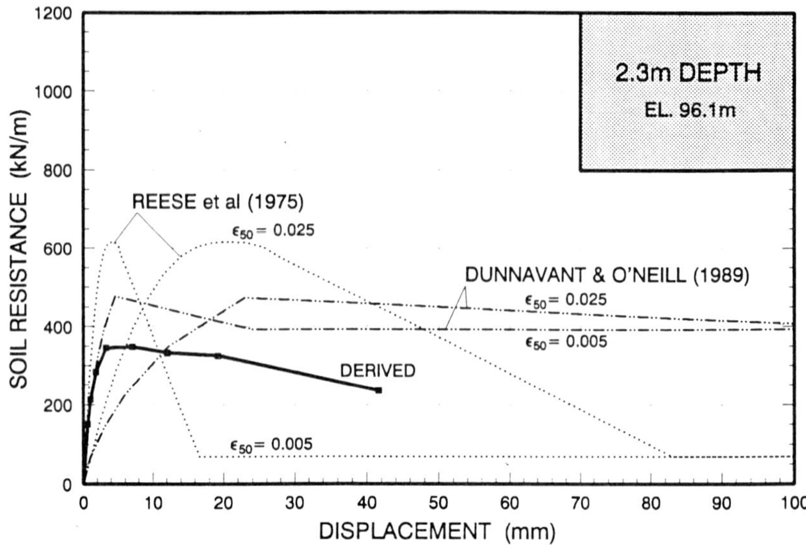

Fig. 7. Comparison of derived p-y curves with published stiff clay cyclic criteria—2.3 m depth below ground surface (pit floor)

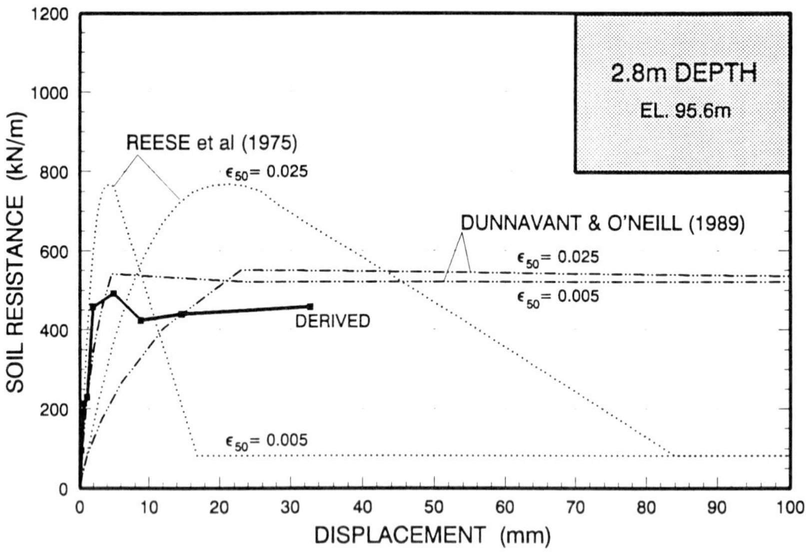

Fig. 8. Comparison of derived p-y curves with published stiff clay cyclic criteria—2.8 m depth below ground surface (pit floor)

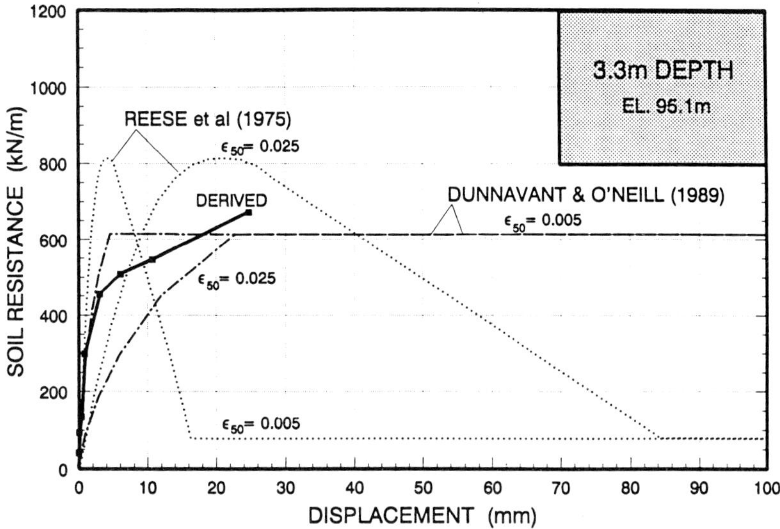

Fig. 9. Comparison of derived p-y curves with published stiff clay cyclic criteria—3.3 m depth below ground surface (pit floor)

Comparison of pile head load–displacement preditions

The effects of using the different *P-Y* curves to predict pile head load displacement behaviour are illustrated in Figs 10, 11, and 12. In Fig. 10 the measured load–displacement curve for the first cycles of loading is compared to predicted curves that were generated using both the derived *P-Y* curves and those developed using various published criteria for static loading. In Figs 11 and 12 the load–displacement curve for the terminal cycles is compared to those generated using cyclic criteria and values of ε_{50} of 0.005 and 0.025, respectively.

The curves were computed using the computer programme LPILE (Ref. 9). The solid line represents measured data points. The curve labelled derived *P-Y* is the predicted load–displacement curve obtained when the derived *P-Y* curves are input into the LPILE programme. The agreement between the actual load–displacement curves and those predicted using the derived *P-Y* curves is very good. Note that because the derived curves are not extracted in the same way they are used, they do not have to provide an exact fit of the load–displacement curve; this is a good check of the *P-Y* extraction process.

Other curves in Figs 10, 11, and 12 indicate the quality of prediction obtained using the Reese *et al.* (1975), Dunnavant and O'Neill (1989), and Matlock (1970) criteria. Note in Figs 11 and 12 the dramatic difference in the predicted curves from the Reese *et al.* (1975) criteria for the two ε_{50} values. The

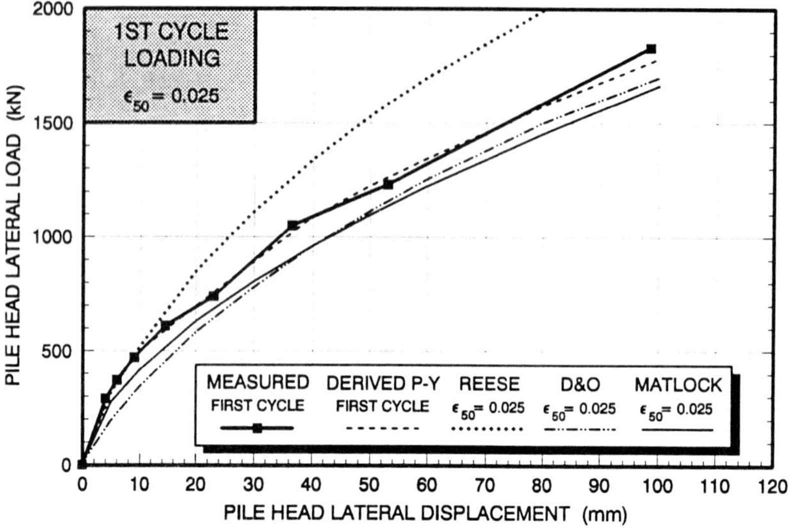

Fig. 10. Comparison of measured pile head load–displacement curve (first cycle loading) with curves predicted using derived p-y curves and various stiff clay criteria

Dunnavant and O'Neill criteria give the closest fit to the measured response. Interestingly, a slightly better prediction is obtained with the curves for ε_{50} of 0.025, even though the *P-Y* curves for that case do not agree as well with the derived ones as do the curves synthesized using ε_{50} = 0.005. We believe the softer initial response and the more gradual rate of degradation of the ε_{50} = 0.025 curves interact fortuitously to produce this result.

Comparison of maximum predicted pile bending moments

In Figs 13 and 14 are shown comparisons between the measured maximum pile bending moments for each displacement increment (terminal cycles) and the maximum moments predicted using the published criteria discussed above. As in the previous figures, Fig. 13 is for an assumed ε_{50} of 0.005, while Fig. 14 is for an assumed value of 0.025. For the case of ε_{50} - 0.005, all criteria give reasonably good estimates of the maximum bending moment (to within 25 per cent). All are conservative as well. The agreement between predicted and measured bending moments for the case of ε_{50} = 0.025 is quite close. We conclude from these figures that any of the published *P-Y* criteria that we examined generally are suitable for predicting maximum moments in free-headed piles. Other researchers (e.g. Ref. 10) also have noted the insensitivity

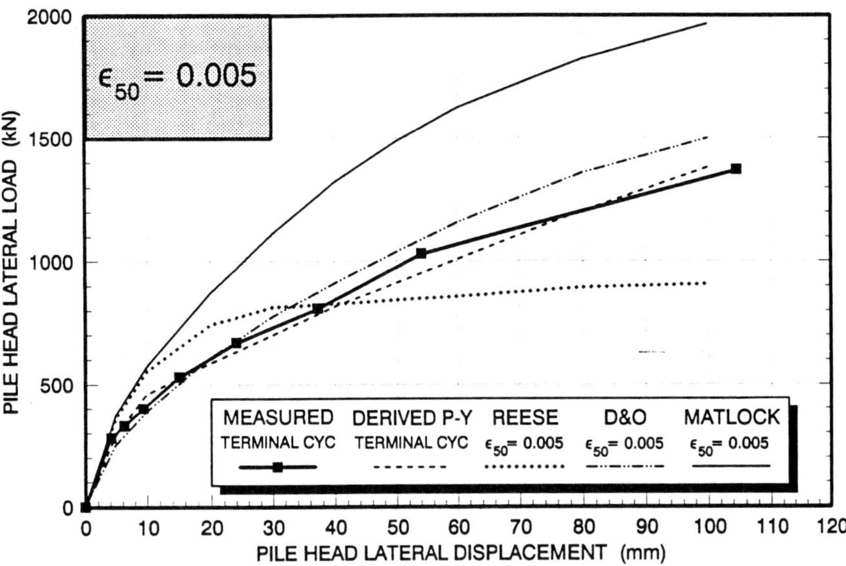

Fig. 11. Comparison of measured pile head load–displacement curve with curves predicted using derived p-y curves and various stiff clay criteria

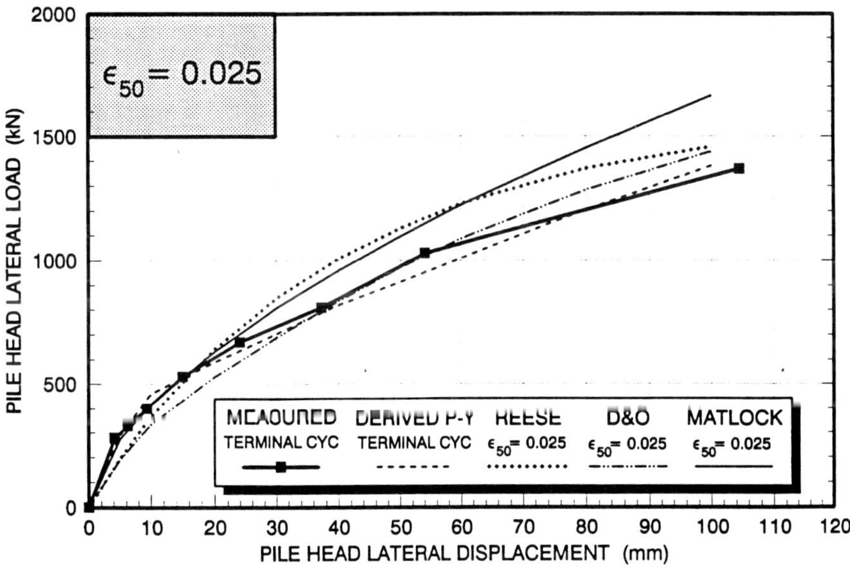

Fig. 12. Comparison of measured pile head load–displacement curve with curves predicted using derived p-y curves and various stiff clay criteria

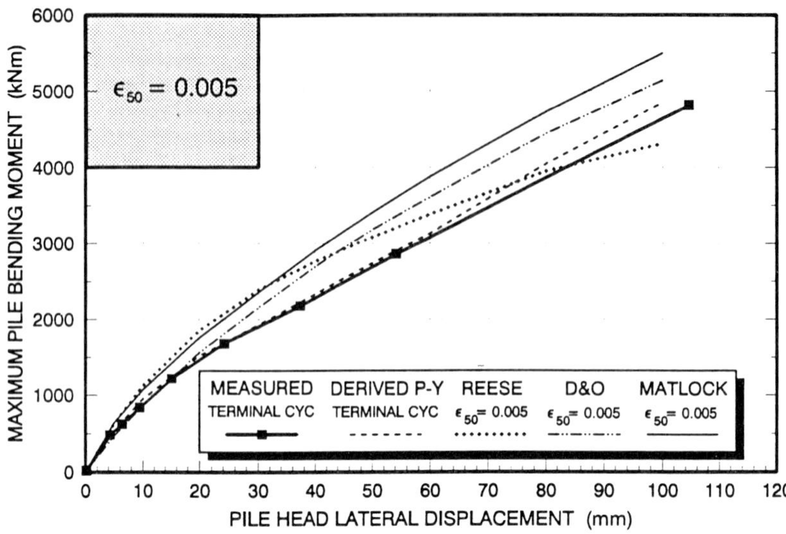

Fig. 13. Comparison of measured maximum pile bending moment with predictions based on derived p-y curves and various cyclic clay criteria

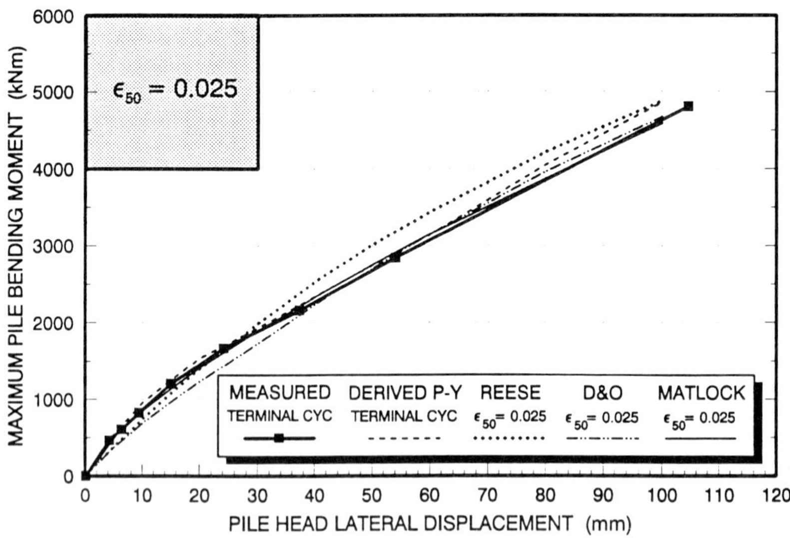

Fig. 14. Comparison of measured maximum pile bending moment with predictions based on derived p-y curves and various cyclic clay criteria

of calculated maximum free-head pile moment to the *P-Y* criterion that is used to obtain the prediction.

Conclusions

Derived *P-Y* curves from the Tilbrook test show a moderate degree of post-peak softening (after cycling) at shallow depths (to 2.3 m or three diameters). No softening was observed below this depth. Large displacements could not be attained at depths beyond 3.8 m.

When applied to the Tilbrook case, the *P-Y* criteria of Reese *et al.* (1975) for cyclic loading of submerged stiff clay overpredict the maximum lateral soil resistance. Based on the measured value of ε_{50} of 0.025, the criteria predicts an initial loading slope that is too soft, a reasonable rate of degradation of lateral soil resistance with displacement, but an excessive total degradation in resistance at large displacement. In the absence of test data, the recommended design value of ε_{50} would be 0.005. Use of this value gives a better match to the early loading curve, but the predicted *P-Y* curves degrade too rapidly (and excessively) and result in a poor prediction of the pile head load–displacement behaviour.

The criteria proposed by Dunnavant and O'Neill (1989) more closely predict the maximum soil resistance and post-peak softening behaviour. Overall, pile head load–displacement behaviour was predicted best using the Dunnavant and O'Neill *P-Y* criteria. Using ε_{50} = 0.005 with this method resulted in good predictions of initial pile head stiffness, stiffness at large displacement (40+ mm) and maximum pile load. A conservative (softer) estimate of initial stiffness and a slightly better prediction of peak load were calculated when ε_{50} = 0.025 was used.

Based on the findings of this study, the Dunnavant and O'Neill criteria are recommended for lateral analysis of cyclically loaded piles in stiff, overconsolidated marine clays.

The best predictions of pile performance were obtained when the laboratory measured values of ε_{50} are used to develop *P-Y* curves. Unfortunately, those values differ by a factor of five (smaller) from the values that are recommended in the absence of laboratory UU test data. The designer may take comfort, however, that maximum pile moments were predicted with acceptable accuracy for both of the ε_{50} values used in the analyses.

Acknowledgements

The authors thank the management of Exxon Production Research Company for permission to publish this paper.

References

1. MATLOCK, H. (1970) Correlations for design of laterally loaded piles in soft clay. Proc. *Second Annual Offshore Technology Conf*, Vol. 1, Houston, Texas, pp. 577-594.

2. REESE, L. C., COX, W. R., and KOOP, F. D. (1974) Analysis of laterally loaded piles in sand. *Proc. Sixth Annual Offshore Technology Conf*, Vol. 2, Houston, Texas, pp. 473-484.

3. REESE, L. C. and WELCH, R. C. (1975) Lateral loading of deep foundations in stiff clay. *J. Geotech. Engng Div., Am. Soc. Civ. Engrs*, **101**, No. GT7, pp. 633-649.

4. REESE, L. C., COX, W. R., and KOOP, F. D. (1975) Field testing and analysis of laterally loaded piles in stiff clay. *Proc. Seventh Annual Offshore Technology Conf.*, Vol. 2, Houston, Texas, pp. 671-690.

5. DUNNAVANT, T. W., and O'NEILL, M. W. (1989) Experimental *p-y* model for submerged stiff clay. *J. Geotech. Engng Div., Am. Soc. Civ. Engrs*, **115,** No. 1, 95-114.

5a. DUNNAVANT, T. W., and O'NEILL, M. W.(1989) Errata – Experimental *P-Y* model for submerged stiff clay. *J. Geotech. Engng Div., Am. Soc. Civ. Engrs*, **115,** No. 11, 1681.

5b. DUNNAVANT, T. W., and O'NEILL, M. W.(1991) Closure – Experimental *P-Y* model for submerged stiff clay. *J. Geotech. Engng Div., Am. Soc. Civ. Engrs*, **117,** No. 5, 827-829.

6. MCCLELLAND, LTD (1988) *Soil and foundation investigation Tilbrook site – final report*. Proprietary report to BP International on behalf of the participants, LDPT, August, 1988.

7. OVE ARUP & PARTNERS (1988) *Report on the axial tension and lateral pile tests at Tilbrook Grange*. Proprietary report to BP International on behalf of the participants, LDPT, Report 18655, November 1988.

8. FUGRO-MCCLELLAND, LTD (1989) *Field and laboratory report – supplemental site investigation – Tilbrook Grange, Cambridgeshire*. Proprietary report to BP International on behalf of the participants, LDPT; Report 89/0530-1, April 19, 1989.

9. REESE, L. C., and WANG, S.-T. (1989) Documentation of computer program LPILE Version 3.0. Ensoft, Inc., Austin, Texas.

10. GAZIOGLU, S. M., and O'NEILL, M. W. (1984) Evaluation of *p-y* relationships in cohesive soils. *Analysis of Design of Pile Foundations*, ASCE, Oct. 1984, pp. 192-213.

20. LDPT as calibration for new pile models

G. SVANØ, Sintef, and T. I. TJELTA and A. EIDE, Statoil, Norway

The performance of a newly developed model for pile response has been verified against the LDPT Tilbrook test results. The model is based on basic elements from triaxial test interpretation and on a simple normalised tangent shear stiffness description on stress-strain level. It is a unified model for axial and lateral pile analysis. It is found that the performance of the model is excellent on an overall level. When going into details, a minor deviation is found between the observed and calculated variation with depth of the axial shaft friction. The best estimate normalized soil shear stiffness was based on a combination of results from in situ tests and triaxial tests. A very good agreement between measured and calculated pile load-displacement response was observed with this best estimate shear stiffness.

Introduction

During the Statoil sponsored project Pile–soil interaction (the PSI project) effective stress models were developed for the analysis of piles under undrained loading conditions. The undrained loading is assumed to take place when load duration is short compared to the time for reconsolidation in the soil around the pile. This is in most cases true for piles in clays, but also for loads of short duration in silty clays (and possibly silts). The fundamental basis is the same as for Janbu's (ref. 1) effective stress models, which essentially were developed for the drained or sustained loading conditions (see also refs 4, 5).

The model development was motivated by the observation that the shaft friction capacity τ_F of piles in overconsolidated (OC) clay as a rule is lower than the undrained shear strength s_u. The higher the degree of overconsolidation, the lower the ratio $\alpha = \tau_F / s_u$. This is reflected in for example the API '91 (ref. 7) and API'87 design rules, where a high $\psi = s_u/p'_0$ gives a low α-factor. In general a high ψ –ratio corresponds to a high overconsolidation degree.

The scatter in s_u is usually larger for overconsolidated than for normally consolidated clays. This is mainly because OC clays dilate after the effective stress path has reached the Mohr–Coulomb failure envelope, and during dilation the sample is rather unstable. Some samples continue to develop uniform strains and therefore follow the Mohr–Coulomb envelope up to large shear stresses, while others develop localized shear bands and fail at lower stresses. Another source of uncertainty is the failure strain definition, which is not standardised, and hence may vary between the different institutions.

A better measure for the shaft friction capacity was therefore sought, and the natural choice was simply to exclude the dilating part of the effective stress path. This choice can be justified as follows: The undrained shear strength in s_u the OC clay relies on dilation after reaching the Mohr–Coulomb envelope, leading to increasing mean effective stresses (Fig. 1) to the extent that this dilation occurs around an axially loaded pile, the dilation induced effective stress increment will be restricted to a relatively narrow zone around the pile, and in this zone, a pore water suction of about the same magnitude will result. The suction gives swelling tendencies, and a localised shear band may therefore easily develop. Hence, the shaft friction is stable up to a shear stress τ_M(Fig. 1), but unstable when loaded further towards the undrained shear strength s_u.

An undrained axial shaft friction model was developed strictly along these lines, on a theoretical basis. The *T-Z* response was obtained from an integration of the shear stiffness measured in the triaxial cell. A similar model was also developed for lateral loading.

The model has been verified against three pile test cases, the Haga test (Refs 13, 14, 15), the Onsoy test (ref. 16) and the LDPT test at Tilbrook. The latter is presented here. The model will further be verified against the LDPT Pentre results.

The developed model

Axial

The shaft friction is equal to

$$\tau_F = r\,\tau_M \tag{1}$$

where the characteristic shear stress τ_M is defined as the shear stress at which an idealized linear effective stress path reaches the Mohr-Coulomb failure criterion. This is illustrated in Fig. 1 for a dilatant clay. The roughness ratio r depends on the type of clay and takes values between 0.7 and 1. For a non-sensitive clay, as the Tilbrook clay, r=1 is recommended. For clays with sensitivity index of S_t=4 and above, r=0.7 is proposed. That is, for most marine clays on land in Norway, r=0.7 should be used, whereas for offshore conditions at the Norwegian continental shelf, the roughness ratio would vary between 0.7 and 1.

The characteristic stress τ_M might in principle be determined directly from the triaxial test, and plotted as a function of depth. However, one objective in the model development was to be able to take into account changes in effective stress (excess pore pressure), and changes in the earth pressure coefficient K'. The shear stress τ_M is therefore expressed as a function

$$\tau_M = T_{vu} (\sigma'_v + a) \tag{2}$$

where σ'_v is the current effective vertical stress, a is the attraction defined as $a = c/\tan\phi$, where c is cohesion (see Fig. 1), and T_{vu} is an undrained shear number. The undrained shear number T_{vu} is developed in the Appendix and is a function primarily of:

D, a dilatancy parameter determined as the slope of an idealised linear effective stress path in p'-q plot (Fig. 1),
K'_a, the earth pressure coefficient against pile, and
ϕ, the internal soil friction angle

The idealised effective stress path is a linearisation of the real effective stress path from consolidation stresses and up to the Mohr-Coulomb failure envelope. For a dilatant clay, the linearised effective stress path may be drawn as a secant through the point on the effective stress path corresponding to a mobilized friction $\tan\rho = 0.9 \tan\phi$, see Fig 1.

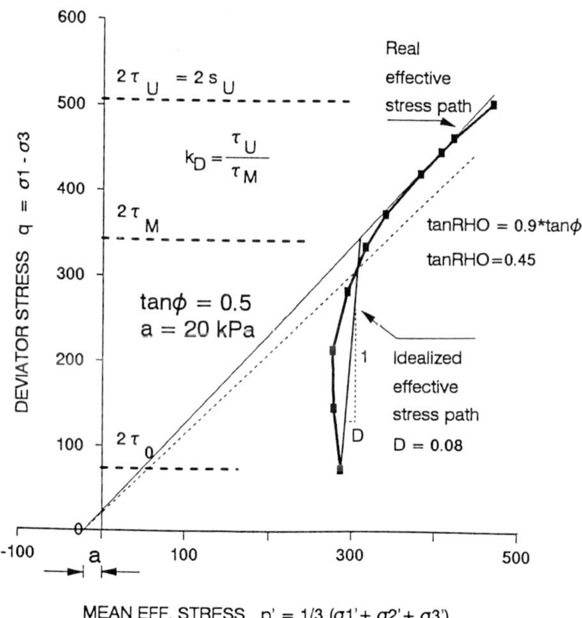

Fig. 1. Effective stress path idealisation

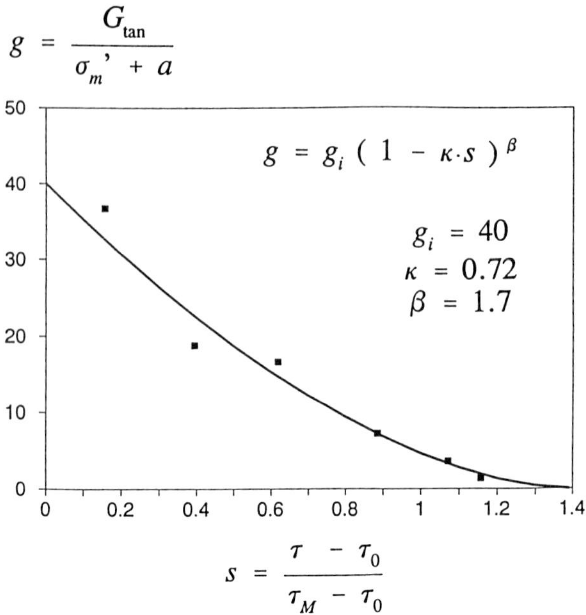

Fig. 2. Normalised tangent shear stiffness (sample)

In the present version of the model the earth pressure ratio is defined as $K'_a = (\sigma'_h + a)/(\sigma'_v + a)$, and this deviation from the standard definition should be observed when parameters are determined for the model. An advantage of this definition is that it immediately reflects the internal friction mobilization due to anisotriopic initial stresses. Furthermore, the value of K'_a will not rise to infinite values at zero depth, as does the standard earth pressure coefficient (K'_0) in an overconsolidated clay.

The axial *t-w* (*t-z*) curve is computed by numerical integration of shear strains over the radius of a soil disk around the pile. Based on these numerical integrations, dimensionless displacement functions are developed. The disk radius is 15 pile radii. The tangent shear modulus G_{tan} is a function of the mean effective stress τ and decreases with increasing shear stress through the equation

$$G_{tan} = g\,(p' + a) = g_i\,(1 - \kappa\, s)^{\beta}\,(p' + a) \qquad (3)$$

where g_i is a shear modulus number and a is the attraction defined above, s is a normalised shear stress, defined as the ratio $s = (\tau - \tau_0)/(\tau_M - \tau_0)$. In the triaxial cell, τ_0 corresponds to the shear stress at consolidation, so that $\tau_0 = 1/2\,(\sigma_a - \sigma_r)$. σ_a is axial and σ_r is radial consolidation stress. For the pile, τ_0 represents the vertical shear stress at the pile surface before loading. The parameters κ and β are determined through interpretation of triaxial test results.

A sample g-function is included in Fig. 2. Note that s is normalised against τ_M, not the ultimate shear stress $\tau_U = s_u$. As for the capacity this choice was motivated by the considerably lesser scatter in τ_M than in s_u for dilating clays. The s-axis on Fig. 2 therefore continues past the value 1.0 defining the κ-parameter.

Lateral

The lateral pile model is based on the ultimate shear strength τ_U. To get an effective stress dependency as for the axial capacity, the ultimate shear strength is determined as $\tau_U = k_D\ \tau_M$ where the k_D-factor is determined from triaxial tests consolidated at the in situ stress level. The lateral capacity is computed as

$$p_{ru} = d\ N_{ru}\ \tau_U \tag{4}$$

where d is the pile diameter and the factor N_{ru} assumes the value 4~5 at the depth range from zero to $5d$, 8~10 below $8d$, and varies linearly for depth ranges in between.

The p-v (or p-y) curve is obtained by numerical integration, and from the results dimensionless displacement functions are developed. The same tangent modulus function is used as for the axial pile model. A unified approach for the axial and lateral pile response is therefore obtained.

From equation 2 it is seen that the capacity is made a function of the current effective stress. Capacity degradation from cyclic loading effects can therefore be taken into account through a cyclic pore pressure generation model, see Appendix.

Material data for the Tilbrook pile tests

The triaxial test results from Tilbrook test site have been interpreted to obtain a set of data which is consistent with the model. The essential data is presented in Figs 3 to 9. Figs 3 and 4 show the interpreted effective strength parameters from the CAU tests. A layer boundary is assumed at 18m depth. Above 18m, there is some scatter in the internal soil friction tan ϕ. However, excluding one freak value of tan ϕ=0.78 at about 8m, an average friction of tan ϕ=0.5 has been interpreted. This value must be seen in combination with the attraction value of a=40kPa. Below 18m, the set a=20kPa and tan ϕ=0.62 has been interpreted.

The dilatancy parameter D is presented in Fig. 5. One will observe that the parameter increases towards the soil surface. This is usually the case, due to increasing overconsolidation degree. The interpreted k_D factor for the hori-

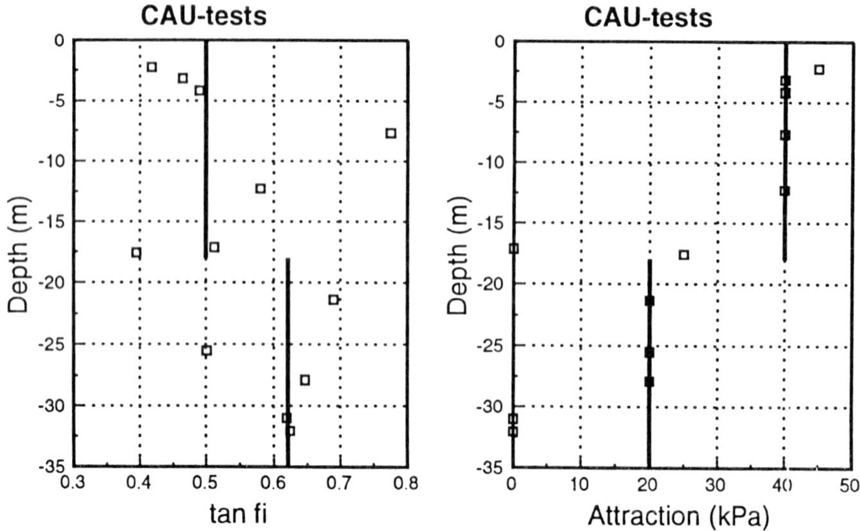

Fig. 3. Interpreted soil friction (tan ϕ)

Fig. 4. Interpreted 'attraction' a

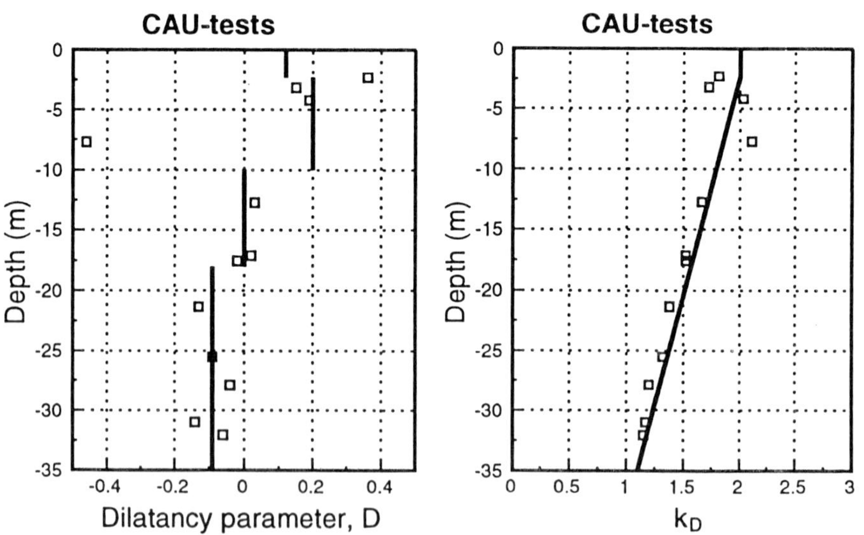

Fig. 5. Interpreted dilatancy parameter D

Fig. 6. Interpreted κ_D

zontal capacity is shown in Fig. 6. Also this factor increases towards the soil surface, up to a value of k_D=2.

Figure 7 shows the measured effective radial stress to effective overburden pressure ratio at 130 days from pile driving. The figure also includes the estimated in situ value. In this plot, an interpretation of the ratio $K'_a = (\sigma'_h + a)/(\sigma'_v + a)$ has been drawn, as a unified plot to be utilised for calculation of both axial and lateral pile capacity.

The shaft friction capacity of the pile is mostly obtained from deeper soil strata. Since the shaft friction is determined by stresses adjacent to the pile, the radial earth pressure after pile installation is given the most weight when assessing the K_a values below 10m. The lateral pile stiffness is governed by the soil response at shallow depth. During lateral loading a larger soil volume is mobilized, and consequently the lateral capacity is less influenced by pile installation effects. Above 10 m the initial, in situ earth pressure (K'_0) estimate is therefore given highest weight in the K'_a assessment.

One additional assumption was made regarding the K'_a coefficient. Prior to the compression test in February 1988, an excess pore pressure of about 200 kPa remained in the soil layers below 20 m. The tension test on the trial pile took place 20 months after its installation, and based on the assumption that most of the excess pore pressure had dissipated, the K'_a coefficient was increased by 25 to 35 per cent below 20 m compared to the observed ratio for the compression pile. This is based on the observation that the total pressure against the pile at this depth was almost constant with time, see Plates 23 and 24 in (ref. 10).

The in situ vertical effective stress is shown in Fig. 8. (From ref. 10).

The tangent shear modulus number g as a function of the shear mobilization *s* has been interpreted from the triaxial test plots in (refs. 9 and 11). The Figs 9 and 10 show the interpreted values above and below 18 m. The shaded regions show the modulus number interpreted directly from the triaxial tests. A line has been drawn, denoted average triaxial. However, due to lack of access to raw data, the stiffness has been interpreted from the rather crude plots in the LDPT laboratory soil data report. In particular in the low strain range the accuracy suffers from this, and an adjusted curve is drawn, denoted best estimate. The best estimate curve lies midway between the average triaxial shear stiffness and the shear stiffness from the normalised in situ shear modulus plot on Plate 88 in (ref.9).

Analysis results

Two piles were installed at Tilbrook Grange, a trial pile and a compression test pile. The compression test pile was originally tested in axial compression in February 1988, 130 days after installation. The trial pile was tested in axial tension in September 1988, approximately 20 months after its installation. At the same time, the piles were tested with horizontal loads. These tests, which are reported in (ref. 12) are back calculated using the pile–soil interaction (PSI) models.

Axial tension

The results from the axial tension load analysis on the trial pile is compared with measurements in Figs 11 to 14. Fig. 11 shows the axial pile force distribution at peak load. The calculated is quite close to the measured, although it must be admitted that there is a little deviation at the middle of the pile.

Figure 12 compares computed and observed load-displacement characteristics of the pile. The solid line gives the observed behaviour, and the dashed lines show the calculated for best estimate and average triaxial stiffness. The calculated pile capacity is very close to the observed, although somewhat higher. The calculated load-displacement curve for the best estimate stiffness lies almost on top of themeasured. The computed response is on the other hand slightly too soft when using the average triaxial stiffness.

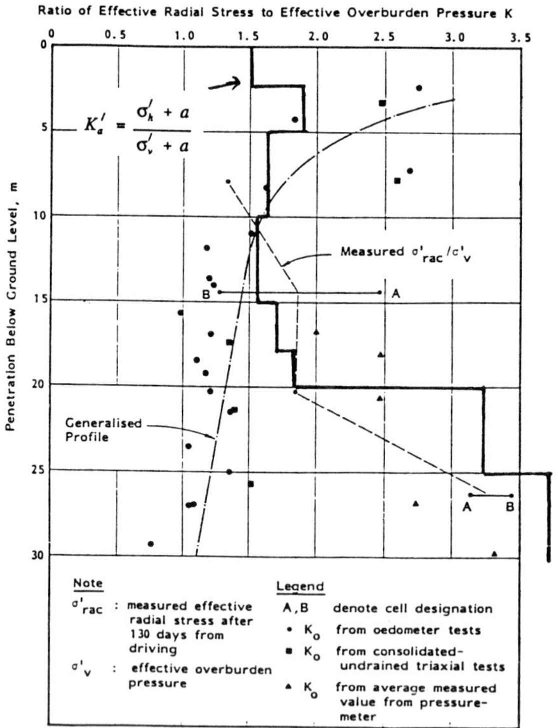

Fig. 7. Assumed K'ₐ coefficient for tension pile

Computed and measured *t-z* or (*t-w*) curves are compared on Figs 13 and 14. At 14 m depth there is a deviation in computed and measured capacity. However, themeasured shaft friction shows a plus/minus variation with depth that a calculation model based on generalised soil profiles cannot be expected to reproduce. At 25 m the agreement between calculated and measured capacity is perfect. The *t-z* stiffness resulting from the best estimate normalised shear modulus matches the observed *t-z* stiffness very well, both at 14 and 25 m.

Lateral pile test

The piles were subjected to cyclic lateral loads. Multistage cycling was applied, in blocks of constant cyclic amplitudes. The cyclic load levels was increased from one block to the next. The envelope around the force displacement for the first load cyclic in each block is taken to represent the static pile response.

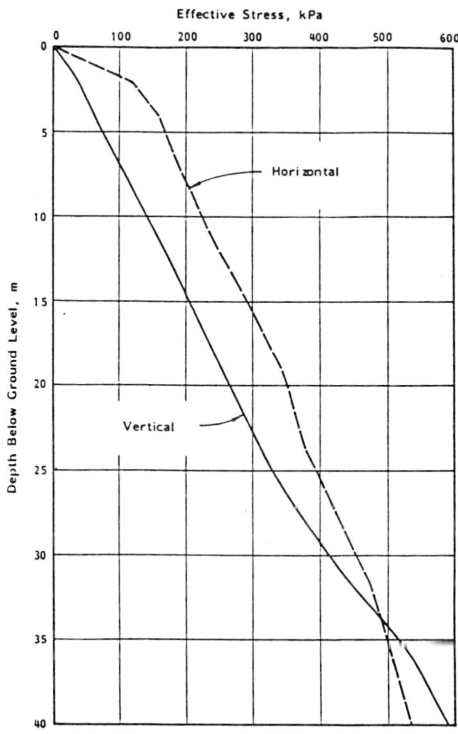

Fig. 8. Vertical effective stress

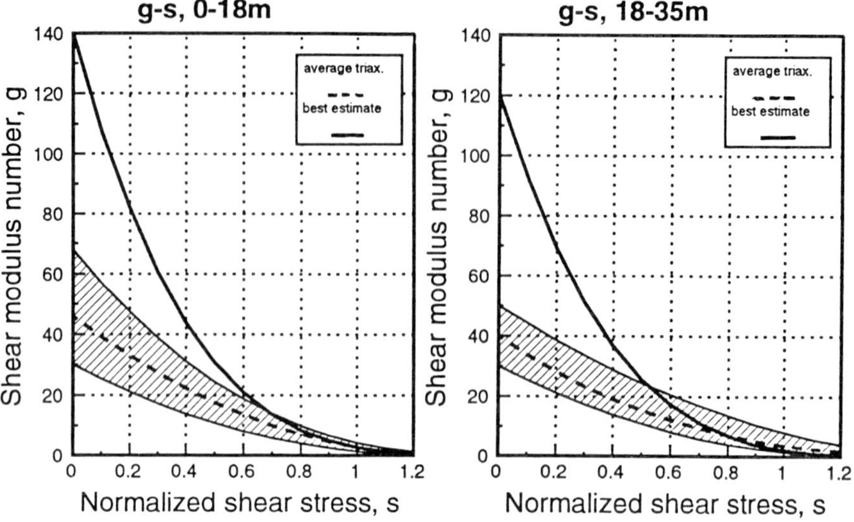

Fig. 9. Shear modulus number above 18 m

Fig. 10. Shear modulus number below 18 m

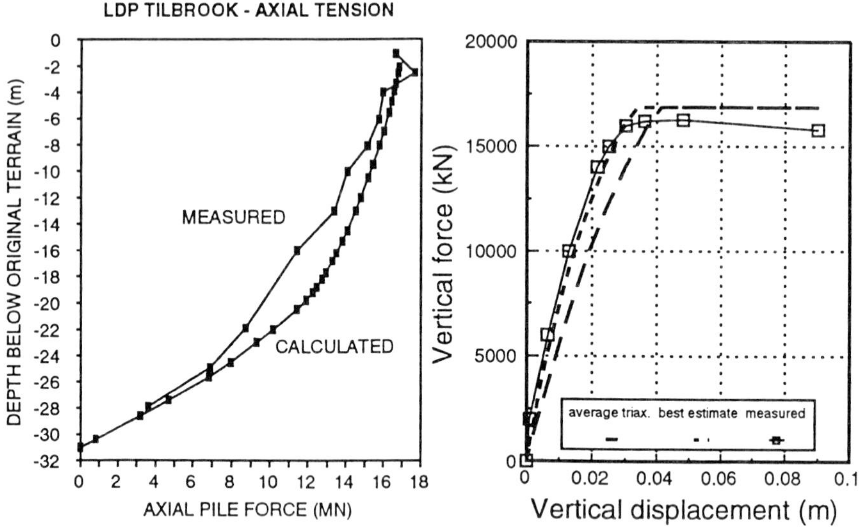

Fig. 11. Calculated and measured axial pile force at tension failure

Fig. 12. Axial force – displacement, tension pile

Figure 15 shows in solid lines the measured force displacement response at the strut-level for the compression pile and the trial or tension pile. The plotted curves represent the first cycle. The dashed curves show the calculated response is slightly to the conservative side, the model performance is fully acceptable.

Figures 16 and 17 show the *p-y* response at two depths. Fig. 16 compares calculated response at 4 m below ground level (dashed lines) with the average of observed shallow depth *p-y* responses (solid lines, first and terminal cycle).

At all the depths within the considered range, a degradation was observed from the first to the terminal cycle. The strain softening type of response that could be expected for the terminal cycles *p-y* curves was seen at some depths, while at other depths it was not so clear as might be expected. The averaged observation curves are drawn horizontal after reaching the peak level. It might be justified, however, to draw the averaged terminal cycle curve with some degree of softening.

The calculated *p-y* curves have a reasonable agreement with observations with regard to stiffness. The capacity is quite close to the degraded capacity for the terminal cycle. On the other hand, no degradation was included in this simulation, and a perfect fit with the first cycle, *p-y* curve would have been a more satisfactory result as far as model verification is concerned.

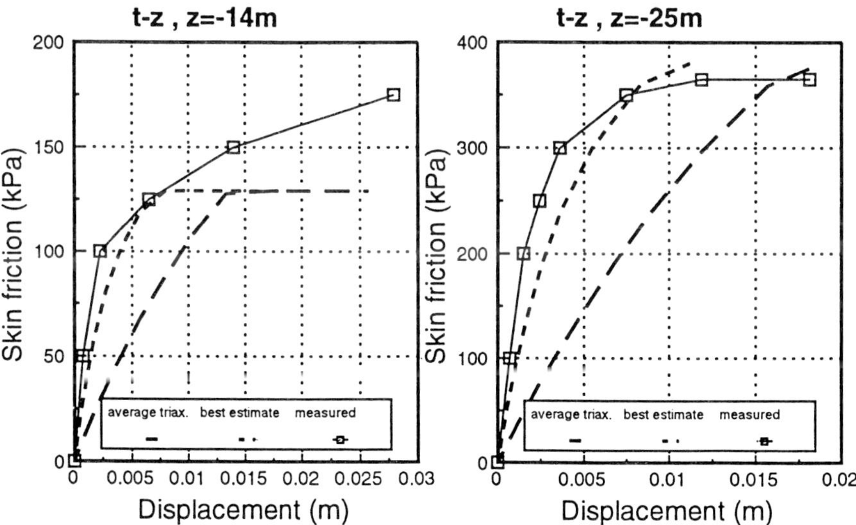

Fig. 13. t–z curves, 14 m depth

Fig. 14. t–z curves, 25 m depth

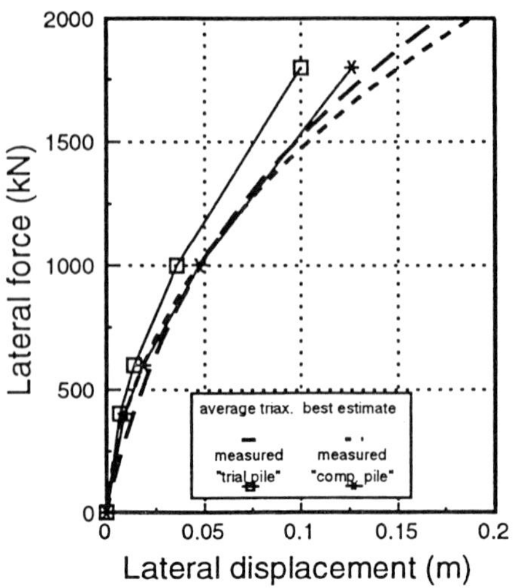

Fig. 15. Lateral load – displacement, tension and compression pile

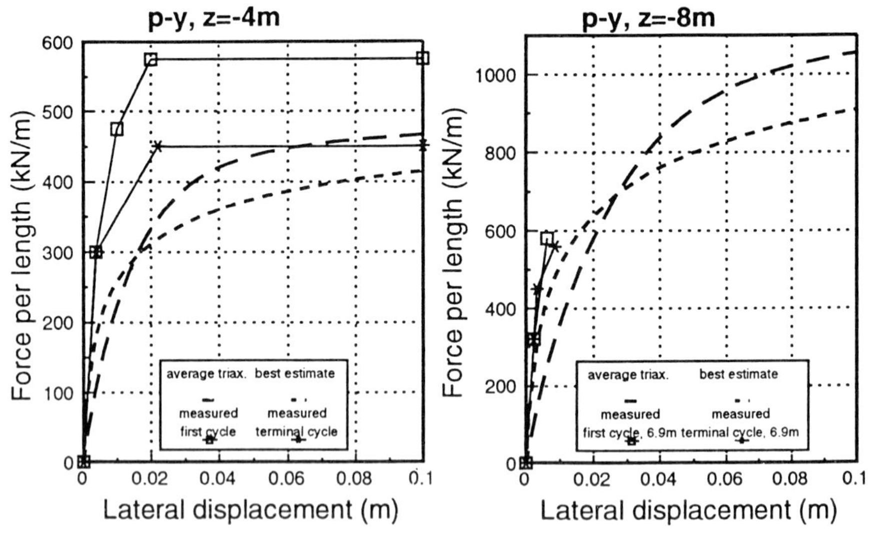

Fig. 16. p–y curves, 4 m depth

Fig. 17. p–y curves, 8 m depth

Compared to this curve, the model underestimates the capacity by some 20 per cent.

Figure 17 shows the results at 7 to 8 m depth. As seen, the model gives acceptable simulation of the stiffness. At this depth, however, the soil is not brought to failure, so a verification of the lateral capacity cannot be made.

Conclusions

Based on the comparison between calculated and measured response for the Tilbrook Grange test piles, it is found that the developed pile/soil model can give a very good prediction of the response of a pile. The calculated overall response of the piles was in very good agreement with the observations. At local depths, there is a deviation between the model and measurements, but not more than must be expected when the analysis is based on averaged soil profiles.

The agreement between computed and observed stiffness in *t-z* as well as in *p-y* response is remarkably good, in particular on the background that the model has a fully theoretical basis. The verification presented in this paper therefore shows that the fundamentals of the soil response on a stress-strain level successfully can be transferred to the rather complex situation around a pile, both for axial and lateral loading.

Appendix

The characteristic shear stress τ_M

The dilatancy parameter D is used to express the tendencies to dilation or contraction before the effective stress path reaches the Mohr-Coulomb failure envelope, and may mathematically be expressed as

$$D = \Delta p' / \Delta q \qquad \text{(A1)}$$

where $q = \sigma_1 - \sigma_2$ is the deviatoric stress and $p' = 1/3\ (\sigma'_1 + \sigma'_2 + \sigma'_3)$ is the mean effective stress. Equation A1 expresses the shear induced change in the mean effective stress. Based on this, the shear stress τ_M can be computed from

$$\tau_M = (M/2)/[(1 - mD)\ /(1 - MD)]\ (p'_0 + a) \qquad \text{(A2)}$$

where p'_0 is the mean effective stress and m is the deviator to mean stress ratio near the pile before loading; m is defined as

$$m = (\sigma_{10} - \sigma_{30})\ /\ (p'_0 + a) \qquad \text{(A3)}$$

The factor M is the deviator to mean stress ratio at the Mohr-Coloumb envelope. It may be expressed as

$$M = [3(N-1)] / [3 + (1+b)(N-1)]$$
$$\text{where } N = (1+\sin\phi) / (1-\sin\phi) \quad \text{(A4)}$$

and b is Habib's σ_2-ratio defined as $b = (\sigma_2 - \sigma_3) / (\sigma_1 - \sigma_3)$. The model verifications have been performed with b=0.3, which may correspond to a plane strain condition; b=0.3 is recommended as default value.

The mean effective stress p'_0 may be expressed through the K' ratio (defined in the text above) and the vertical effective stress σ'_v as $p'_0 + a = 1/3(1+2K'_a)(\sigma'_v + a)$. The undrained shear number T_{vu} can hence be expressed

$$T_{vu} = (M/2)\,[(1 - mD) / (1 - MD)]\,[(1 + 2K'_a) / 3] \quad \text{(A5)}$$

The theoretically based equations are basically taken from (ref.6).

The axial displacement function

Shear strains in a soil disk with radius equal to ξ times the pile radius are integrated numerically into the axial pile displacement w, and from the numerical results, a displacement function is fitted.

$$w = (1/\kappa)(\tau_F/G_i)\,R_0 \ln(\xi) \cdot Q(f) \quad \text{(A6)}$$

where R_0 is the pile radius. The function $Q_a(f)$ is given as

$$Q = (1 - A)\,[(\kappa f) / (1 - \kappa f)] + A\kappa f + B(\kappa f)^2 \quad \text{(A7)}$$

f is the ratio between the current axial shear stress and the shaft friction at failure; $f = \tau / \tau_F$. The factors A and B are functions of the modulus exponent β.

$$A = 1 - 0.00196\beta^{4.518}$$

$$B = 0.3 + 0.25\,(\beta - 1.05)(3 - \beta)$$

for β values from 1 to 4.

The lateral displacement function

From numerical integrations of strains, a displacement function is also fitted for the lateral pile model:

$$v = [(p_{ru}) / (5G_i)]\,(1/\kappa)\,H(f) \quad \text{(A8)}$$

$$H = (1 - A)\,[(\kappa f) / (1 - \kappa f)] + A\kappa f + B(\kappa f)^2 \quad \text{(A9)}$$

For the lateral model, f is the ratio between the current and the ultimate lateral soil reaction; $f=p_r/p_{ru}$. For β-values between 1 and 4, A and B for the lateral model may be estimated from

$$A = 1 - 0.0152\beta^{4.26}$$

$$B = -2.35 + 4.05\beta - 1.2\,\beta^2$$

Cyclic degradation of capacity

A cyclic capacity degradation model is developed and implemented based on a cyclic pore pressure generation concept (refs 2, 3). The excess pore pressure at any depth along the pile is computed according to

$$\Delta u = [(4\tau_{cy}) \;/\; (r_u)] \ln [(1 + N_c + N_{c0}) \;/\; (1 + N_{c0})] \tag{A10}$$

τ_{cy} is the cyclic shear stress and N_c is the number of cycles. The material parameters r_u (the pore pressure resistance number) and N_{c0} (the reference number of cycles) are functions of the static and cyclic shear stress mobilization ratios $\tau_{cy} \;/(p'_0 + a)$ and $\tau_a \;/(p'_0 + a)$ as well as the degree of overconsolidation. τ_a is the average or ambient shear stress at the pile.

A systematic evaluation of the parameters has so far only been made for a normally consolidated marine clay, and it was found that while N_{c0} varied only between 2 and 3, the values of r_u covered a range from about 4 to 30, see Table A1. A high value gives a low rate of pore pressure generation.

The accumulated excess pore pressure is taken into account by reducing the effective stress in calculation of the characteristic shear stress τ_M

$$\tau_M = T_{vu} \, (\sigma'_v - \Delta u + a) \tag{A11}$$

For a normally consolidated clay, the model typically gives a shaft friction capacity under symmetric gwo way cyclic loading of 60 to 65 per cent of the static capacity (100 to 500 cycles to failure).

Table A1 Pore pressure resistance number r_u for NC clay

	$\tau_{cy} /(p'_0 + a) = 0.3$	$\tau_{cy} /(p'_0 + a) = 0.3$
$\tau_a /(p'_0 + a) = 0$	$r_u = 23$	$r_u = 4$
$\tau_a /(p'_0 + a) = 0.34$	$r_u = 33$	$r_u = 4$

References

1. JANBU, N., Simple pile foundation analysis. The Arne Selberg symposium, Trondheim 1980. Reprinted in bulletin no. 14 NTH, Geotechnical Division, Trondheim 1982.
2. JANBU, N., Soils under cyclic loading, Chairman's report, spec. sess. no 1 on soils, *Proc. Behaviour of offshore structures*, NTH, Trondheim, 1976
3. JANBU, N., Soils in offshore engineering. 25th Rankine Lecture. *Géotechnique* **35,** No. 3 Sept. 1985.
4. GRANDE, L. O., *Samvirke mellom pel og jord* (Unoff. title trans, pile soil interaction), Dr. ing. thesis NTH, 1976.
5. GRANDE, L. and NORDAL S., Pile soil interation analysis on effective stress basis. *Proc. Recent developments in the design and construction of piles*, ICE 1979, London.

6. SVANO, G. *Undrained effective stress analysis*. Dr. Ing thesis, NTH, 1981.
7. AMERICAN PETROLEUM INSTITUTE. *Recommended practice for planning, designing and constructing fixed offshore platforms* API RP2A, 19th Edition, August 1, 1991.
8. SINTEF. *Pile-soil interaction. Theory development for computer implementation*, STF69 F91007, 1991-04-26, 102p (G. Svano, T. Lieng).
9. MCCLELLAND LTD, London, *Large diameter pile test programme. Soil and foundation investigation. Tilbrook site*. Fianl report 2 volumes. August 1988.
10. MCCLELLAND LTD, London. *Large diameter pile tests programme project report*. Volume III of V, Tilbrook pile tests results, analysis and interpretations. August 1988.
11. FUGRO-MCCLELLAND LTD, *Field and laboratory report. Supplementary site investigation, Tilbrook Grange, Cambridgeshire*. Report No. 89/0530-1 Rev 00.
12. OVE ARUP & PARTNERS, *Report on the axial tension and lateral pile tests at Tilbrook Grange*, November 1988
13. KARLSRUD, K. and HAUGEN, T. *Cyclic loading of piles and pile anchors — field model tests, final report, summary and evaluation of results*. Norwegian Geotechnical Institute, rep. no. 40010-28, April 1983.
14. KARLSRUD, K and HAUGEN T. Axial static capacity of steel model piles in overconsolidated clay. *Proc. 11th Int. Conf on Soil Mech. and Fndn Engng*. San Francisco 1985.
15. KARLSRUD, K., NADIM, F. and HAUGEN T. Piles in clay under cyclic axial loading – field tests and computational modelling. *Proc. 3rd int. conf on num. methods in offshore piling*, Nantes, May 1986
16. KARLSRUD, K., HANSEN B., KALSNES, B., NADIM, F., and NOWACKI, F., *Design of offshore piles in clay – field tests and computational modelling – summary, interpretation and results of the pile load tests at the Onsoy test site*. Norwegian Geotechnical Institute. Report no. 52523-23, December 1987.

Discussion

Professor T. J. POSKITT, Queen Mary and Westfield College, University of London

At the instrument selection stage the possibility of using piezoresistive devices was examined. Preliminary tests revealed that under the testing regime all devices then available were not sufficiently robust to survive.

The problems arose due to the blocks which must be used in order to attach the transducer to the pile wall. A wide variety of such blocks were investigated all of which had natural frequencies which lay in the range 5–20 kHz. Resonance of these blocks during the passage of a stress wave appeared to be the problem. This could be so vigorous that the signal resulting from it dwarfed that which came from the pile wall. A discussion of this problem can be found in *Ground Engineering*, May 1991.

Acceleration measurement problems are apparent in the signal given in Fig. 7 of Randolph's paper. The area of the curve above and below the axes can be seen by casual inspection not to be equal, thereby implying a terminal velocity. The signal must therefore be suspect.

Professor M. F. RANDOLPH, University of Western Australia

I agree with Professor Poskitt that, at the time the project was planned, today's range of modern accelerometers were not available. Nowadays, robust piezoresistive accelerometers exist that weigh a few grammes, have resonant frequencies of 180 kHz or more and can withstand accelerations of 10 000*G* for thousands of blows. These transducers may be used to obtain reliable measurements of dynamic velocities and displacements during each hammer blow.

Dr U. MIRZA, Kvaerner, Earl & Wright

I seem to have got the indication that the authors are attaching too much significance to the K_0 values we in fact measured with the pressuremeter. What we said yesterday is that by the end of the consolidation, 130 days, we were getting values of K_c which are substantially higher than the K_0. We are talking about Kc values of about 3 compared to a K_0 of about 1.2. But the authors seem to be saying that the K_0 of the site at that depth should actually be higher than what the design profile implied.

G. SVANOE, SINTEF, Norway

What I actually used in the calculations was the Kc value as measured from the compression pile, but I increased it slightly because I assumed that the trial pile would be fully consolidated. That is what I used but the comment was that the pressuremeter K_0 is higher than the design, and it is systematically higher, so I meant to say that perhaps the

K_0 is a low estimate.

Dr D. J. HENKEL, Arup Geotechnics

The thing that struck me particularly about Professor Randolph's paper was the radical change in pile resistance due to what he called "friction fatigue". This seems to me to be primarily a phenomenon associated with the development of pore pressure and is it possible that we can predict this sort of behaviour from laboratory tests? We all know that in silts "shake down" and the generation of high pore pressures can take place and I would like to have his comments on this.

As far as the *p-y* curves are concerned, there did not seem to be a link in the discussion between pile behaviour and the laboratory tests I would have expected to have been carried out. The horizontal displacement of the pile produces stresses similar to those in a triaxial extension test, in which the horizontal stress is increased. Would it not be more appropriate to look at extension tests to derive the stress-strain curve of the material concerned?

The shape of the stress-strain curve used by Reese is rather severe, most materials I have come across there is a gradual build up of stress followed by a gradual decrease, and not the sudden collapse as suggested by Reese. I am therefore not surprised that Hamilton and Dunnavant get better results with more reasonable stress-strain curves.

I find the effective stress approach very interesting because I think that the only way in which we can understand the problem is through effective stress. However, without having worked through an example I am not able to offer a comment on how valuable the rather complicated approach suggested is likely to be in the future.

Dr K. L. YIP-WONG, Maunsell & Partners

With reference to Professor Randolph's point about length of the signal needed for analysis. By double integrating the acceleration signals to obtain the displacement, we need to work out the area under the acceleration and the velocity, say between time *T* and time *T*-1. If we have got errors on the acceleration signals between time *T* and time *T*-1, by double integrating the acceleration signals we would have an accumulative type of error on the displacements. Therefore the longer the signal to analyse, the more error would occur, so the best length to work on to do the analysis is $2\,^{L}/_{C}$. Another consideration is the parameter sensitivity by using the signal length $2\,^{L}/_{C}$. We did some sensitivity checks by using the wave equation, by varying the parameters damping, quake and resistance. Compared with the damping, resistance and quake are more sensitive. For the purpose of predicting soil resistance, working on $2\,{}^{L}/_{C}$ is sufficient.

K. KARLSRUD, Norwegian Geotechnical Institute

Concerning the lateral pile behaviour, I was pleased and surprised to see how close the Tilbrook test came to predictive models. The way I understand the way Dr Murff analysed it, was that Reese's *p-y* curve for cyclic loading is a final sort of degraded value, it is not a complete *p-y* curve that follows each cycle, but it just gives you the *p-y* curve for the end of the situation. At NGI in 1982 we did a lateral load test on some piles at Haga. Some observations are very similar to those at Tilbrook. We saw the gap effect that also has been observed previously. The Haga pile was 15 cm in diameter. The maximum displacement put on the pile which was about 20-25 mm, and we had a gap that went down 60 cm along the pile, so it is the same type of observation that the author has seen. We also had earth pressure gauges down along the pile, at very close intervals along the top of the pile, so we could see exactly when the pile parted from the soil, because the earth pressures dropped to zero at the moment the crack formed behind the pile.

Also interesting, if you really want to make a correct theoretical analysis of the lateral behaviour, was that, until we got the crack formation, we had as much suction at the back side of the pile as pressures built up at the front side of the pile. This is not usually accounted for in analytical models. Thus, there is complete shear failure going on at the front face as well as the back face until the crack forms at the back side. If you really want to go into details of the lateral pile behaviour I think these are elements that one should consider in future modelling. In any case, I think we can still use the simplified forms of *p-y* curves because the load-displacement response of the pile top is, after all, rather insensitive to some of these assumptions.

I think I agree with Mr Svanoe up to the point that the effective stress at onset of pile loading is one very important parameter. I have no problems with analysing the response of the clay as you load it, you can use either an undrained type of approach or you can use an undrained effective stress type of approach. Basically, if you have the right type of soil modelling in there, it should give the same result. I think you should recognise that we cannot use the stress-strain strength behaviour of the undisturbed clay. I think you should consider putting into the concept the properties of the material that is closest to the pile wall and, as you will see from this diagram, this is a typical type of strain that we impose by the pile installation. We have, for a closed ended pile, strains within one and a half pile radii from the pile wall that goes from 20% to much more than 100% strain. Of course, soils that have been strained to that sort of level do have different shear strain properties from the material further out that has hardly been strained at all. So we should take into account how this stress-strain behaviour has been influenced by the pile installation. If we have those concepts in place, I think there are various ways. We understand the effective stress and its influence on the shear strength. One can make various assessments. If you assume that, very simply, the shear strength close to the pile wall is proportional to the mean effective stress and the normalised undrained shear strength properties of the remoulded material, you can show exactly what was suggested here: that the shear strength is related to effective stress, or you can also show exactly as we showed in the paper to OTC in 1990, how alpha will reduce with overconsolidation ratio as a result of such a model which includes the effect of the mean effective stress and the shear strength properties of the remoulded clay. There are various ways of explaining the reduction in alpha with undrained shear strength ratio or overconsolidated ratio, but I think we should recognise the effect of pile installation on the soil properties.

M. ENGLAND, Cementation Piling and Foundations

I hope the message that I have put across is that the time element of foundation behaviour should not be ignored. Fig. 1 Shows a modelled pile behaviour characteristic. This is a driven cast in situ pile of 0.425 m diameter and the lowest contour is the unique drained, consolidated behaviour of the pile determined by the ten pertinent parameters. When we talk about effective stresses and pore water pressures, one of the main elements to consider is time. This illustration is generated from an analysis in time of each load applied during a static load test. If one considers what happens when we do not allow for the behaviour of the pile changing until infinite time, and one considers what happens under shorter applications of load, one can follow declining load hold period and contours until one arrives in the extreme at the elastic shortening of the pile.

This means that unless time is taken into account in all the measurements that are being made, you really cannot easily paste together the whole picture of how a pile behaves under static load conditions.

One other thing that may be added when we are talking about lateral loading, is that exactly the same thing is going to apply. If one does not let the pore water pressures dissipate, misleading results may be generated, and if tests are performed progressively faster, one will actually find that the response will be closer to the modulus of elasticity of water than that of the material surrounding the pile.

Dr J. D. MURFF, Exxon Production Research Company, USA

In response to a couple of the questions that have been asked, I would like to get back to Karlsrud's point about the *p-y* curves that were presented. Those *p-y* curves and all the *p-y* curves that I am familiar with that have been

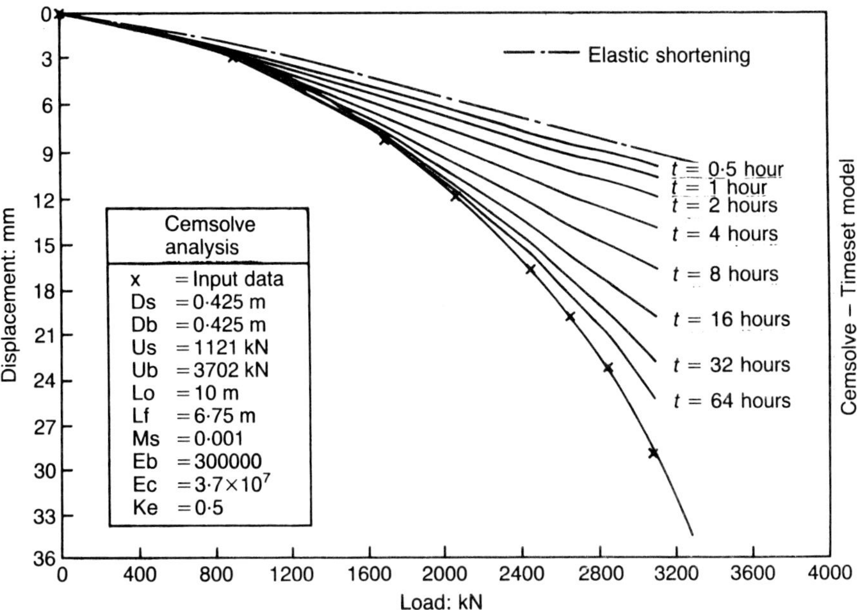

Fig. 1.

presented in literature, are really not monotonic curves representing behaviour, but are really a locus of a set of points measured at a particular point in time, and we used them as if they were load displacement curves, but they in fact are not derived in that manner. I welcome the comments raised by Dr Henkel with regard to the extension test. The reason, I believe that the tests were used as they were was because of convention. They simply conform with the standard practice, so there was not an attempt here to try to do any sort of stress analysis. It is really an empirical procedure. My suspicion is that the stress state that affects the pile varies throughout the soil. It includes extension but I suspect that there are a lot of complex things going on, including shear stress rotations that probably affect the behaviour. Dr Henkel mentioned that the Reese criteria was a rather unusual one and he was not surprised that these results did not agree, and I would agree with that. I think the relevance of that criterion is that it has been around a long time, it is referenced in API and it has been used, or at least proposed, by a number of investigators.

Professor M. RANDOLPH, University of Western Australia

I would like to respond to the comment on matching displacements and how much of the time signal you need to match. My own view is that matching velocities is preferable and more direct than matching displacements. Certainly, if you wish to deduce information regarding the distribution of soil resistance with depth, I think it is crucial to match the detailed velocity response rather than the integrated pile head displacement. If displacements are going to be matched, then I think it is essential to match both the peak displacement, the temporary compression, but also most importantly, the final permanent set and to tie that in with the actual measured blow counts. Something which is not often done in practice in stress-wave analysis, is to make sure that the final model yields the measured permanent set of the pile. In order to estimate the final displacements, it is necessary to extend the stress-wave match to at least three or four times the travel time ($^{L}/_{c}$), or say, at least two return waves, to get any real confidence. Estimation of the viscous component of soil resistance, over and above the ordinary static friction, is best achieved by matching the data in the latter part of the blow, for example between $2^{L}/_{c}$ and $4^{L}/_{c}$. If you confine your attention to the first part of the blow, for times less than $2L/c$, then it is very difficult to distinguish how much of the resistance is due to a viscous, velocity dependent, enhancement of the static friction.

21. The impact of axial pile load tests at Pentre and Tilbrook on the design and certification of offshore piles in clay

R. HOBBS, Lloyd's Register, Offshore Division

A review is presented of the compressive axial pile load tests carried out at Pentre and Tilbrook Grange. The tests were designed to improve the data base for pile design, and provided the opportunity to verify procedures currently used in the design and certification of offshore piles in clay. As a result of the review, it is recommended that: (a) the frictional capacity of offshore piles in normally and overconsolidated clay should be based on the criteria given in the Commentary to API RP2A (1991) (unit friction varying from 1.0 to 0.5 times undrained shear strength depending upon shear strength alone), with unit friction in near surface soils being limited to that sustainable by passive pressure against the pile wall; (b) the criteria given in the main text of API RP2A (1991) should only be used for normally consolidated clays with an appropriate length effect correction. The method is, however, recommended for heavily overconsolidated clays as a useful check on the procedure given above.

Introduction

This paper reviews axial pile load tests carried out, under the direction of BP, to confirm and extend the data base for the frictional capacity of driven piles in clays representative of North Sea conditions.

High quality compression tests on well-instrumented 762 mm diameter pipe piles were undertaken at a stiff, essentially normally consolidated clay site at Pentre, near Shrewsbury, and at a very hard, heavily overconsolidated clay site at Tilbrook Grange, near Huntingdon.

As a participant in the project, Lloyd's Register attended major activities, including site investigations, laboratory work, instrument testing and the compression pile load tests, together with all participants' meetings.

Details of the tests are given in reports prepared by the project manager (McClelland, 1987, 1988a, b) and in other papers to this conference but are briefly discussed herein for completeness. This paper concentrates on recommending unambiguous and practical methods of calculating the frictional capacity of offshore piles in clay which take account of the Pentre and Tilbrook tests. Particular attention is paid to examining the validity of codified methods used in established offshore practice. Other methods are not reviewed in detail.

Background to tests

Established practice for design of offshore piles in clay is to use criteria given in the American Petroleum Institute's document RP2A (API, 1986). In the North Sea the practice has been to adopt API Method 2 (API, 1986 para 2.6.4b2 and API, 1991, Commentary) in which unit friction, f, varies from 1.0 to 0.5 times undrained shear strength, S_u, depending upon shear strength alone ($f = \alpha S_u$). Effects related to pile length or stiffness are not explicitly considered.

The method is an empirical procedure (Drewry *et al.* 1977) based on the results of a number of pile load tests (Vijayvergiya and Focht, 1972, Olson and Dennis, 1982) which, however, were limited with regard to relatively long pipe piles in stiff normally consolidated clays and in heavily overconsolidated clays with undrained shear strength in excess of about 400kPa. Thus design of piles in the North Sea involved extrapolation beyond the data base.

Several authors suggest that stress history, expressed as overconsolidation ratio or the ratio of undrained shear strength to effective overburden pressure, has an important influence on the frictional capacity of piles in clay (Wroth, 1972, McClelland, 1974, Semple, 1979, Semple and Rigden, 1984, Randolph and Murphy, 1985). Indeed, after the present test programme commenced, API revised the recommendations in the main text of RP2A to reflect this, whilst retaining the AP12 method in the Commentary (API 1987, 1991).

To fill the perceived gaps in the API data base, in 1982 the UK Department of Energy instigated an extensive feasibility study and site investigation programme to find suitable locations for pile tests in stiff, essentially normally consolidated and very hard, overconsolidated clays representative of North Sea soils. Pentre and Tilbrook Grange were selected by participants as the most suitable sites for the Large Diameter Pile Tests which followed.

Site investigation

Site investigation techniques

Detailed site investigations were made at both sites (Lambson, *et al.* 1993). Drilling, sampling, in situ testing and sample handling procedures all closely matched current offshore site investigation techniques.

Scope of site investigation

Site investigation at Pentre consisted of three near-continuous sampled borings to 60 m using 76 mm hydraulically operated piston and thin walled push samplers, one borehole with vane, *T-Z* probe and piezocone, one with self boring pressuremeter and three piezocone tests from ground level to 50 m.

Site investigation at Tilbrook Grange was similar in scope, with three near-continuous sampled borings to 40 m, using predominantly thin and thick wall 76 mm push samplers, one borehole with *T-Z* probe and piezocone, one borehole with *T-Z* probe and one with self boring pressuremeter.

Laboratory testing

Testing similar to that currently adopted for offshore site investigation was undertaken, namely: unconsolidated undrained triaxial tests (which give the reference strengths used with the API procedures), isotropically consolidated undrained triaxial tests, consolidation, classification and index tests. Additional tests were performed to provide information on fundamental soil behaviour for future calibration and on load transfer characteristics, namely: anisotropically and K_0 consolidated undrained triaxial tests, consolidated drained direct shear and ring shear tests on soil–soil and soil–steel interfaces, consolidated constant volume simple shear tests, resonant column and permeability tests.

Pile design and installation

Piles for both sites were 762 mm pipes with 15 mm (Pentre) or 30 mm (Tilbrook) wall thickness. Because of anticipated (but not realised) high driving and static stresses, the piles were fabricated in RQT steels which have higher yield stress but similar stiffness to steels used offshore. Surface condition was similar to offshore piling, namely lightly rusted with regular paint markings to indicate penetration.

At Pentre the pile was installed through a 15 m sleeve, vibrated in place, from which soil had been removed by augering. The piles were driven vertically with hydraulically operated drop hammers, a BSP HH7 (rated energy 82 kNm) at Pentre and a BSP HA40 (471 kNm) at Tilbrook Grange. These are similar in operation to hammers currently used in the North Sea. Pile guidance frames were used during driving. Soil plug movements were monitored throughout driving. At Pentre the plug moved up 1.7 m from original augered level while at Tilbrook it moved down 17.1 m below ground level. Installed penetrations were 55 m at Pentre, i.e. 40 m embedment below the bottom of the sleeve, and 30 m at Tilbrook. At Pentre, the test pile was installed within 5 m of sampled boring 101, while at Tilbrook Grange, the pile was installed approximately equidistant from, and within 5 to 7 m of, the three sampled borings.

Instrumentation

The piles were well instrumented with strain gauges and strain modules, total pressure and pore pressure cells together with accelerometers (Cox, Solomon and Cameron, 1993). Instruments were internally mounted or flush with the pile outside wall and cables were run in channels on the inside wall. Prior to selecting instruments, extensive testing was conducted to ensure that they could survive pile driving and to minimise zero drift, etc. (Solomon *et al.* 1993). There was a good level of redundancy. Piezometers and standpipes were also installed in the surrounding soil to measure pore pressures prior to and during pile driving and to monitor their dissipation after driving.

Pile testing

Piles were loaded by hydraulic jacks reacting against a load frame anchored by diaphragm walls at Pentre and bored anchor piles at Tilbrook Grange. These foundations were over 5 pile diameters away from the test pile to minimise interaction effects. Lateral support was provided by a ground slab. Initial static tests were carried out at a constant rate of penetration of 1 mm/min. In addition, creep and one-way cyclic tests were performed (Cox, Cameron and Clarke, 1993).

Review of pile test at Pentre

Soil conditions

Soils at the Pentre site (Fig. 1) are considered to be reasonably similar to many essentially normally consolidated, but stiff, clays in the North Sea. While the clay-size content is relatively low, clay minerals are found in the silt-size fraction, so that the material is less silty than the grading curves would suggest. Measured plasticity indices (10–25 per cent, locally 15–35 per cent) and liquid limits (30–40 per cent, locally 50–60 per cent) show the material to be a clay. Cone friction ratios generally range 2–4 per cent, also indicative of clay. Detailed descriptions of the soils are given by Lambson *et al.* (1993).

Shear strength profile

The profile of undrained shear strength, S_u, selected by McClelland (1987) (Fig. 2) is considered to be a reasonable overall fit to unconsolidated undrained triaxial test data from all boreholes below the 15 m pile sleeve. An alternative profile based on boring 101, which is closest to the as installed pile location, was also used by the author. The average shear strengths for the two profiles are 87.5 kPa and 89.1 kPa respectively. These are considered to be the profiles which would be used for design and certification were the site the location for an offshore platform.

Pile test

At the time of the initial static load test, 29 days after installation, instrumentation indicated that excess pore pressures generated during driving had more or less fully dissipated.

The test provided a well-defined unequivocal failure load (Fig. 3), although estimation of the end bearing involved extrapolation of (limited) strain gauge data over the lower section of the pile. This should not have significantly affected interpretation of friction, however. The measured peak capacity was 6.03 MN, achieved at 36 mm pile head movement. This reduced

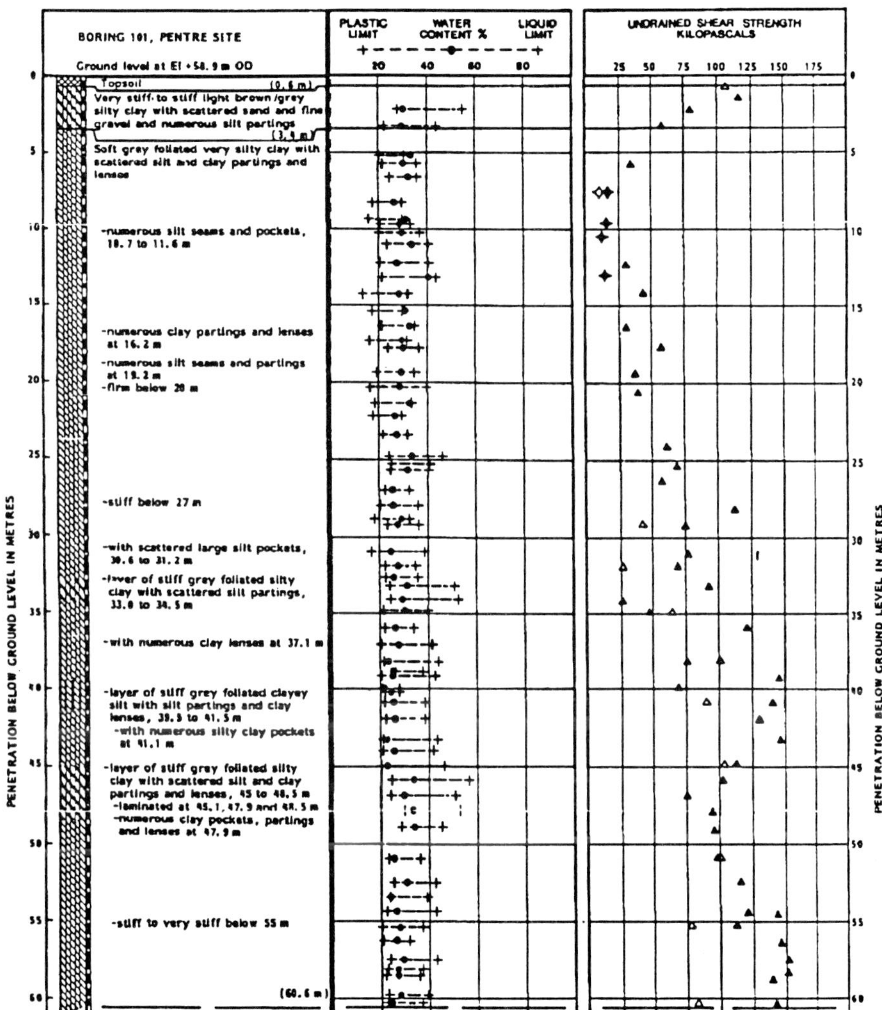

Fig. 1. Typical borehole log, boring 101, Pentre

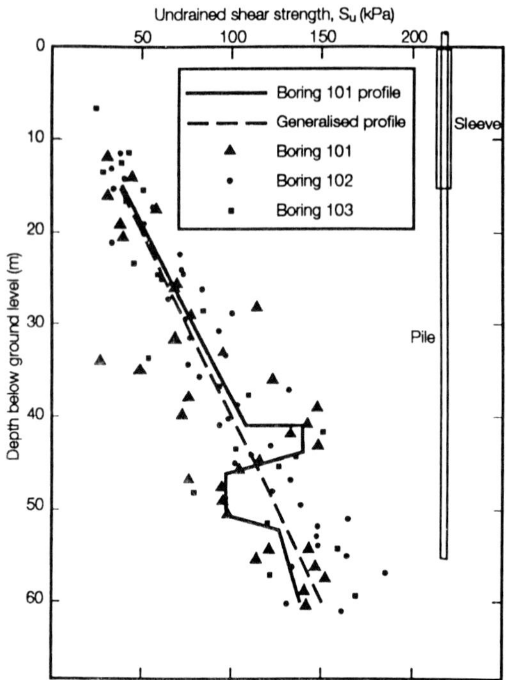

Fig. 2. Shear strength profile, Pentre

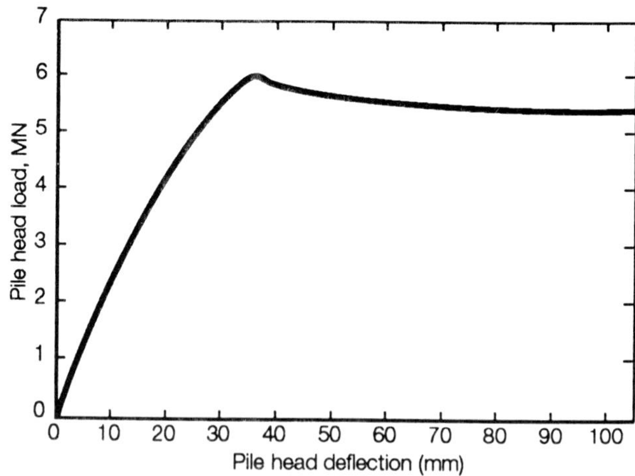

Fig. 3. Load–deflection curve, static test, Pentre

to 5.47 MN at 115 mm. At the peak, the end bearing component was calculated to be 0.86 MN and at 115 mm, 1.22 MN. The end bearing at peak capacity corresponds to an N_c of 10.5–11 (where $q = N_c \, . \, S_u$) and at residual 14.8–15.6, compared to the value of 9 normally assumed. A subsequent static test at the same rate of loading gave a peak capacity of 5.63 MN.

Comparison of results with predictions, Pentre

Mobilised friction distribution

Figure 4 shows the interpreted (measured) distribution of unit friction along the pile shaft at peak load compared to that predicted employing the API-2 method. This figure suggests that mobilised friction is somewhat lower than calculated in the upper part of the pile. McClelland (1988 b) explained this as the result both of pile whip reducing the lateral stress on the pile and of negative friction locked-in at the start of the test, albeit these two explanations would appear to be, to some extent, mutually exclusive.

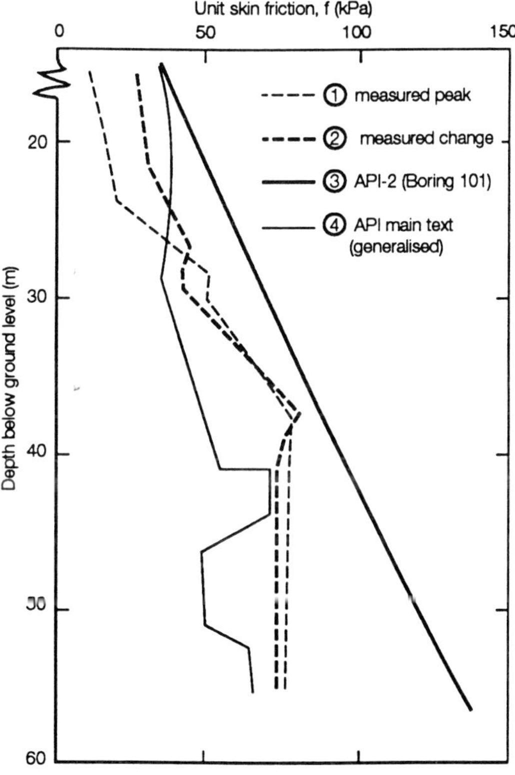

Fig. 4. Comparison of predicted and measured friction distribution, Pentre

If only the *changes* in friction from start of test to peak load are considered, however, there is fairly good agreement between measurement and prediction to about 30 m below ground level. This represents the friction actually mobilised as the load was increased from zero to peak (i.e. as if strain gauges had been zeroed at the start of the test). Below 30 m the measured frictions exceed those predicted.

Mobilised α

Mobilised average unit friction at peak load corresponds to an average α of 0.606 (boring 101 profile) to 0.617 generalised profile), while the average a calculated using the API-2 method is 0.535 (boring 101) to 0.540 (generalised profile) (Table 1). These values are slightly higher than the value of 0.5 which corresponds to the average S_u since the method is used incrementally, layer by layer. It is also the author's practice to place the transition from $\alpha = 1.0$ to 0.5 at S_u=25 to 75 kPa instead of the 24 to 72 kPa recommended by API (Lloyd's Register, 1989).

In contrast, strict application of the API (1991) main text procedure on an incremental basis gives $\alpha = 0.948$ (generalised profile), that is, an overprediction of 56 per cent, which is clearly very unconservative. Fig. 4 shows that this overprediction occurs over the full pile penetration. In order that better agreement be reached, a significant length effect reduction would be required either as an overall factor, or as a defined falling branch *T-Z* curve. No clear indication is currently given in RP2A to the length reduction factor, if any, to be applied with this method.

It is noted that the capacity dropped to 5.47 MN by a pile head displacement of 115 mm and that a subsequent static test at the same rate of loading gave a peak load similar to this residual value, which corresponds more closely to that predicted by API-2.

Table 1. Comparison of measured peak capacities with API predictions, Pentre

	Friction (kN)	End bearing (kN)	Total peak capacity (kN)	Average unit friction (kPa)	Average α
API-2	4527	564	5091	47.3	0.540
API (1991) main text	7939	564	8503	82.9	0.948
Measured	5170	860	6030	54.0	0.617

(a) Generalised profile

	Friction (kN)	End bearing (kN)	Total peak capacity (kN)	Average unit friction (kPa)	Average α
API 2	4562	538	5100	47.6	0.535
Measured	5170	860	6030	54.0	0.606

(b) Boring 101 profile

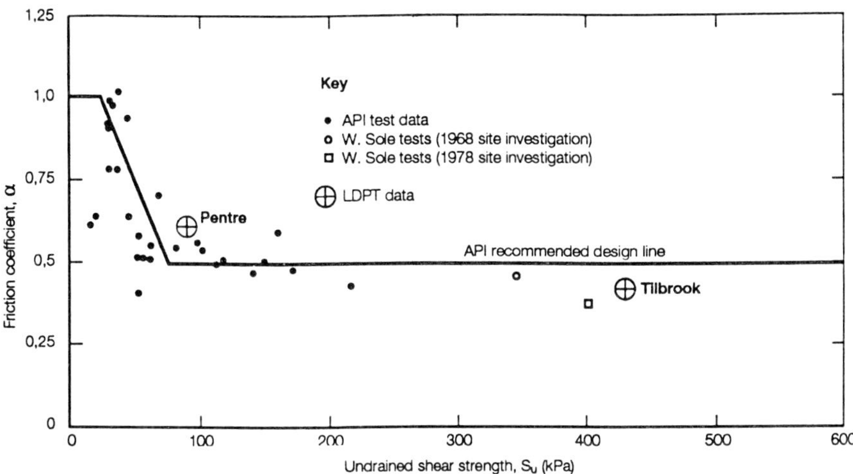

Fig. 5. Comparison of Pentre and Tilbrook pile test results with API data base ($\alpha - S_u$)

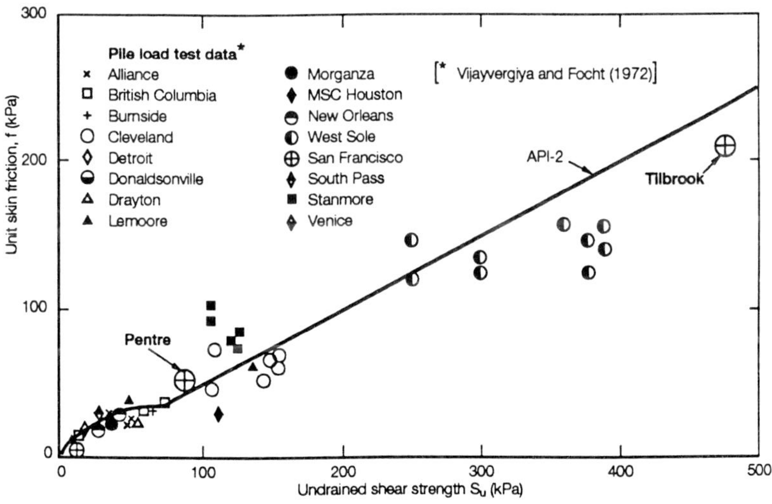

Fig. 6. Comparison of Pentre and Tilbrook pile test results with API data base ($f - S_u$)

API data base

Comparison of the Pentre test with the API test data and the API-2 method (Figs 5 and 6) shows that the result is within the upper scatter of previous pile tests supporting API-2. While the author's review of the tests in the data base indicates that some appear to have limitations with regard to several factors, including: sampling, shear strength measurement, set-up times, presence of oversized closure plates, re-testing, re-driving, underdrainage or artesian pressure, this agreement suggests that they cannot be rejected. The review also suggests no clear correlation with soil plasticity for tests in the current API data base.

Most of the sites in the data base had shear strengths measured by unconfined compression tests, whereas unconsolidated undrained triaxial (UU) tests are used in current UK North Sea site investigations. Re-interpretation of results using shear strengths measured in UU tests would be expected to lead to an increase in strength and a consequent decrease in α. For the Beta site (Pelletier and Doyle, 1983) UU tests gave significantly higher strengths than UC tests and calculated α values using UU strengths would consequently be lower than those previously quoted for that site.

Pile length and scale effects

Pile length (or stiffness) has been identified as an important factor in frictional capacity and it is not clear how much of this effect is due to pile stiffness and soil *T-Z* response and how much is related in some way to the amount of pile driven past any given element of soil.

As a qualitative indication of possible length effects the ratio of mean α values predicted for Pentre (0.762 m piles embedded 40 m) and for a North Sea site with similar plasticity and shear strength (2.438 m piles to 85 m), have been compared. These gave ratios of 1.12 for the Kraft *et al.* (1981) α–length relationship, 1.09 for the λ–length relationship (no limits applied to λ) and 1.09 for the Flaate and Selnes (1977) method. Prototype North Sea piles may

Table 2. McClelland's (1988b) predicted frictional capacities, Pentre

Method	Friction capacity (MN)
API-2	4.47
Kraft et al	6.59
Meyerhof β	6.92, 8.66
Randolf and Murphy	7.09
Semple and Rigden	8.41
Esrig and Kirby	9.46
T-Z probe	5.37
Measured	5.17

therefore have slightly lower α values than the test pile. (Methods using stiffness rather than length do not give such a clear difference, reflecting one aim of the test piles' design).

The North Sea piles above were required to achieve a capacity of the order of 43 MN, while a current design for another site calls for an ultimate friction capacity of 60 MN at a penetration of 100 m in normally consolidated clay. In contrast the Pentre piles achieved just 6 MN. Thus considerable extrapolation is still required from the test to actual North Sea piling.

Other methods

Good correlation was achieved by using the peak frictions measured in the limited number of in situ *T-Z* probe tests and further work may be useful to see whether this has general applicability. This agreement is perhaps surprising as the probe is a short, small diameter, relatively stiff device compared to a pile and it might be expected that, in general, it would give higher frictions. The test takes some time to perform, if sufficient pore pressure dissipation is allowed, and its use offshore on a routine basis could therefore lead to increases in site investigation costs, particularly as the likelihood of borehole abandonment would increase.

McClelland (1988b) presented a number of predictions of total friction using alternative methods (Table 2). All these methods were unconservative compared to the measured frictions. The unconservatism was reduced if an analysis using theoretical, falling branch axial load–deflection *T-Z*) curves was carried out. However, this is considered to be inappropriate for empirically based methods and, in particular, the lambda method (Kraft *et al.*, 1981) where length effects are explicitly included already. Further, the assumed shape of the *T-Z* curve has a significant effect. The opportunity to select parameters to fit the desired result is seen as something to avoid in codified procedures. The Randolph and Murphy (1985) method, without *T-Z* curve analysis, gave $\alpha = 0.85$ and it is noted that the API (1991) main text procedure differs in detail from this method, upon which it is based. There are various methods for selecting the ratio of normally consolidated shear strength to effective overburden pressure and these lead to different results. In using effective stress methods, whether simple such as Meyerhof (1976) or sophisticated, there appears to be considerable scope for selection of parameters. This is evidenced by the two alternative Meyerhof predictions presented in Table 2. Generally, only a limited number of effective stress tests are carried out for offshore site investigations, which again may limit their application. While effective stress theories undoubtedly aid understanding of pile behaviour, the most practical approach to predicting pile capacity is the use of simple, empirically based methods which utilise well-defined parameters and result in unique solutions. It is interesting to note that the back-calculated

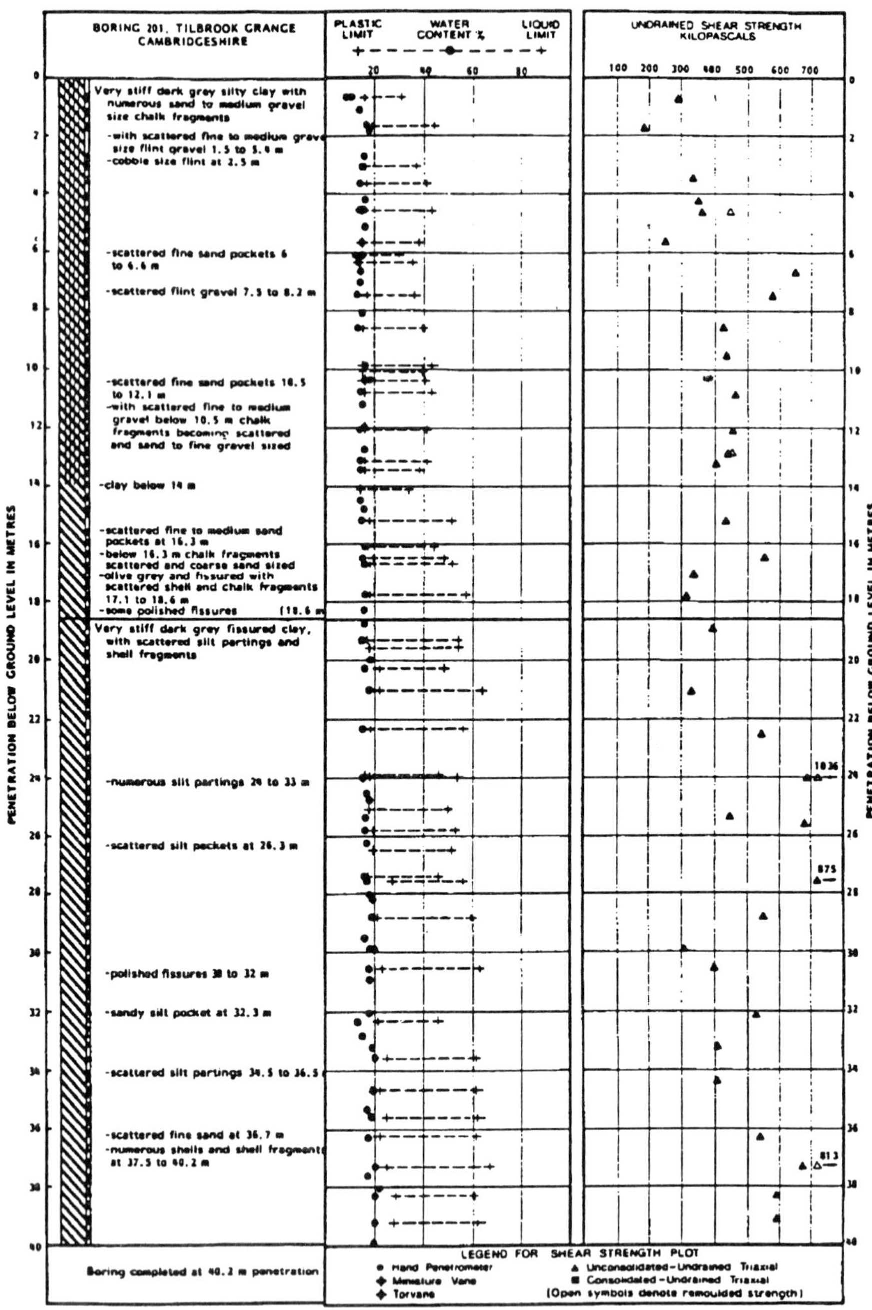

Fig. 7. Typical borehole log, boring 201, Tilbrook Grange

λ value of 0.111 is significantly lower than the arbitrary minimum of 0.14 set by Kraft *et al.* (1981).

Recommendations

The above review has not identified a clear case for a change in established practice and has highlighted the potential for non-unique predictions using some alternative non-codified procedures. While further analysis of the tests with a view to producing more robust procedures is encouraged, it is considered prudent at this time to continue to use the well-established API-2 method of predicting frictional capacity of piles in normally consolidated clays. While theoretically not as satisfying as effective stress methods, it is simple and unequivocal. The only interpretation involved is in fitting a shear strength profile to measured UU results (with due regard to geology and to trends established from other laboratory and in situ tests). It appears that the procedure given in the main text of API RP2A (1991) is unconservative for Pentre and its use for similar soils is therefore not recommended without appropriate, probably large length effect reductions, over which there is likely to be considerable debate.

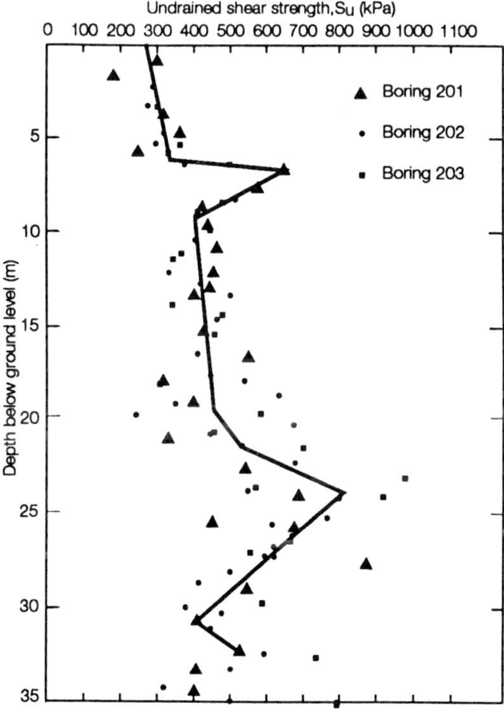

Fig. 8. Shear strength profile, Tilbrook Grange

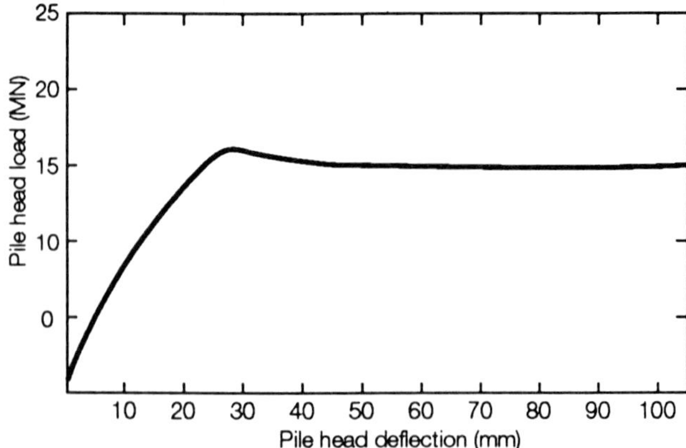

Fig. 9. Load–deflection curve, static test, Tilbrook Grange

Pile test at Tilbrook Grange

Soil conditions

Soils at the Tilbrook site (Fig. 7) are considered to be reasonably similar to many heavily overconsolidated clays in the North Sea.

Two distinct strata have been identified, namely Lowestoft Till (to about 18 m) and Oxford Clay. The Lowestoft Till has a relatively high carbonate content (unlike most North Sea clays) but this is present predominantly in discrete particles and fragments of chalk in a non-carbonate matrix and is therefore not expected to have a significant effect on pile behaviour. Plasticity indices generally range 20–30 per cent in the Lowestoft Till and 30–40 per cent in the Oxford Clay. Natural water content is generally at or below the plastic limit. Full descriptions of the soils are given by Lambson *et al.* (1993).

Shear strength profile

The profile of undrained shear strength selected by the author (Fig. 8) differs only slightly from that of McClelland (1988a). The average shear strength over the embedded pile length is 478 kPa. Again, the selected profile is considered to be that which would have been used in design and certification save that, for offshore sites, where boreholes are more widely spaced, borehole – specific profiles are usually developed and the worst profile then used to calculate design pile capacity.

Pile test

At the time of the load test, some 130 days after installation, instrumentation indicated that excess pore pressures generated during driving had not completely dissipated, particularly in the Oxford Clay, despite a delay in the programme. It was estimated that 10 per cent excess pore pressures remained in the Lowestoft Till and 20 per cent in the Oxford Clay (McClelland, 1988b).

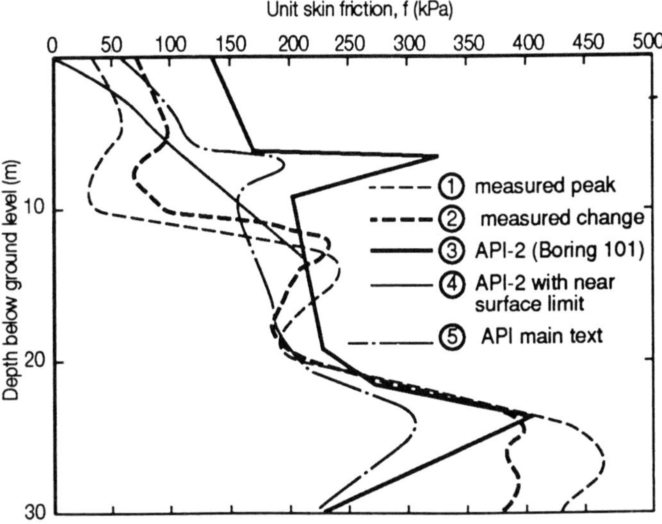

Fig. 10. Comparison of predicted and measured friction distribution, Tilbrook

As at Pentre, the initial static test provided an well defined failure load (Fig. 9). The measured peak capacity was 16.13 MN, achieved at 28 mm pile head movement. This reduced to 14.2 MN at 129 mm. At peak load the mobilised end bearing was calculated to be 1.45 MN (corresponding to N_c = 7.1) while at 129 mm it was 2.07 MN (N_c = 10.2). Subsequent static tests achieved peak capacities of 14.6 MN and 13.6 MN.

Comparison of results with predictions, Tilbrook

Mobilised friction distribution

Figure 10 shows the interpreted distribution of unit friction down the pile at peak load compared to that predicted using the API-2 method. As at Pentre, mobilised friction is somewhat lower than calculated in the upper part of the pile. Agreement is improved if only the changes from start of test to peak load are considered, but the measured frictions are still much lower in the upper part of the pile.

In applying the API-2 method to offshore sites where heavily overconsolidated clays occur near seabed, it is Lloyd's Register's current practice to limit skin friction to that which can be sustained by passive pressure acting on the pile wall, on the basis that, following disturbance due to pile driving (including whip) and lateral loading, the previous in situ soil stresses cannot be re-established, since the source of overconsolidation has, by definition, been removed. The passive pressure represents a theoretical maximum to the lateral pressure which can be mobilised. Some judgement has to be exercised in the selection of parameters (see previous comments with regard to effective stress methods) for example, what angle of friction to use and whether to include an effective cohesion term in the formulation for K_p. It does, however, appear to be a reasonable modification to API-2.

CIU triaxial tests in the Lowestoft Till (McClelland, 1988) give $\phi' = 26°$ while, for a mean PI of 25 per cent, empirical), correlations give $\phi' = 30°$ (Department of the Navy, 1971) and $\phi' = 25°$ (Mayne, 1980). In the underlying Oxford Clay, both CIU tests and department of the Navy (1971) give $\phi' = 28°$.

Figure 10 shows the effect on calculated friction of such a reduction based on the above data. ϕ' has been taken as 28° for $K_p = \tan^2(45° + \phi'/2)$ and the interface friction angle, δ, has been taken as ϕ'–5°. The overall fit between measured change and calculated friction is good.

Mobilised α

The measured average mobilised friction at peak load (204.4 kPa) corresponds to an average α of 0.428 (Table 3). The average friction calculated using the API-2 method is 238.9 kPa (α= 0.5). However, using the reduction in capacity for shallow soils as discussed above, the average friction is 204.1 kPa and the average α is 0.427. Thus, there is remarkable (and probably fortuitous) agreement between the measured pile friction and that calculated using the modified API-2 approach for shallow soils, for the assumptions made about ϕ' and δ and the formulation for K_p. (Using the measured ϕ' of 26° and δ of 21° gives an average unit friction of 194.1 kPa and α of 0.406, while the empirical correlation ϕ' of 30° together with a δ of 25° gives 211.7 kPa and 0.443). The API (1991) main text procedure used incrementally gives α = 0.396, which is slightly conservative. Use of a *T-Z* curve approach would worsen this agreement. Alternative interpretations of the test, while not giving as good agreement of friction distribution, are a local friction limit of about 225 kPa or a uniform α of 0.425.

API data base

Comparison of the Tilbrook test with the API test data (Figs 5 and 6) show that the result has extended the data base, but that the unmodified API-2 method is optimistic for this test.

Table 3. Comparison of measured peak capacities with API predictions, Tilbrook

	Friction (kN)	End bearing (kN)	Total peak capacity (kN)	Average unit friction (kPa)	Average α
API-2	17160	1826	18986	238.9	0.5
API-2 with limit	14655	1826	16480	204.1	0.427
API (1991) main text	13601	1826	15427	189.4	0.396
Measured	14681	1450	16131	204.4	0.428

Other methods

According to McClelland's (1988b) calculations (Table 4) the Randolph and Murphy method gives $\alpha = 0.39$. One Meyerhof prediction overestimates and one underestimates the measured friction, the spread between the results again demonstrating the scope for non-unique solutions. Use of *T-Z* curve analysis worsens the fit for the Randolph and Murphy method and one of the Meyerhof methods.

As for Pentre, fairly good correlation was obtained between measured frictions and the results of (limited) *T-Z* probe tests, again using peak values.

Table 4. McClelland's (1988b) predicted frictional capacities, Tilbrook

Method	Friction capacity (MN)
API-2	16.97
Meyerhof β	12.84, 18.87
Randolf and Murphy	13.34
T-Z probe	13.10
Measured	14.68

Recommendations

Notwithstanding the results of the tension tests, which are described for the first time in the public domain by Clarke *et al.* (1993) (and in which Lloyd's Register did not participate), it is recommended that design of piles in overconsolidated clays be based on the API-2 method, with friction in surficial soils limited to that which can be mobilised by a lateral stress equal to passive pressure. Tentatively it is proposed that the passive pressure coefficient K_p be defined by the expression $K_p = \tan^2 (45° + \phi' /2)$ and the limiting friction by $f = K_p\ \sigma'_v \tan (\phi' - 5°)$. Where effective stress test results are not available, use should be made of empirical correlations with plasticity. The API main text (1991) procedure is considered to be a valid additional method for heavily overconsolidated clays.

Caution should be exercised in extrapolating the above findings to high shear strength sites which differ significantly from Tilbrook, for example, those with much higher average strength, significantly different strength profile or stratification, or much longer piles. Further, it is noted that current North Sea jackets at overconsolidated clay sites have piles designed to ultimate capacities of up to 95 MN compared to Tilbrook Grange's 16 MN. Thus, as at Pentre, while the load tests have extended the data base, extrapolation is still necessary in terms of total capacity (and diameter).

Conclusions

The main conclusions of this review of the compressive pile load tests carried out at Pentre and Tilbrook, as part of the Large Diameter Pile Tests project, are as summarised below.

(*a*) The tests were carried out on driven piles installed in soils reasonably representative of stiff to very hard North Sea clays.
(*b*) Failure loads were well-defined.
(*c*) The interpreted unit friction, f, from the initial static tests corresponds to average α values of 0.61 for Pentre and 0.43 for Tilbrook Grange, where $f = \alpha S_u$.
(*d*) The average α values predicted for the two sites using the two RP2A criteria (API, 1991) were 0.54 and 0.50 respectively (API-2), and 0.95 and 0.40 (main text method). For Tilbrook, limiting the API-2 friction to that sustainable by a lateral stress equal to passive pressure leads to a predicted average α of 0.43.
(*e*) The in situ *T-Z* probe was found to give reasonable predictions of friction at both sites, however, practical considerations may limit its use.
(*f*) Effective stress methods of calculating friction, while aiding understanding of pile behaviour, appear to allow considerable scope for selection of parameters and thus can result in non-unique solutions.
(*g*) Similarly, incorporating *T-Z* curves into the calculation gives scope for non-unique solutions.
(*h*) The most practical approach to calculating pile capacity is the use of simple, empirically based methods which utilise well-defined parameters.
(*i*) It is recommended that:

- For normally consolidated clays, unit friction be calculated using the API-2 method;
- For overconsolidated clays, unit friction be calculated using the API-2 method with a limit of $f = K_p \sigma'_v \tan(\phi' - 5°)$ where $K_p = \tan^2(45° + \phi'/2)$;
- The API (1991) main text method should only be used for normally consolidated clays with application of an appropriate length effect correction. However, for overconsolidated clays the method is recommended as a useful additional check;
- The above methods should be employed incrementally with depth.

Caution should be exercised in extrapolating the above findings to high shear strength sites which differ significantly from Tilbrook; for example, those with much higher average strength, significantly different strength profile or stratification, or much longer piles.

(*k*) The tests have greatly extended the data base in terms of soil type or strength and have increased confidence in the empirical procedures used in established offshore practice. It is worth noting, however, that in terms of pile diameter, penetration and ultimate axial load, extrapolation is still required for typical North Sea piles in clay, already installed or currently being designed.

Acknowledgements

The author wishes to thank Lloyd's Register for permission to publish this paper and colleagues in Offshore Division for comments on the draft.

References

API (1986, 1987, 1991). *RP2A: Recommended Practice for Planning, Designing and Constructing Fixed Offshore Platforms*, 16th, 17th, 19th editions.

CLARKE, J., LONG, M. M., and HAMILTON, J. (1993). Tilbrook Grange tension test. This volume.

CLARKE, J., RIGDEN, W. J. and SENNER, D. W. F. (1985). Reinterpretation of the West Sole platform 'WC' pile load tests. *Géotechnique*, Vol. 35, No. 4, 393-412.

COX, W. R., CAMERON, K., and CLARKE, J. (1993). Static and cyclic axial load tests on two 762 mm diameter pipe piles in clay. This volume.

COX, W. R., SOLOMON, I. J. and CAMERON, K. (1993). Instrumentation and calibration of two 762 mm diameter pipe piles for axial load tests in clay. This volume.

DEPARTMENT OF THE NAVY (1971). *Design Manual, Soil Mechanics, Foundations and Earth Structures*. NAVFAC DM-7.

DREWRY, J. M., WEIDLER, J. B., and HWANG, S. T. (1977). Predicting axial pile capacities for offshore platforms. *Petroleum Engineer*, May 1977, 41-44.

FLAATE, K. and SELNES, P. (1977). Side friction of piles in clay. *Proc. 9th Int. Conf. Soil Mech. & Fndn Engng*, Tokyo, Vol. 1, pp. 517 -522.

KRAFT, L. M., FOCHT, J. A. and AMERASINGHE, S. F. (1981). Friction capacity of piles driven into clay. *J. Geotech. Engng Div., Am. Soc. Civ. Engrs*, Vol. 107, No. 11, 1521-1541.

LAMBSON, M. D., CLARE, D. G., SEMPLE, R. M. and SENNER, D. W. F. (1993). Investigation and interpretation of Pentre and Tilbrook Grange soil conditions. This volume.

LLOYD'S REGISTER (1989). *Rules and Regulations for the Classification of Fixed Offshore Installations*, Pt 3, Ch. 2.

MAYNE, P. W. (1980). Cam-clay predictions for undrained strength. *J. Geotech. Engng Div., Am. Soc. Civ. Engng*, Vol. 108, No. 6, 851-872.

MCCLELLAND, B. (1974). Design of deep penetration piles for ocean structures. *J. Geotech. Engng Div., Am. Soc. Civ. Engrs*, Vol. 100, No. 7, 705-747.

MCCLELLAND LTD (1987). *Large Diameter Pile Tests programme, soil and foundation investigation, Pentre site – final report*, UK 86-800-1, Vols I-IV.

MCCLELLAND LTD (1988a). *Large Diameter Pile Tests programme, soil and foundation investigation, Tilbrook site – final report*, UK 86-800-2, Vols I-IV.

MCCLELLAND LTD (1988b). *Large Diameter Pile Tests programme, test programme – final report*, UK 86-800-3, Vols I-V.

MEYERHOF, G. G. (1976). Bearing capacity and settlement of pile foundations, *J. Geotech. Engng Div., Am. Soc. Civ. Engrs*, Vol. 102, No. 3, 1-29.

OLSON, R. E. and DENNIS, N. D. (1982). Review and compilation of pile test results, axial pile capacity, final report to API on project PRAC 81-29, University Texas at Austin.

PELLETIER, J. H., and DOYLE, E. H. (1982). Tension capacity in silty clays – Beta pile test. *Proc. 2nd Int. Conf. on Numerical Methods in Offshore Piling*, Austin, Texas.

RANDOLPH, M. F., and MURPHY, B. S. (1985). Shaft capacity of driven piles in clay. *Proc. 17th Offshore Technology Conf*. Houston, Vol. 1, pp. 371-378.

SEMPLE, R. M. (1972). Discussion on Session 7 on the effective stress analysis of piles. *Proc. Conf. on Recent Developments in the Design and Construction of Piles*, ICE, London, pp. 397-399.

SEMPLE, R. M., and RIGDEN, W. J. (1984). Shaft capacity of driven pipe piles in clay. *Proc. Symp. on Analysis and Design of Pile Foundations*, ASCE, San Francisco.

SOLOMON, I. J., COX, W. R., CLARKE, J. and POSKITT, T. J. (1993). The development of instrumentation for drivability testing and load testing of large diameter piles. This volume.

WROTH, C. P. (1972). Discussion on the design and performance of deep foundations. *Proc. Conf. on the Performance of Earth and Earth-Supported Structures*, ASCE., Vol. 3, pp. 231-234.

VIJAYVERGIYA, V. N., and FOCHT, J. A. (1972). A new way to predict capacity of piles in clay. *Proc. 4th Offshore Technology Conf.*, Houston, pp.. 865-740.

22. Comparison of recent tests on OC clay and implications for design

F. NOWACKI, K. KARLSRUD and P. SPARREVIK, Norwegian Geotechnical Institute

The aim of physical testing is to learn and improve the understanding of the mechanisms involved in the pertinent problem. Being engineers, there is also a strong motivation to as soon as possible implement the results of the tests in practical design of similar structural elements.

This paper gives a brief summary of recent high quality pile tests performed in OC clays and intends to provide conclusions from these tests which may have practical implications for design of piles.

Recent pile load tests in OC clays

The criteria for the selection of load tests of interest were that:

(*a*) the soil conditions and the main soil parameters should be reasonably well defined
(*b*) attempts should have been made to measure the effective stresses acting on the pile shaft and the distribution of the side resistance along the shaft.

The need to fulfil (*a*) is obvious for any meaningful interpretation of a test. Criterion (*b*) is important to avoid serious misinterpretation (e.g. due to residual stresses along the pile shaft) and provides also a key element for any improvement of the basic physical understanding of the problem in question.

Only few pile test programmes have included attempts of measurements of effective stresses acting on the pile shaft. A literature review revealed that such tests have been performed in stiff, OC clay at six different locations. Key data from the test piles, the ground conditions and the test results are presented in Table 1.

The range of OCR varies from 4 to more than 30 at the various sites. The variation in OCR is also large within some of the soil profiles. Also the range of undrained strengths at the various test sites is very large from $s_u = 40$ kPa to about $s_u = 700$ kPa. Although the information on index data is limited for some of the sites, one can conclude that the tests were performed in a variety of soil types.

University of Houston Central Campus (UHCC) pile tests

The tests were described by O'Neill *et al.* (1982a and b). The objective of the tests was mainly to study the behaviour of a pile group but also two reference piles were studied.

The clay was described as a very stiff, saturated clay preconsolidated by desiccation. This is different from the Tilbrook, Canons Park and Cowden sites where the soil was truly preloaded. There is ambiguity in the available documentation in respect to the in situ pore pressure condition.

The piles were closed-ended 0.273 m diameter steel pipes driven to a penetration of 13.1 m. They were heavily instrumented but the piezometer filters were de-saturated during driving. No reliable pore pressure data was obtained during and shortly after installation. The total pressure data were highly scattered but some clear trends could be deduced.

Both the inferred load transfer pattern during driving and the lateral earth pressure measured after driving suggested that the pile installation produced a slightly oversized cavity along the top half of the pile, after which the soil swelled back against the piles. Peak side resistance in uplift was approximately equal to that in compression although the observed distribution was different.

Canons Park piling project

The tests were described by Bond and Jardine (1988 and 1989). The soil consists of a stiff overconsolidated London Clay. The piles were mainly closed-ended (except one) with a diameter of 0.102 m penetrated to a depth close to 6 m. The upper 2 m was blinded off by a casing. The piles were equipped with an instrument cluster at three levels consisting of axial load cell, surface stress transducer (capable to measure both σ_h and τ_v and pore pressure probes.

The main parameters varied were:

(a) The method of installation (slow-jacked, fast-jacked, or driven piles)
(b) The equalisation time (short, medium, or long time after installation)
(c) The direction of first loading (compression or tension).

Many highly interesting results were produced from the tests. However, the amount of variable testing conditions may create a problem for interpretation of the tests because more than one parameter was varied from one test to another. The point raised by Bond and Jardine (1988) that the skin friction is influenced by the rate of displacement during installation will not be pursued herein. All driven piles are in practice installed at a high rate of soil displacement.

Cowden pile tests

A summary report of this research project was prepared by Ove Arup & Partners (1986a). Nine piles of either 305 or 203 mm diameter were driven to a depth of 10m in an overconsolidated glacial clay. No satisfactory measurements of the horizontal stresses acting on the pile surface were obtained at Cowden. Neither was a satisfactory measurement of the distribution of side shear stresses along the pile shaft obtained. The piles were covered with shiny mill varnish. The shearing resistance of clay against this surface was found to be much lower than that of clay against rusted steel. The Cowden data therefore provides lower bound values of side friction.

Haga pile tests

Detailed information from Haga can be found in Karlsrud and Haugen (1985a and b) or Karlsrud *et al.* (1986). Ove Arup & Partners (1986b) prepared a comparison of the Cowden and the Haga pile tests in a brief summary report. In the Haga pile test programme, 5 m long piles, 153 mm diameter, were jacked into a sensitive overconsolidated clay, tested and extracted at thirty-one different locations.

In contrast to the soil profiles at the other sites considered in this paper, the water contents of the Haga soils are close to the liquid limit. All the other sites show water contents close to the plastic limit. The Haga clay is a leached and sensitive ($S_t \approx 4$). Nevertheless, there are many important conclusions which can be drawn from the Haga tests.

Field studies of an instrumented model pile at Madingley (Coop and Wroth, 1989) presented results with an in situ model pile. The soil at the test site Madingley consisted of overconsolidated clay. The model pile was closed-ended, 80 mm in diameter and 1135 mm long. The model pile segment was instrumented to measure radial total stress, pore pressure and side friction. The pile was installed by jacking. Very little load test data were presented by Coop and Wroth (1989). One may argue that the segment is too short to produce data which are comparable with pile data. The data are, however, a valuable supplement to the very scarce data from piles in overconsolidated clays.

Tilbrook Grange tests

The series of pile load tests carried out at this site are thoroughly described elsewhere in papers in this volume. The soil condition is described by Lambson *et al.* (1993).

Two open-ended 762 mm OD steel piles were driven to about 30m. One pile, denoted the McClelland pile, was heavily instrumented and subjected to an axial static compression test, succeeded by a compression creep test and finally a one-way cyclic compression test. The other pile, which originally was used in a driving trial, was instrumented by strain gauges after installation and plug removal and tested in tension. This other pile is denoted the Arup pile. Three closed-ended 219 mm OD steel piles and an open-ended 273 mm OD steel casing were also installed and tested in static and cyclic tension at the site by NGI. A summary of the pile outlines and instrumentation is shown on Fig. 1.

The ground consists of a very stiff silty clay (Lowestoft Till) above about 18 m and a very stiff fissured clay (Middle Oxford Clay) below this depth. The soil profile is heavily overconsolidated.

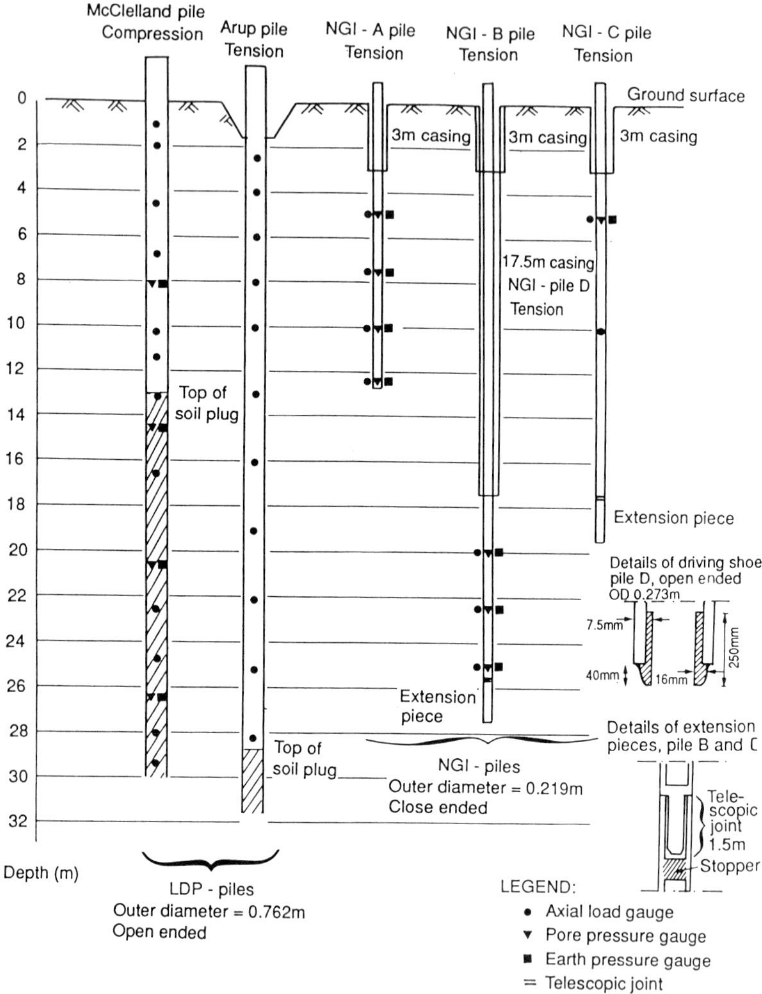

Fig. 1. Tilbrook Grange. Pile outlines and instrumentation.

The quality of the data from the series of tests at Tilbrook Grange is judged as very high. Two piles were close to full scale. The results of the tests with different piles with different instrumentation provides a high degree of confidence in the data.

Comparisons of the pile load test results

Installation earth and pore pressures

Figure 2 summarizes measured pore pressures at end of pile installation of all the piles at Tilbrook Grange. There are considerable variations between the piles and with depth. The excess pore pressures must, however, generally be characterized as low. Normalized with respect to the assumed in situ undrained shear strengths, the excess pore pressure ratio $\Delta u_i/s_u$ ranges from

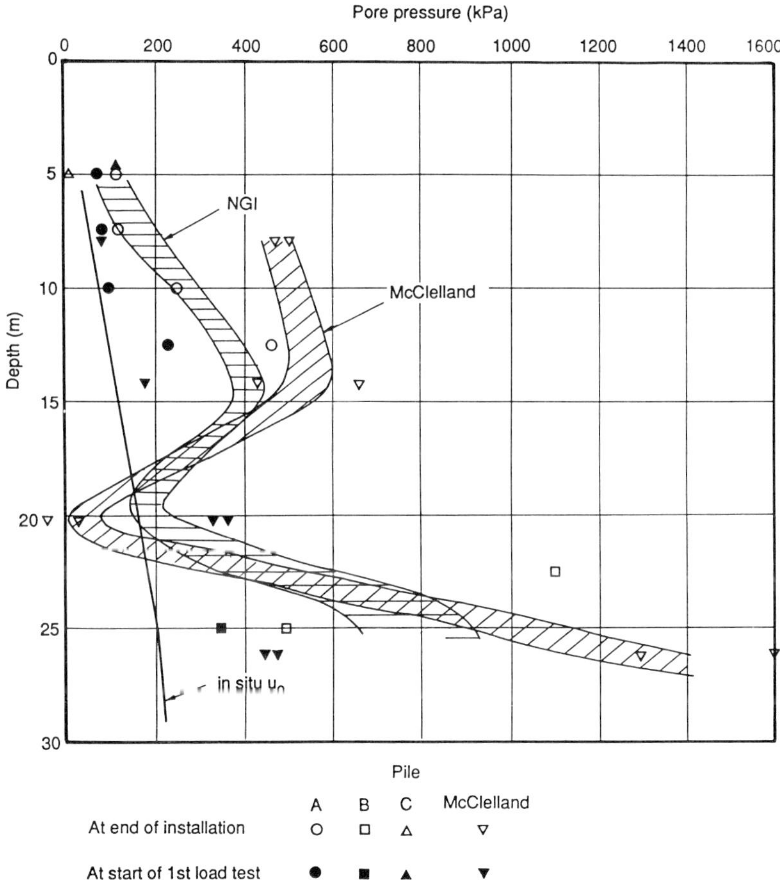

Fig. 2. Pore pressures immediately after pile installation and at onset of initial static tests

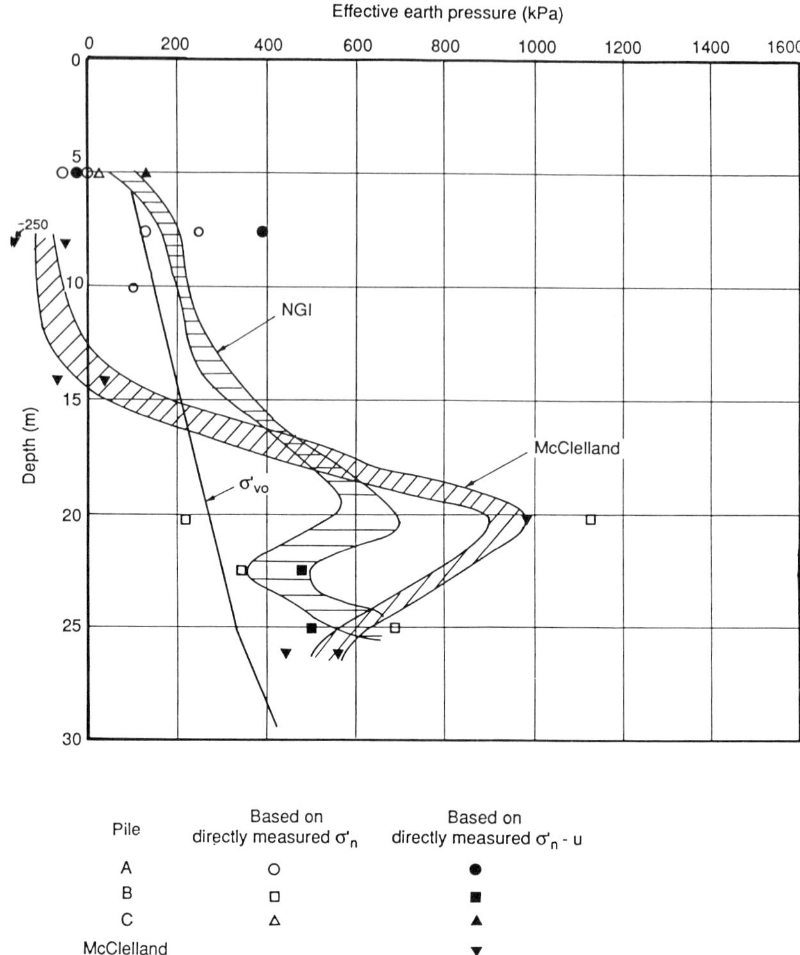

Fig. 3. Measured effective horizontal stresses immediately after pile installation

-0.32 to 1.85, with an average around 0.70. For comparison, predictions based on Cavity Expansion Methods (CEM) with a simple elasto-plastic soil model gives $\Delta u_i/s_u$ in the range 3.4 to 3.7 for closed-ended piles, or a factor of about 4 times larger than the average measured. This is probably due to shear induced dilation of the stiff clay.

It is surprising that along the top 14 m, the McClelland pile generated significantly larger excess pore pressures than the closed ended NGI pile A. Because the McClelland pile penetrated down to this depth in almost open mode, one would have expected smaller pore pressures than for the closed-ended NGI pile.

The results from Canons Park and Madingley, where pore pressures were successfully measured during installation, reveals similar results to those obtained at Tilbrook. At Haga relatively higher pore pressures were induced.

The installation effective earth pressures, σ'_{hi}, measured at Tilbrook show considerable scatter, even more than the pore pressures, see Fig. 3. At instrument levels within the Lowestoft Till (5 to 14 m depth), σ'_{hi} ranges from –250 to +380 kPa. Negative values of effective stresses are difficult to accept, and are probably due to too low measured total earth pressures.

At instrument levels within the Middle Oxford Clay the σ'_{hi}-values range from about 0.8 to 4.0 times the in situ vertical effective stress, σ'_{vo}. The extremely high values near 20 m are mainly due to the low installation pore pressures at this level (Fig. 3).

The installation effective earth pressures at the other test sites showed also a considerable scatter although at some of the sites less than observed at Tilbrook.

Bond and Jardine (1988) emphasized the systematic fall in magnitude of the total and effective radial stresses observed in Canons Park tests as the pile passes any particular soil horizon (*Z/R* effect or strain fatigue effect.) A similar effect was observed at UHCC. The inferred load transfer pattern during driving as well as the lateral earth pressure measurements after driving suggested that pile installation at UHCC produced a slightly oversized cavity along the upper half of the pile after which the soil swelled back on the piles. Such effects were not directly observed during installation of the piles at the other test sites although length effects may be inferred from the tests at Tilbrook as discussed later.

Large positive pore pressures were generated shortly after installation at the leading instrument levels at Canons Park. Much smaller pore pressures were generated at a higher position. The NGI A Pile at Tilbrook shows similar results for the lowest instrument unit which was situated close to the pile tip for this particular pile (Fig. 1).

Reconsolidation

Most theories predict the total radial stress to be constant or decrease during the reconsolidation process. There are test results in OC clays which contradict this prediction.

The total radial stresses tended to rise over the top half of the pile and reduce over the lower part at Canons Park during equalization of stresses. Also the UHCC tests revealed this effect as mentioned above. Most theories also predict a monotonic decrease in the pore pressures with time. Again the test results may show a different behaviour. A fairly monotonic decrease in pore pressure was observed at several pore pressure gauges at Tilbrook where also the measured response with respect to pore pressure dissipation for the various piles reflected the mode of soil displacement during installa-

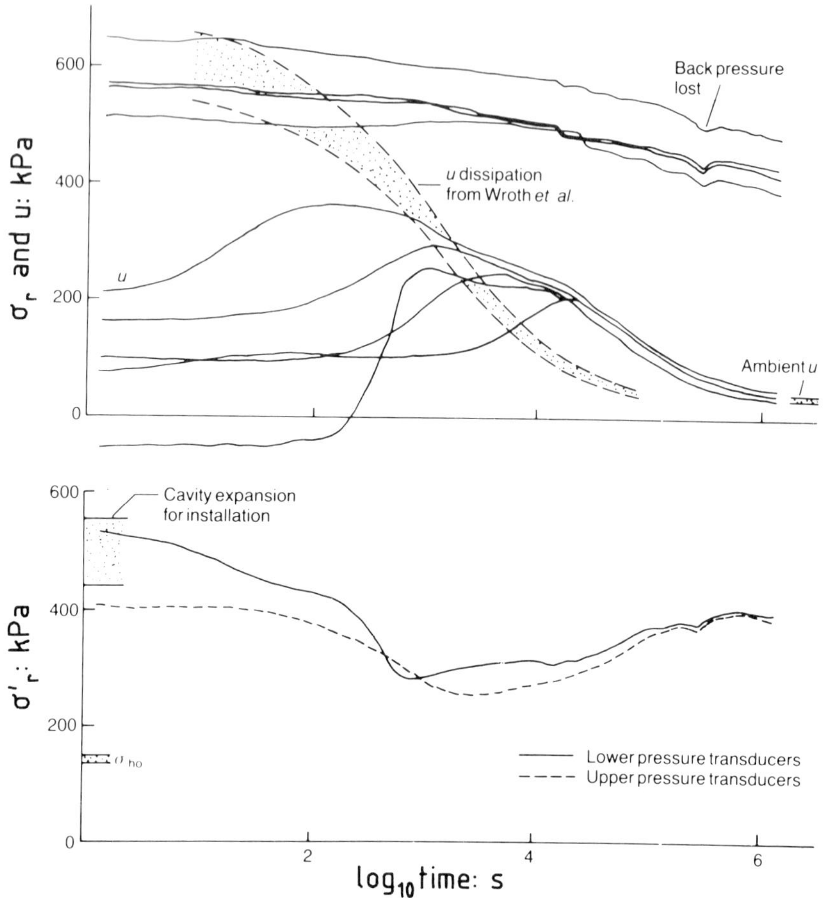

Fig. 4. Madingley reconsolidation data

tion (closed-ended or plugged versus open-ended). A completely different typical reconsolidation process is shown in Fig. 4 which is taken from Coop and Wroth (1989). Soon after halting penetration at Madingley, the pore pressures built up followed by a period of stabilization. Coop and Wroth (1989) suggest that there is a pore pressure maximum during penetration which is radially remote from the shaft. This pore pressure is spreading to the shaft during reconsolidation.

Very similar reconsolidation results to those from Madingley were observed at Canons Park for the upper probes and at one particular depth (20.4 m) for the McClelland pile at Tilbrook.

The average radial effective stresses decreased during the equalization process at Canons Park. This observation is also contradictory to most theoretical predictions. The final values at Madingley were also slightly lower than the initial effective stress shortly after installation. The data from Tilbrook indicate an increase in effective stresses except for the McClelland pile at a depth of 20.4 m. The Haga tests showed close to zero initial effective stresses after driving. The effective stress increase during equalization was consequently relatively higher at Haga and more similar to experience from normally consolidated clays.

Earth and pore pressures at onset of pile loading

The load tests at Tilbrook were carried out about 60 and 120 days after installation for the NGI piles and the McClelland pile, respectively. At that time the pore pressures were still somewhat larger than in situ u_o-values (Fig. 2) (75 to 95 per cent dissipation).

Figure 5 summarizes the measured horizontal effective stresses, σ'_{hc}, at onset of pile loading. At this stage there is also a fair amount of scatter or variation, but to a lesser extent than during installation (compare to Fig. 3). There is little scatter in measured pore pressures at this stage (Fig. 2) and the remaining scatter or variation is mainly due to total earth pressure variations.

For the NGI piles, k'_c vary from about 2.3 to 3.0, generally increasing with depth. The k'_c values for the McClelland pile increase with depth from 1.2 to 2.9. The significantly lower k'_c values for upper portion of the McClelland pile compared to the NGI piles could be due to the open (almost non-plugged) mode of penetration of the McClelland pile down to 13 m penetration. It is also possible that the McClelland pile was more influenced by pile whipping effects during driving. The best estimated k'_c values at Tilbrook are plotted versus s_u/σ'_{vo} in Fig. 6 where also data from the other sites are presented.

The horizontal effective stress ratio k'_c measured in the UHCC test series was similar to that observed at Tilbrook. The Haga tests revealed $k'_c = 1.2$ which is consistent with a lower OCR. The horizontal effective stress ratio measured for the Canons Park piles and on the segment pile at Madingley were more than twice as high as those measured at Tilbrook and at UHCC. The values of k'_c increased from about 5 to about 11 over the depth range 3.25 m to 5.5 m at Canons Park. The variation with depth was less at Madingley where a typical value was $k'_c = 7$. The Z/R effect which was observed at Canons Park may not have been developed at Madingley due to the short pile segment used.

The substantial higher k'_c values observed at Canons Park and at Madingley cannot be explained by a higher OCR at these sites. The main differences between these test series and the Tilbrook and UHCC tests are that the pile diameter is smaller and that the piles were installed by jacking instead of

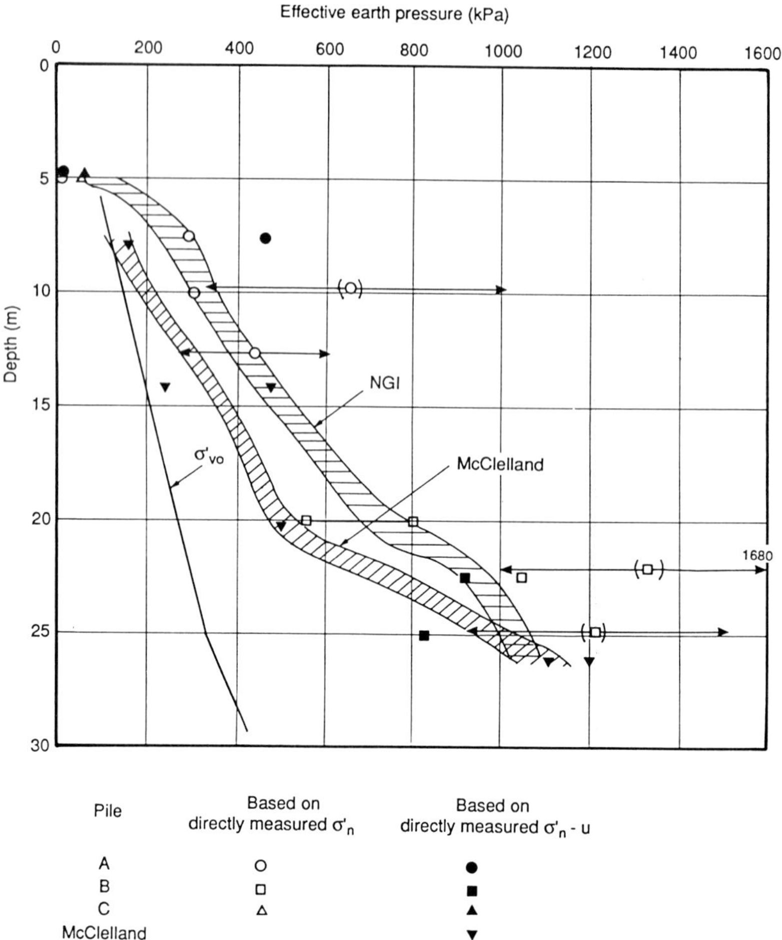

Fig. 5. Measured effective horizontal stresses at onset of pile loading (first initial static tests)

driving. Since very little scale effect was observed for the tests at Tilbrook, one might suspect that installation by driving is the main reason for less observed horizontal effective stresses. However, two uninstrumented piles were also installed by driving at Canons Park. These driven piles showed very similar load displacement characteristics to the piles installed by fast-jacking. Unless the friction characteristics along a driven pile is different from a fast-jacked pile, one would therefore expect the horizontal stresses to be similar for the driven and fast-jacked piles at Canons Park.

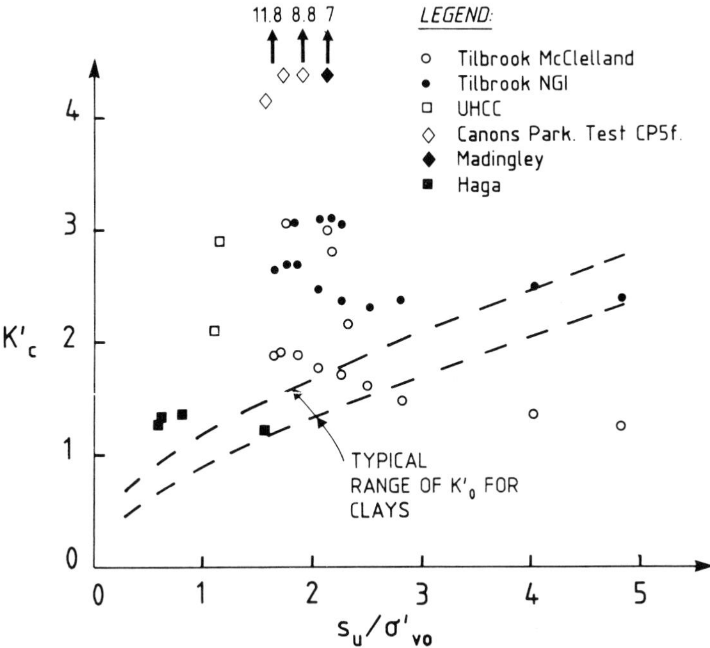

Fig. 6. $k'_c = \sigma'_{hc}/\sigma'_{vo}$ *versus* s_u/σ'_{vo} *from pile tests in overconsolidated clays*

It should also be noted that the Haga piles were installed by jacking. A moderate $k'_c = 1.2$ was measured which is slightly higher than the estimated $k'_0 \approx 1.0$. These tests do not support the hypothesis that installation by jacking *per se* induce very high horizontal stresses.

Ultimate static skin friction

The ultimate skin friction values at Tilbrook are presented in Fig. 7. To arrive at the skin friction values for the Arup pile, which was instrumented after driving, it was assumed the same residual skin friction distribution at onset of loading as measured for the McClelland pile.

The following main aspects are noted from Fig. 7:

- The Arup pile and the NGI piles A, B and C (all loaded in tension) gave very similar ultimate skin friction values at all depths. The peak skin friction value coincides with the zone of peak undrained shear strength.

Table 1. Summary of pile and soil data and key results from sites studied

		Tilbrook	UHCC (3)	Canons Park test CP5f/L1T	Cowden	Haga	Madingley
Pile penetration depth	m	32	13	≈6	10	5	8 (4)
Pile diameter	m	0.219-0.762	0.273	0.102	0.203-0.305	0.153	0.08
Water content	%	15	25/20	30	16	≈40	-
Plasticity index	%	30	45/15	45	20	15	-
s_u (1)	kPa	300-700	38/69	70-120	100-150	43	≈125
s_u/σ'_{vo}	-	1.3-5	1.13/1.10	1.8-1.9	0.7-5	0.6-1.65	≈2.4
OCR	-	6-30	≈6/≈4	10-30	2-20	3.4-10	-
K'_o	-	1.2-3.0	1.8/0.9	1.6-2.3	0.8-1.8	≈1.0	2.5
K'_c	-	1.4-3.0	2.9/2.1 (5)	5-12	-	1.2	≈7
$\Delta U_i/s_u$	-	-0.32-1.85	-	≈0	-	2-6	≈0.9
$\Delta U_i/\sigma'_{vo}$	-	-0.25-5	-	≈0	-	3-4.5	≈1.9
σ'_{hi}/s_u	-	-0.5-2.0	-	2.5-9	-	≈1.2	≈3.5
$\sigma'_{hi}/\sigma'_{vo}$	-	-5 -4.0	-	5-15	-	≈1.3	≈9
Time to reach 90% cons.	days	60-300	≈4	1.5	-	≈4	≈4
Appr. time to failure	min	≈30	500	30?	-	20	1.5
τ_{us}	kPa	50-400	22/35	45-200	40-59	18-50	≈65
$\alpha = \tau_{us}/s_u$	-	0.44 (2)	0.58/0.52	0.64-1.70	>0.36-0.54	0.43-1.0	≈0.5
$\beta = \tau_{us}/\sigma'_{vo}$	-	1.02 (2)	0.65/0.56	1.0-3.0	>0.53-0.79	0.35-0.67	-
τ_{us}/σ'_{hc}	-	0.35-0.7	0.22/0.27 (5)	0.20-0.26	-	0.25-0.55	-
$\tau_{us}/\sigma'_{hf} = \tan\delta$	-	0.3-0.7	0.24/0.28 (5)	0.2-0.24	-	0.35-0.07	≈0.17
Z_c/D	-	0.01-0.03	≈0.005	0.02-0.03	≈0.03?	0.015-0.020	-

(1) Generally based on s_u from UU or CU triaxial tests, except Haga which is based on DSS tests
(2) McClelland pile, mean
(3) Data from strata B&D. Hydrostatic in situ pore water pressure assumed
(4) Model pile segment of length 1.135 m
(5) Lateral stresses taken as measured in the pile group

Table 2. Methods used to predict static skin friction for the Tilbrook Grange test series

α-methods ($\tau_{us} = \alpha \cdot s_u$)
1) API-RP2A (1982) $\alpha = f(s_u)$
2) API-RP2A (1989) $\alpha = f(s_u/\sigma'_{vo})$
3) Semple & Rigden (1984) $\alpha = f(s_u/\sigma'_{vo}, L/D)$
4) Randolph & Murphy (1985) $\alpha = f[s_u/\sigma'_{vo}, (s_u/p'_o)_{nc}]$
5) Randolph & Wroth (1982) $\alpha = f(s_u/\sigma'_{vo}, \phi')$

β-methods ($\tau_{us} = \beta \cdot \sigma'_{vo}$)
6) Meyerhof (1976) $\beta = k'_c \tan\phi'$, $k'_c = 1.5\, k'_o$
7) Parry & Swain (1977) $\beta = f(\phi', \sigma'_{vf}/\sigma'_{vo})$
8) Randolph & Wroth (1982) $\beta = f(s_u/\sigma'_{vo}, \phi')$

Other methods
9) λ-method (Kraft & Focht, 1981) $\tau_{us} = \lambda(\sigma'_{vave} + 2\, s_{u\,ave})$, $\tau = f(\pi_3)$
10) Flaate & Selnes (1977) $\tau_{us} = 0.4\sqrt{OCR} \cdot \mu_L \;\; \sigma'_{vo}$
$\mu_L = f(L)$
11) NGI-method (Karlsrud & Nadim, 1990) $\tau_{us} = \xi\, (\tau_f/\sigma'_{ac})^{DSS}_{RR}\;\; k'_c\, \sigma'_{vo}$
where ξ and $k'_c = f(OCR)$
12) Janbu (1976) $\tau_{us} = S_v\, (\sigma'_{vo} + a)$
$S_v = k_{ap} \cdot r \tan\phi'$

- In the depth range of about 5 to 12 m, the McClelland pile (loaded in compression), gave considerably lower skin friction values than the other piles. In average it corresponds to 50 kPa in the depth range 5 to 10 m or about 30 per cent of the values for the other piles.
- Between 12 and about 23 m depth, the McClelland pile gave values agreeing very closely with the other piles.
- Below ≈23 m the McClelland pile gave apparent skin friction around 440 kPa, which seems to be about 120 kPa larger than indicated by the Arup and NGI B piles.

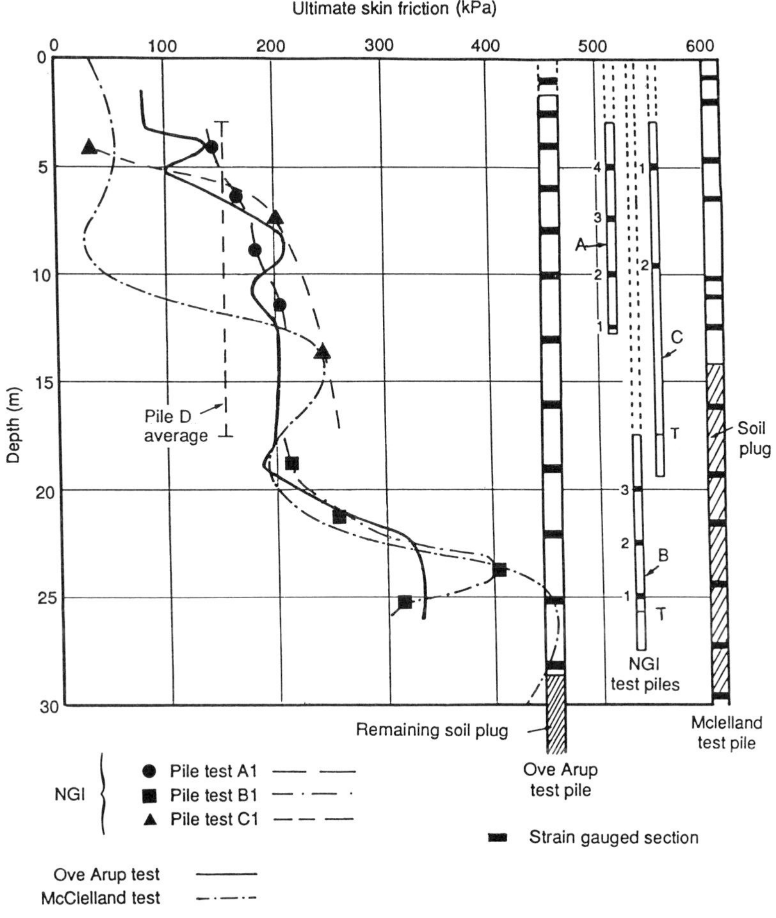

Fig. 7. Tilbrook Grange. Ultimate skin friction

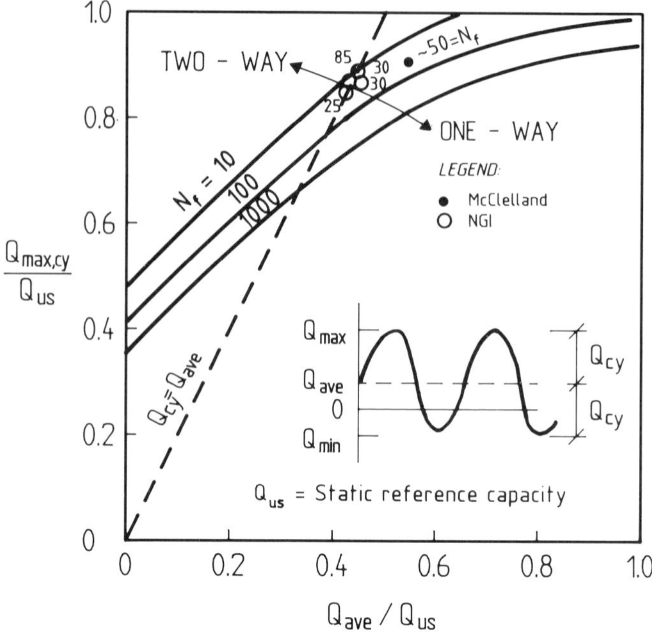

Fig. 8. Number of load cycles to failure (N_f) in relation to maximum cyclic load and average bias load. Solid lines infrerred from Haga tests. Data points from Tilbrook Grange.

When considering and comparing the ultimate skin friction values one should bear in mind that there still remained some excess pore pressures along the piles at onset of pile loading (Fig. 2). It is estimated that, after 100 per cent pore pressure dissipation the ultimate skin friction could have increased on average by 5 per cent to 12 per cent.

The Arup pile was installed 20 months prior to load testing (although redriven 0.5 m 9 months after installation) and thus, can be assumed to have reached full pore pressure dissipation.

No definite explanation has been found for the very low skin friction along the upper 12 m of the McClelland pile, but the following aspects have been considered:

- It could be an effect of pile driving, either a *Z/R*, strain fatigue or pile whipping effects. It is then surprising, that the Arup pile did not show similar results. It should be noted that the driving sequences of the piles were not identical.

- The measured σ'_{hc} values are for the upper 14 m of the McClelland pile on average 65 per cent of what was measured for the NGI pile A (Fig. 5). This can only partly explain the observed low skin friction.
- The tensile and compressive shear stresses along the bottom and top part of the piles at onset of pile loading implies a reversal of shear stresses (tension to compression) along the top part of the McClelland pile (loaded in compression). Such shear stress reversal could lead to somewhat reduced local skin friction, but there has been no indications from previous pile load tests or laboratory tests that a single reversal of shear stresses could cause the strength to drop to 30 per cent.
- Open versus closed (plugged) penetration could contribute to the difference between the McClelland and NGI piles, but again the Arup pile does not confirm this.
- There are no indications from the measured local shear stress versus displacement (*t*-*z*) data, that strain softening plays any significant role.

Regarding the relatively high skin friction indicated along the lower 5 m of the McClelland pile compared to the other piles, this is mainly an artificial effect caused by a contribution to the measured stresses in the pile wall of inner skin friction from the soil plug. If one assume the plug tip resistance is transferred to inner skin friction over a 5 m length above the pile tip, the inner skin friction becomes 140 kPa. This corresponds closely to the suggested difference in apparent skin friction for the McClelland pile and the NGI B pile and Arup pile below 25 m depth.

Pursuing the issue of the difference between open and closed (plugged) pile penetration, it may be noted that the average skin friction along the non-instrumented NGI pile D corresponds to 155 kPa (Fig. 7). Pile D, which was the casing for pile B, did not plug during driving, partly because it was equipped with a driving shoe with 6 mm inside clearance. The average skin friction along pile D corresponds to 78 per cent of the average skin friction for the closed-ended pile C with identical penetration. This could be a pure effect of open versus closed pile penetration. However, it is more likely that subsequent driving of pile B inside the casing pile D caused some whipping effects on pile D, reducing the horizontal stress and skin friction.

The load capacity of the closed-ended pile at Canons Park was 6 per cent higher than the capacity of the open-ended driven pile. Also the test series at Cowden showed very similar capacities for open-ended and closed-ended piles.

Variation in horizontal effective stress during pile loading up to failure were small for the test series at all sites, in general not greater variation than ±15 per cent.

The measured τ-z response at Tilbrook shows that the peak ultimate skin friction was reached at a local displacement of 1 to 3 per cent of the pile diameter. Similar relative displacements were observed at all other sites except at UHCC where the peak friction was reached at about 0.5 per cent. Strain softening effects were observed in some of the local τ-z curves observed

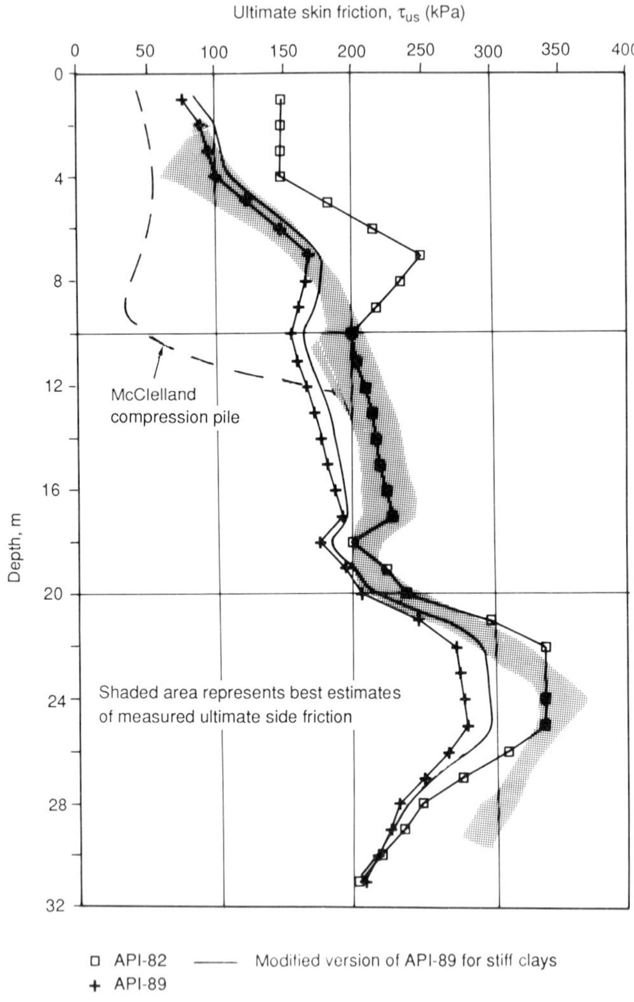

Fig. 9. Tilbrook Grange. Plot showing predicted and measured ultimate side frictions

at Tilbrook. A minimum observed value of residual to peak values of local skin friction was about 75 per cent. Very little or no strain softening was observed at some levels. To quantify the effects of local strain softening in the ultimate (peak) pile head resistance, the pile capacities were computed by integration of the peak values of local skin friction (e.g. assuming that the piles were completely rigid). This gave only 3 per cent higher capacity for the McClelland pile and less than 1 per cent higher for the Arup pile than the measured capacities. Thus, strain softening has not significantly influenced the ultimate (peak) capacities of these piles.

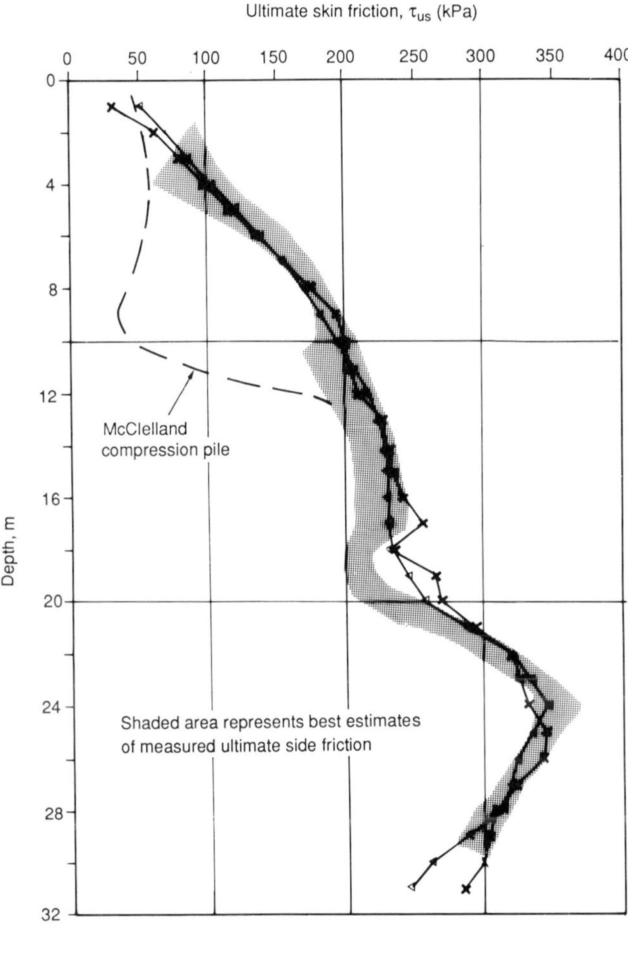

Fig. 10. Tilbrook Grange. Plot showing predicted and measured ultimate side frictions

Cyclic pile capacities and response

All the NGI-piles and the McClelland pile were subjected to cyclic loading following the static reference tests at Tilbrook. All the tests were so-called one-way cyclic loading tests, e.g. both the maximum and minimum loads were applied in the same direction (tension for NGI's tests, compression for McClelland). All the piles reached failure by an increase of accumulated upward (NGI's tests) or downward (McClelland's test) pile top displacement. The number of cycles to failure, N_f was defined as when the rate of accumulated displacement exceeded 1.0 mm/min, which is consistent with the rate of displacement at failure in the static tests. The number of cycles to reach failure according to this criterion are very consistent and typically suggest $N_f \approx 30$ for a cyclic peak load corresponding to 86 per cent of the static capacity. The results from each test is plotted in Fig. 8, which shows that the one-way cyclic capacities for the Tilbrook tests also agree well with the Haga tests (Karlsrud *et al.*, 1986). Since only one-way cyclic tests were carried out at Tilbrook, it may be worthwhile to notice that the Haga pile tests as well as

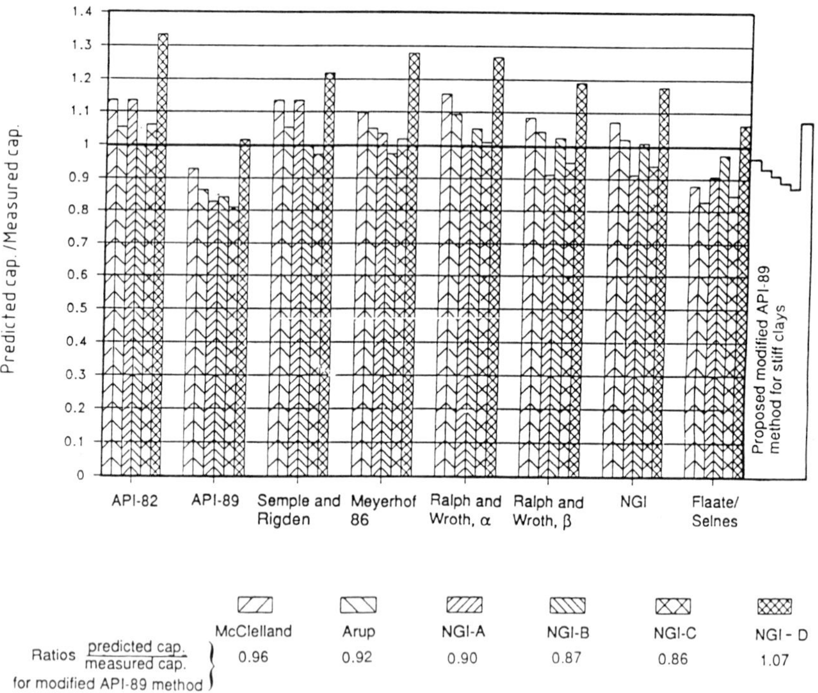

Fig. 11. Tilbrook Grange. Predicted and measured capacities

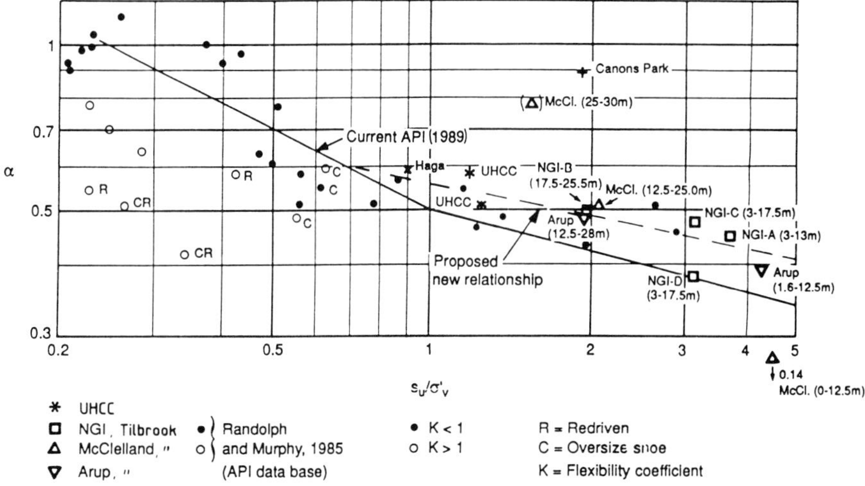

Fig. 12. Comparison between $\alpha - s_u/\sigma'_{vo}$ relations from new tests and load test data from API data base

cyclic DSS tests and theoretical analyses show that the cyclic capacity of piles subjected to two-way type cyclic loading (e.g. load reversed tension–compression) is generally much lower than in one-way cyclic loading. For symmetrical two-way loading the cyclic capacity may be as low as 50-60 per cent of the static capacity. The results of the cyclic tests at Cowden supported the results obtained at Tilbrook and Haga. No cyclic data exist from the other test sites.

Predicted compared to measured ultimate static skin friction

The ultimate pile capacity and limit skin friction were predicted and compared to the measured values for all piles tested at Tilbrook, using twelve different methods, see Table 2.

With regard to grouping of the methods in Table 2, the use of α and β definitions may be a bit misleading if one generally thinks of β-methods as more fundamentally effective stress based compared to α-methods. Definition of β is just a way of normalization. Methods 10, 11 and 12 could actually also have been classified as β-methods in this respect.

Selection of input parameters is a question of uncertainty and judgement regarding several of the methods. This applies for instance to the assessment of undrained strengths (all α-methods), to estimating k'_0- and k'_c-values (methods 6 and 11), effective friction angle and attraction (methods 6, 7, 8, 12) and overconsolidation ratio (methods 10 and 11). As for design purposes, the predictions for the Tilbrook tests were generally based on selection of lower bound values for the input parameters, except the undrained shear strengths, where best estimate values in terms of mean values from UU tests over selected depth intervals were used.

Figures 9 and 10 present predicted distributions of ultimate skin friction for some methods. Measured values are given as a shaded range representing NGI piles A, B, C, the middle part of the McClelland pile, and the Arup pile. The top part of the McClelland pile is shown separately, whereas the bottom 5m is neglected due to the effect of inner soil plug.

Most of the methods in Table 2 give in general good correlation to measured values represented by the shaded area (neglecting the upper part of the McClelland pile). The exception was mainly methods 7 and 12 which underpredict by more than 50 per cent.

Methods with best correlations to measured ultimate skin friction were used to compute the total pile head capacities, taking into account tip resistance and weight of piles and plug as appropriate. The predicted values are compared to measured in the bar chart in Fig. 11. Neglecting the NGI pile D casing, which for all methods are overpredicted by approximately 20 per cent when compared to the other piles, the average and extreme deviations of the ratio predicted/measured capacities for the methods shown in Fig. 11 are within +8 per cent and -14 per cent.

For practical design purposes the outcome of all these methods would be regarded as very satisfactory. Considering that there have been no previous pile tests in such stiff clay, it is actually remarkable that they all do so well. It seems like the β methods 8, 11 and 6, based to a various extent on the state of effective stress in the ground, give the best predictions. One should bear in mind, however, that selection of input parameters for these β methods were more uncertain than selection of undrained shear strengths from the UU-tests used as basis for the α method predictions. The methods are currently based on a broader statistical study of pile tests in other clays than the β methods used in this study.

On the other hand, determining a design profile for undrained shear strength also involves large uncertainties. Undrained strengths obtained from CU tests at Tilbrook was much lower than UU strengths used in the predictions. It is also common to apply α-methods on the basis of undrained shear strengths from UC tests, fall cone tests, pocket penetrometer and torvane tests. Such tests may give considerable scatter and deviation from UU tests. Thus, in future application of α methods a better understanding of the difference between various strength tests is warranted.

The reasons why the measured skin friction along the upper and lower part of the McClelland pile and the NGI D pile deviated from the general trend of the measured values have been discussed previously.

Table 1 shows typical back calculated mean values of α and β for the other pile test sites. Except for the Canons Park data, all tests produce reasonable values in accordance with the predictive methods.

Mobilized apparent friction on the pile shaft

Table 1 also shows typical ratios between the limit skin friction and the measured horizontal effective stress against the piles at onset of loading (σ'_{hc}) and at failure (σ'_{hf}).

It should also be noted that the apparent friction do not necessarily represent the maximum or true mobilized effective friction within the clay. This can only be determined if one knows the normal effective stresses and shear stresses acting on a plane in addition to the plane defined by the pile wall.

There is a fair amount of scatter in the data from Tilbrook. The typical values of τ_{us}/σ'_{hc} at Tilbrook were within the range 0.4 to 0.6. Consolidated drained Direct Shear (DS) testing of samples from Tilbrook also showed a large scatter in the results. However, the mobilized apparent friction in direct shear tests at peak load was typically in the similar range 0.4 to 0.6 for both undisturbed and remoulded samples, as also observed on the piles.

The apparent residual friction as measured in the direct shear tests were typically 60 per cent of the peak value and is therefore less than the apparent friction observed on the piles. Residual apparent friction from Ring Shear (RS) tests were also less than the values observed along the piles.

It is seen from Table 1 that the ratio τ_{us}/σ'_{hc} (or τ_{us}/σ'_{hf}) typically in the range 0.2 to 0.25 for the tests at UHCC, Canons Park and Madingley while the Haga tests revealed a range more similar to the observations at Tilbrook. One of the conclusions from the Haga tests was that the side shear resistance along a pile shaft can be reasonably modelled by undrained Direct Simple Shear (DSS) testing on remoulded reconsolidated clay and taking into account the difference that may exist in mean effective consolidation stress in the DSS apparatus compared to an element against the pile shaft at onset of loading.

Residual apparent friction as measured under *drained* conditions in direct shear apparatus was 20 to 50 per cent higher than the measured apparent friction along the piles in the UHCC tests.

Peak apparent friction as measured in a ring shear apparatus under *drained* conditions was very similar to the observed apparent friction on the piles at Canons Park.

Residual apparent friction as measured under *drained* conditions in a ring shear apparatus was very similar to the observed apparent friction at Madingley.

No rational explanation has been found for the differences in measured apparent friction on the piles at the various sites. The attempts made in the various pile testing projects to relate the observed apparent friction on the pile wall to the apparent friction measured in the laboratory in DS and RS tests are also difficult to evaluate. There is a mixture of drained or undrained testing conditions and reported peak or residual friction values which makes a rational interpretation difficult.

Conclusions and design recommendations

Installation and reconsolidation

The excess pore pressures induced by pile installation at all sites except at Haga were relativly low compared to experiences with softer clays and cavity expansion predictions.

The variation of the total stress and pore pressure during the reconsolidation period may be significantly different from most theoretical predictions. Effects of dilatancy and pile installation (Z/R, strain fatigue, whipping) may be decisive in overconsolidated clay. We are not aware of theories which take such effects into account.

The horizontal effective stress ratio $K'_c = \sigma'_{hc}/\sigma'_{vo}$ is expected to increase with increasing OCR. However, the measurements along the piles do not directly support this assumption. Both OCR and installation effects (*Z/R*, strain fatigue, whipping) decrease with depth and tend to cancel each other out. The net effect is a relatively uniform K'_c observed for each test pile. The variation in observed K'_c is large ($K'_c = 1.2$ to more than 10 for the range of from 0.6 to 5).

Ultimate skin friction and capacity

Peak value: A number of different methods gave remarkably good predictions of the capacity of the Tilbrook Grange tests. For practical design purposes the current API(89) method seems generally well founded, and is recommended as a main design method. The results of the recent research programmes suggest, however, that the method may be slightly adjusted.

Figure 12 presents back-calculated α values versus normalized undrained shear stength ratio, s_u/σ'_{vo}, for all the Tilbrook Grange tests and the other recent tests reviewed. The figure also presents the data from the API data base presented by Randolph and Murphy (1985), and the current API (1989) design line proposed on this basis.

It appears that API(89) lies somewhat on the low (conservative) side of the test results for s_u/σ'_{vo} greater than about 0.7. It is therefore proposed that for $s_u/\sigma'_{vo} > 0.7$ API(89) could be upgraded as shown by the dashed line in Fig. 12. This corresponds to an increase in α-value compared to API(89) increasing from 11 per cent at $s_u/\sigma'_{vo}=1$ to 21 per cent at $s_u/\sigma'_{vo}= 5$.

The proposed new α-relationship is given by

$$\alpha = 0.50\ (s_u/\sigma'_{vo})^{-0.5} \text{ for } s_u/\sigma'_{vo} < 0.7$$

$$\alpha = 0.56\,(s_u/\sigma'_{vo})^{-0.2} \text{ for } s_u/\sigma'_{vo} < 0.7$$

In relation to application of this method it should be noted that selection of the undrained shear strength is an aspect of some uncertainty. In the analyses of the Tilbrook Grange tests mean values of the UU shear strength results were used. The shear strengths from CU tests were significantly lower than UU strengths. Other tests such as fall cone, pocket penetrometer and torvane tests may also give considerable deviation from UU tests. The review of the other tests also clearly show that the basis for determination of the undrained strength may vary significantly.

Attempts were made to determine the relationship between effective stresses against the pile surface and the skin friction and investigate whether the skin friction was governed by a drained or undrained soil response during pile loading. The results clearly show a correlation between skin friction and effective stresses against the pile surface at all test sites. However, one can not conclude whether the skin friction is governed by undrained/drained or peak/residual strength characteristics on the basis of the available data. The scatter is high when extensive shear test data exist and the reported laboratory test results in the different test series are quite different.

Although there is a clear relationship between the ultimate skin friction and the horizontal effective stresses along one particular test pile, the scatter in horizontal stresses and friction characteristics among the various test piles and test sites is high. The simple empirical methods seem to cancel out some of these scatter in a remarkable way.

Methods which are based on a more sound geotechnical approach, for example the NGI method, gave very close agreement with the Tilbrook Grange tests. Such methods can serve as independent verifications of the suggested modified API method.

Length effects

It should be noted that the API(89) method is intended for use in computation of the peak local ultimate skin friction. Reservations are made in the commentary for taking into account possible pile length effects. The pile length effect has been indicated by Kraft *et al.* (1981), Semple and Rigden (1984) and Randolph and Murphy (1985). The length effect has primarily been related to the following three main factors:

(1) Length of pile driven past a particular level in the ground
(2) Lateral movements of a pile (whipping) during driving
(3) Progressive failure due to strain softening effects (peak to residual skin friction)

The Tilbrook tests suggest that strain softening or progressive failure has insignificant effects on the pile capacity. When the peak pile head load was reached, strain softening effects only reduced the capacity compared to a completely rigid pile, by maximum 1 to 2 per cent. Even at very large pile head displacements, corresponding to about 10 per cent of the pile diameter, the capacity did not reduce by more than 10 per cent. However, strain softening may be more pronounced in soils which exhibit high strain softening effects.

The locally very low skin friction measured along the top 12 m of the McClelland pile (at Tilbrook) could be an effect of driven pile length or pile whip. Stress reversal effects could also have contributed. None of the other Tilbrook pile tests showed any similar low skin friction along the pile top. It is, therefore, difficult to draw any definite conclusion regarding length effects.

It is concluded that some possible length effects should be accounted for in combination with the proposed revised α–s_u/σ'_{vo} relationship. It seems presently very difficult to tie length effects to specific physical mechanisms, to which a method and input parameters can be defined. It is therefore proposed, simply to reduce the computed ultimate skin friction values by 50 per cent from the pile top and down to a depth of 15 pile diameters or 25 per cent of the driven pile length, whichever is smallest. In the extreme case this will not reduce the total pile capacity by more than 10 per cent, and in most practical cases by less than 3 per cent.

Tension versus compression loading

The tests at Tilbrook, UHCC, Canons Park, Cowden and Haga all confirm previous experiences that there seems to be no significant difference in ultimate pile capacity arising from skin friction, between piles in clay loaded in compression compared to tension.

Open versus closed piles

The Tilbrook tests are not quite conclusive concerning the difference in skin friction between a pile penetrated in an ideal open mode, or in a closed plugged or partially plugged mode. Although NGI's open pile D showed about 22 per cent lower skin friction than the corresponding closed NGI pile C, this was conceivably mostly due to whipping during driving of the pile B inside pile D. The top part of the Arup pile showed 5–10 per cent lower skin friction than the trend from the closed piles, but this could also be a length effect.

The tests at Canons Park and Cowden also showed that closed-ended and open-ended piles had very similar capacity.

Although there is room for different interpretations with respect to effect from open versus closed piles, it is concluded that as long as length effects are accounted for one can use the proposed design curve in Fig. 12 for both open and closed piles.

Pile set-up

Prediction of the developement of pile capacity with time after installation may be an important design consideration. In many practical cases it is required that the piles shall be able to carry their full design load only 3–4 months after driving. For the large diameter offshore piles it will not be possible to reach full dissipation of the excess pore pressures set up during pile installation whithin such a short time period, even if the piles do not plug during driving. It may be particularily important for the set-up predictions to assess pile plugging. For the time being no well funded theoretical procedures for doing this are found, and one must resort to practical experiences. Set-up effects are generally less significant in OC clays compared to piles in softer NC clays. This follows from the observation that induced installation pore pressure ratio $\Delta u_i/s_u$ decreasing with increased OCR. Observations of negative set-up effects exist, for example from the Canons Park tests (skin friction increased during the equalization period).

Load–displacement response

For most of the pile test series the peak ultimate skin friction was reached at a local displacement corresponding to 1 to 3% of the pile diameter. This is somewhat on the high side compared to experience from softer clays.

Effects of cyclic loading

The response of the Tilbrook piles to one-way cyclic loading was quite as expected. For a peak cyclic load of 85–90 per cent of the static capacity, failure was reached after 25 to 85 load cycles.

A North Sea storm may correspond to an equivalent number of load cycles under the peak load of about 50 (Karlsrud and Nadim, 1990). Thus, the Tilbrook test results may be more or less directly applicable to offshore piles in similar clays subjected to pure one-way cyclic loading.

For cyclic loading where the pile top will undergo load reversal, both previous pile tests (Karlsrud *et al.*, 1986) and cyclic simple shear tests show that the cyclic capacity may be significantly lower. Values down towards 50 per cent of the static capacity may be relevant for piles in stiff clays subjected to symmetrical two-way loading. Work at NGI clearly suggests that the general behaviour of piles subjected to cyclic loading can be reasonably predicted on the basis of data from cyclic simple shear tests and a numerical model as presented by Karlsrud and Nadim (1990).

Acknowledgement

NGI and the authors would like to thank Amoco, BP and the Department of Energy, UK for their technical and financial support to this pile test review study.

References

API RP2A. *Recommended practice for planning, designing and constructing fixed offshore platforms*. American Petroleum Institute, Dallas, Texas, Thirteenth Edition, 1982.

API RP2A *Recommended practice for planning, designing and constructing fixed offshore platforms*. American Petroleum Institute, Dallas, Texas, Eighteenth Edition, 1989.

BOND, A. J. and JARDINE, R. J. *Canons Park piling project. The behaviour of displacmeent piles in overconsolidated clay*. Final report, December 1988.

BOND, A. J. and JARDINE, R. J. *Research on the behaviour of displacement piles in an overconsolidated clay*. Prepared by the Imperial College, London for the Department of Energy. Offshore Technology Report. Department of Energy. OTH 89296, 1989.

COOP, M. R. and WROTH, C. P. Field studies of an instrumented model pile in clay. *Géotechnique* **39,** 1989 No. 4, pp. 679-696.

FLAATE, K. and SELNES, P. Side friction of piles in clay. *Proc. Ninth Int. Conf. on Soil Mech. & Fndn Engng,* Tokyo, Vol. 1, 1977 pp. 517-522.

JANBU, N. Static bearing capacity of friction piles. *Proc. European Conf. on Soil Mech. & Fndn Engng,* Vol.12, 1976, pp. 479-88.

KARLSRUD, K. and NADIM, F. Axial capacity of offshore piles in clay. Proceedings. *Proc. Offshore Technology Conf.,* Houston, May 1990. Paper OTC 6245.

KARLSRUD, K., NADIM, F. and HAUGEN, T. Piles in clay under cyclic loading: Field tests and computational modelling. *Proc. 3rd Int. Conf. on Numer. Meth. in Offshore Piling,* Nantes, France, May 21–22, 1986, pp. 165-190.

KRAFT, L. M., FOCHT, J. A. and AMERASINGHE S. F. Friction capacity of piles driven into clay. *J. Geotech. Engng Div., Am. Soc. Civ. Engrs* **107,** No. GT11, 1981 pp. 1521-1541.

LAMBSON, M. D., CLARE, D. G., and SEMPLE, R. M. Investigation and interpretation of Pentre and Tilbrook Grange soil conditions. This volume, 1993.

MEYERHOF, G. G. Bearing capacity and settlement of pile foundations. *J. Geotech. Engng Div., Am. Soc. Civ. Engrs* **102,** No. GT3, 1976, pp. 1-29.

O'NEILL, M. W., HAWKINS, R. A. and AUDIBERT, J. M. E. Installation of pile group in overconsolidated clay. *J. Geotech. Engng Div., Am. Soc. Civ. Engrs* **108,** No.GT11, November 1982, pp. 1369-1386.

O'NEILL, M. W., HAWKINS, R. A. and MAHAR, L. J. Load transfer mechanisms in piles and pile groups. *J. Geotech. Engng Div. Am. Soc. Civ. Engrs* **108,** No. GT12, Dec. 1982, pp. 1605-1623.

OVE ARUP & PARTNERS. *Research on the behaviour of piles as anchors for buoyant structures.* Summary report prepared by Ove Arup & Partners for the Department of Energy. Offshore Technology Report. OTH 86215, 1986.

OVE ARUP & PARTNERS. *Comparison of British and Norwegian research on the behaviour of piles as anchors for buoyant structures.* Report prepared by Ove Arup & Partners for the Department of Energy. Offshore Technology Report. OTH 86218, 1986.

PARRY, R. H. G. and SWAIN, C. W. Effective stress methods of calculating skin friction on driven piles in soft clay. *Ground Engineering* **10,** No. 3, 1977, pp. 24-26.

RANDOLPH, M. F. and WROTH, C. P. Recent developments in understanding the axial capacity of piles in clay. *Ground Engineering*, Oct. 1982, pp. 17-32.

RANDOLPH, M. F. and MURPHY, B. S. Shaft capacity of driven piles in clay. *Proc. 17th Offshore Technology Conf.*, Houston, Vol.1, 1985, pp. 371-378.

SEMPLE, R. M. and RIGDEN, W. J. Shaft capacity of driven pipe piles in clay. American Society of Civil Engineers. Analysis and design of pile foundations. *Proc. symposium and session in connection with Am. Soc. Civ. Engrs Annual Convention*, San Francisco, 1984, pp. 59-79.

23. Degradation of skin friction for driven piles in clay

T. R. ALDRIDGE, and Dr F. SCHNAID, Fugro-McClelland Ltd

This paper presents the results of a joint industry study perpormed to determine the form of T-Z data to be used for assessing the axial load–deflection behaviour of driven steel pipe piles in clay. In particular it addresses the degradation of skin friction after peak shear is reached. The usefulness of in situ T-Z probe data and laboratory drained direct shear tests in determining the form of the T-Z data is evaluated. The study was primarily based on the results of the two Large Diameter Pile Tests (LDPTs), but the conclusions were also tested against the results of 11 other pile tests in clay. The main conclusion of the study is that the T-Z data follow very distinct forms over three different zones down the pile length, and a form of T-Z curve which incorporates the observed variations is recommended for use.

The pile tests

The LDPTs were performed at Pentre, in a lightly overconsolidated very silty clay, and at Tilbrook, in a heavily overconsolidated clay. The soil conditions at the two sites are very briefly summarised in Fig. 1.

Both the piles were nominal 762 mm OD steel pipe piles, driven by hydraulic hammers 30 to 40 m into clay deposits. The piles were loaded at a constant rate of displacement in static compression after excess pore water pressures generated during driving had dissipated. The load deflection curves from the tests, derived at the top of the section of pile/soil contact, are presented on Fig. 2. The curves show a clear peak load followed by a reduction in capacity of the order of 10 per cent.

The measured residual unit friction distributions over the embedded pile lengths, following driving and consolidation, are presented on Fig. 3. In general, the curves indicate upward stresses applied by the pile to the soil over the upper 30 per cent of the pile at Pentre and 43 per cent of the pile at Tilbrook. It is also notable that, at both sites, the residual unit friction distribution results in a small zone where upward stresses are applied to the soil, at depths of between 0.5 and 0.7 Z/L, where Z is the depth from the top point of pile/soil contact and L is the embedded pile length.

Large-scale pile tests in clay. Thomas Telford, London, 1993

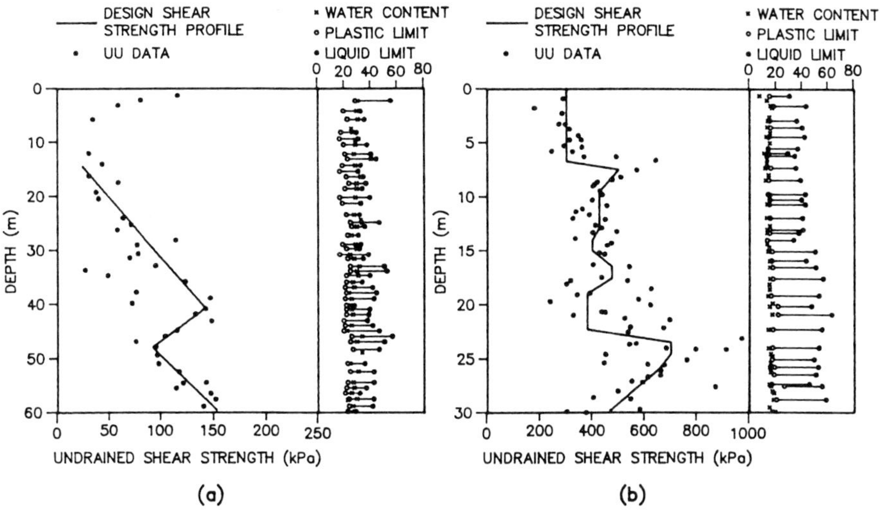

Fig. 1. Basic soils data for (a) Pentre (left) and (b) Tilbrook (right)

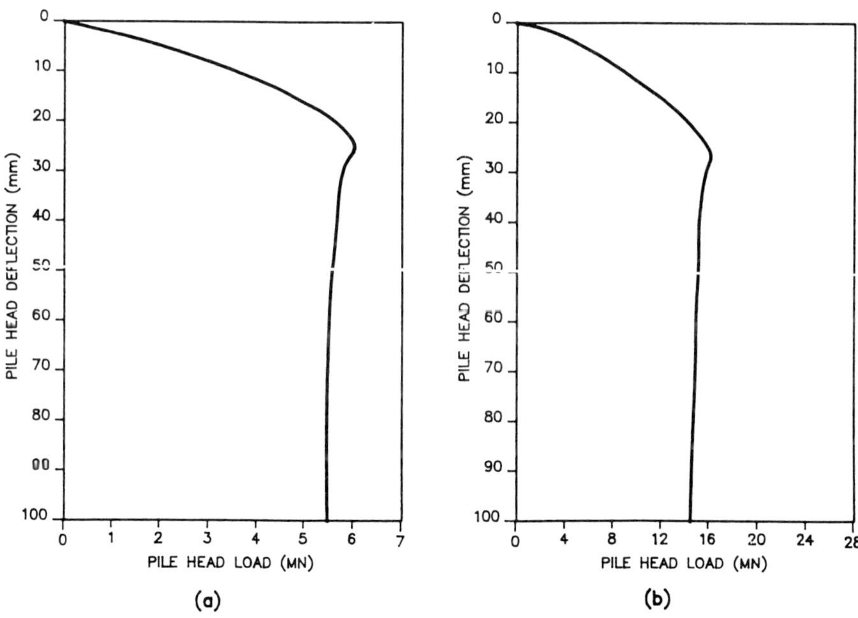

Fig. 2. Pile axial load–deflection for (a) Pentre (left) and (b) Tilbrook (right)

T-Z curves

T-Z curves relate the development of frictional resistance at any point on the pile to the absolute movement of that part of the pile. The *T-Z* data used in standard beam–column programs are therefore Winkler springs, attached to a fixed point in space. The displacement values on the curves must therefore incorporate the effect not only of the localised slip of the pile past the soil at any depth, but also of the elastic movement of the soil mass around the pile at that depth (Toolan and Horsnell, 1980). A Winkler *T-Z* curve typically consists of the following parts:

(a) An initial linear elastic section
(b) A non-linear section up to the peak residual resistance
(c) A reduction from the peak to residual resistance
(d) A constant displacement st residual shear stress.

This paper reviews the appropriate form of each part of the *T-Z* curve, and relates it to the LDPT data and in-situ and laboratory soil tests.

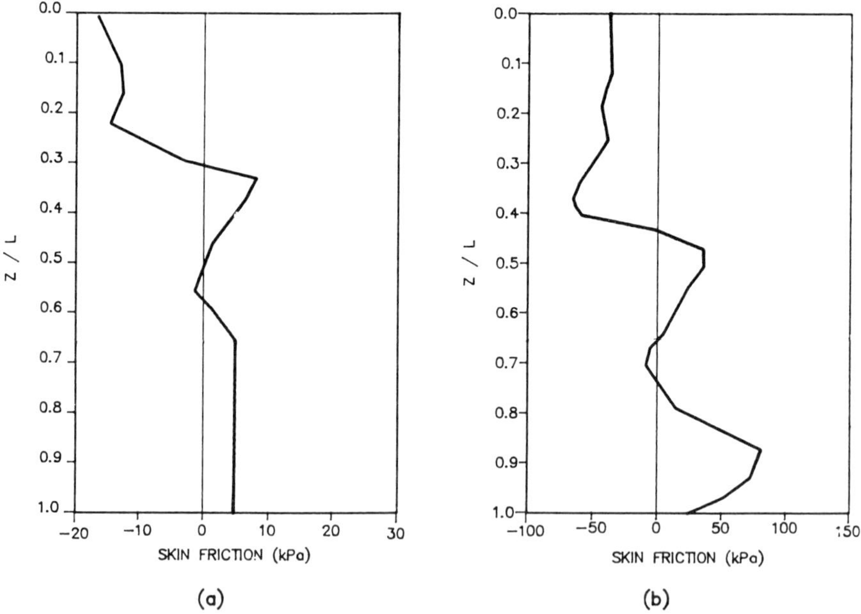

Fig. 3. Residual friction on pile for (a) Pentre (left) and (b) Tilbrook (right)

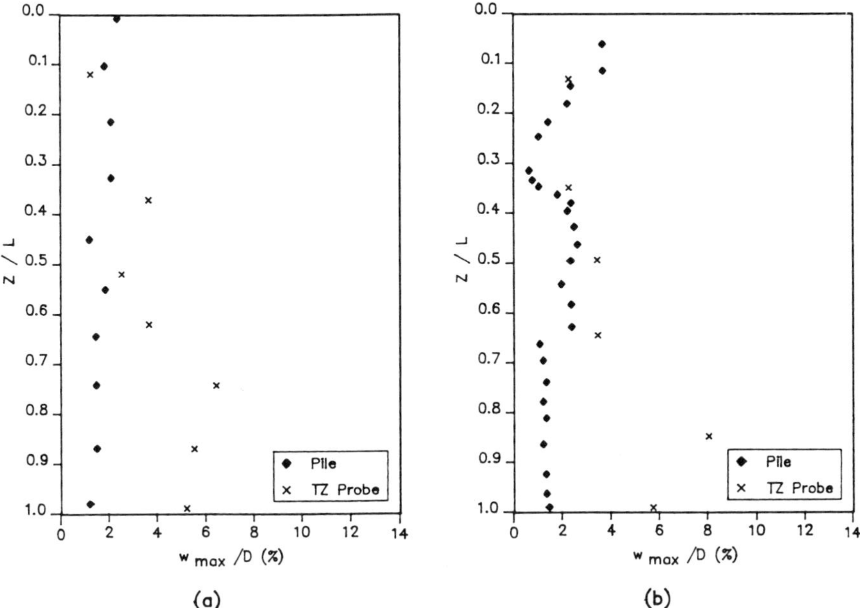

Fig. 4. Displacement to peak shear from pile an T-Z probe for (a) Pentre (left) and (b) Tilbrook (right)

Pre-peak form of curve

Quake at peak shear

The values of quake at peak shear, w_{max}, divided by pile diameter, D derived from the load tests are presented on Fig. 4. The ratio of w_{max}/D is of the order of 1.0 to 1.5 per cent in the lower third of both piles and increases above this region by a factor of two to three by the to of the pile, with some significant local variations with depth. The w_{max}/D values from the *T-Z* probe, which are also presented on Fig. 4, show reasonable agreement with the values obtained from the pile test in the top half of the pile, being of the order of 2 per cent. In the lower half of the pile, however, the results from the *T-Z* probe are approximately 4 times higher than the values derived from the pile test results.

The ratio between the diameter of the *T-Z* probe and the LDPT piles is 762/43.7, i.e. 17,4. From this data it appears that the displacement to maximum resistance is essentially diameter dependent, rather than an absolute value. There are differences between the results from the pile and *T-Z* probe tests, however, and there may therefore be some minor effects of scale.

In comparison with the *T-Z* probe results, which should reflect in situ soil parameters, the pile test results indicate a comparatively softer response in the top 25 per cent of the pile than over the lower part of the pile. It is

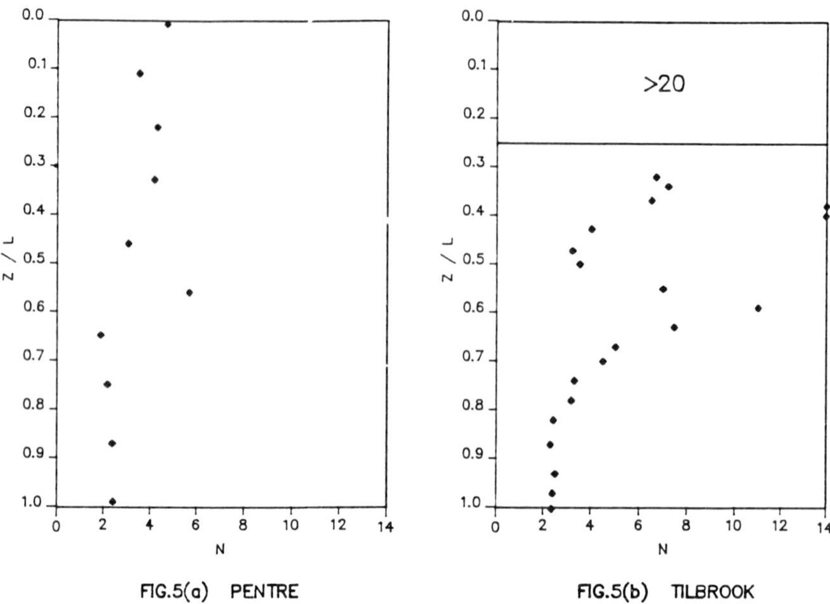

Fig. 5. Normalised initial slope of T-Z curves for (a) Pentre (left) and (b) Tilbrook (right)

considered that this softening of the soil response is probably a function either of the damage done to the soil in this region during pile installation or of the residual soil stress distribution prior to loading.

It is also notable that there is only a small variation in displacements to maximum shear between the two pile test sites, despite the large variation in overconsolidation ratio, and therefore in situ lateral stresses.

Initial T-Z slope

The initial slope of the normalised *T-Z* curves $(d\tau/\tau_{max})/(dw/w_{max})$ is presented on Fig. 5. The data appear to fall into two similar zones at both sites, being of the order of 2.5 in the lower 40 per cent of the pile, and twice this value in the top 60 per cent of the pile. In the upper 25 per cent of the pile at Tilbrook the values are in excess of 20. Theoretical initial *T-Z* curve stiffness values may be computed using methods proposed by Randolph snd Wroth (1978) or Kraft *et al.* (1981). Based on the work of Randolph and Wroth, it may be shown that, for a pile in a uniform isotropic elastic soil of Poisson's ratio 0.5, the initial slope of the *T-Z* curve may be estimated from:

$$\frac{d\tau}{d(w/D)} = 0.5\ G \qquad (1)$$

where τ is shear stress; w is displacement; D is pile diameter; and G is the shear modulus applicable for the soil strains around the pile.

The following general pre-peak T-Z formulation has been proposed by Randolph (1991):

$$\frac{\tau}{\tau_{max}} = 1 - \left(1 - \frac{w}{w_{max}}\right)^N \qquad (2)$$

where τ_{max} is maximum shear stress; w_{max} is displacement at maximum shear stress; and N is the initial slope of the normalised curve. The initial slope $(d\tau/\tau_{max})/(dw/w_{max})$ of the curve given by equation (2) is equal to the value of N. A value for w_{max} may be derived from equations (1) and (2) using the adhesion factor, α, and undrained shear strength, s_u, as follows:

$$\tau_{max} = \alpha . s_u \qquad (3)$$

From equations (1) and (2)

$$0.5\ G = \frac{\alpha\ s_u}{(w_{max}/D)/N} \qquad (4)$$

Thus:

$$w_{max}/D = \frac{2\ \alpha\ N}{G/s_u} \qquad (5)$$

If an overall average initial stiffness value of 3.0 is adopted, based on observations in the LDPTs from the lower two thirds of the pile, equation (5) then gives:

$$w_{max}/D = \frac{6\ \alpha}{G/s_u} \qquad (6)$$

Using equation (6) it is possible to calculate theoretical values of w_{max}/D from the shear modulus values inferred from the in situ pressuremeter testing and the α values derived from the pile test measurements of skin friction. This has been done for the average response in the lower two thirds of the piles, where the values are generally uniform in both pile tests. The measured average w_{max}/D values are of the order of 1.30 per cent at Pentre, and 1.5 per cent at Tilbrook (Fig. 4). The calculated w_{max}/D values are presented in Table 1, and are 1.46 and 1.15 times higher than predicted from the pressuremeter G/s_u values, respectively. This could indicate that the higher shear strain levels around the pile under load result in effective G/s_u values which are on average approximately 1.4 times lower than measured by the pressuremeter. If this result were generally applicable it would be possible to modify equation (6) to give:

$$w_{max}/D = \frac{6\ \alpha\ F_G}{G_p/s_u} \qquad (7)$$

where G_p is the shear modulus value measured by the by pressuremeter and F_G is the ratio of shear modulus measured by the pressuremeter to the average value mobilised around the pile. A review of pile test and shear modulus data indicates, however, that the uncertainty in the applicable range of F_G is such that for practical design purposes the simple adoption of a w_{max}/D value in the range 1.0 to 1.5 per cent will yield equally acceptable results.

Table 1. w_{max}/D derived from measured G_p/s_u and values

PARAMETER	PENTRE	TILBROOK
OCR	1.75	20
G_p/s_u	445	250
α (measured)	0.66	0.54
w_{max}/D (%)	0.89	1.30

Peak and residual resistance

The pile test data are presented in terms of the ratio $(\tau_{max} - \tau_{res})/\tau_{max}$ versus Z/L on Fig. 6. Whilst there is some scatter, it is clear at both sites that no degradation is observed near the top of the pile, but that the average of the ratio $(\tau_{max} - \tau_{res})/\tau_{max}$ is of the order of 20 per cent over the lower third of the pile. The equivalent data from the drained direct simple shear tests are also plotted on Fig. 6. The results show a remarkable agreement with the pile test data over the lower two thirds of both piles. In the upper third of the piles, the direct shear data at both sites indicate degradation of the order of 20 per

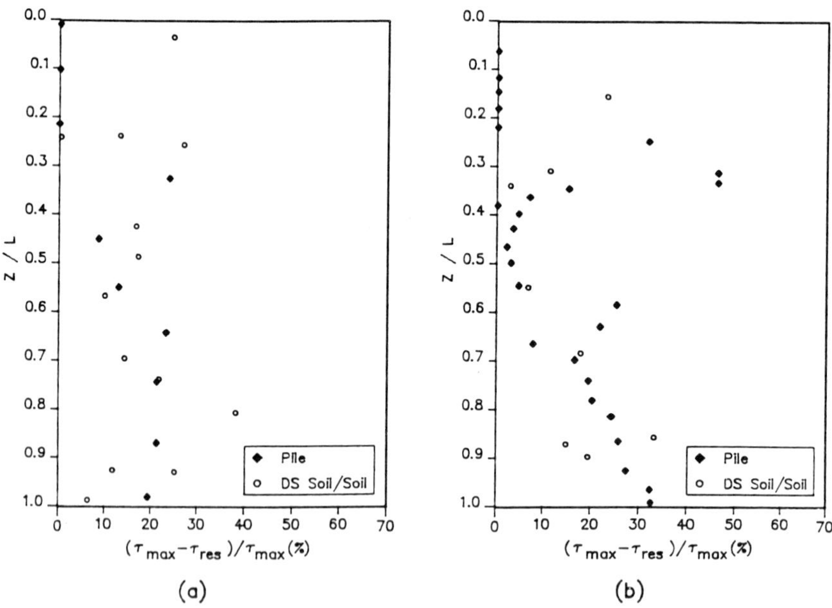

Fig. 6. Post-peak degradation from pile and direct shear test for (a) Pentre (left) and (b) Tilbrook (right)

cent, in strong contrast to the complete lack of degradation in the pile test results.

Observations from the LDPTs and from a study by Bond and Jardine (1989) indicate that only minor changes in horizontal effective stress may take place during loading. If this is generally the case, the ratio of peak to residual friction values might be expected to be a function primarily of the peak and residual friction angles of the soil, rather than the actual values of horizontal stress in the soil either prior to or following pile installation. The peak friction angle, for normally consolidated soil has been related to plasticity index, PI, by Kenney (1959), who presents data which result in the expression

$$\sin \phi'_p = 0.807 - 0.23 \log_{10} PI \qquad (8)$$

The residual friction angle, ϕ'_r has been related both to clay plasticity and to particle size by Kenney (1967) and Lupini *et al.* (1981). The residual friction angle is typically of the order of 27 degrees to 33 degrees for clays of low plasticity and of the order of 8 degrees to 12 degrees for clays of high plasticity. If the horizontal effective stress remains reasonably constant, the post-peak and peak friction values should be in the ratio $\tan \phi'_r / \tan \phi'_p$. The ratio of $\tan \phi'_r / \tan \phi'_p$ may be calculated using equation (8) and the typical ϕ'_r values given above, to be of the order of 0.75 to 0.85 for clays of low plasticity, and in the range 0.30 to 0.40 for clays of high plasticity. Whilst the post-peak degradation is generally in the range estimated for soils of low plasticity at both LDPT sites, as expected, there are local variations shown by the direct shear tests which are not shown by the plasticity data.

The lack of degradation in the upper third of the pile is not shown by the direct shear test data or indicated by considerations of peak to residual friction angle. It must therefore be concluded that the lack of degradation is due to the installation process, in which the soil around the upper third of the pile at both LDPT sites has already been reduced to residual conditions by the installation process.

Post peak form of curve

Displacement to residual shear

The displacement from peak to residual shear value normalised by the pile diameter as $(w_{res}/w_{max})/D$, is plotted on Fig. 7 for the two LDPT tests. Also on this Figure are the equivalent data obtained using the *T-Z* probe, normalised by the probe diameter. The plotted values are based on a straight line fit to the post peak behaviour to allow a simple numerical comparison to be made. The values follow different rrends for the two pile tests, but are remarkable at both sites for the agreement between the pile and the *T-Z* probe below the uppermost sections of the piles. Given the factor of 17 between the diameters

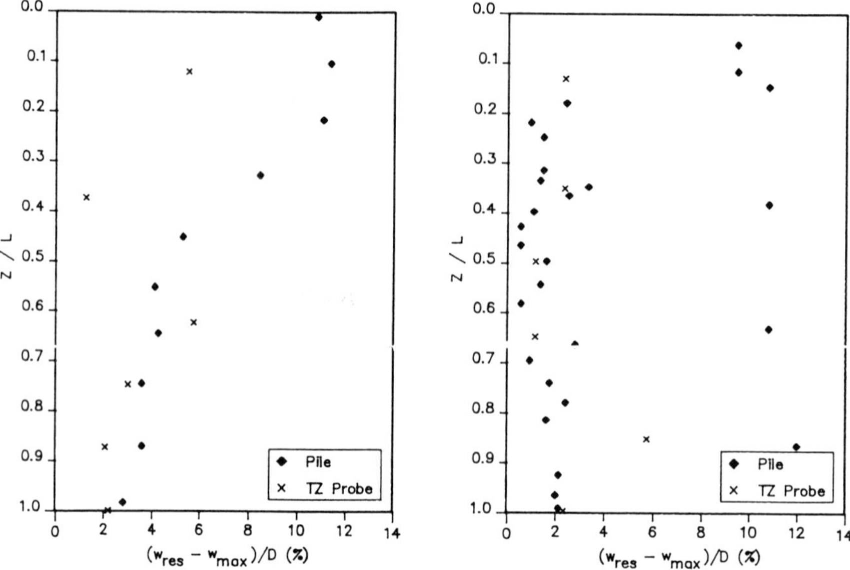

Fig. 7. Displacement to residual shear from pile and T-Z probe

of the pile and *T-Z* probe, this is a clear indication that the displacement form peak to residual shear is a function of diameter.

In the upper 20 to 30 per cent of the pile the displacement from peak to residual increases suddenly, to values of the order of 10 per cent of pile diameter. This trend is not shown by the *T-Z* probe, indicating it not to reflect in situ soil properties, but presumably to be a function of the effect of the pile installation on the soil.

Shape of post-peak curve

For comparison, the following curves are presented on Fig. 8

(a) a typical observed *T-Z* post-peak shape
(b) the inital straight line assumption
(c) a bi-linear post-peak reduction from peak to residual
(d) the recommended curve of Randolph (1985)

$$\tau = \tau_{max} - 1.1\ (\tau_{max} - \tau_{res})\ (1 - e^{-2.4\ (\Delta w/\Delta w_r)^{\zeta}}) \qquad (9)$$

where: $\Delta w = w - w_{max}$, $\Delta w_r = w_{res} - w_{max}$ and ζ= an exponent defining the rate of shear stress reduction.

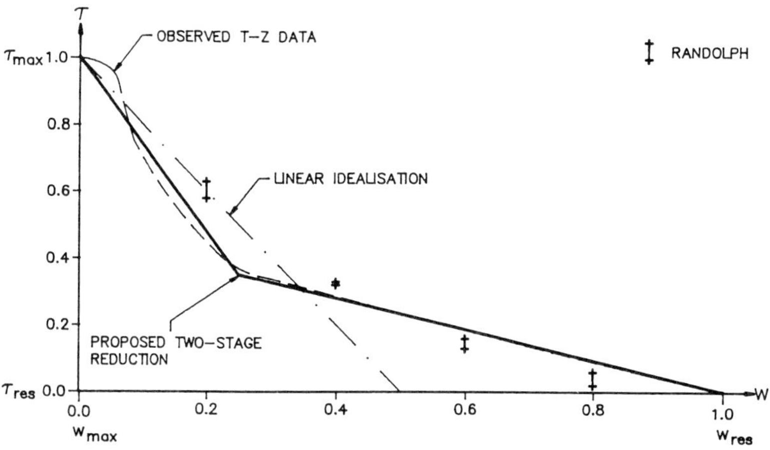

Fig. 8. Post-peak form of T-Z curve

Table 2. Co-ordinates of post peak T-Z curve

Parameter	Peak	Intermediate	Residual
τ	τ_{max}	$\tau_{max} - 0.65\ (\tau_{max}-\tau_{res})$	τ_{res}
w	w_{max}	$w_{max} + 0.015D$	$w_{max}+0.06D$

The best overall straight line fit to the LDPT data reached residual at $0.03D$ after peak, assuming a linear idealisation after peak displacement. The bi-linear curve presented on Fig. 8 is similar to that given by equation (9), and very close to the observed data. The co-ordinates of this part of the curve are presented in Table 2.

Discussion

Zones of T-Z response

It is clear from the LDPT data that the *T-Z* response around a driven pile essentially divides into three zones:

- Zone A: The upper 25 per cent of the pile, in which the results of degradation prior to static loading are clearly observed, and the *T-Z* response is comparatively softened.
- Zone B: The upper half of the remaining 75 per cent of the pile (i.e. from 0.25 to 0.625 *Z/L*) in which *T-Z* springs have a very high initial stiffness.

- Zone C: The lower half of the remaining 75 per cent of the pile (i.e. from 0.625 to 1.0 *Z/L*) in which *T-Z* springs form a smooth curve up to peak.

Various possible causes have been considered por the degradation in Zone A. The zone of upward soil stresses around the pile prior to loading (Fig. 3) covered the upper 30 per cent of the pile at Pentre and the upper 43 per cent of the pile at Tilbrook, and is therefore not quite the same depth range as the measured degraded zone at either site.

Whilst it is not considered the most probable cause, the possibility of pile whip defining the length of the degraded zone has previously been suggested, and could be investigated further. If this mechanism is considered possible, then note should be taken of the confining effect of pile sleeves when applying such an analysis for piles to be driven offshore.

It is considered that the most probable cause for the reduction in both soil stiffness and strength in Zone A is the two-way cyclic stresses applied to the soil during driving, and it is considered that analyses to determine the relevant percentage of the pile length affected should be made on this assumption for other sites.

The top of Zone C would appear to correlate very well with the section below the point at which the residual skin friction crosses the zero line for a second time (Fig. 3). It is therefore possible that the dividing line between zones B and C is related to the direction of the residual soil stresses, although more research would be needed to confirm this.

Optimum form of T-Z curve

For the LDPT sites, the use of a *T-Z* formulation with an initial slope of 3.0, the Bogard and Matlock (1990) curve up to peak and a two-stage degradation to residual as defined in Table 2 resulted in the calculated load–deflection response presented on Fig. 9. The quake to peak was taken as 1.3 per cent of pile diameter at Pentre and 1.5 per cent at Tilbrook. A reduction of 20 per cent from peak to residual friction was also assumed. This may be seen to give an acceptable fit, for most engineering purposes, to the observed load–deflection behaviour.

The ratio of predicted to observed displacement at any given load is plotted for both piles on Fig. 10. The use of one overall *N* value of 3.0 resulted in an extremely good match to the observed initial stiffness at Pentre. It is notable at Tilbrook that the proposed formulation does not match the initial stiffness of the load–deflection response very well, with the calculated response being approximately 50 per cent softer than the observed behaviour. There are probably two major reasons for this mismatch:

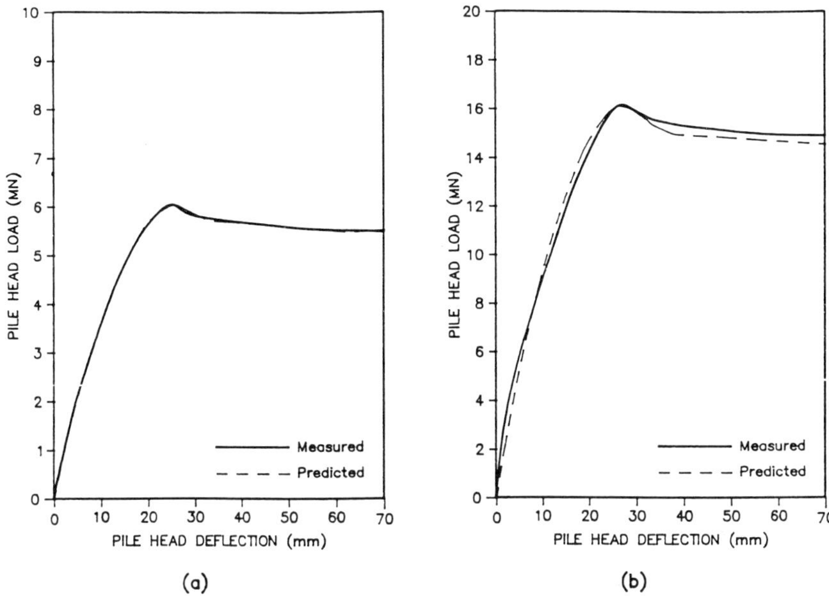

Fig. 9. Predicted pile head load–deflection for (a) Pentre (left) and (b) Tilbrook (right)

(i) The initial stiffness of the *T-Z* data in Zone B (0.25 *Z/L* to 0.625 *Z/L*) is extremely high. Typical *N* values (Fig. 5) are of the order of 5 in this zone. At sites where the τ_{max} values in this zone are high, this could obviously affect the load-deflection response. A test run performed for Tilbrook using *N* values of 5 in Zone B, and 2.5 in Zone C, showed a significantly better match to the observed behaviour at low load level.

(ii) The axial load–deflection analyses have all been performed assuming the *T-Z* curves to be initially unstressed, in accordance with normal design practice. The *T-Z* springs in the upper part of the pile are initially in tension – the shear stress applied by the pile to the soil being upwards. The very stiff response measured during stress reversal will result in an even stiffer initial response than predicted, since the upper springs actually move through a larger stress range than given by the τ_{max} value.

It is clear that both of the major factors considered to cause the mismatch on initial stiffness could be addressed in the *T-Z* formulation, although more data would be required at a range of sites to confirm that the observations at Tilbrook could be generally applied.

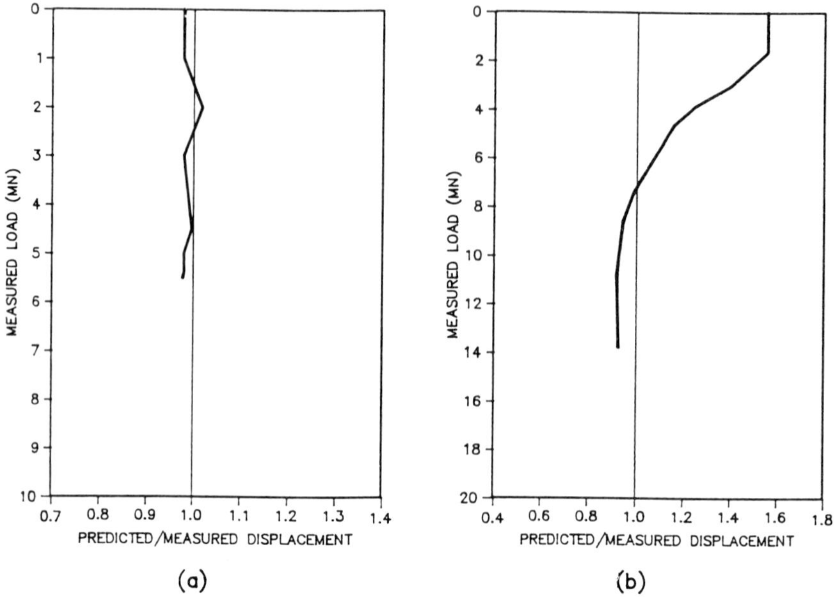

Fig. 10. Ratio of predicted to observed deflection for (a) Pentre (left) and (b) Tilbrook (right)

Pile capacity

Due to the degradation of the *T-Z* response in the upper 25 per cent of the pile, the peak pile capacity calculated using API RP2A (1991) peak friction values in the proposed *T-Z* formulation will be lower than calculated using the current API method, even for a rigid pile. Since the behaviour observed for the LDPT piles presumably occurred for all the driven piles in the pile load test data-base used to develop the API recommendations, it is recommended that after the peak values are reduced in the top 25 per cent of the pile length, the unit friction values at all depths down the pile should be increased by a constant factor, so that the sum of all the side springs plus the end bearing spring is equal to the total capacity given by the API pile capacity method.

Conclusions

Observations of the LDPT piles indicate that the *T-Z* response around the pile depends not only on the in situ soil conditions but also on the state of the soil around the pile following installation. Three zones have been identified at a range of depths down the pile. In each zone the *T-Z* response may be expected to be different, even in uniform soil conditions. If the best prediction

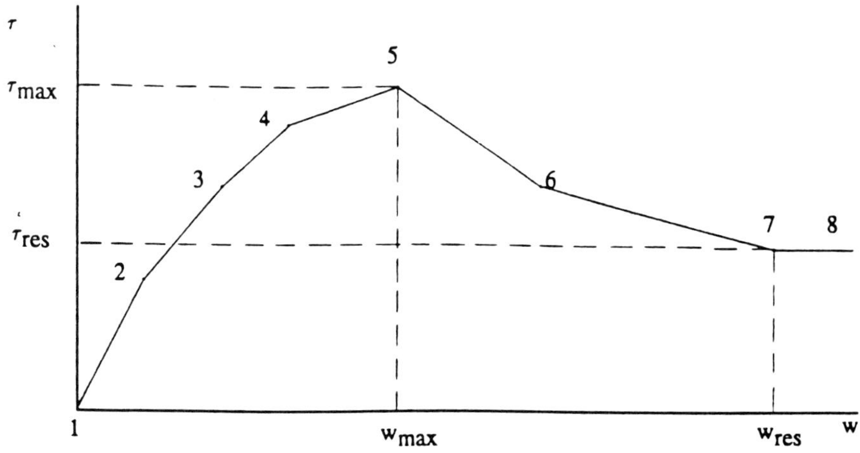

POINT	w	τ
1	0.0	0.0
2	$0.125w_{max}$	$0.375\tau_{max}$
3	$0.48w_{max}$	$0.75\tau_{max}$
4	$0.72w_{max}$	$0.90\tau_{max}$
5	w_{max}	τ_{max}
6	$0.75w_{max} + 0.25w_{res}$	$0.35\tau_{max} + 0.65\tau_{res}$
7	w_{res}	τ_{res}
8	Large displacement	τ_{res}

Zones B and C : 0.25 to 1.0 Z/L

w_{max} = 0.01 to 0.015 D
w_{res} = w_{max} + 0.06 D
τ_{max} : see text
τ_{res} = 0.8 τ_{max}

Zone A : 0.0 to 0.25 Z/L

w_{max} = 0.02 to 0.03D
w_{res} = Large displacement
τ_{max} = τ_{res}
τ_{res} : see text

Explanation of Terms:

τ = shear stress
w = displacement
D = pile outside diameter
L = pile length in soil
Z = depth from pile head

Fig. 11. Recommended form of T-Z curve

of pile stresses and pile response is required, then it is essential that the *T-Z* analyses incorporate the initial soil stresses prior to the static loading, and that the three *T-Z* zones are incorporated into the analysis.

The recommended overall *T-Z* curve formulation is presented on Fig. 11. It should be noted that this curve was applied to 17 other pile test results during the course of the joint industry study, covering a range of pile stiffnesses and soil conditions, and generally resulted in a very close match to the observed behaviour.

The following points summarise the conclusions from the study, and may be used in conjunction with the proposed form of *T-Z* curve on Fig. 11 to define *T-Z* data for use in determining the axial load–deflection and capacity of driven steel pipe piles in clay:

1. The w_{max} values are generally calculated to fall in the range 1.0 to 1.5 per cent of pile diameter. If the pile or jacket design could be sensitive to this value, it is recommended that a sensitivity analysis should be performed.
2. The w_{max} value adopted in Zone A should be twice the value used for Zones B and C.
3. The value of $(w_{res} - w_{max})/D$ should be obtained directly from the results of in situ *T-Z* probe tests, if these are available. The default value of 0.06 should otherwise be used.
4. The reduction from peak to residual friction should be determined primarily from drained direct shear tests. In situ *T-Z* probe tests may also be used to indicate appropriate values. In the absence of specific data, a reduction factor of 0.8 is recommended.
5. The τ_{max} values should be reduced to residual values in Zone A. The τ_{max} values over the full pile length should then be factored up uniformly to give the original total frictional capacity for a rigid pile.
6. If the pile response stiffness at low load is critical to the design, then a full analysis should be performed incorporating the residual stresses in the pile prior to loading, and an increase in *T-Z* stiffness in Zone B. In this case, the definition of the extent of Zones A, B and C should be determined from considerations of the soil response during installation and the residual stresses prior to static loading.

Acknowledgements

This joint industry study was funded by the Shell, BP, Exxon and Amoco oil companies, and by the UK Health and Safety Executive, all of whom were sponsors for the initial Large Diameter Pile Test programme. Dr M. F.

Randolph of the University of Western Australia also gave technical input to the study.

References

AMERICAN PETROLEUM INSTITUTE (1991) *Recommended practice for planning, designing and constructing fixed offshore platforms.* API RP2A.

BOGARD, J. D. and MATLOCK, H. (1990) Application of model pile tests to axial pile design. *Proc. Offshore Technology Conference*, OTC 6376, pp. 271-278.

BOND, A. J. and JARDINE, R. J. (1989) *Research on the behaviour of displacement piles in overconsolidated clay.* Offshore Technology Report, OTH 89 296, Department of Energy, UK.

KENNEY, T. C. (1959) Discussion. *Proc. Am. Soc. Civ. Engrs*, Vol. 85, No. SM3, pp. 67-69.

KENNEY, T. C. (1967) The influence of mineral composition on the residual strength of natural soils. *Proc. Geotech. Conf.*, Oslo,Vol. 1, pp. 123-129.

KRAFT, L. M., RAY, R. P. and KAGAWA, T. (1981) Theoretical *T-Z* curves. *J. Geotech. Engng. Div., Am. Soc. Civ. Engrs*, **107,** No. GT 11, pp. 1543-1561.

LUPINI, J., SKINNER, A. and VAUGHAN, P. R. (1981) The drained residual strength of cohesive soil. *Géotechnique*, **31**, No. 2, pp. 181-213.

RANDOLPH, M. F. (1985) *RATZ – load transfer of axially loaded piles.* Cambridge University Engineering Department.

RANDOLPH, M. F. (1991) *Review of Fugro-McClelland Limited.* Report No. 91/ 1843(01).

RANDOLPH, M. F. and WROTH, C. P. (1988) Analysis of deformation of vertically loaded piles. *J. Geotech. Engng. Div., Am. Soc. Civ. Engrs,* **104,** GT12, pp. 1465-1488.

TOOLAN, F. E. and HORSNELL, M. R. H. (1980) Analysis of load deflexion behaviour of offshore piles and pile groups. *Proc. Int. Conf. Numerical Methods in Offshore Piling*, ICE, London.

Discussion

M. SWEENEY, BP Engineering

At this conference, Dr Hobbs showed some photographs of the Miller Platform at floatout. It occurred to me that these are useful to put our discussions into the context of real costs. During the Miller studies, we tried to save one pile per group in the large pile groups. For such a large jacket, a saving of one pile per group amounted to £2–£4 million per pile, which for a four-legged jacket amounts to £8–£16 million per jacket. The Miller Platform is a few years old now, but those figures are still roughly current in our cost models. The R&D seems expensive, but that puts it into a realistic context.

When my colleagues were setting up the LDPT programme I remember teasing them that they had two options. First, they could have a very tight, well controlled test programme with incontrovertible data at the end of the day. To achieve that there was one obvious course of action, they should omit all instruments from the piles. Alternatively, they could have all the joys of conflicting data from different types of instrument, new and unexplained phenomena would emerge, there would be more questions than answers and there would be years of debate over what it all meant. We have clearly got something closer to the latter than the former — thank goodness! Not only is it a lot more fun, but I believe that worrying away at all the ambiguities in fundamental data is the only way forward. In this 'Design implications' session, and indeed throughout this conference, we saw many new ideas. Some were simple and accessible to designers. Many were immature, complex and inaccessible to designers. Dr Semple discusses these issues in a structured way with the panel. Here, comments refer to the last three papers.

Dr A. J. BOND, Geotechnical Consulting Group

Mr Nowacki showed a correlation between K_c and the undrained strength ratio of the clays included in his study, which shows the data from Canons Park and Madingley off the scale (Paper 22, Fig. 6). At Imperial College, we have found a strong correlation between K_c and overconsolidation ratio (OCR) as indicated by Fig. 1. I have included on this figure the results from Pentre and Tilbrook (denoted by the letters P and T). The Tilbrook data are shown separately for the two different clays.

There is a degree of agreement between Fig. 1 and Mr Nowacki's Fig. 6. In particular, the data for Pentre fall well below the other points in the database.

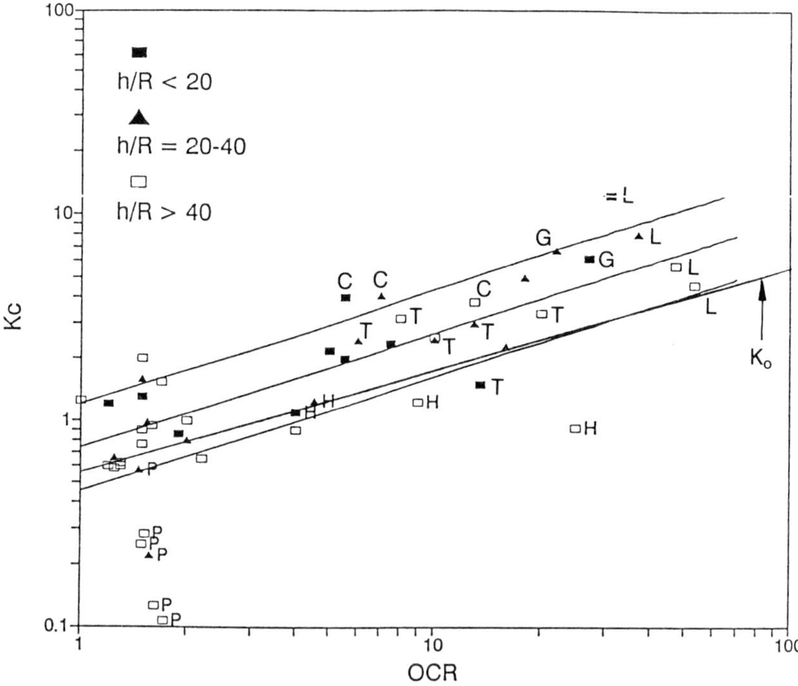

Fig. 1.

I should mention that the data shown on Fig. 1 are from instrumented pile tests on closed-ended piles. At high OCRs, we have data from Canons Park in London Clay (points L); Coop and Wroth's data from Madingley in the Gault (G); and Barry Lehane's data from Cowden (C). The OCRs in these particular clays were measured in high pressure oedometer tests. Conventional oedometer tests would give misleadingly low OCRs, so the way you measure OCR for heavily overconsolidated clays is very important. Other things to note are that at Tilbrook the excess pore pressures were not fully dissipated at the start of pile loading, and hence at this site the K_c values will be slightly lower than if the ground had been fully equalised.

Figure 1 uses OCR as a measure of the clay's tendency to dilate or contract when sheared. High OCR clays generally dilate when sheared and hence the K_c value is very much higher than the K_o. This is shown on Fig. 1 by the line marked "K_o". You can see that at Canons Park the values of K_c are generally many times greater than K_o. Similarly, the Cowden and Tilbrook results are well above the K_o -line.

There are exceptions, of course. In particular, the data for Haga does not fit the trend of the other high OCR clays. This is because the Haga clay has a high apparent OCR owing to leaching, rather than to the removal of overburden. The Haga clay contracts when sheared, and hence gives a low K_c.

Factors that affect K_c other than OCR include the height above the pile tip at which measurements are taken - the nearer the tip, the larger the values of K_c - and of course, the clay's sensitivity. It would be interesting to see where the NGI's Lierstranda results fit on this figure.

References

BOND A.J., JARDINE R.J. and LEHANE B.M. (1992), Factors affecting the shaft capacity of displacement piles in clays. *Proc. Int. Conf. on Offshore Site Investigation and Foundation Behaviour*, Society of Underwater Technology.

LEHANE B. M. (1992), *Experimental investigations of pile behaviour using instrumented field piles*. PhD thesis, University of London (Imperial College).

Dr R. J. JARDINE, Imperial College of Science, Technology & Medicine

Firstly, a point concerning the correlation of K_c values with OCR and or c_u/σ'_{vo}. Fig. 14 of Paper 4 shows that for soils like Magnus till, there is a systematic trend for the effective stress paths (followed by identical samples swelled back to different OCRs) to tend towards critical state undrained shear strengths after shearing to large strains. It can be shown experimentally, or theoretically, that the ratio c_u/σ'_{vo} has a simple power law relationship with OCR for such soils. But not all soils do this. Many more plastic, overconsolidated clays bifurcate when they undergo large strain yielding and develop shear bands. Within those bands water contents increase and the soil fabric aligns to produce a low strength residual fabric; this is shown diagrammatically on Fig. 2. Soils that behave in this way include the Gault Clay (found at Madingley), the London Clay (from Canons Park), the Lias Clay (from West Sole) and, I expect, the Oxford Clay from Tilbrook Grange. In this case, c_u/σ'_{vo} and OCR no longer follow the simple power law relationship. One result is that the K_c -c_u/σ'_{vo}. plot becomes inconsistent, as shown in Paper 22.

Professor A. J. WHITTLE, Massachusetts Institute of Technology, USA

I would like to bring to your attention some of our experiences in predicting the set-up behaviour of piles in clays which were referenced previously by Mr Karlsrud of NGI. This work has involved a combination of analytical modelling and experimental measurements on an instrumented model pile shaft (the PLS cell). Our analysis methods are fairly widely known and include the strain path method for quantifying the disturbance effects caused by pile driving, and generalized soil models for predicting the effective stress changes through successive phases in the life of the pile.

Figures 3 and 4 compare predictions and measurements of effective stress changes during consolidation at the pile shaft. Fig. 3(a) shows results for tests performed in the lightly overconsolidated, plastic Empire clay which is a

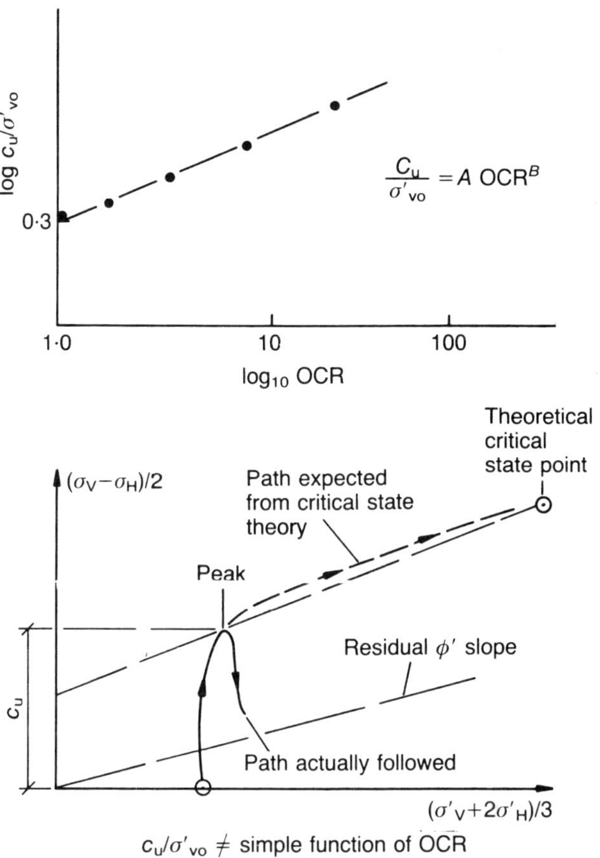

Fig. 2.

typical Gulf of Mexico clay. In this case there are relatively high effective radial stresses on the pile shaft during installation and very pronounced set-up such that the final effective stress K_0 is much higher than the initial K_0 stress conditions in the ground. In contrast, Fig. 3b shows the set-up measured for the lightly overconsolidated Boston Blue Clay, which is a sensitive, low plasticity marine clay. In this case, there are very low effective stresses during installation, the set-up is significant and the final stress conditions are close to the original K_0 conditions in the ground. The measurements tie in well with predictions made using the strain path method and MIT-E3 soil model as shown in both figures. However, it is important to

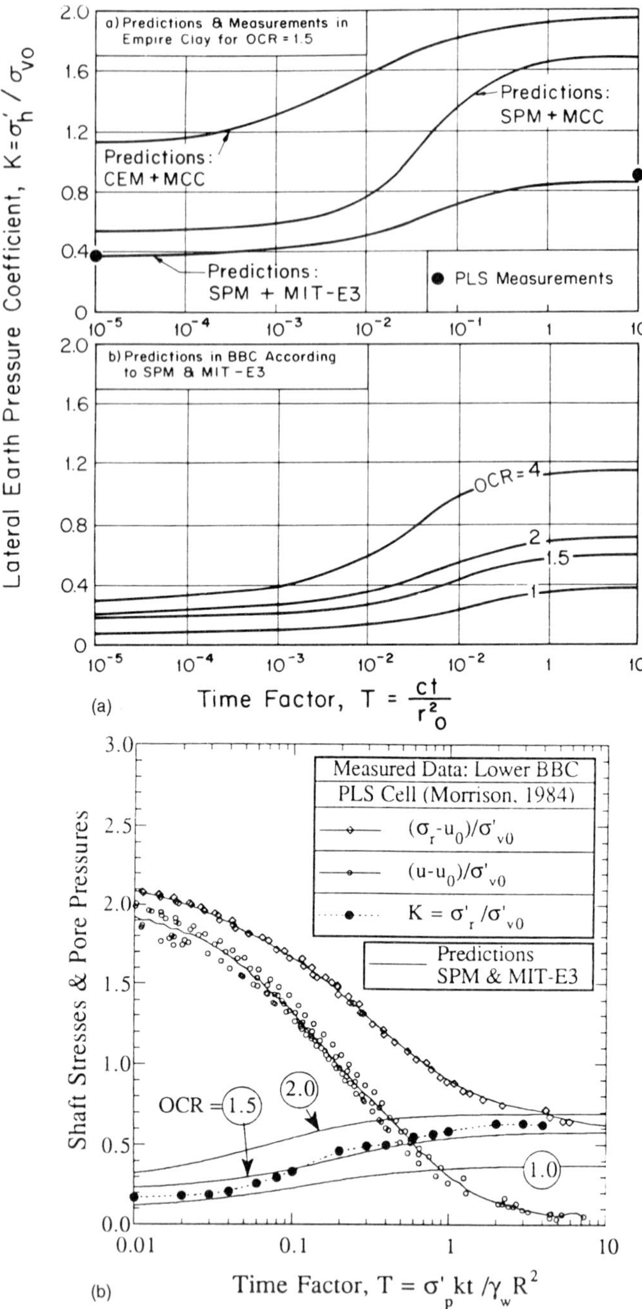

Fig. 3. Set-up predictions (a) at the pile shaft (after Azzouz et al. 1990); (b) in Boston Blue Clay (after Whittle, 1992)

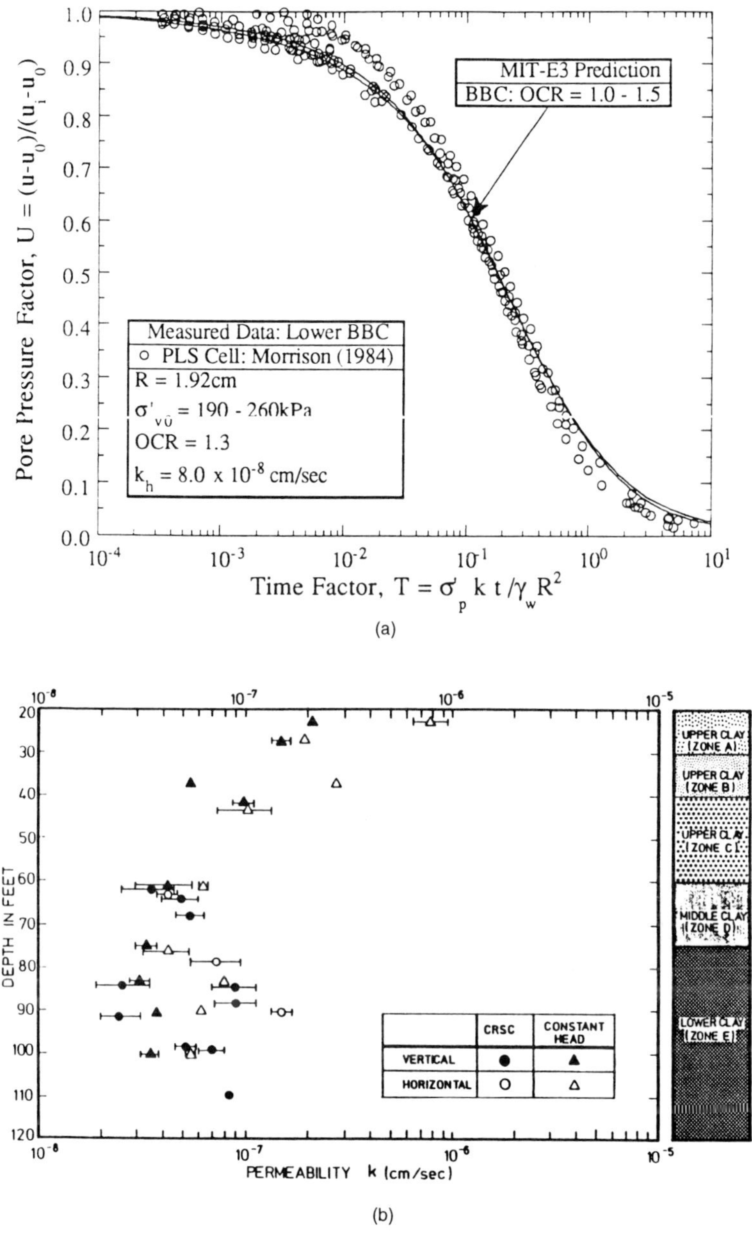

Fig. 4. (a) Evaluation of pore pressure dissipation predictions in BBC (after Whittle, 1992); (b) summary of laboratory permeability measurements in BBC (after Morrison, 1984)

emphasis that there are a large number of soil properties which can affect the set-up predictions.

We have now extended our studies to look at some of the data from the NGI tests on large-scale model piles and recent measurements from the Imperial College instrumented pile. The NGI data from the Haga and Onsøy sites seem to fit rather closely our understanding of behaviour from the Boston Blue Clay site, while the data from the Bothkennar test site are in line with previous experience at the Empire site. The data we have seen from the Pentre site certainly do not tie in with our predictions. I think that one of the main reasons is the effect of partial drainage during the installation process. The pore pressure data suggest significant dissipation of pore pressure during pile installation. This type of behaviour implies that there are significant volume changes within the clay and reduces the subsequent changes in effective stresses during consolidation. The low set-up effective stresses at Pentre are not due to the sensitivity/strain softening of the soil in undrained shearing, as was the case in Boston Blue Clay. The net result is that the fundamental mechanisms of pile capacity in the Pentre soils are very different to lower permeability soft clays but may be more similar to the NGI experience at the Lierstranda site.

Mr Karlsrud has stated that predictions of set-up times are fairly well understood. This is an important issue, of course, as the pile tests are relatively small-scale compared to the real prototype piles. Our understanding of set-up rates comes largely from the comparisons with the PLS cell data. Our analysis of the consolidation process assumes a constant coefficient of permeability in the soil, but non-linear stress-strain response of the soil. The predictions of set-up in a given soil are controlled by the coefficient of horizontal permeability (k_h). We have performed systematic studies to compare predictions and measurements of pore pressure dissipation in Boston Blue Clay. Fig. 4 shows that there is very good agreement between the estimated values of permeability and laboratory measurements of k_h (from CRS and constant head tests). We have found that for piles installed in low permeability clays, the key to dissipation is to have a good model of the initial pore pressure distribution caused by pile installation. However, it is not possible to rely on this type of information when there is partial drainage occurring during pile installation. This is an important issue for future pile design in silts.

I would like to see a regrouping or clustering of the existing database for piles in clays. For example, several people have stated that the tests performed at the Haga site were performed in a highly overconsolidated clay and have compared this data with results obtained in London Clay, Madingley Clay, etc. This appears inconsistent with the geological history of the Haga clay which is a recent clay deposit. This highlights a serious limitation in using the OCR (or plasticity index or undrained shear strength) of the soil for identifying and comparing the test results from different sites. It seems that we need to consider other physical or mechanical properties in relating the pile performance.

References

AZZOUZ, A. S., BALIGH, M. M. and WHITTLE, A. J. (1990). The shaft resistance of friction piles in clay. *J. Geotech. Engng Div., Am. Soc. Civ. Engrs*, **116**, No. 2, 205-221.

MORRISON, M. J. (1984). *In-situ measurements on a model pile in clay*. PhD Thesis, Dept of Civil Engineering, MIT, Cambridge, MA.

WHITTLE, A. J. (1992). Assessment of an effective stress analysis of predicting the performance of driven piles in clays. *Proc. on Offshore Site Investigation and Foundation Behaviour*. Society for Underwater Technology, London, September.

J. BEERS, IHC Hydrohammer, The Netherlands

I am employed by the fabricator of the IHC pile drivers, and so I am involved with a lot of pile driving. I notice that on the Tilbrook pile that a driving head was used. My question is why was a driving head used?

Dr W. R. COX, Fugro McClelland, Marine Geoscience Inc., USA

In Paper 10 (Cox, Solomon and Cameron) it was stated that the bottom 54 m of the NC pile was 15 mm thick and top 4.5 m was 20 mm thick. The bottom 27 m of the OC pile was 30 mm thick and the top 5 m was 40 mm thick. The analyses that were made early in the design phase of the programme indicated that high stresses could be expected during driving. The principle reason for adding the thicker sections to both piles was to minimize damage to the top of the piles.

J. BEERS, IHC Hydrohammer, The Netherlands

I would like to put forward that normal practice with 30" diameter piles and 1" wall thickness - the commonly used conductor pile - is to drive without any driving head. I would suggest that in the future we try to get away from the driving head and also the driving toe. Then you will not get any reflection. I cannot understand completely when you say a driving head is needed because you have only a certain length of 40 mm wall thickness before you get back again to 30 mm wall thickness.

Dr W. R. COX, Fugro McClelland Marine Geoscience Inc., USA

It is correct that the increase in wall thickness was restricted to only the top 4.5 m in the NC pile and the top 5 m in the OC pile, but it is not uncommon to have seating problems with hammers and this can cause much higher stresses than can be predicted by dynamic analyses. Although the piles were each 30"

(762 mm) in diameter, which is a common size for offshore conductors, the intent of the tests was to simulate a driven pile for jacket support and not a conductor. For this reason we were not inhibited in having a change in wall thickness since this is common practice for many load-bearing piles used offshore.

K. KARLSRUD, Norwegian Geotechnical Institute

Maybe I should reserve ammunition for the Panel discussion.There are quite a few points that have been brought up that I would like to address, but I will limit myself to very few comments at this stage. I fully take the point brought up by Dr Jardine, and further elaborated on by Professor Whittle. Certainly we have to be cautious about treating all clay materials in the same way and, relating the properties of clay materials only to the OCR. You have to show great respect for differences in clay behaviour due to different geological origins, mineralogies and so forth. So I fully agree with the fact that there can be a very large amount of variability that is not related to the OCR itself.

When you imply that the Haga clay is very special I do not agree. If you look at the stress paths and normalized shear strengths for Haga clay they follow almost precisely the SHANSEP concepts which give you precisely the type of dilatant behaviour in relation to OCR you would expect for that type of normalized behaviour. So, for example, for over consolidated normally plastic clays, the Haga clay is a normal deposit rather than an extraordinary deposit. How can we explain K_c values as high as 10 to 12 as observed by tests in some of the materials? K_c ratios as large as that are mind boggling. Of course that does not fit with any sort of passive earth pressure theory. We have to consider again the overall equilibrium of that soil element in the ground and possible changes in vertical effective stress close to the pile.

F. NOWACKI, Norwegian Geotechnical Institute

In response to Dr Bond, I am a little sceptical as to what you said was a strong correlation between two parameters when you use a plot in a double logarithmic diagram. I think you may have a factor of three, maybe more, between the upper and lower line you showed us. However, I basically agree with Dr Bond and Dr Jardine about using the ratio C_u/sé$_{vo}$ and, I think I said that the OCR may probably be a much better parameter to use from a theoretical point of view. The problem is that in practice, OCR is not a straightforward parameter. If you are not able to correlate your OCR to the C_u/σ'_{vo}, which was showed by Dr Jardine, then you are in trouble. So from a practical design point of view, I do not see it is a big step forward to correlate between K_c and OCR.

Panel discussion

Introduction

Dr R. M. SEMPLE, Dames & Moore, USA

The panel members have all been introduced previously during the conference. Mark Randolph, Kjell Karlsrud and Richard Jardine have their own independent lines of research on piling and, as such, have knowledge and data to bring to bear on the subject. Don Murff of Exxon Production Research helps to pay for it all. Don Murff, in fact, has had many years of involvement in the foundation sub-committee responsible for the recommendations that are in APIRP2A. He is the past Chairman of that committee, and from my own involvement, I can say that the committee is a place of lively discussion and that Don is influential in that.

The panelists do not know what questions I am going to ask them. I will pose a question and ask a particular panelist to respond before throwing it open to the rest of the panel. They may choose to answer or not — they are not obliged to answer each question. Then we will throw it open to the floor.

I would like to quote from the overview paper by Colin Mullis, indicating the objectives that the consortium had in undertaking this work and he said, "data from the tests were to be used to review the interpretation of the existing database with the objective of recommending more fundamental and rigorous approaches to the computation of the ultimate axial capacity of pile foundations in clay". A more fundamental and rigorous approach.

Dr Hobbs indicated that, although these piles are fairly large in terms of test piles, they are still a long way away from North Sea offshore piles. That places a great emphasis on procedures that capture the physics of what is going on as opposed to merely empirical correlations, when we make those extrapolations to full size piles. This discussion is structured in four areas. I would like to have some discussion on the site characterisation and the site data, and then move on to the instrumentation. Then we will deal with the behaviour of the piles before moving on to design implications.

The investigations of the sites were comprehensive for design purposes, and the scatter in the data would present no unusual problem to an offshore pile designer. There is no limit to the amount of testing that could be done on sites but are the site investigation and the published results reasonably complete? I would like to ask Dr Jardine to address this.

Site characterisation and site data

Dr R. JARDINE, Imperial College of Science, Technology & Medicine

The site investigations reflect offshore design practice. As an academic researcher, one would have probably included some different types of tests in addition to those that were performed. At our own research sites, for example, we have tended to include suites of stress path tests to define parameters that might be needed, for example, if we asked Professor Whittle to try and model it for us. So there we would be looking at perhaps programmes of SHANSEP type tests under K_0 consolidation, and also tests on intact samples taken with the greatest possible care to try and look at the in situ properties of the soils.

Dr J. D. MURFF, Exxon Production Research Company, USA

I asked the question about driven samples and, as far as I have been able to determine, there were some driven samples taken, but I am not sure exactly what happened to them. I think it is a relevant question and one that is particularly important in terms of characterising or coming up with a design procedure. There are some very carefully chosen words in the design practice RP2A, and I would like to cite those. The specification I have heard cited here several times has been that API calls for UU tests on push samples, and what it really calls for is the following, "unconsolidated, undrained triaxial compression tests on high quality samples, preferably taken by pushing a thin wall sampler, the diameter of 3" or more are recommended for establishing strength profile variations because of their consistency and repeatability. In selecting the specific shear strength values for design, however, consideration should be given to the sampling and testing techniques used in comparison to the techniques used to correlate the procedures to available relevant pile load test data." I think there is a difference between literally using UU tests with pushed samples and conforming to what RP2A says.

K. KARLSRUD, Norwegian Geotechnical Institute

If we are working with empirical design rules, I think we have sufficient knowledge of the general properties. If you want to go into more fundamental approaches on the shaft friction - on some of these materials I think we missed the most relevant test which is to do tests on material that has been severely remoulded as happens next to the pile surface and reconsolidated to appropriate levels of effective stresses representative of the state of effective stress that we have around the pile surface. On the Pentre material, we at NGI did a few tests of that kind. We had a very limited number of samples to do that, so we have some indications of the normalised shear strength properties of remoulded, reconsolidated material that I used to fit in to our theoretical model. But on the stiff clays, we have very little or no detail of that kind, so in that case I think we have to divide, the issues between purely empirical approaches like the API, and then we should use all the relevant methods as Dr. Murff says, that have formed the basis for that empirical method. To use the empirical method, you have to use all steps in that method. If we are going to use theoretical approaches that are more fundamental, we have to make sure that we measure those fundamental properties on the correct type of material.

Dr M. D. LAMBSON, BP Engineering

Dr Jardine referred to the extent of the laboratory testing reported in the paper on soil conditions. A comprehensive laboratory testing programme was undertaken to evaluate the pertinent physical properties of the foundation soils. For both the Pentre and Tilbrook Grange sites soil strength and behaviour in terms of effective stresses were assessed from triaxial compression testing of isotropically consolidated, anisotropically consolidated and K_0 consolidated 'undisturbed' specimens. At Pentre a series of isotropically consolidated compression tests were also carried out following the Shansep approach. Unfortunately, not everything could be detailed in the paper due to the need for brevity. Some stress paths are reported in our paper for the Pentre site since we felt the behaviour of the material warranted it. However, complete results are available in the site investigation reports. These can be made available to researchers.

In reply to Dr Murff's question regarding the use of driven samples, the site investigation practice generally followed standard North Sea based procedures where push samples are usually taken. A few driven samples were recovered at the Pentre site, although these were between 6 m and 11 m penetration, and therefore above the zone of interest since the pile was sleeved to a depth of 15 m.

Professor A. J. WHITTLE, Massachusetts Institute of Technology, USA

I feel compelled to respond to Dr Lambson. You mention the importance of Shansep testing in your laboratory test programme. Shansep is the acronym introduced by Professor Ladd and his co-workers (at MIT in the 1990's) for test procedures used to estimate the stress-strain-strength properties of natural clay deposits which exhibit normalised behaviour. In the Shansep method, clay samples are consolidated under one dimensional (K_0) conditions which are essential for reproducing the shear stress history of soil elements in the ground. In contrast, the data reported by Lambson *et al* in Fig. 25 of their paper use hydrostatic consolidation prior to undrained shearing. In our experience, there are large differences in the measured stress-strain-strength properties of samples consolidated under hydrostatic and K_0 consolidation conditions.

Dr R. HOBBS, Lloyds Register

I think it is unfortunate that the Norwegian pile tests did not include some UU triaxial tests. UU give the reference strength which is used with the empirical database and it would have been useful to have included sufficient of these tests to define a shear strength profile, in addition to the other tests which NGI may have considered more representative.

Instrumentation

Dr R. M. SEMPLE, Dames & Moore, USA

These piles were intensively instrumented and there was clearly a very big effort put forth by the consortium and by the contractor to instrument the piles. It is a heroic task to try and make total stress measurements in piles which are going to be so severely driven. There have been questions about the total pressure measurements and there have been questions about the residual stress distribution, and also what we know or have been able to infer about the distribution of load between end bearing and shaft friction. I would like to ask Mr Karlsrud if he would comment on whether he thinks that it is a good practice to instrument piles as comprehensively as these were, and if he accepts the data as presented.

K. KARLSRUD, Norwegian Geotechnical Institute

First of all, I think that if we are going to improve on the theoretical approaches we have to know effective stress, so we definitely have to do that type of instrumentation to understand what is going on. I have not gone into the details of the LDPT Pentre results in terms of the total and effective stresses measured there. They seem a bit funny but they had a bit of a problem. The Tilbrook data looked more reasonable, but I think that probably one issue that one should look at in the future, is to have greater confidence in those total stress earth pressure measurements. Great efforts were taken and we have very severe driving stresses on these piles. One should also look at how these things perform. For Tilbrook we also have a natural variability in the soil which makes interpretation a little difficult and we have, as I mentioned, the stones, and chalk fragments, that could also influence the stresses locally. So earth pressure measurements are the most difficult part, and maybe different types of sensors could have been put on to that pile to make sure that we had some independent types of instruments actually to do the same job. As for the residual stresses, again the problem you have when you drive a pile is you risk changing the zero points of the instruments. For the NGI piles, we were very fortunate to be able to extract all the piles after pile testing, so that we could recheck the zero frequencies of all the sensors and, although there were not, for the NGI piles, very large differences in zero frequencies, there were some. That allowed us to adjust the values that we had used originally. It is a question of how we also measure the strains in the pile. The type of system that was used, seems to give good values. There are some doubts about what happened in the top. Whether the data had measurement problems or other problems, I do not know. I think when you compare the LDPT piles with the NGI piles, we have a fairly consistent picture, I think, from two very independent types of measurements, which is comforting. By and large, I would say that I have good confidence at least with the practical use of these results in LDPT actual load measurements. I think I will stop there.

Dr R. M. SEMPLE, Dames & Moore, USA

Before going on to discuss the behaviour of the piles—the test results—I would like to make a few general points that may help us to focus on fundamentals. I suggest that the key considerations are as shown in Table 1.

Table 1.

Soil	Pile	Load	Events
Fabric	Diameter	Direction	Installation
Plasticity	Length	Rate	Consolidation
Stress history	Plug	Repetition	Loading
Density			

Fabric seems to be important in affecting the soil response to pile installation. Two otherwise similar soils can provide different skin frictions if their fabric differs. Information presented in this volume seems to confirm that plasticity is a controlling variable. The role of stress history is recognised in the current pile capacity procedure, i.e., API RP2A. I think soil unit weight - clearly fundamental to the effective stress controlled soil behaviour - is also an independent variable to be considered. We know that there are pile length effects in the data that we see. Pile diameter, and for a pipe pile, the wall thickness are important, so we also have the stiffness involved. The behaviour of the pile plug will have an effect. In terms of loading, the variables are the direction of the load, the rate at which the load is applied and the repetition or cyclic loading. The primary events in the history of the pile are the installation, the consolidation and the loading phases, and we have models of installation, models of consolidation and of loading. With regard to models, I would like to try and just focus on a very simple aspect of the behaviour and that is the skin friction being related to the effective radial stress at the time of failure, and some type of interface friction coefficient.

$$\tau f = \sigma'_{rf} \cdot \tan \delta \tag{1}$$

So that's a fairly straightforward little equation. The problem is, what is the value of the variables? Let's concentrate on the radial effective stress. If I think about that, then it would be helpful if I knew what the final radial stress was at the end of consolidation. So, then all I need to be concerned about is the change in effective stress due to loading and Richard Jardine would probably be able to tell me what that loading effect is. So let me recast the equation in terms of that radial consolidation stress at the onset of pile loading,

$$\tau_f = \left(\frac{\sigma'_{rf}}{\sigma'_{rc}}\right) . \sigma'_{rc} \tan \delta \tag{2}$$

However, we still do not know what that radial consolidation stress is. On the other hand, when the pile is driven, we will set up pore pressures and those will dissipate. That is a consolidation process. And we heard Mr Karlsrud saying that we understood that pretty well, so that's great. We can simplify life further and we now have the same equation recast in terms of ratios between effective stresses.

$$\tau_f = \left(\frac{\sigma'_{rf}}{\sigma'_{rc}}\right)\left(\frac{\sigma'_{rc}}{\sigma'_{ri}}\right)\sigma'_{ri} . \tan\delta \tag{3}$$

The second ratio, the new one that has been added, is between the radial stress at the end of consolidation and the radial stress initially before the pile was installed. I still have to estimate what that initial radial stress is, but, of course, we can relate that to the initial vertical stress that was in the ground. There is information available for that, and we have an even simpler equation to deal with here.

$$\tau_f = \left(\frac{\sigma'_{rf}}{\sigma'_{rc}}\right)\left(\frac{\sigma'_{rc}}{\sigma'_{ri}}\right)\left(\frac{\sigma'_{ri}}{\sigma'_{vo}}\right)\sigma'_{vo} \tan \delta \tag{4}$$

With regard to the initial radial stress, it is very often the case that the stresses immediately after installation are related to the undrained shear strength.

$$\tau_f = \left(\frac{\sigma'_{rf}}{\sigma'_{rc}}\right)\left(\frac{\sigma'_{rc}}{\sigma'_{ri}}\right)\left(\frac{\sigma_r - U}{S_{uo}}\right)\left(\frac{S_{uo}}{\sigma'_{vo}}\right)\sigma'_{vo} \tan \delta \tag{5}$$

Things are really going very well now! We have also broken the radial effective stress into the total value and the pore pressure because people measure those things, or attempt to. We have related or normalised that quantity to the undrained shear strength. Then we have introduced the $^{C}/_{P}$ ratio, as it was once called, which is a well understood parameter. We started off with an equation that could have been cast in terms of a lateral earth pressure coefficient. So, I have taken a bit of a liberty here

$$\tau_f = K_l - K_c \, . \, K_i \, . \, \frac{S_{uo}}{\sigma'_{vo}} \tan \delta \tag{6}$$

$$K_l = \left(\frac{\sigma'_{rf}}{\sigma'_{rc}}\right); \; K_c = \left(\frac{\sigma'_{rc}}{\sigma_{ri}}\right); K_i = \left(\frac{\sigma_r - U}{S_{uo}}\right)$$

and rewritten the equation using three terms which are K_l for the loading effect, K_c for the consolidation effect and K_i for the installation effect, and then the stress ratio or $^C/_P$. Professor Burland, some years ago defined the term "beta", in which the skin friction at failure was related to the initial vertical stress, and that clearly comes out of the equation quite simply.

$$\frac{\tau_f}{\sigma'_{vo}} = \beta = (K_l \, . \, K_c \, . \, K_i)\left(\frac{S_{uo}}{\sigma'_{vo}}\right) \tan \delta \tag{7}$$

The other interesting thing is that we can also define "alpha" in this way,

$$\frac{\tau_f}{S_{uo}} = \alpha = (K_l \, . \, K_c \, . \, K_i) = \tan\delta \tag{8}$$

and in fact, it is a simpler equation with less terms in it, So we have alpha related to beta is equal to the stress ratio,

$$\frac{\beta}{\alpha} = \frac{S_{uo}}{\sigma'_{vo}} \text{ or } \frac{C}{P}$$

and in a simple expression like this we have indeed captured the main aspects of the behaviour.

Behaviour of piles

I would now like to try and discuss installation-consolidation-loading. We will attempt to focus on the salient features. The procedure that is currently used in API RP2A for assessing skin friction is based on effective stress concepts, although it does not appear that way in the actual document. It appears simply as a correlation, but it represents a very considerable step forward from an alpha-S_u correlation, which is totally empirical and gives us no way we can progress. So I would like to ask Professor Randolph if he would comment on the observations from the LDPT tests, that there were very low excess pore pressures and initial radial stresses at Pentre, which is the normally consolidated site. The excess pore pressure normalised to undrained strength was less than 3, and the initial radial effective stress was very small. Also, in the Lowestoft Till at the overconsolidated clay site, the ratio of pore pressure to shear strength was less than unity. How these accord with our theories and if we have adequate tools to explain these observations?

Professor M. RANDOLPH, University of Western Australia

I think that no one has really tackled the high permeability aspect of the Pentre soils and the fact that, apart from the slightly more plastic zone at the 40-45 m depth, there was significant pore pressure dissipation during installation of the piles. I think that this is a crucial reason why, in the upper part of the Pentre site, the excess pore pressures were lower than expected. It also explains the low shaft friction and lack of any peak-residual behaviour, since significant damage at the pile-soil interface would have occurred during pile installation. This will always be a problem with silty clays, particularly for relatively small diameter piles where the consolidation times are short. Dr Semple's proposed series of *K* factors to quantify radial stress changes during installation, consolidation and loading will, of course, break down if the radial effective stresses become zero or negative after installation! Whether the recorded negative radial effective stresses are real, or due to small errors in the instrumentation, analyses using the strain path method show that very low radial effective stresses are induced adjacent to the pile immediately after installation. This is consistent with low shaft resistance during driving of piles into lightly overconsolidated or sensitive soils. To some extent, low effective stresses along the shaft imply correspondingly high excess pore pressures which will yield high radial effective stress after consolidation. Thus an approach based not so much on ratios of radial effective stress, but on incremental changes of radial stress and pore pressure may be preferable. However, at this stage it is true that our current theories do not allow full quantification of the stress changes that occur during the pile installation and loading history, particularly for intermediate soil types such as at Pentre.

Dr J. D. MURFF, Exxon Production Research Company, USA

I would like to comment on one of your factors: the loading factor. Quite a number of years ago, I think it was 1969 when Hurricane Camile came through the Gulf of Mexico, one of the platforms in the South Pass 70 area failed due, presumably, to a mud slide. During that failure, the platform completely fell over on its side. The piles were bent over broadside and in four of the piles the steel actually parted. That gave the investigators of the post-mortem team a point to tag the capacity to. The capacity presumably was at least enough to hold the pile while it was failing and that capacity turned out to be 2-3 times the calculated capacity using standard procedures. I think that event significantly influenced the opinions of Bob Bea, who has been a champion of rate effects for many years now. The API Resource Group recently commissioned Professor W. Tang, University of Illinois, to take a look at the available data. He has looked at all sorts of influences, and one of the large ones he believes is this rate effect factor. He believes that although we observe a fairly low bias based on slow rates of loading, the biases in clay are conservative and are in the neighbourhood of 1½ to 3.

Dr K. KARLSRUD, Norwegian Geotechnical Institute

To your question on Pentre and the effect of partial drainage during loading. I think you can make an argument that during the loading phase of the Pentre piles we do have a partially drained situation. If you look at the pore pressure dissipation time, it was a matter of hours due to the very large shear strains imposed by the pile installation. I think we should consider the possibility that during the pile loading we also have a partially drained situation for part of the Pentre piles. How important that is in relation to the fact that we start off with a very low effective stress is another question. I think that it is maybe of the order or 10-20% influence.

K. KARLSRUD, Norwegian Geotechnical Institute

As I have said the two most important elements in predicting the shaft friction are: (a) the effective stress state for this element close to the pile wall and (b) what is the shear strength properties of that material that has been so severally remoulded and is close to the pile surface. Let us look at this conceptually first. If we install a pile in, say, a normally consolidated or moderately over-consolidated clay, if this material reconsolidates back to the in situ effective stress ratio, the shear strength properties in this zone is some normalised shear strength times effective stress, which may not be too different from the shear strength of the intact material if it is normally consolidated. If we are in a stiff over-consolidated clay, the shear strength of this zone is definitely going to be quite a bit lower than further from the pile wall. Most soil models - and this goes for the MIT soil models and others - predict that the shear strength of remoulded clay is proportional to the mean effective consolidation stress. At NGI, we have done lots of tests on re-moulded and reconsolidated clays, that have had very large variation in undrained shear strength and OCR. They all show that the shear strength is essentially, related to the mean effective consolidation stress, irrespective of the previous stress history. But the normalised shear stress ratio varies somewhat with OCR or the absolute value of the intact material. Note that I am dealing with direct simple shear test but it could have been a triaxial test. The basic thing is, that if you look at the normalised shear strength of this re-moulded, re-consolidated material (Fig. 1), we start with a shear strength ratio typically for a material that was originally normally consolidated in the range of 0.25–0.30.

There is some tendency in these data, for this normalised shear strength ratio to increase with the original OCR or $^{Su}/_{p}$ ratio of the intact clay. If you know, or can determine this normalised shear strength ratio of remoulded, reconsolidated clay, we have one element for getting at the skin friction. The other element is the mean effective stress. The horizontal consolidation stress we have against the pile can be determined on the basis of empirical relations from pile tests. Now we have to find a way of relating this horizontal

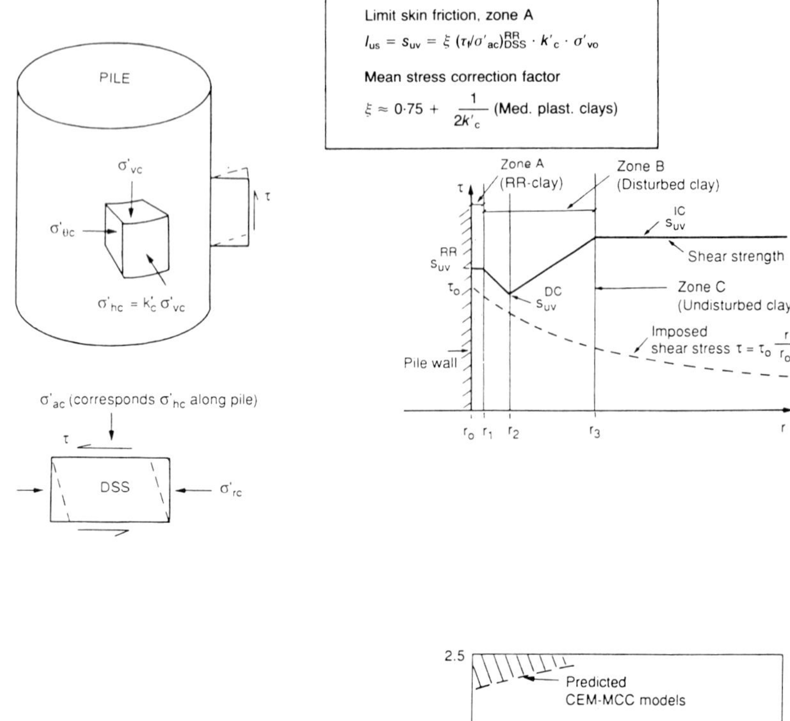

Limit skin friction, zone A
$l_{us} = s_{uv} = \xi\,(\tau_f/\sigma'_{ac})^{RR}_{DSS} \cdot k'_c \cdot \sigma'_{vo}$
Mean stress correction factor
$\xi \approx 0{\cdot}75 + \frac{1}{2k'_c}$ (Med. plast. clays)
PILE
σ'_{vc}
$\sigma'_{\theta c}$
$\sigma'_{hc} = k'_c\,\sigma'_{vc}$
τ
σ'_{ac} (corresponds σ'_{hc} along pile)
DSS
σ'_{rc}
Zone A
(RR-clay)
Zone B
(Disturbed clay)
s^{IC}_{uv}
Shear strength
s^{RR}_{uv}
τ_o
s^{DC}_{uv}
Zone C
(Undisturbed clay)
Imposed
shear stress $\tau = \tau_o \frac{r}{r_o}$
Pile wall
r_o r_1 r_2 r_3 r

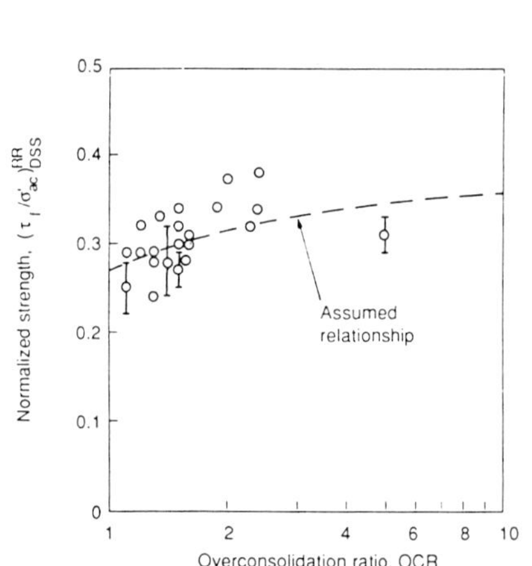

Normalized strength, $(\tau_f/\sigma'_{ac})^{RR}_{DSS}$
Assumed relationship
Overconsolidation ratio, OCR
0.5
0.4
0.3
0.2
0.1
0
1
2
4
6
8
10

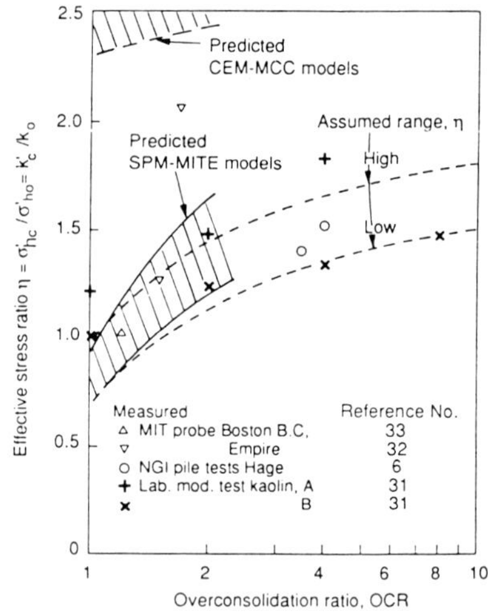

Predicted
CEM-MCC models
Predicted
SPM-MITE models
Assumed range, η
High
Low
Effective stress ratio $\eta = \sigma'_{hc}/\sigma'_{ho} = k'_c/k_o$
Measured
Reference No.
△ MIT probe Boston B.C. 33
▽ Empire 32
○ NGI pile tests Hage 6
+ Lab. mod. test kaolin, A 31
× B 31
Overconsolidation ratio, OCR
2.5
2.0
1.5
1.0
0.5
0
1
2
4
6
8
10

consolidation stress to the overall mean effective stress against the pile. We have proposed a factor here that just accounts for the difference in mean effective stress in the simple shear apparatus, compared to the overall mean effective stress along the pile. Thus, you can develop a procedure that really ties in all the three main important elements, the horizontal effective stress ratio, how that is different in the laboratory tests compared to the mean effective stresses against the pile and the shear strength ratio of that re-moulded material. That is at least a physical or theoretical way of modelling things that follows all the primary elements, and we have shown that it gives very good prediction of shaft friction in most of these plastic clays. The principles outlined are discussed in more detail in my paper and in a paper to OTC, 1990.

Dr R. JARDINE, Imperial College of Science, Technology & Medicine

Many measurements of interface failure parameters have been made in our recent laboratory and field research. Instrumented pile tests in London Clay, Cowden till and Bothkennar clay all indicate that shaft failure is not controlled by a general failure of the soil continuum, but is governed by a simple Coulomb effective stress relationship. The operational delta parameters reflect the degree of soil fabric re-allignment developed during installation at the interface. In some materials a highly polished shear surface exists. In others, the soil remains matt in texture, but still fails on a single surface with peak δ falling below ϕ' and decreasing post-peak. If shaft failure was controlled by continuum failure, then would increase post-peak and would also depend on OCR and the other stress components (σ'_z and σ_θ) developed in the soil close to the pile. Our tests show that this is not the case: δ hardly changes in soil profiles where K_c and OCR undergo large variations with depth. In the same way, OCR *per se*, does not appear to affect the δ values measured in ring shear tests on most clays.

As mentioned in the earlier discussions, peak δ can depend on the interface type (roughness and hardness) and on the displacement rate at which the soil was presheared during pile installation. In most cases ring-shear interface tests provide realistic estimates of the δ angles that are mobilised in pile loading tests.

Fig. 1. (a)(top left) Stress states along pile and in direct simple shear tests; (b)(top right) variation of undrained shear strength away from the pile wall; (c)(bottom left) normalized undrained shear strength values from DSS tests on remoulded reconsolidated clay; (d)(bottom right) horizontal effective stress ratio $\eta = K'_c/K'_o$ for closed-end piles

E. TOOLAN, Fugro McClelland Limited

I would like to address the question of K_i, which I think is the K on installation. It is that it is quite clear from the papers by Randolph, Jardine and Aldridge, that the skin friction is not constant with pile penetration. This is not new. There was a design method which was standard in the industry maybe 20 years ago, which was called the Lambda method. The Lambda method had a length effect in it so that in uniform soil, as you drove down deeper, the average friction reduced. This was nothing to do with elastic compression, it was to do with the formulation of the way the friction was calculated. This fell out of favour for the simple reason that the friction that was being lowered was the friction at the bottom of the pile and this struck most people as nonsence, that the friction of the shear strength of a clay of 200 kPa at 50 m had a much lower friction than 200 kPa at ground level. The method, although it was based on empirical data, and there was evidence that friction did reduce as you drove down, fell out of use. Now we appear to have data that indicates that the concepts of friction in uniform clay reducing as you drive down is, in fact, correct. The only thing is that when Vijayvigera formulated his Lambda method, he had it at the wrong end. He should have been reducing it at the top. I think that unless we take this on board, we are in danger of constantly coming up with empirical relationships that may work through sheer chance. If we had shorter piles, at the Pentre site or Tilbrook site, we would have got quite different alpha factors.

We probably would have got alpha factors for shorter piles that would have plotted well above the API line and we would have all said we should draw the line higher. But in fact, we know that the friction was reducing as the pile was driven, I would say, below a depth of 10 m in both cases. I wonder how the panel react to that concept which I think is more important than anything else that has come out of these discussions.

Theoretical approaches that require re-moulding the clay to the in situ conditions, then sampling it and testing it have the problem that clients come to me and say ‘do a site investigation, the structure designer is onboard, we want the piles designed by 6 weeks’ time’. To do an enormous amount of work on reconsolidating stiff soils, we have found from a bitter experience, it takes three years. You have to start going that theoretical route on a wide range of soils. But you could not possibly adopt that approach on a site-specific basis.

Dr J. D. MURFF, Exxon Production Research

One comment on the Lambda method. My recollection of the Lambda method was that it was based on average shear strengths and average mean stresses over the total pile.

E. TOOLAN, Fugro McClelland Limited

In the company where I work, Vijayvigera produced a manual on how to use the Lambda method and he did apply it incrementally because a lot of sites have layered soils.

Dr J. D. MURFF, Exxon Production Research

But he did not derive it incrementally.

E. TOOLAN, Fugro McClelland Limited

It is the way the method was applied which gave very peculiar results in layered soils.

Dr R. HOBBS, Lloyds Register

Does the size of the pile driving hammer affect the final pile capacity?

Dr R. M. SEMPLE, Dames & Moore, USA

I don't know. Good question.

A. R. BIDDLE, Independent Consultant

I think the results that we have seen on both the Pentre site and at Tilbrook Grange are very interesting. I have not seen them before, but, particularly on Tilbrook Grange, I would like to make two points. One is, that on that site I think the reality of trying to achieve ultimate capacity, came home. We have to wait a significant amount of time to get our pore pressure dissipation in strong clays, in clays which are particularly plastic and heavily overconsolidated. We cannot afford to base design on that theoretical ultimate capacity. We have got to take a realistic capacity which can be achieved by the pile as we consider in the design office, by the first winter that the structure has to sustain storm wave loading. This is because the 100 year wave could come along at any time and it could be in the first winter. Therefore, we have to take capacities which are not necessarily the ultimate capacity. They are those which can be achieved at the maximum within a 6 month period, and I think we should not forget that. The other point is that we have a problem in the design process that we, as geotechnical engineers, would like to think that our input is extremely valuable in the end process for getting an economically designed structure. However, I know from my own experience that this is not so. Most of the decisions are taken on a pragmatic basis by engineers who have got a lot of factors to consider and the foundation is but one small part of that. Until we get experienced geotechnical engineers into the design process in both the design contractors and into every oil company's decision process, I do not think we are going to be able to achieve a lot of efficiency in the pile design. It is still going to be done very conservatively and it is going

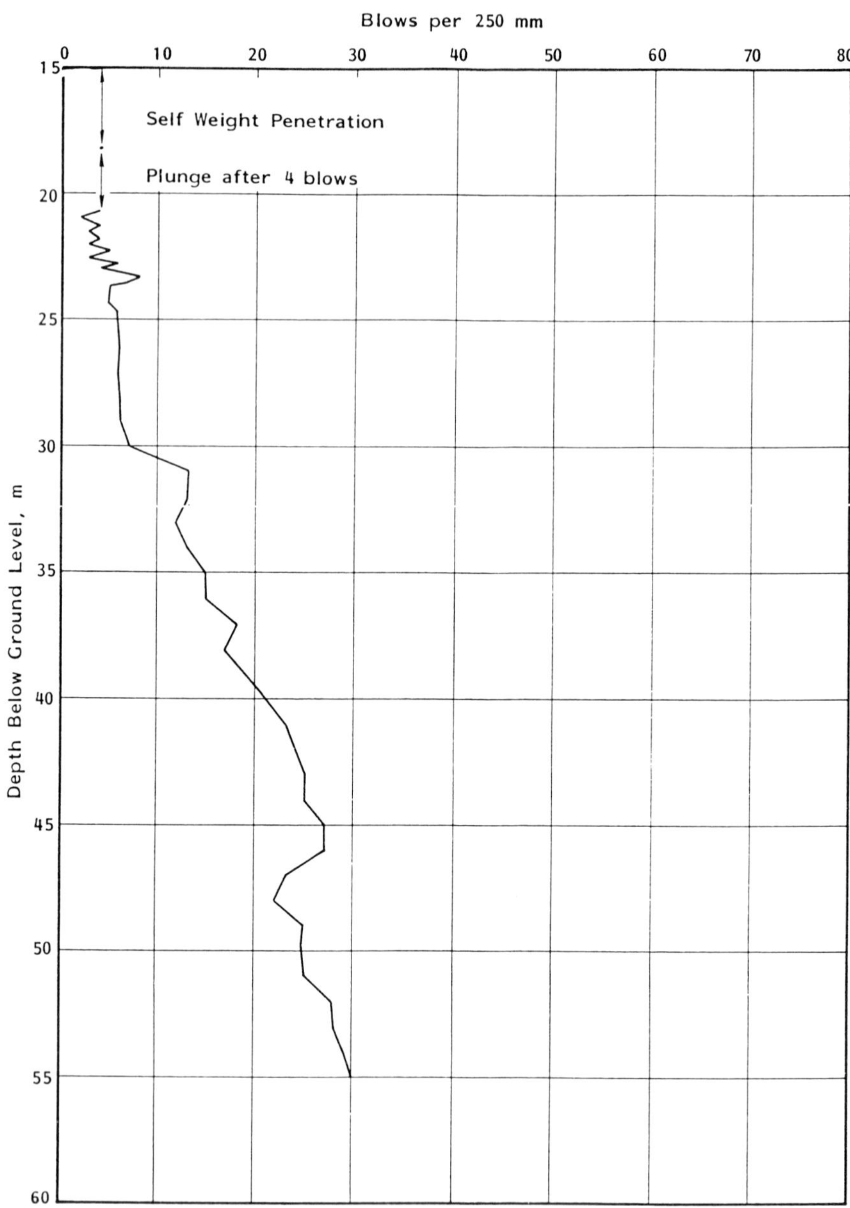

Fig. 2. (a) Blow count against penetration, NC pile, Pentre

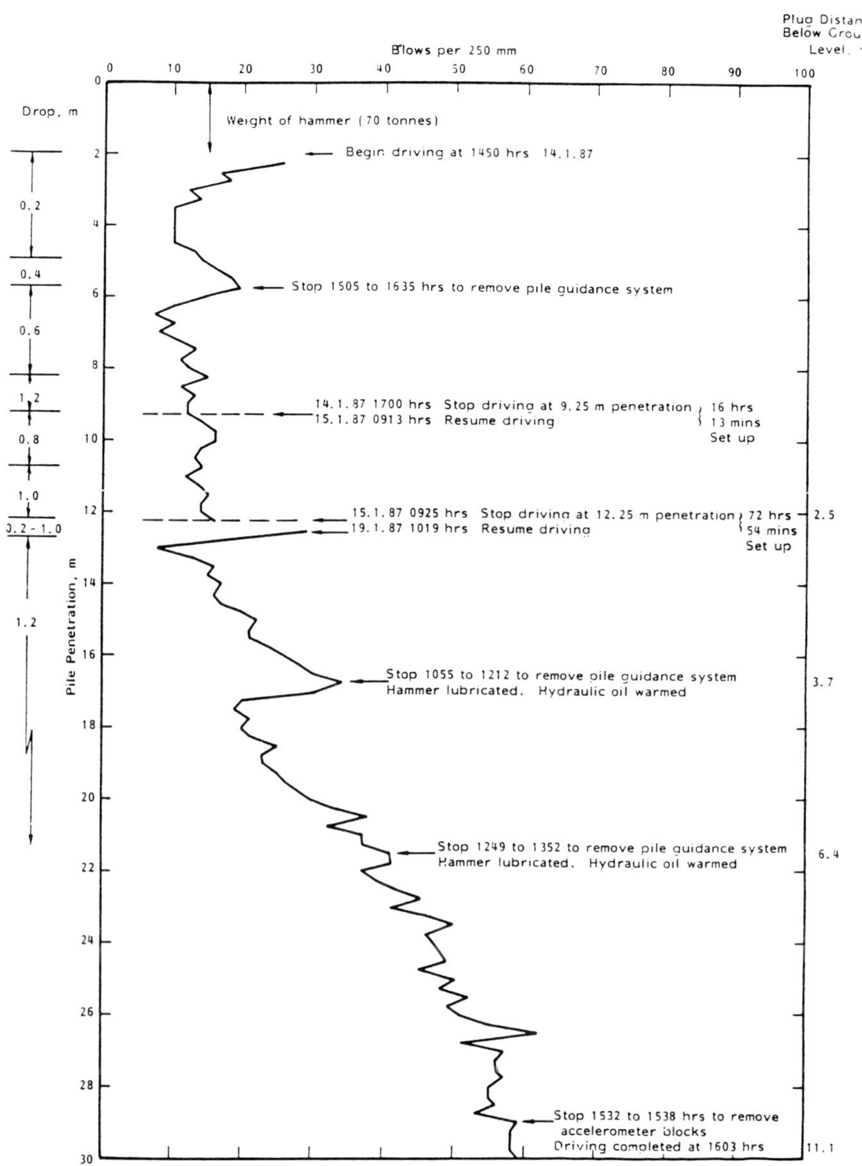

Fig. 2. (b) Blow count against penetration, pile driving trial, Tilbrook Grange

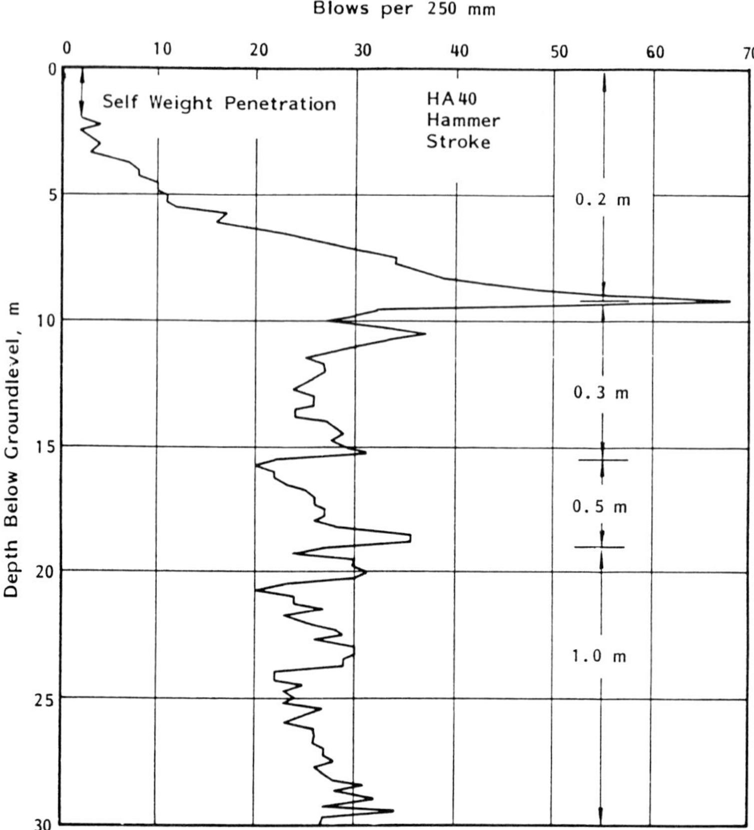

Fig. 2. (c) Blow count against penetration, OC pile, Tilbrook Grange

to be done according to such a simple empirical method as the API code which the average structural engineer can understand.

J. L. COLLIAT, Elf Aquitaine

I just wanted to add a factual comment on installation. We have a huge amount of data. I think we have missed one data set which is the driving curves. For the friction fatigue phenomenon, I think at least it should be obvious from the blow count curves (Fig. 2).

Dr R. M. SEMPLE, Dames & Moore, USA

It is apparent that there was a weak zone or zone of low load transfer at the tops of the two piles, and there are various possible reasons, that have been put forward, some mutually exclusive that appear in the same publication. But clearly we can have a gap due to pile whip. There are residual stresses that are claimed might affect things. There can be strain softening and there can be friction fatigue, that is what I consider to be a cyclic reduction in the radial effective stress as the pile is driven past a point. I would like to ask Professor Randolph what he thinks is the reason for the low skin friction at the top of the two large diameter piles?

Professor M. RANDOLPH, University of Western Australia

I do not think that the low skin friction at the top of the two piles is due to pile whip. I would really like to kill pile whip off as far as possible, especially for the offshore arena where the piles are fairly well constrained at their heads. For lateral loading under static conditions, at moderate mud-line displacements, the effects decay rapidly with depth, becoming negligible within 3 or 4 diameters. My intuitive feeling, without having done the calculations, is that under dynamic conditions, the soil inertia will increase the attenuation of those movements, and the effects of any pile whip will be confined to the top 2-3 diameters at the very most. That region is very often discounted anyway as regards shaft friction. As far as I understand it, the actual piles at Pentre and Tilbrook were reasonably well constrained without gross lateral movements during installation, which would further preclude pile whip.

On the other hand, I do think that there is a very severe effect of two-way cyclic loading during installation. From dynamic analysis of the piles, positive and negative limiting friction occurs with each blow, right down to the bottom maybe 10 m of the pile. In other words, the upper two thirds of the Tilbrook pile and the upper 75% of the Pentre pile, are being cyclically loaded. But I also think that there are other effects. At the Pentre site in particular, there is a very marked change at about the 15 m level. If the low friction values are just due to cyclic loading effects, for example, I would expect to see a gradual transition from what appears to be fully residual conditions towards peak and strain softening behaviour. The transition at Pentre appears to be more abrupt than that, and potentially linked with the change in plasticity index of the soil at that level.

K. KARLSRUD, Norwegian Geotechnical Institute

For the Pentre site I think the relatively low shaft friction along the upper 15 m or so is just reflected in the low horizontal effective stresses that we are measuring. Also, as I have shown, we are getting almost identical shaft friction on our piles at the same depth levels, so that correlates very nicely. Pentre is, in that sense, no mystery.

Dr R. M. SEMPLE, Dames & Moore, USA

Yes, but then why is the effective radial stress so low in the upper portion of the pile?

K. KARLSRUD, Norwegian Geotechnical Institute

Conceptually we could model what is happening regarding the low effective stress. Professor Whittle referred to his consolidation model for modelling what happens during the reconsolidation phase. The current limitation of the model he is using is, as far as I understand, is to consier the influence on the soil properties inherently on the medium to small strains that have occurred some distance from the pile surface. But he has not really reflected the very large remoulding close to the pile wall and the very considerable reconsolidation we get in that zone, and that is what causes the whole unloading of the total stresses.

If you recall the comparison of the stress reduction for Onsøy and Pentre they both showed very large total stress reduction. If you compare the percentage change, the difference in total stress change in those two cases was not at all that large. the one dropped by 50%, the other dropped by 60%. But when you are taking differences between two large numbers, you are getting there. So these are rather minute details in the consolidation properties we assume, because we are going to have a radial variation in consolidation properties of the soils due to the remoulding action of the pile installation. A number of people have done this kind of test as well. At the Haga site we remoulded the clay very severely, reconsolidated in an oedometer and looked at the volume change characteristics of the remoulded material. In the block samples we took next to the Haga pile, we measured the water content and back figures the volume change in that zone close to the pile wall. I plotted this in a diagram and it agreed exactly with what we measured in an oedometer. We could predict that volume change in that zone close to the pile wall precisely based on oedometer tests on remoulded material. Now, of course, then we have to remodel all the consolidation characteristics at all radial distances from the pile wall and how it is influenced. What are the consolidation properties of a soil that has been prestrained to 50% or 100% and how would you change the oedometer curve of that material. We have a differential variation in consolidation pressure in relation to radial distance from the pile surface. If you can couple this into an analytical model I think we are going to begin to see some of these reductions in horizontal effective stress.

Dr R. JARDINE, Imperial College of Science, Technology & Medicine

I think that the mechanisms that Mr Karlsrud has just been discussing are in not doubt acting. The question is why should they be acting so powerfully near the top of the pile. Why should you end up with such low radial effective stress? I think one thing we have to ask ourselves, is what are we expecting. If we are working on some kind of alpha method, we are expecting the stresses towards the top of the pile to be some factor, 0.5 or 0.7 of the undrained shear strength. But if we look at it from an effective stress point of view, then we tend to think of K_c values and we expect those to be perhaps constant, the same as alpha values. But if, of course, the effective vertical stress is low, as it is near the surface, then you are already starting at a low effective stress state and if you compound with that, the effects that we found of $^{h/R}$ or the length effect on K_c, we find that K_c reduces still further. So what we are actually seeing is that the soil near the top of the pile, as a result of this large number of accumulations of cycles of load that it experiences, plus the reductions in stress that take place that can be explained by the strain path method, ends up in a very low state of effective stress. In fact, in our experience at some overconsolidated clay sites, the radial total stresses may even go up with time, very close to the top of the pile. We found that at Canons Park for example. But in fact they started from a very low value, there is generally a very great fatigue of either effective stress or of total stress as you penetrate further in the stress layers.

Dr R. HOBBS, Lloyds Register

It is also worth thinking about the effect of lateral loading on piles, on their capacity near the surface. Obviously offshore piles do experience cyclic lateral loads and, as we saw with the rather extreme case of the lateral pile test (at Tilbrook), you can get postholing and loss of friction.

Dr R. M. SEMPLE, Dames & Moore, USA

There were two tests performed at Tilbrook Grange on large diameter piles. One was the compression test and the other was the subsequent tension test. There was a very different distribution of skin friction that was seen in those two tests and the low load transfers near the surface did not appear int he tension test. I would like to ask if Dr Jardine would comment on what he thinks is the reason for that behaviour.

Dr R. JARDINE, Imperial College of Science, Technology & Medicine

I do not know; I have not studied the data long enough to have a definite opinion. The upper stratum at Tilbrook is a glacial till; tests in dilatant dense tills can give some surprises with regard to time-dependent phenomena and to tension loading. Our experiments in Cowden till showed that capacity dipped dramatically shortly after installation and then recovered with time.

The same programme of experiments showed that the tension shaft capacity was significantly lower than that in equivalent compression tests. These two features were absent in our experiments at other sites and seem to be related to the till's capacity to dilate, and its greater susceptibility to induce anisotropy. Perhaps the same factors were influential at Tilbrook Grange; more details on the Cowden tests are given in Lehane and Jardine at the BOSS Conference held at Imperial College on 1992.

Professor M. RANDOLF, University of Western Australia

The difference in apparent skin friction distribution in the compression and tension tests is intimately connected with the problem of residual stresses. Dr Hobbs, in his paper, did in fact plot both deduced peak skin friction and deducted changes of skin friction during test - in other words, taking, as I understand it, zero residual stresses in the pile to start with. The magnitude of assumed residual stresses makes a big difference in the interpreted friction profile in the upper region of the Pentre site, which may partially explain the differences between the two tests.

Professor J. BURLAND, Imperial College of Science, Technology & Medicine

Listening to the discussion, one thing that occurs to me is that we have no had a discussion about the effects of the reaction system on the pile behaviour. A lot of trouble was taken for the compression tests to use barettes and bored piles as to minimise interaction with the test pile. However, for the tension tests reaction was by means of surface pad footings. Could that have increased the shaft resistance higher up the pile?

Dr R. M. SEMPLE, Dames & Moore, USA

Good observation.

The low transfer zones at the top of the pile is one issue that had received quite a lot of attention, but I think another somewhat surprising finding was the low skin friction at the lower region of the Pentre pile, which came out significantly less than an API prediction for a normally consolidated clay. I would like Dr Murff to comment on this.

Dr J. D. MURFF, Exxon Production Research Company, USA

I agreed that it is a little surprising. I think there is a possibility that the full end bearing of the pile had not been mobilised and that some of that end bearing that is assumed to be across the plug was skin friction.

Professor M. RANDOLPH, University of Western Australia

I was just trying to make a quick calculation of how much compression of the internal soil plug might be necessary to mobilise the end bearing pressure of 1.4 MPa that was assumed for the Pentre pile at peak load. An approximate calculation, based on compression of the plug over a length of a few metres, leads to estimates of the order of 200 mm. On that basis, the full end bearing capacity was probably not mobilised, which would affect the deduced skin friction profile significantly. Thus, a reduction in mobilised end-bearing of 75% would suggest a corresponding increase in skin friction of 20-25 kPa in the lower part of the pile shaft.

Design criteria

Dr R. M. SEMPLE, Dames & Moore, USA

Unless there are nay other comments on this point from the floor, I would like to move on to a couple of questions on design criteria. The results from Pentre, both from the large diameter test programme and the results that NGI have obtained, indicate that we have relatively low skin friction in very silty clays compared to some expectations at the APR RP2A criteria. There was a high plastic band running through that site and that gave high skin frictions relative to the undrained shear strength of the material. So how should these test results in very silty clay be recognised in design practice and should they influence what is in API RP2A?

Dr J. D. MURFF, Exxon Production Research Company, USA

This is an issue that we of course have been struggling with, having seen the Pentre results some time ago. I have learned a lot at this meeting, getting some of the insights of various people and seeing some new data that I do not think we have seen before, I think there seems to be a ground swell of plasticity proponents, so maybe we are going back to the future. There are several ways we could look at this. I thing that it is important to realise that in the pile load test database there is something like a 30-40% coefficient of variation in the load test results. So in one sense the Pentre test, even though it is quite low, fits within a reasonable error. It is not that far out the scatter band of the data. We have not reached any conclusions yet on how to address these tests results, but this conference will be of great value in our future deliberations.

J. H. P. PELLETIER, Shell Development Company, USA (current Chairman on the API Geotechnical Resource Group)

My first thought on this would be to handle it in a similar way to the way we currently handle carbonate soils: in that case, we put in some extra test about sites which may lie outside the database from which the API method was developed. I would like to throw the question back to the panel, or maybe to some of the people who have commented from the audience, as to the applicability or the relationship of the Pentre site to the other sites that are in our API database. This seems to me to be a very silty clay. I liked Professor Whittles comments about partial drainage on the lateral stress during testing. As Chairman of our group I would like to have some other comments as to how typical this site is to some of the other sites that are currently in the database, and is it typical of North Sea normally consolidated conditions.

E. TOOLAN, Fugro McClelland Limited

I think that one thing that people need to be aware of about API RP2A current method, is that it is based on Randolph and Murphy's paper which is incremental. In other words you are not supposed to treat the pile as a single element. You should go down the pile and, possibly following American practice, input some *t-z* curves This would certainly bring better agreement at the Pentre site - the results if you did that would tie in, I would think, to probably withinb 10%. So the overall result at Pentre is not perhaps as depressing as it seems. Certainly at Tilbrook, if you put *t-z* curves in, I think you get almost a perfect match provided you put in some degration. I think Dr Hobbs's paper alluded to this, I am not quite sure. It contained a little bit about *t-z* curves but said no more about it. As far as how relevant Pentre is to North Sea clays, they are as relevant as to clays worldwide. We have clays in the North Sea with PIs of as high as 60 and we have some clays down at about the same level PI at Pentre of 15. So there is no such thing, I think, as a typical North Sea clay. I would say that it seems to be - leaving out PI, and looking at the silt content - a lot siltier. I was dealing with a site with very silty clays in Scotland I can tell you when you drive piles through there, they go through the clay as if it were butter. My own feelings is that the material could behave somewhat brittley. In other words, as you are remoulding it, you are getting a considerable loss of strength. I think the stiffness is because of the high silt content alone, and therefore there is hardly any rebound effect. If you start looking at soil behaviour using pressuremeters, particularly my own company's type of pressuremeter, the cone pressuremeter, you will start picking up the very great importance of horizontal stiffness in this type of material with regard to pile design. If you have a material that has a low horizontal stiffness then you are going to get hardly any rebound and as you drive the pile you are just going to fall to something like the active pressure as far as K_0. I am just looking at K_0 values quoted, they seem quite close to the active pressure in this particular case. But I would not say it was irrelevant to the North Sea, it would be relevant to some situations, but as a general application to the North Sea, I would say it is not.

Dr R. M. SEMPLE, Dames & Moore, USA

I think it was comforting that the high plastic zone in the Pentre site behaved more or less as we would have expected. The need if anything to recognise the low skin friction would be applicable to very silty clays.

R. M. DUTT, Furgo McClelland Marine Geoscience Inc, USA

I would like to take Mr Pelletier's point a little bit further and ask the panel about data sets: the Pentre site data is closer to the Beta pile load test site that was in clayey silts/silty clay. I believe that Pentre and Beta data should be considered separately from the data set for the normally consolidated clay. Otherwise I think we will be comparing apples with oranges.

Dr R. M. SEMPLE, Dames & Moore, USA

I would like to comment that if this site came up as a design problem that I would hope that we would spot the particular characteristics of this material. The highly dilatant behaviour in triaxial shear for example. And that, if anything, the issue may be one of correctly characterising the shear strength rather than the method of calculating skin friction from it. The shear strength profile for Pentre is very, very high for a normally consolidated clay. Anyone who is used to dealing with normally consolidated material in the Gulf of Mexico would balk at using such high shear strength values. It corresponds to a 16 pound line, i.e. 16 psf per foot of penetration, which is 50% more than designers in the Gulf of Mexico are used to. I expect alarm bells would go off when that was spotted.

K. KARLSRUD, Norwegian Geotechnical Institute

First of all, lets go back to the top part of the Pentre pile which is the part that shows the lowest skin friction. If the entire Pentre deposit had been like that we would have had a low alpha value for the entire pile. I think that would be the case. At Lierstranda we had a 30 m or even 40 m deposit of silty clay. We got skin frictions of 10 kPa and if you normalise that with respect to the in situ, vertical effective stresses, it reaches a value of 0.05 times the vertical effective stress. So you cannot argue that the reason for this is that because we misinterpreted the in situ undrained shear strength. Have you ever heard about a material having a normalised in situ undrained shear strength ratio of 0.05? No. And the same applied for Pentre. The undrained shear strength ration that we used for the interpretation of $S_u/_{p'} = 0.3$, is not an unreasonably high value for a material that is claimed to have an overconsolidation ratio of 1.6 and that has the total unit weight and in situ effective stresses that we have at the site. So the question then is: are these Pentre deposits as well as the Lierstranda deposits abnormal in offshore context? Do you find them in other places? Indeed we have shown that on land in Norway we do find similar marine deposits that give similarly low capacities, if you go back and look at the test results.

You can argue that the Pentre soil is a far more silty material with much higher permeability, but still the results, to some extent at least, tie in with what we see at Lierstranda. I think we must have to watch out for the very silty clays. The Fig. presented earlier shows how sensitive it was to the plasticity index. Of course there is statistical variation here if you change the plasticity index by 2% and you get a 30% increase in shaft friction, according to that curve, then you have a problem. Plasticity index is of course not the only parameter that determines this. We have done at these sites a large number of CPTU tests and we have done dilatometer tests. From the CPTU tests, we get no indications that Liestranda is anything special compared to Onsøy. But the dilatometer tests at Lierstranda show us something interesting. We got a material index that was lower than anything we had measured anywhere in other deposits and we have been in many deposits with the dilatometer. We allowed the dilatometer to stay in the ground to allow for consolidation. At Lierstranda, compared to Onsøy, we measured very low effective stresses on the dilatometer, relative to more plastic clays. So I think we have to group our future design rules in to two. I am quite comfortable with the plastic clays based on these results, as well as previous results. We have a problem with the silty clays. All of these tests were done after a relatively short period of time and as I indicated earlier, I am fairly confident that we cannot stay with those very low effective stresses for a very long period of time. Observations that we have on earth pressures against sheet pile walls where we have measured very low active stresses when we get the main deformations, they tend to restore. I think in these silty materials that show these very low effective stresses against the piles after a relatively short period of time, they are probably going to partially restore after a long period of time. The question in them, when do I need my pile capacity?

Dr R. M. SEMPLE, Dames & Moore, USA

That means there is hope for the future. I would like to raise just one more question that I think is important. I am going to put Dr Murff on the spot again. At Tilbrook Grange and at Pentre we did have this low load transfer zone at the surface. The discussion we have just had could be applied to the lower part of the Pentre pile, and the question is should this low load transfer zone at the surface be recognised in our design practice.

Dr J. D. MURFF, Exxon Production Research

I think if we can understand what is driving it, it probably should be, but the reality is that for most offshore piles it is probably very inconsequential. Particularly piles 300 feet long. If you throw away the tip 10 m of soil, it makes no difference. That is of some consolation to me in worrying about this. I think you have to ask the question does this effect cover 10 pile diameters, 10 m or a third of the length of the pile. It depends upon how you characterise your results as to what conclusion you reach. We really do not know the answer to this.

Dr R. JARDINE, Imperial College of Science, Technology & Medicine

I spend a few moments earlier looking at the routine SI data from Pentre, searching for indications that the Pentre soils might not be typical of North Sea low plasticity clay. We discussed dilation, grading and permeability in Session 2. Another feature seemed to be the ratio of CPT cone resistance to triaxial UU shear strength, *Nk*. Perhaps we should expect in situ penetration tests (piezocone, dilatometer, full displacement pressuremeter, etc.), which have some degree of geometrical similitude with displacement piles, to give the most direct indications of non-standard behaviour.

We must also be aware that there are potential problems at the other end of the spectrum. High plasticity clays can show a very wide range of δ values. Soils with a high proportion of active clay minerals (such as London, Gault, Oxford or Lias clays and some North Sea tills) tend to give very low angles (down to 8° at Canons Park). However, very silty materials like the Bothkennar clay that contain only small amounts of active clay mineral can appear to be very plastic if they contain modest organic contents. Such materials can show rather high δ values (30° at Bothkennar).

The best way of distinguishing between such soils is to perform interface ring-shear tests; I would be happy if these experiments became a standard part of offshore site investigations.

Dr R. M. SEMPLE, Dames & Moore, USA

I think Mr Karlsrud is on record in a paper as indicating that NGI were unable to distinguish between sites that would produce very low skin frictions and sites that would be more like a normally consolidated clay on the basis of the CPT. He could not see any difference in them.

Dr R. HOBBS, Lloyd's Register

About Dr Murff's point regarding not having a great effect: there are many overconsolidated sites in the North Sea where we have very high shear strengths near the sea bed with relatively short piles. In those particular sites, I think it would make a very great difference to the pile capacity if we excluded some of the friction rather than just using the new API procedure.

Dr J. D. MURFF, Exxon Production Research

From the results I have seen, in the designs I have been involved in, I do not think it would make a lot of difference. I would agree that is some cases where you have particularly short piles and are trying to take advantage of very hard clays, discounting the top 10 m would make a difference.

A. LUSTED, BP Engineering

Interesting characteristics of the Pentre site are that in cone penetration tests (CPTs) the sleeve friction ratio, f_s, in the near surface soil is low in the range 2-4% and the c/p' ratio was relatively high at around 0.3. I have been working on a site in Scotland where a clay has broadly similar characteristics and piles have penetrated easily, perhaps in a similar manner to those described by Mr Toolan at another Scottish site. The f_s provides a measure of the performance of the remoulded zone of soil around a pile during its installation whilst the c/p' ratio can provide a measure of the stiffness of the elastically deformed zone surrounding the plastically deformed zone during pile installation and upon reconsolidation. At the extreme end of performance where the remoulded zone has a very low shear modulus and the elastic zone is relatively very stiff, such properties have a similarity to many types of calcareous soils, where unusually low pile shaft friction values have been observed. We have been discussing how we might identify unusual soils in design codes. A first step might be to think more carefully about the combination of f_s and c/p'.

Professor A. J. WHITTLE, Massachusets Institute of Technology, USA

Given that K_c is a very important parameter in estimating pile capacity and that there is good agreement between K_c measures by large displacement piles, model piles and even miniature shafts (such as the PLS cell), might it not be a good idea to consider disposable instruments? For example, we could simply install the cone in the ground and use remote monitoring of the instruments to estimate the lateral effective stresses on the pile shaft. I am actually involved in a joint project with geologists at the Woods Hole Oceanographic Institute who are trying to monitor pore pressures in the abyssal plain using disposable probes. Maybe I have been living in a throw away society for too long, but it seems to me that if there are no significant scale effects, then this might be a cost-effective solution for the design of offshore pile installations.

Dr R. M. SEMPLE, Dames & Moore, USA

I would just make a comment about Magnus that one of the things that has come out of that, is that it appears that the designers have been over-estimating the loads that are applied to our structures. I can imagine we will see a reduction in some of the loading parameters, drag co-efficients and so on, and the consequence of that will be an increase in the actual loads being applied to the piles. So I think these test data from Pentre and Tilbrook Grange are particularly important. There has been, I think, a hidden margin of safety in the past due to the over-estimation on the loading side of the problem, and I think we are going to lose that, so we have to be very, very careful that we have our soil information right.

J. H. P. PELLETIER, Shell Development Company, USA

In the 20th edition of RP2A, which is coming out in 1993, wave loads will increase by approximately 40% over the minimum values that are in RP2A. The minimum values that are in RP2A do not include such things as current. For particular operators, wave loads may go up.

D. E. SHARP, BP Exploration

I would like to add to Mr Pelletier's comments, I think it makes the question, 'what are the foundation stiffnesses', which is the question structural designers look to us to answer, even more important than perhaps the last 5% of the ultimate pile capacity. A number of methods predict the ultimate capacity with reasonable accuracy. I wonder if we should not be devoting more time and effort to modelling the foundations as seen by the structure.

Dr R. HOBBS, Lloyds Register

One comment about the siltiness of the site, and whether Pentre is exceptional. I remember very clearly that, in the early days of this project, great lengths were gone to persuade me that this site was representative of the North Sea, indeed that both were, and that Lloyds Register should not, having seen the very high alpha values which were going to come out, then object to them because the sites were odd. It seems that we have turned full circle and are looking from the other viewpoint now, trying to find some way to explain why they were lower than anticipated.

Closing address

Professor J. B. BURLAND

The importance of full-scale field measurements in civil engineering has been emphasised many times. This conference was of unique importance since it provided a forum for the presentation and discussion of measurements on the Magnus oil production platform and recent large-scale instrumented pile loading tests. My closing remarks must, of necessity, be highly selective and cannot possibly do justice to the many important contributions that were made.

The instrumentation for the Magnus foundations had to withstand extremely hostile conditions and the implementation required complex organisation and sophisticated management and control. Ultimately the success of the measurements made at Magnus and for the large-scale instrumented piles depended on the resourcefulness and dedication of many individuals working under very trying conditions. For this reason I particularly liked Bill Cox's presentation which brought out so well the human angle.

The major findings from the Magnus Foundation Monitoring Project has been well summarised in a number of presentations. They include the following:

(i) For a known discrete wave the loads coming on to the pile group can be calculated with reasonable accuracy.
(ii) To date the foundations have only experienced about 30 per cent of the design load.
(iii) The contribution of the mud mat, both due to dead load and wave loading, is significant. It remains to be seen whether this result will be used in practice.
(iv) Pile fixity is important.
(v) The foundations are 2 to 4 times stiffer than assumed in design. I understand that, at the time, the design stiffness was felt by some to be too high!
(vi) Interaction between piles was less than assumed in design.

Both the higher foundation stiffness and the reduced interaction effects are attributable to the small strain non-linear stiffness properties of the ground. Our studies of small strain stiffness carried out at Imperial College coincided with the construction of the Magnus platform. It is both gratifying and important that the results of these studies are consistent with the measured response of the foundations.

It has been noted by Don Murff that the higher foundation stiffness does not entirely account for the measured natural period of the structure. There appear to be effects within the structure itself which require further investigation.

The large instrumented pile tests

The point has been made that the API database is for a limited range of soil types and pile sizes. The results from the large instrumented pile tests add significantly to this data base. I have long been an advocate of an effective stress approach to the analysis and design of piles and it is gratifying to see the recognition that this approach is beginning to receive. Before examining the contribution of the large diameter pile tests to the data base, it is important to examine how the effective stress approach can be used in the analysis of data from routine test piles.

It is now clear that the shaft friction τ_{sf} is controlled by the simple Coulomb equation:

$$\tau_{sf} = \sigma_{rf} \, . \, \tan \delta' \qquad (1)$$

where σ_{rf} is the shaft radial effective stress at failure and δ' the effective angle of interface friction.

Research by Bond, Jardine and Lehane (1992) suggests that tan δ' tends to decrease with increasing plasticity and generally lies between about 0.4 and 0.2.

The magnitude of σ_{rf} depends principally on the initial vertical effective stress σ'_{vo}, the overconsolidation ratio (OCR), installation effects and loading. The value of σ'_{vo} can be assessed with considerable accuracy so that it is logical to re-write equation (1) as:

$$\tau_{sf}/\sigma'_{vo} = \beta = (\sigma_{rf}/\sigma_{vo}) \, . \, \tan \delta' = K_{sf} \, . \, \tan \delta' \qquad (2)$$

The determination of K_{sf} requires the measurement of σ_{rf} and this has only recently been achieved. However a dominant controlling variable is the OCR. Hence we might expect there to be a reasonable correlation between τ_{sf} and σ_{vo} for various OCRs. The OCR is itself related to the undrained strength ratio s_u/σ_{vo} which is routinely determined at most sites. We should note that for most routine pile tests only the average values of the shaft friction τ_{sf} can be determined. However, for the large diameter pile tests the measurement of the axial load distribution allows the determination of local values of τ_{sf}.

The above reasoning suggests that it would be logical to relate τ_{sf} to σ_{vo} and s_u/σ_{vo}. In Fig. 1 the API data base given by Semple and Rigden (1984) has been plotted on axes of τ_{sf} against σ_{vo}. The open circles are for normally and

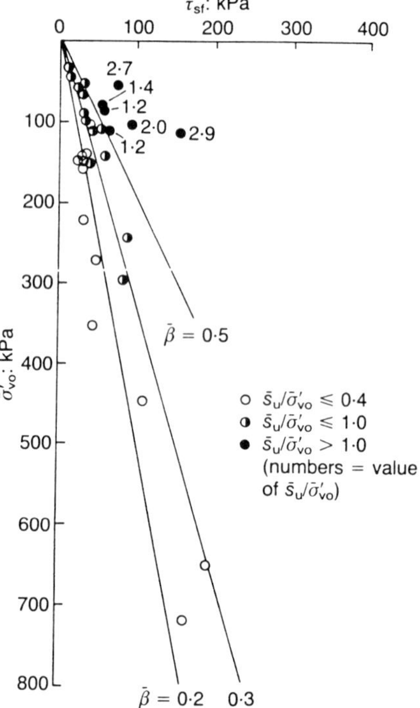

Fig. 1.

lightly overconsolidated soils ($s_u/\sigma_{vo} \leq 0.4$), the half circles are for intermediate overconsolidated soils ($s_u/\sigma_{vo} \leq 1.0$) and the black circles are for heavily overconsolidated soils.

It can be seen that the results for normally and lightly overconsolidated soils generally lie below the line corresponding to β=0.3 with most of the results lying between β=0.2 and 0.3. The results for intermediate overconsolidation generally lie between β=0.3 and 0.5 and the results for heavily overconsolidated soils all lie above β=0.5.

A logical next step, which is consistent with equation (2), is to plot β against s_u/σ'_{vo} and this is done in Fig. 2. It can be seen that the results lie within a reasonably narrow band with the lower bound given by the line corresponding to α=0.45. In view of the number of factors influencing K_{sf} and tan δ' in equation (2) and the uncertainties associated with the determination of s_u, one could not reasonably expect the spread of results to be less than that shown in Fig. 2.

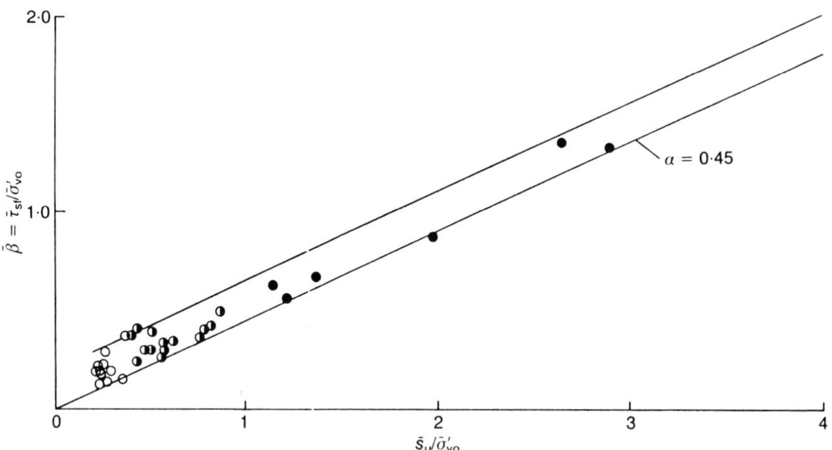

Fig. 2.

We can now see how the results from the large diameter instrumented piles compare with the API data base. Fig. 3 shows a plot of the local values of τ_{sf} against σ'_{vo} for the Pentre test (open circles). The black circles are from the paper by Karlsrud *et al.* For values of less than about 225 kPa the local values of β are well below 0.2. However at higher values, and hence greater penetrations, the values lie around the β = 0.2 line. These results are strongly suggestive of arching around the upper part of the pile due to the highly contractant behaviour of the loose silt. The average values of τ_{sf} and σ'_{vo} are shown by the point labelled P corresponding to a value of β = 0.17 for the whole pile.

Fig. 4 shows the local values of β plotted against s_u/σ'_{vo} for the test at Tilbrook Grange (open points). The scatter is larger than for the average results from the API data base but the α=0.45 line again forms a reasonable lower bound. The average value for the whole pile lies just below this line as shown by the point labelled T. The black circles are for the NGI results reported in the paper by Karlsrud *et al.* and it can be seen that the scatter is much less. Also shown in Fig. 4 are the range of β values obtained from the Pentre test.

τ_{sf}: kPa
100
200
300
100
200
300
400
500
600
σ'_{vo}: kPa
LDIP Pentre
Karlsrud *et al.*
P
$\beta = 0{\cdot}2$

Fig. 3.

2·0
1·0
0
$\beta = \bar{\tau}_{sf}/\sigma'_{vo}$
1
2
3
4
s_u/σ'_{vo}
$\alpha = 0{\cdot}45$
T
P
LDIP Tilbrook Grange
Karlsrud *et al.*

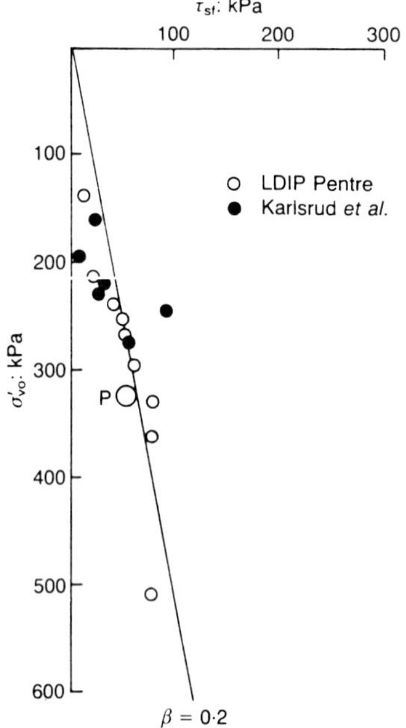

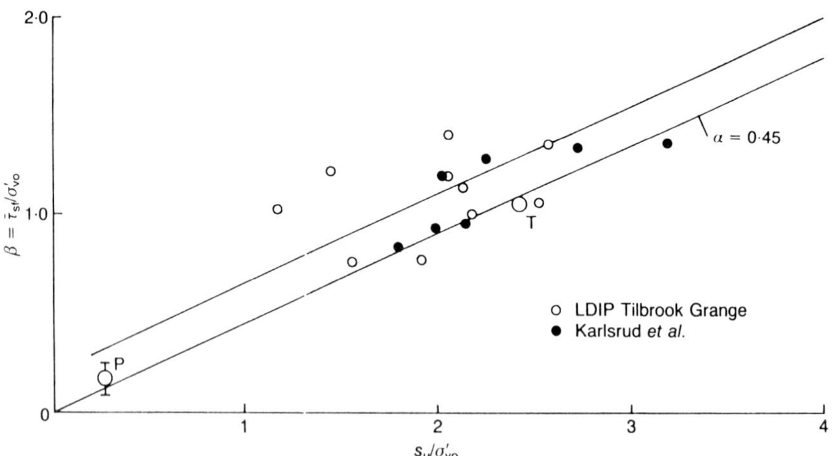

Fig. 4.

It is clear from the above that the large diameter instrumented pile tests have added considerably to the API data base and the results will continue to be analysed in detail for a long time to come. I have no doubt that the way ahead is via effective stress approaches. At the level of research and development, the reliable measurement of local values of σ'_r and τ_s are essential, as well as the most advanced techniques of in situ and laboratory testing to determine σ'_{ro}, δ', etc. For routine design we should be aiming to replace the use of s_u, which is highly test dependent, with a reliable measure of OCR which should correlate better with β.

This conference brought much new data into the public domain and it afforded us the opportunity to understand a wide range of perspectives on offshore piling. The proceedings are a most valuable addition to knowledge and experience on this important subject.

References

BOND, A. J., JARDINE, R. J. and LEHANE, B. M. (1992). Factors affecting the shaft capacity of displacement piles in clay. *Proc. Int. Conf. on Offshore Site Investigation and Foundation Behaviour*. Society for Underwater Technology, London.

SEMPLE, R. M. and RIGDEN, W. J. (1984). Shaft capacity of driven pipe piles in clay. *Proc. Symp. on Analysis and Design of Pile Foundations*. ASCE, San Francisco.